MATERIALS

MATERIALS

Introduction and Applications

WITOLD BROSTOW
University of North Texas
Denton, Texas, USA

HALEY E. HAGG LOBLAND
University of North Texas
Denton, Texas, USA

Published by John Wiley & Sons, Inc., Hoboken, New Jersey
Published simultaneously in Canada

Library of Congress Cataloging-in-Publication Data:

Names: Brostow, Witold, Hagg Lobland, Haley E.
Title: Materials : introduction and applications / Witold Brostow, Haley E. Hagg Lobland.
Description: Hoboken, New Jersey : John Wiley & Sons, Inc., [2016] | Includes bibliographical references and index.
Identifiers: LCCN 2016006351| ISBN 9780470523797 (cloth) | ISBN 9781119281009 (epub)
Subjects: LCSH: Materials science–Study and teaching. | Surfaces–Study and teaching. |
Thermodynamics–Study and teaching.
Classification: LCC TA404 .B76 2016 | DDC 620.1/1–dc23
LC record available at http://lccn.loc.gov/2016006351

Cover image courtesy of the authors.

Set in 10/12pt Times by SPi Global, Pondicherry, India

Printed in the United States of America

10 9 8 7 6 5 4 3 2 1

"Keep away from people who try to belittle your ambitions. Small people always do that, but the really great make you feel that you, too, can become great."—Mark Twain

CONTENTS

Foreword by Ulf W. Gedde xv

Preface xvii

Acknowledgments xix

Part 1 Foundations 1

1 Introduction 3

1.1 History of Materials Science and Engineering (MSE), 3
1.2 Role of MSE in Society, 4
1.3 Teaching MSE, 5
1.4 Basic Rules of MSE, 5
1.5 States of Matter, 6
1.6 Materials in Everyday Life, 7
1.7 How to Make New Materials, 8
1.8 How to Use this Book, 9
1.9 Self-Assessment Questions, 9
References, 9

2 Intermolecular Forces 11

2.1 Interactions: The First Vertex of the Triangle, 11
2.2 Primary Chemical Bonds, 12
2.3 Physical Interactions, 12
2.4 Force and Energy, 15
2.5 Interactions and States of Matter, 16
2.6 Contactless Transport, 18
2.7 Self-Assessment Questions, 19
References, 19

3 Thermodynamics and Phase Diagrams **21**

3.1 What is Thermodynamics and Why is it Useful? 21
3.2 Definitions, 22
3.3 Zeroth Law of Thermodynamics, 23
3.4 First Law of Thermodynamics, 23
3.5 Second Law of Thermodynamics, 24
3.6 The So-Called Third Law of Thermodynamics, 25
3.7 Still More Laws of Thermodynamics? 26
3.8 Thermodynamic Potentials, 26
3.9 Thermodynamic Stability Criteria, 28
3.10 Unary Phase Diagrams and Supercritical States, 29
3.11 Liquid-Vapor Equilibria, 32
3.12 Liquid-Liquid Equilibria, 37
3.13 Solid-Liquid Equilibria, 38
3.14 Self-Assessment Questions, 42
References, 43

4 Crystal Structures **45**

4.1 The Nature of Solid Phases, 45
4.2 Formation of Solid Phases, 48
4.3 Crystal Structures, 50
4.4 Defects in Crystals, 60
4.5 Self-Assessment Questions, 65
References, 66

5 Non-Crystalline and Porous Structures **67**

5.1 Quasicrystals, 67
5.2 Mineraloids, 68
5.3 Diffractometry, 69
5.4 The Binary Radial Distribution Function, 70
5.5 Voronoi Polyhedra, 73
5.6 The Glass Transition, 76
5.7 Glasses and Liquids, 79
5.8 Aging of Glasses, 81
5.9 Porous Materials and Foams, 82
5.10 Self-Assessment Questions, 86
References, 86

Part 2 Materials **89**

6 Metals **91**

6.1 History and Composition, 91
6.2 Methods of Metallurgy, 94
6.3 Alloys, 104
6.4 Phase Diagrams of Metal Systems, 105

6.5 Ferrous Metals: Iron and Steel, 105
6.6 Non-Ferrous Metallic Engineering Materials, 107
6.7 Structures of Metals in Relation to Properties, 109
6.8 Glassy Metals and Liquid Metals, 110
6.9 Self-Assessment Questions, 116
References, 116

7 Ceramics **119**

7.1 Classification of Ceramic Materials, 119
7.2 History of Ceramics, 120
7.3 Crystalline Ceramics, 121
7.4 Network Ceramics: Silicates and Sialons, 127
7.5 Carbon, 129
7.6 Glassy Ceramics, 133
7.7 Glass-Bonded Ceramics, 136
7.8 Cements, 139
7.9 Advanced and Engineering Ceramics, 141
7.10 General Properties of Ceramics, 146
7.11 Self-Assessment Questions, 147
References, 148

8 Organic Raw Materials **151**

8.1 Introduction, 151
8.2 Natural Gas, 152
8.3 Petroleum, 154
8.4 Coal and Coal Tar, 158
8.5 General Remarks, 160
8.6 Self-Assessment Questions, 161
References, 162

9 Polymers **163**

9.1 Polymers among other Classes of Materials, 165
9.2 Inorganic and Organic Polymers, 166
9.3 Thermoplastics and Thermosets, 167
9.4 Polymerization Processes, 172
9.5 Molecular Mass Distribution, 177
9.6 Molecular Structures of Important Polymers, 178
9.7 Spatial Structures of Macromolecules and Associated Properties, 178
9.8 Computer Simulation of Polymers, 183
9.9 Polymer Solutions, 184
9.10 Polymer Processing and the Role of Additives, 185
9.11 Applications of Specialty Polymers, 187
9.12 Self-Assessment Questions, 188
References, 188

10 Composites 191

10.1 Introduction, 191
10.2 Fiber Reinforced Composites, 193
10.3 Cermets and other Metal Matrix Composites (MMCs), 196
10.4 Ceramic Matrix Composites (CMCs), 198
10.5 Carbon–Carbon Composites, 199
10.6 Polymer Matrix Composites (PMCs), 199
10.7 Hybrid Composites, 200
10.8 Laminar and Sandwich Composites, 200
10.9 Concretes and Asphalts, 202
10.10 Natural Composites, 205
10.11 A Comparison of Composites, 208
10.12 Self-Assessment Questions, 209
References, 209

11 Biomaterials 211

11.1 Definitions, 211
11.2 Overview of Biomaterials and Applications, 213
11.3 Joint Replacements, 214
11.4 Dental Materials, 218
11.5 Vascularization in Cardiac and other Applications, 219
11.6 Intraocular Lenses and Contact Lenses, 222
11.7 Drug Delivery Systems, 224
11.8 Biological and Natural Materials, 226
11.9 Bio-Based Materials, 231
11.10 Other Aspects of Biomaterials, 233
11.11 Self-Assessment Questions, 236
References, 236

12 Liquid Crystals and Smart Materials 241

12.1 Introduction, 241
12.2 Liquid Crystals, 242
12.3 Field-Responsive Composites, 248
12.3.1 Magnetorheological Fluids, 249
12.3.2 Electrorheological (ER) Fluids, 252
12.3.3 Electrorheological and Magnetorheological Elastomers, 254
12.4 Electrochromic Materials, 255
12.5 Piezoelectric and Pyroelectric Materials, 256
12.6 Shape-Memory Materials, 260
12.7 Self-Assessment Questions, 263
References, 263

Part 3 Behavior and Properties 267

13 Rheological Properties 269

13.1 Introduction, 269
13.2 Laminar and Turbulent Flow and the Melt Flow Index, 270

13.3 Viscosity and How it is Measured, 273
13.4 Linear and Nonlinear Viscoelasticity, 277
13.5 Drag Reduction, 281
13.6 Suspensions, Slurries, and Flocculation, 285
13.7 Self-Assessment Questions, 287
References, 288

14 Mechanical Properties **289**

14.1 Mechanics at the Forefront, 289
14.2 Quasi-Static Testing, 290
14.3 Properties: Strength, Stiffness, and Toughness, 298
14.4 Creep and Stress Relaxation, 299
14.5 Viscoelasticity, Dynamic Mechanical Analysis, and Brittleness, 302
14.6 Fracture Mechanics, 305
14.7 Impact Testing, 309
14.8 Hardness and Indentation, 312
14.9 Self-Assessment Questions, 315
References, 316

15 Thermophysical Properties **319**

15.1 Introduction, 319
15.2 Volumetric Properties and Equations of State, 320
15.3 Differential Scanning Calorimetry (DSC) and Differential Thermal Analysis (DTA), 323
15.4 Thermogravimetric Analysis, 326
15.5 Thermal Conductivity, 327
15.6 Negative Temperatures, 330
15.7 Self-Assessment Questions, 333
References, 334

16 Color and Optical Properties **335**

16.1 Introduction, 335
16.2 Atomic Origins of Color, 335
16.3 Color and Energy Diagrams, 339
16.4 Light and Bulk Matter, 344
16.5 Optical Properties and Testing Methods, 345
16.6 Lasers, 348
16.7 Electro-Optical Effects and Luminescence, 348
16.8 Photoinduction, 351
16.9 Invisibility, 352
16.10 Self-Assessment Questions, 355
References, 355

17 Electronic Properties **357**

17.1 Introduction, 357
17.2 Conductivity, Resistivity, and Band Theory, 358
17.3 Conductivity in Metals, Semiconductors, and Insulators, 363

17.4 Semiconductors: Types and Electronic Behavior, 364
17.5 Superconductivity, 371
17.6 Phenomena of Dielectrical Polarization, 371
17.7 Self-Assessment Questions, 375
References, 375

18 Magnetic Properties **379**

18.1 Magnetic Fields and their Creation, 379
18.2 Classes of Magnetic Materials, 383
18.3 Diamagnetic Materials, 384
18.4 Paramagnetic Materials, 384
18.5 Ferromagnetic and Antiferromagnetic Materials, 384
18.6 Ferrimagnetic Materials, 386
18.7 Applications of Magnetism, 386
18.8 Self-Assessment Questions, 389
References, 389

19 Surface Behavior and Tribology **391**

19.1 Introduction and History, 391
19.2 Surfaces: Topography and Interactions, 393
19.3 Oxidation, 395
19.4 Corrosion, 399
19.5 Adhesion, 400
19.6 Friction, 404
19.7 Scratch Resistance, 411
19.8 Wear, 418
19.9 Lubrication and Nanoscale Tribology, 419
19.10 Final Comments, 421
19.11 Self-Assessment Questions, 422
References, 423

20 Materials Recycling and Sustainability **427**

20.1 Introduction, 427
20.2 Water, 428
20.3 Nuclear Energy, 430
20.4 Energy Generation from Sunlight, 432
20.5 Energy Generation from Thermoelectricity, 435
20.6 Degradation of Materials, 437
20.7 Recycling, 438
20.8 Final Thoughts, 439
20.9 Self-Assessment Questions, 440
References, 441

21 Materials Testing and Standards **443**

21.1 Introduction, 443
21.2 Standards and Metrics, 443

21.3 Testing, 444
21.4 Microscopy Testing, 445
21.5 Sensors in Testing, 447
21.6 Summary, 448
21.7 Self-Assessment Questions, 448
References, 448

Numerical Values of Important Physical Constants **449**

Name Index **451**

Subject Index **455**

FOREWORD

There are many textbooks in the Materials Science and Engineering (MSE) field, why bother about one more? An important objective of this Foreword is to try to answer this question. We know that for thousands of years, materials—whether stone, iron, bronze, or clay—have largely influenced the ways people live. Matter, what makes up materials, is an integral part of society. This is no less true in the 21st century than it was in the 1st century or at any time before that. There is, therefore, a need for scientists, engineers, and laymen alike to understand the nature of materials.

One develops new materials and solves problems in MSE following three kinds of approaches (alone or in different combinations): experiment, theory, and computer modeling and simulations. All three approaches are covered in this book. A reader thus obtains not only a description of experimental facts but also improved explanation of what has been observed and above all indication of how one moves towards prediction—which provides direction for the development of new materials.

MSE includes Metallurgy which has been in existence for thousands of years, as well as, for instance, elements of Solid State Physics, Microelectronics, and Polymer Chemistry created only in the last half-century. MSE is based on real life and has to be connected to it repetitively—this is a very important feature of this textbook. The book has definitely been written to serve readers with various possible backgrounds: Materials Engineering, Materials Science, Chemical Engineering, Mechanical Engineering, Physics, Chemistry, and more. A reader who already happens to be familiar with a particular topic (or who is not interested in a particular class of materials, properties, or explanations) can safely skip sections and even entire chapters. I find the flexible structure of the text, as well as frequent cross-references, convenient from this point of view.

This is a two-track textbook. Apparently the first basic track can be covered in one semester or term; a two-semester course would include the second track in combination with more time spent on difficult elements of the main track. I believe that this provides enhanced choice and flexibility for the reader: student, professor, or engineer in industry.

I like in this book the fact that several classes of materials have "equal rights": glassy metals, biomaterials, petroleum, smart materials, mineral and polymer concretes.... An important advantage are comparisons between different kinds of materials such that common features of different kinds of materials are pointed out.

There are several other very attractive features that make this book special. It is written in a very positive and enthusiastic style. It simply makes me happy and smiling while reading it. Every chapter ends with a section presenting questions for us to solve. Have I understood the content? This crucial question will for most serious readers be replied by a clear yes, because explanations are clear and simple to follow.

ULF W. GEDDE
Stockholm

PREFACE

There are many textbooks of Materials Science and Engineering (MSE). Almost *none of them* can be covered in a one semester course; what you are getting now is adapted to covering all basics, all classes of materials, and all classes of their properties inside of one semester or one term.

Extant books contain extensive tables of parameters characterizing materials. This is not needed; such data can be found mostly via the Internet and additionally in reference books. We do include some data, but this is done specifically to illustrate a point. Memorization of data is not a goal of this textbook. Consider the following scenario. The management of Company X has decided one morning that their future is not in materials class Y, which they have manufactured exclusively until that day, but in materials class Z. Suddenly the extensive knowledge accumulated by Company X's engineers and scientists working on specific properties of specific materials in class Y becomes useless. This has really happened to some people we know and is just one illustration of why understanding is much more important than memorization.

Certain topics to be discussed in this textbook are more advanced. These will be presented in shaded gray boxes—a two-level approach. Depending on whether you are a student of Mechanical Engineering, Electrical Engineering, Engineering Technology, MSE, Chemistry, Physics, etc., you can decide for yourself whether a topic presented on a more advanced level is not important for you—or else essential for you given your professional profile. You will find some specific instructions on using this book in Chapter 1.

Various elements of this book were taught as courses, parts of courses, or seminars at a number of locations. Our hosts and participants have provided precious input. At least the following institutions that hosted one or both authors of this book need to be named: Abdus Salam International Centre for Theoretical Physics, Trieste; AGH University of Science and Technology, Cracow; Almaty Technological University; Bartin University; Center of Applied Physics and Advanced Technology of the National University of Mexico, Queretaro; Center for Research and Advanced Studies of the National Polytechnic Institute, Mexico City; Chulalongkorn University, Bangkok; Drexel

University, Philadelphia; Imperial College London; Indian Institute of Technology Delhi; Institute of Macromolecular Chemistry of the Academy of Sciences of the Czech Republic, Prague; Institute of Polymer Science and Technology, Madrid; Ivane Javakhishvili University, Tbilisi; Johannes Gutenberg University, Mainz; Kaunas University of Technology; Lvivska Politechnika National University; National University of San Luis, Argentina; Royal Institute of Technology, Stockholm; Tokyo Institute of Technology; Uladimir Belyi Institute of Metal and Polymer Mechanics of the Academy of Sciences of Belarus, Homel; Technical University of Berlin; Technical University of Warsaw; University of Antioquia, Medellin; University of Helsinki; University of Minho, Braga and Guimarães; University of Montreal; University of North Texas, Denton; University of Riga; University of Rouen; University of Vienna; University of Zagreb. Thanks to colleagues and students at these institutions, this book is better than it would have been without them.

ACKNOWLEDGMENTS

The authors wish to thank those who have helped in various ways and during various stages of the preparation of this textbook. We thank our spouses, whose support has made the completion of this book possible. We thank our parents, who were our first teachers, setting us on the road with aspirations for learning and discovery. We gratefully acknowledge Peter Steiner, our graphics artist, for his excellent work in creating our illustrations. We also thank colleagues and associates worldwide who have offered helpful comments and suggestions for improvement of the text. Among these we recognize with special thanks the following: Jan Barcik, Aneta Bialy, Karem Boubaker, Michael Bratychak, Bruna Camporezi, Chin Han Chan, Hyoung Jin Choi, Michael J. Demkowicz, Volodymyr Donchak, David Eglin, Jenny Fagerland, Sarah Forester, Daniele Foresti, Ulf W. Gedde, Anne-Sophie Gillot, Andrei Grigoriev, Elazar Gutmanas, Mikael Hedenqvist, Michael Hess, William L. Johnson, Ramaz Katsarava, Anna Krzton-Maziopa, Josef Kubát, Werner-Michael Kulicke, Rimantas Levinskas, Jonna Lind, Elizabete F. Lucas, Lilian Medina, Samantha Micciulla, Lawrence E. Murr, Nikolai Myshkin, Patrick R. Onck, Raymond H. Pahler, Marianna Pannico, Christian Paulik, Marcos Rizzotto, Piotr Rusek, Shannon Shipley, Michael Schorr, Olena Shyshchak, Michael Silverstein, Ricardo Simoes, Zbigniew H. Stachurski, Chunye Xu, and Jacques L. Zakin.

PART 1

FOUNDATIONS

1

INTRODUCTION

Hic mortui vivunt et muti loquuntur.

—Latin inscription on the Library building of the Lvivska Politechnika National University; it means: here the dead are alive and the mute speak.

1.1 HISTORY OF MATERIALS SCIENCE AND ENGINEERING (MSE)

Unlike many other branches of science and engineering—such as mathematics, chemistry, physics, or civil engineering and architecture—MSE has not been formally recognized for a very long period of time. Mathematics has been growing for some thousands of years. Contemporary chemistry is based on alchemy developed by Chinese, Indians, Egyptians, Greeks, Romans, and Arabs. Compared with these, MSE did not even exist in the middle of the 20th century. It was only in 1959 that von Hippel and Landshoff [1] started to talk about molecular engineering as a way to create materials to order. Some historians of science claim that this was the birth of MSE. Others claim just the opposite: MSE is thousands of years old! This is because one of the constituents of MSE is metallurgy—which indeed is several thousand years old. Likewise ceramics as pottery and stoneware have been used for millennia. Presently, MSE absorbs almost instantly new and useful elements, for instance, from solid state physics or physical chemistry, created in the 21st century. Thus, in MSE we have a blend of very old and brand new.

Materials: Introduction and Applications, First Edition. Witold Brostow and Haley E. Hagg Lobland.

1.2 ROLE OF MSE IN SOCIETY

Materials surround us, comprising the clothes we wear, the tools we operate, the luxuries we enjoy, and the toys with which we entertain ourselves. Even food has now been recognized by the Materials Research Society (MRS) as a class of materials akin to commonly categorized metals, ceramics, and plastics [2]. Historians divide the history of humanity into ages: Stone Age, Bronze Age, Iron Age. Some historians have declared that we now live in the Plastics Age. The way people live and function is determined by how they utilize the available materials. From the Stone Age onwards, some people have been industrious with the materials available: maximizing their uses, inventing new uses, and sometimes discovering or developing new materials better than the old ones. Thus, moving from the Stone Age to the Bronze Age meant a dramatic improvement; people were able to furnish a desired shape more easily to bronze (alloys based on copper) objects. After—strictly speaking—the Bronze Age was over, people continued to make intricate objects from bronze. For instance, in the British Museum in London there is bronze statue of Nataraja, the lord of the dance, created ca. 1100 BC during the Chola Dynasty in Tamil Nadu, India.

In the same way, materials are constantly evolving, improving in processing and performance, finding their way into or even driving new applications. The adaptation of existing materials for new uses is driven mostly by economic factors: what manufacturers and consumers want or need. An early bicycle builder, for instance, wanted a material that was tough like iron but would dampen the shock of rocks and ruts in the road. Consequently, the process of creating fiber-reinforced materials was invented and pneumatic tires were designed, greatly improving the bicyclist's experience.

In the present as in all of history, materials are limiting. Imagine that suddenly everything made from polymeric materials disappeared. Cars would not be able to move because instead of every tire there would be a heap of carbon black dust (a tire consists of some 70% rubber which is a polymer and of 30% carbon black). Little girls would be crying because of disappearance of their plastic dolls. These are just two examples...

Alternatively, consider the future in light of the role of polymeric materials: civil engineers are accustomed to working with mineral concretes based on cement, but more and more polymer concretes and/or polymeric fiber composites are used instead [3]. Many car components that have long been made from metals are now being fabricated from polymers and/or composites (whose lower density than metals results in lower car weight, thus more miles per gallon or kilometers per liter). We now have all-composite airplanes and are not completely reliant on the aluminum fuselage. The 787 Dreamliner first rolled out by Boeing in Summer 2007 is the first large commercial airplane with fuselage, windows, wing boxes, control surfaces, tail, and stabilizers all made from carbon-fiber containing composites.

Above we said that MSE absorbs (steals?) elements of other disciplines such as physical chemistry or solid state physics, using them for its own purposes. However, this is not all one-sided. To give just one example, synthetic organic chemists create new polymers because they know that MSE people will find applications for their polymerization products. Materials Science and Engineering thus also stimulates the growth of other disciplines. Moreover, the development of novel materials and applications often enables fabrication of new equipment and techniques that serve to advance other fields. To state the obvious: the fact that materials determine the way people live is not a self-serving declaration by materials scientists and materials engineers—we did not bribe the historians. The activities of everyday life involve humans interacting with materials.

1.3 TEACHING MSE

Now a skeptic can ask the question: if MSE is so important, why do we have in high school boring classes—of chemistry, for instance—with much memorization but MSE is not taught? There are at least two answers to this. First, educational systems are slow. Because in the middle of the 20th century MSE did not yet formally exist, educational authorities such as the Ministries of Education in various countries have not yet figured out that it should be taught. A more discomfiting fact is that at the university level, one can often still get a degree in Physics, Chemistry, or Biology without taking even one class in MSE. The second answer is that materials are like air: all around us and necessary for life, but paid attention to very little. Thus, we tend to pay no attention to materials—until objects made from those materials begin to disappear.

If somehow you are not yet convinced that materials are as important as air is, consider the implications of a few simple questions about materials:

- Why does one use gasoline as a fuel, but one cannot use water instead?
- Why can we elongate a rubber band seven times its original length, but this is not doable for steel?
- Why can we skate with metal blades on ice but not do so on a wooden floor?
- Can water blow up a five-story house?
- Can water be non-transparent and iron transparent?
- What is a mirror made of, apart from glass?

We are not going to answer these questions right now, but answers will be provided later at appropriate locations in the text.

1.4 BASIC RULES OF MSE

Materials Science and Engineering provides us with description, explanation, and prediction of materials behavior. Our basic tool is the following statement: The macroscopic properties of physical systems are determined by structures and interactions at lower levels, such as microscopic, molecular, and atomic. Thus, there is a constant traversing back and forth between *fundamental theories* and *prediction of macroscopic properties*. This principle is represented schematically in Figure 1.1.

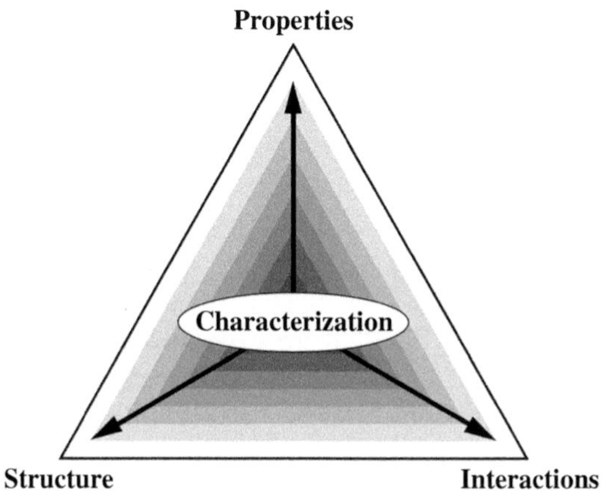

FIGURE 1.1 The basic triangle of Materials Science and Engineering.

The triangle in Figure 1.1 has the following important property: complete knowledge of two vertices of the triangle provides us with knowledge of the third vertex. Thus, if we had available full knowledge of structures and interactions at the atomic and molecular level, we would not need to go to a laboratory to determine macroscopic properties; they would be calculable from the other two vertices. In practice, we typically have fractional knowledge of structures and fractional knowledge of interactions, while some properties of interest but not all have been measured in a laboratory. Thus, we "run around the triangle" to acquire missing information, and the acquisition is called characterization. The word characterization is applied to the experimental techniques used in determining the three elements of the triangle: (1) structure (such as spatial location of phases in multi-component systems seen in microscopy), (2) interactions (such as hydrogen bonds in water which are quite important for water properties), and (3) properties (such as the melting temperature, impact strength, dynamic friction, electrical conductivity—and all other properties).

Some people add processing to the triangle, converting it into a tetrahedron. We think the triangular structure has more perspicuity—but this does not mean that processing is not important. Processing affects all vertices of the triangle. For a material, important events in its history occur during processing; a small change in processing can affect a large number of properties in the final product. In essence, knowledge of a material's history helps us to understand better its current behavior. Certain phenomena occur also during, say, sitting on a shelf, but changes caused by such phenomena are usually slow and not drastic.

The business of characterization is conducted by experimentation. There is no single correct approach to experimentation, but framing the right question is key to success. Theory-oriented experimentation, which was typical of the work of Isaac Newton, is also the classic approach in the philosophy of science. Experiments "are designed with previously demonstrated theories in mind and serve primarily to test or demonstrate them" [4]. On the other hand, an approach known as exploratory experimentation concentrates less on a global explanation of isolated experiments and more on the connections between closely related experiments. Such a method involves "the systematic and extensive variation of experimental conditions to discover which of them influence or are necessary to the phenomena under study" [4]. Clearly, both kinds of experimentation can take us towards deeper understanding. At the core lies keen observation, to see what is the structure of a material and how does it interact with its environment.

To appreciate the importance of this simple triangle diagram, realize that by starting with knowledge of structure and interactions as a foundation, one is enabled to appraise and work with any type of material. A materials scientist or materials engineer who understands this approach does not limit him or herself to a narrow class of materials but is prepared with reasoning skills to analyze all kinds of materials and has a good chance of developing new ones.

1.5 STATES OF MATTER

Ancient Greek philosophers claimed that the whole world consists of just four elements: Air, Water, Stone (or Earth), and Fire. Scientists, particularly in the 18th and 19th centuries, made a much different claim: there are many elements such as copper, silver, oxygen, carbon.... This resulted in the generally held opinion that ancient Greeks simply made wild guesses since they did not have the necessary knowledge. Now, however, it is generally acknowledged that the ancient Greek philosophers had more knowledge than

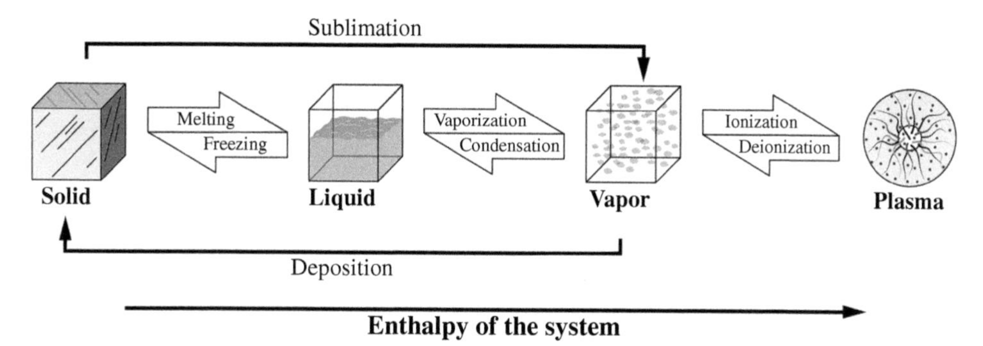

FIGURE 1.2 Basic states of matter and the transitions between them. Enthalpy of the system increases to the right. The melting transition is known also as fusion.

has historically been credited to them. They simply used a different terminology. There are four basic states of matter: gas, liquid, solid, and plasma. These correspond with the Greek elements: Air with gas, Water with liquid, and so on. Recognition of plasma as a separate state of matter is fairly recent; but investigations and experiments involving plasma now abound. An example of a process in the plasmatic state is described in Section 6.7. Many surface-modification techniques also utilize plasma processes. Transitions between the basic states of matter are shown schematically in Figure 1.2. There are also several kinds of mesophases, intermediary between solids and liquids (e.g., colloids, micelles, liquid crystals), that are discussed in subsequent chapters.

1.6 MATERIALS IN EVERYDAY LIFE

We have already pointed out that materials determine the way people live. This statement deserves reinforcement by some examples:

- In the 1980s, the best racing canoes and kayaks were made of wood by a small number of European craftsmen. The best paddlers could cover 1000 meters in about 4 minutes and 5 seconds, under ideal conditions. At the 2012 Olympics, Sebastian Brendel of Germany earned the gold medal with a winning time of 3 minutes and 47.176 seconds. The speed improvement was largely due to innovations in boat and paddle design: what was formerly a craft had turned into an industry, and new composite materials had made that innovation possible.
- A Ukrainian Olympic ice skating champion Viktor Petrenko said at Dartmouth College about electronic brakes built into skis and snowboards: "The change in friction you get is equivalent to going from being on ice to dry pavement" [5].
- If you do not care about racing in a canoe or skating on ice, think about this situation: you are driving a car on an icy and slippery road. You start skidding, and getting out of skidding depends on *friction* between the car tires and the road. This is a material property that can—literally—make the difference between life and death.
- If the above example seems to you overly dramatic, then consider a very simple situation at home: you are frying a couple of eggs on a frying pan. You can lift the eggs after they are fried because the pan surface is covered with Teflon, a low friction material. There is a connection between the story of the skidding car and the story of

frying eggs. On the road you wish as high friction as possible. On the frying pan you wish as low friction as possible. Just one—and the same in both cases—material property is involved.

1.7 HOW TO MAKE NEW MATERIALS

Some students might later go on to designing new materials. How does one do that? Well, we already noted that past history, including processing, determines the properties. Thus, the first stage before creating a new material is investigation of the way the current ones are made. Often one is *not* told by the manufacturer how a given material was made. We have to play a detective, analyzing a structure to figure out what happened before the present structure materialized (a technique known as reverse engineering). This is a capability which is useful also outside of MSE. For instance, look at the scene in Figure 1.3. What occurred before the situation shown in the artwork materialized?

Once we have some knowledge of the materials already used for a desired application, we can begin to think about better ones. Do not assume that you have to follow the path already established; **there might be better options, just nobody thought of them before**. **Originality** and **innovation** are the keys. Do *not* assume something because it seems obvious.

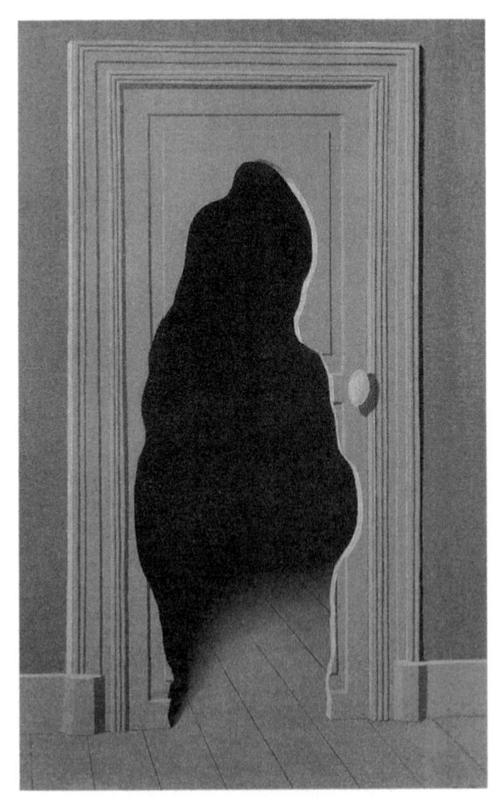

FIGURE 1.3 *La Réponse imprévue* by René Magritte. *Source*: René Magritte: © 2013 C. Herscovici, London/Artists Rights Society (ARS), New York. Reproduced with permission.

1.8 HOW TO USE THIS BOOK

There is a problem typical for practically all students: "I have listened to the lecture, participated in a laboratory since there was one, read this chapter, but I am not sure whether I am getting enough knowledge. I am taking several courses this term/semester. Should I put more effort in the MSE course and less in my Y and Z courses, or the other way around?"

There is a way to deal with this problem. Assessment questions are listed at the end of each chapter. If you can answer them all without consulting this chapter (and your lecture notes if such exist), then you better focus on the Y and Z classes you are taking. If you cannot answer any, this means you were asleep in class and also went through a chapter in a perfunctory way; you now need to read the chapter with some attention. If—a frequent situation—you are able to reply to some questions "off the bat" but not to all questions, this tells you which parts of the chapter you need to return to.

There is a temptation to use wording from the chapter itself, the cut-and-paste procedure. Avoid it! When you use your own words, your understanding gets better. Also note that there might be questions such that more than one answer is a good one. There is an infinity of good answers to the first assessment question in this chapter.

The two-level approach described in the Preface will help you also to manage the material you need to learn. You as a student have to make some choices there—and so has your instructor. Consult your instructor when in doubt; he or she wishes you to acquire useful knowledge and on this basis to be proud of you when the course is completed.

Finally, pay attention to italics and boldface, they convey extra information. Throughout the text, boldface is mostly used whenever important terms are first defined.

1.9 SELF-ASSESSMENT QUESTIONS

1. Describe your favorite material. Give reasons why it is your favorite.
2. Why do historians tell us that the way people live is determined by MSE?
3. Explain the meaning and purpose of the basic triangle of MSE.
4. Discuss transitions between the states of matter known to you.

REFERENCES

1. A.R. von Hippel & R. Landshoff, *Physics Today* **1959**, *12* (10), 48.
2. A.M. Donald, *MRS Bulletin* **2000** December, 16.
3. G. Martínez-Barrera, E. Vigueras-Santiago, O. Gencel, & H.E. Hagg Lobland, *J. Mater. Ed.* **2011**, *33*, 37.
4. N. Ribe & F. Steinle, *Physics Today* **2002** July, 43.
5. V. Petrenko, Skis to Get Electronic Brakes (by Ian Sample), *New Scientist*, February 6, **2002**.

2

INTERMOLECULAR FORCES

daß ich erkenne, was die Welt/in Innersten zusammenhält.

—from a research proposal of Johann Wolfgang von Goethe, as formulated by his hero *Dr. Faustus*; translated from the German by T.E. Webb: ... *that I may discern what inner force/Holds the world together in its course.*

2.1 INTERACTIONS: THE FIRST VERTEX OF THE TRIANGLE

We already know from Figure 1.1 that the macroscopic properties of materials are determined by interactions and structures at the atomic and molecular level. This situation determines the order of presentation in this course. In the present chapter we discuss interactions between particles such as atoms, molecules, ions, or polymer chain segments. There is a direct connection between the interactions and thermodynamic (also called thermophysical) properties of materials such as the heat capacity or the melting temperature; these will be discussed in Chapter 3. Then we shall deal with materials structures: crystalline structures in Chapter 4 and non-crystalline ones in Chapter 5. That will conclude our unit on Foundations, the first of three large units of the course. The second unit will be on classes of materials, the third on properties.

The objective of the present chapter can be summarized simply: to obtain a basic understanding of the main types of intermolecular forces affecting materials behavior. Thus, we aim to get a picture of the effects these forces have on the **properties** of materials in all states of matter.

There are actually two types of interactions in materials. The first type, called chemical bonds or **primary interactions**, deals with interactions inside of molecules. Primary interactions are also sometimes referred to as **intramolecular** interactions or forces.

Materials: Introduction and Applications, First Edition. Witold Brostow and Haley E. Hagg Lobland.

The second type, called **intermolecular**, secondary, or **physical interactions**, entails interactions between separate molecules, ions, or polymer chain segments.

2.2 PRIMARY CHEMICAL BONDS

How can we distinguish primary interactions from secondary (physical) interactions? First, the chemical bonds are stronger and interatomic distances are shorter in primary interactions than in *inter*molecular interactions.

There are three types of primary bonds, distinguished largely by how the participating atoms share their electrons. In **covalent** bonds, the electrons are localized and equally shared between partner atoms. Examples of covalently bonded compounds are: O_2, Cl_2, CO_2, SiO_2. The positively charged nuclei of two or more atoms simultaneously attract the negatively charged electrons that are being shared between them. In **ionic** bonds, as in covalent bonds, electrons shared between participating atoms are localized to the molecule. However, in contrast to the "equal sharing" of covalent bonds, there is in ionic bonds a charge transfer between partner atoms owing to electrons preferentially residing around a particular atom (or atoms). Examples of ionically bonded materials are: NaCl, $CaCO_3$, Fe_2O_3, MgO. **Metallic** bonds differ in that the electrons are delocalized. If one would succeed in capturing an electron, one would *not* be able to tell which atom it comes from. Individual electrons roam freely in the whole material. For that reason, some use the term "cloud of electrons" or even "electron gas" in reference to metallic bonding. Examples of materials with metallic bonding are: Al, Fe, Cu, Ag.

2.3 PHYSICAL INTERACTIONS

There are several categories of secondary interactions. We now discuss them in turn, in the order of increasing strength. First are **dispersion forces**, known also as **London dispersion forces** or **van der Waals dispersion forces**. Now we describe the nature of these forces by considering a scenario. Electrons belonging to two neighboring molecules might have, at a given moment, a situation wherein most electrons on one molecule, say, the left one, are far away from the right molecule; at the same time, most electrons on the right molecule are close to the left one. Thus, there are temporary dipoles on both molecules and a resultant attraction between them. This lasts only a small fraction of a second! Then the electrons are dispersed, hence the name. However, a similar situation occurs at other locations in the material. Thus, at any given time there are enough of these instantaneous short-living dipoles to hold together, say, helium monoatomic molecules (in Figure 2.1) in either liquid or solid state. The other two names for dispersion forces come from the fact that Johannes D. van der Waals was concerned with this issue already in 1873 [1], although he did not understand the mechanism of appearance of these forces. That mechanism was explained in 1930 by Fritz London (nothing to do with the city of London) jointly with Robert Eisenschitz [2] using quantum mechanics. These interactions appear in all materials, in addition to any other interactions present.

Induced dipole forces, also known as **Debye forces** (so named after Peter J.W. Debye), are characterized by a permanent multi-pole (such as dipole, quadrupole, octupole, or hexadecapole, involving, respectively, 2, 4, 8, or 16 poles with electrical charges) exercising its influence on a neighboring non-polar molecule, causing formation of an

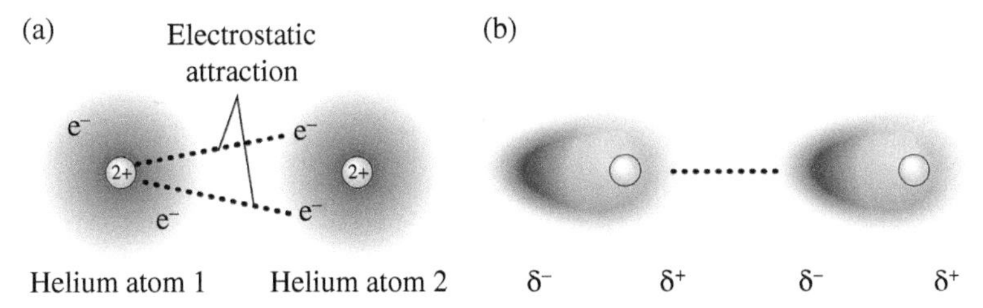

FIGURE 2.1 Dispersion forces. Part (a) shows the electrons of atoms 1 and 2 momentarily concentrated near atom 1. The partial charges of each atom—owing to the situation in (a)—are denoted in (b).

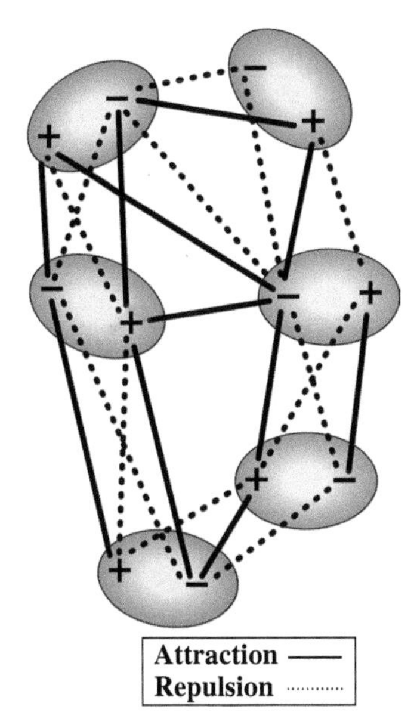

FIGURE 2.2 Dipole-dipole forces. Polar molecules are shown as elongated spheres with permanent charge separation. Attractive and repulsive forces are indicated by solid (—) and dashed (---) lines, respectively.

induced dipole on that molecule. A good example is HCl with Ar. An atom of Ar experiences a momentary dipole as its electrons are attracted (to the H side of HCl) or repelled (from the Cl side) by HCl. This type of interaction can occur between any polar molecule and symmetrical non-polar molecule.

A third type of physical interaction results from **dipole-dipole forces**. In this case we are talking about interactions between permanent dipoles. Such forces are shown schematically in Figure 2.2. As the molecules tumble, there is a mix of attractive and repulsive forces. If we consider HCl alone, there will be such interactions. Likewise, interactions of HCl with chloroform ($CHCl_3$) will also be affected by dipole-dipole forces.

Ion-dipole and **ion-induced dipole forces** are somewhat similar to the dipole-dipole forces just described—only necessarily one of the interacting partners is an ion while the other is a dipole, permanent or induced. These interactions are illustrated in Figure 2.3.

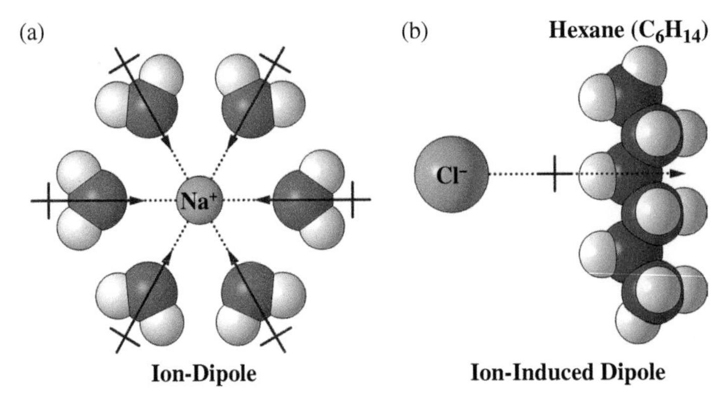

FIGURE 2.3 (a) Ion-dipole and (b) ion-induced dipole forces.

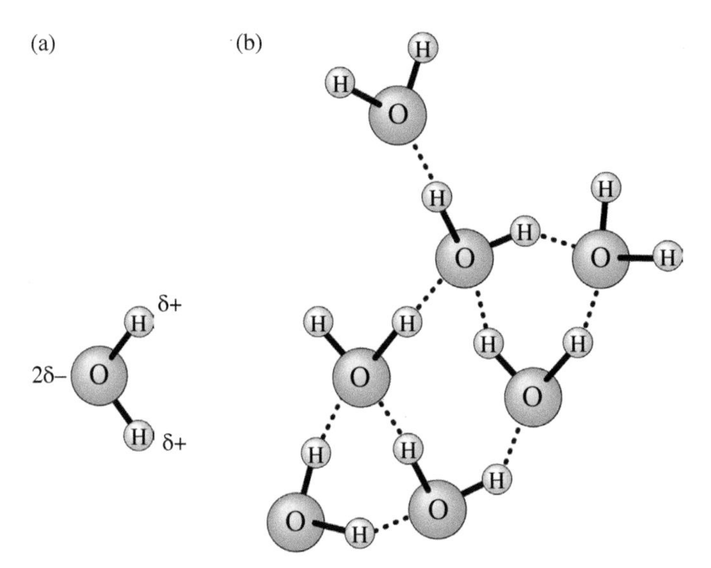

FIGURE 2.4 (a) Schematic of H_2O (water) showing the partial negative ($\delta-$) and partial positive ($\delta+$) charges on the atoms. (b) Dotted lines represent hydrogen bonding, which occurs between H and O atoms on neighboring molecules.

The last and strongest of the physical interactions between molecules is **hydrogen bonds**: the name is misleading since these are not "real" primary chemical bonds. However, hydrogen bonds are the strongest interactions other than chemical bonds. Hydrogen bonding consists of an attraction between an electronegative atom's lone-pair electrons and a hydrogen atom that is bonded to either nitrogen, oxygen, or fluorine. Electrons in the molecule H—X (where X = electronegative element N, O, or F) lie much closer to X than H. H comes with only one electron, so in the H—X bond, the partially positive H presents an almost bare proton to a partially negative X. This is illustrated for water (H bonded to oxygen) in Figure 2.4.

Hydrogen bonding in water is responsible for the phenomenon of ice floating. In solids, molecules are usually more closely packed than in liquids, therefore solids are denser than liquids. Ice, however, is typically ordered with an open hexagonal structure that optimizes hydrogen bonding. Consequently, ice is less dense than liquid water. Because ice floats, it forms an insulating layer on top of lakes, rivers, etc., *thereby allowing aquatic life to survive in winter.* The uniqueness of water is highlighted by an experiment substituting deuterium ("heavy hydrogen", denoted ^2H or D) for hydrogen. Without the influence of hydrogen bonding, D_2O (heavy water) solidifies in a different structure than H_2O; the result is that solid D_2O sinks. Besides this, hydrogen bonding plays a critical role in determining the molecular arrangements of proteins and DNA (deoxyribonucleic acid).

Note that electrical charges are involved (i.e., electrostatic interactions) in each type of intermolecular interaction just described. Thus, these physical interactions are associated with electrostatic forces between atoms and molecules. They are related to physical properties (recall the MSE triangle, Figure 1.1) such as melting point and boiling point because intermolecular attractions must be overcome in order for a material to change state. Physical interactions also determine the solubility of solids, gases, and liquids in various solvents.

In talking about interactions, one sometimes uses the concept of **electronegativity**. It is defined as the tendency of an atom or a functional group in a molecule to attract electrons towards itself, that is to increase the electron density. Electronegativity cannot be directly measured and must be calculated from other atomic or molecular properties. There are various numerical scales of electronegativity, beginning with the first one formulated by Linus Pauling [3]. On any of these scales, the higher the electronegativity number, the more an element, a group of atoms, or a compound attracts electrons towards it.

2.4 FORCE AND ENERGY

We have been using the terms force, attraction, and repulsion in our discussion thus far. The connections between these can be defined more concretely. For a distance R between atoms (ions, atom groups) we have

$$u(R) = -\frac{dF(R)}{dR} \tag{2.1}$$

where u is the potential energy of interaction between the two particles while F is the force acting between them. If the distance R between particles is equal to infinity, clearly there is no interaction and $u=0$. This defines the energy scale: repulsion is represented by positive values of $u(R)$; for attraction $u(R)$ is negative. When we are at $u(R)=0$, the distance between the particles is called the collision diameter R_σ; the particles then touch each other. Any attempt to make the distance smaller than the collision diameter means trying to push one atom or particle inside the other (i.e., repulsion). This situation is described by the curves in Figure 2.5, which shows the force $F(R)$ and interaction potential $u(R)$ as a function of R. The shape of the curve applies to two separate atoms interacting by van der Waals forces as well as to two atoms forming a chemical bond to one another. However, the bottom of the potential well will be much deeper in the latter case. The distance R_m marks the bottom of the potential well. In the case of ion-dipole forces, the force acting between the particles depends on the electronic charges and on the distance between the particles. The larger the charge and the smaller the ion, the larger is the ion-dipole attraction.

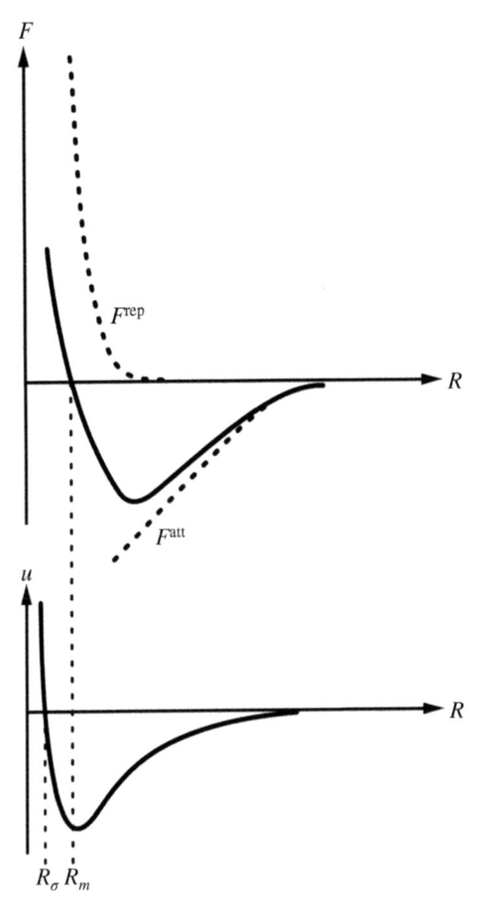

FIGURE 2.5 Forces F and interaction energy u of a pair of particles as a function of interparticle distance R. F^{rep} and F^{att} refer to repulsive and attractive forces, respectively. R_σ is the collision diameter, and R_m signifies the bottom of the potential well.

2.5 INTERACTIONS AND STATES OF MATTER

Throughout this course we will be highlighting connections between interactions, structure, and properties. We shall better understand the importance of the interactions just described by looking at some of those connections now. Consider the liquid state. **Viscosity** is the resistance of a liquid to flow; and a liquid flows by the sliding of molecules over each other. The stronger the intermolecular forces, the higher the viscosity.

Surface tension is also a consequence of intermolecular forces. A familiar example is being able to float a needle or other small dense object on water provided the object is placed on the surface gently, without breaking the tension. How does this happen? Bulk molecules, that is those within the interior of the liquid, are equally attracted to their neighboring molecules. By contrast, surface molecules are only attracted inward towards the bulk molecules since the attraction to air molecules above is so small. Therefore, surface molecules are packed more closely than bulk molecules, while a tensile force is generated across the surface. See Figure 2.6. A small insect called a

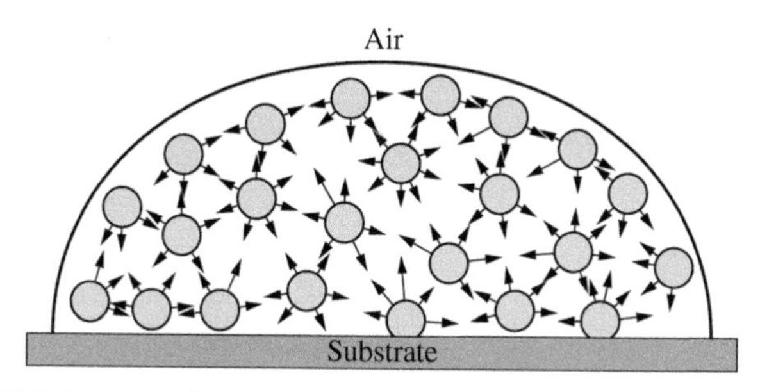

FIGURE 2.6 Origins of surface tension: schematic shows directionality of forces in a liquid drop on a solid substrate.

water strider can actually run on the surface of water; because of the bug's low mass and its leg geometry, the water surface is not pierced through.

The surface tension can be thought of as the energy required to break through the surface of a substance or to spread a drop of the substance out into a film. Consider the example of tent walls. Common tent materials are rainproof—to some extent. The surface tension of water will bridge the pores in the finely woven material. However, if one touches the tent material with a finger, one breaks the surface tension and the rain will drip through. An important example is **soaps** and **detergents**. How do they work? These materials lower the surface tension of water so that it more readily soaks into pores (including the skin pores) and soiled areas.

Surface tension has the dimension of force per unit length, which is equivalent to energy per unit area; one talks about either surface tension or surface energy. Increasing the surface area leads to exposure of more molecules and a higher overall energy. The extent of surface tension is governed by the intermolecular forces; for instance, water has a much higher surface tension than alcohol. What happens when we put a drop of a liquid on the surface of a solid? Well, either the liquid will "like" the solid or it will not. In the first case the angle θ between the solid surface and the drop will be small; see Figure 2.7. The less the liquid likes the solid surface, the larger will θ be. If it seems we somehow got away without considering the substrate surface itself, now we are back to solid surfaces—for surface tension applies to the solid phase as well as to the liquid. Van Oss, Chaudhury, and Good [4] have developed a method of determination of surface tension of a solid from three measurements of contact angles of that solid with liquids, one nonpolar and two polar. One obtains three contributions to the surface tension of that solid: acidic, basic, and nonpolar; the latter corresponds of course to dispersive interactions (van Oss *et al.* call it Lifshitz–van der Waals interactions). The method has been used for a variety of solids [5–7].

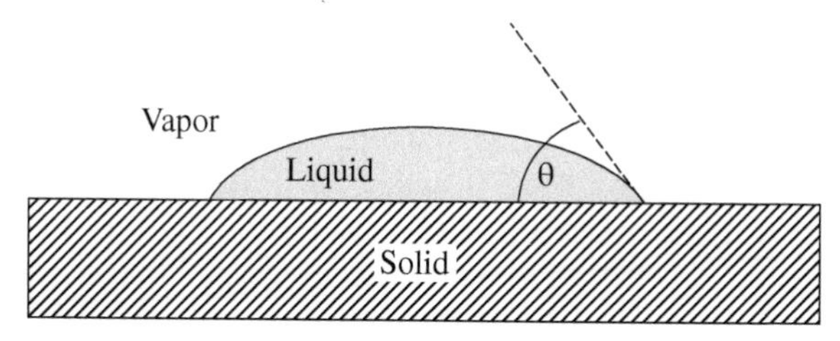

FIGURE 2.7 Schematic of the experimental setup for determination of the contact angle θ of a sessile drop of liquid on a solid.

TABLE 2.1 Characteristic Interactions and Structures of Solid, Liquid, and Gaseous Materials

Material Type	Structural Unit	Forces/Interactions	Examples
Ionic	ions	ionic bonds	NaCl, MgO
Metallic	metal atoms	metallic bonds	iron, copper
Molecular	molecules	dispersion, dipole-dipole, H-bond	H_2O, CO_2
Network	ordered atoms	covalent bonds	diamond, graphite
Amorphous	disordered atoms or molecules	covalent bonds	glass, plastics

A different scenario is also possible. We can measure the wetting angle of a liquid on a solid at an elevated temperature. After cooling to room temperature, the liquid drop solidifies. While the main use of the combined materials may be at the ambient temperature, determination of wetting angles at higher temperatures is still useful, especially, for instance, in coatings that must be applied at elevated temperatures. If the wetting angles are small, then as we have said there is good adhesion between the components; in this scenario *adhesive forces—binding molecules to a surface—are stronger than cohesive forces inside each component.* The reason behind weakening of cohesive forces at higher temperatures is higher free volume. In turn, higher free volume means lower viscosity. These phenomena are important, for example, for encapsulation of thermoelectric materials by high temperature polymers [8].

It is helpful to think of the different classes of materials in terms of their characteristic interactions and associated structural units. That information is provided in Table 2.1. Network solids are characterized by long-range order, although they do not necessarily have a regular repeating lattice. Silicates are materials in which the atoms are bonded covalently and form a network structure. These are contrasted with amorphous materials, which have no long-range order to their structure. Clearly Table 2.1 is not an exhaustive list. There may be multicomponent systems and composite materials that involve two or more types of materials and an increasing variety of interactions. Yet by this we get an idea of the forces that hold molecules together, the forces that act between molecules, and the general behavior of such forces.

2.6 CONTACTLESS TRANSPORT

Except in space exploration, atoms, molecules, ions—all particles constituting materials—are subjected to the gravitational field of the Earth. Staying on the Earth, can this field be eliminated? In other words, what if we toss an object into the air and it does *not* fall? We are not talking about a fantasy here, it can be done. In 2013 Daniele Foresti and his colleagues at the Swiss Federal Institute of Technology in Zurich have shown a way to do it [9]. The device they have constructed is called an **acoustic levitator**. One causes an explosive mid-air reaction between a small water droplet and a grain of sodium metal. The levitator sends powerful ultrasound waves between an emitting surface and a reflector, the reflector bounces the waves back. In the small region where the emitted and the reflected waves meet, the interference creates stable areas that can trap and keep small pieces of a material.

One could mix this way water droplets with instant coffee granules. Apparently even a toothpick can be levitated. Foresti and his colleagues stress that acoustic levitation might affect magnetic, optical, or electrical properties of materials. Contactless processing is important in biochemistry and in dealing with pharmaceuticals.

2.7 SELF-ASSESSMENT QUESTIONS

1. Which interactions are stronger, chemical bonds or physical interactions? In which of these two cases are the distances between atoms smaller?
2. Explain why ice forms on rivers and lakes first at the surface and not at the bottom. Laymen talk about density; you have to talk also about interactions. Compare to behavior of liquid copper during freezing.
3. Imagine bringing together two atoms separated by a large distance. Describe what happens to the electrostatic forces as the distance between the atoms gets smaller and smaller all the way to the collision diameter (R_σ).
4. Hot water has lower density than room temperature water. Which of the two has better wetting capabilities?

REFERENCES

1. J.D. van der Waals, *Over de Continuiteit van den Gas-en Vloeistoftoestand*, A.W. Sijthoff: Amsterdam **1873**.
2. R. Eisenschitz & F. London, *Z. Physik* **1930**, *60*, 491.
3. L. Pauling, *J. Am. Chem. Soc.* **1932**, *54*, 3570.
4. C.J. van Oss, M.K. Chaudhury, & R.J. Good, *Adv. Colloid & Interface Sci.* **1987**, *189*, 361.
5. R.J. Good, Contact angle, wetting and adhesion: a critical review, in *Contact Angle, Wettability and Adhesion*, ed. K.L. Mittal, VSP: New York **1993**.
6. K. Zukiene, V. Jankauskaite, G. Buika, & S. Petraitiene, *Mater. Sci. Medziagotyra* **2002**, *8*, 266.
7. W. Brostow, P.E. Cassidy, J. Macossay, D. Pietkiewicz, & S. Venumbaka, *Polymer Internat.* **2003**, *52*, 1498.
8. W. Brostow, J. Chang, H.E. Hagg Lobland, J.M. Perez, S. Shipley, J. Wahrmund, & J.B. White, *J. Nanosci. & Nanotech.* **2015**, *15*, 6604.
9. D. Foresti, M. Nabavi, M. Klingauf, A. Ferrari, & D. Poulikakos, *Proc. Natl. Acad. Sci. USA* **2013**, *110*, 12549.

3

THERMODYNAMICS AND PHASE DIAGRAMS

I cannot ever remember having sold a book, but I once burned one. It was a textbook of thermodynamics. I have felt a little guilty about that ever since, but only because the particular book that so incensed me at the time was in fact not much worse than nearly all the others in its field. Thermodynamics is incredibly badly presented, for the most part by people who do not understand it.

—Maxwell L. McGlashan [1], University of Reading, University of Exeter and later University College London.

3.1 WHAT IS THERMODYNAMICS AND WHY IS IT USEFUL?

Let us now answer the two questions asked in the title of this section and then deal with the "environment" of these questions. The definition: **Thermodynamics** is the science dealing with effects of temperature on properties of physical systems (i.e. materials). Why is thermodynamics useful? *Because some quantities are more difficult to measure than other ones.* Thermodynamics helps us make predictions or draw conclusions from principles and theory when experimental measurements are hard to obtain.

We shall discuss in Section 3.9 the stability criteria which tell us whether a given process will take place if we leave a material alone. One of those stability criteria deals with entropy S. There is no entropy meter, that is no device that allows to measure S directly. The situation is even worse if we need to know how entropy changes with pressure P. However, thermodynamics tells us that the change of S with P can be calculated from the change of volume V with temperature T, which is a simple measurement (see more in Section 3.8).

There are two types of thermodynamic properties. First, there are **equilibrium** properties. When we measure an equilibrium property such as density, nothing changes in the material.

Materials: Introduction and Applications, First Edition. Witold Brostow and Haley E. Hagg Lobland.

However, thermodynamics deals also with **transport** properties such as diffusivity, viscosity, or thermal conductivity. Transport properties cannot be measured unless flow occurs.

Why did McGlashan make the statement that we include as a motto? He was right when he made it; unfortunately his statement is largely true still. It seems that the problem is related to the age of thermodynamics. Nobody sane would work or teach on electronic devices without scanning the latest literature. However, thermodynamics, which is believed to have been created in 1824 when a French revolutionary Sadi Carnot wrote a book about the motive power of fire [2], is taught still far too often in much the same fashion as it was almost two centuries ago. Many do not realize that the contents of thermodynamics have undergone a large evolution since 1824. Consequently, information is used from old literature, a process that seems to be ongoing for generations. For instance, one talks at length about Carnot cycles—only to admit at the end that such cycles cannot be realized in practice. Some students infer from this that thermodynamics deals with fictitious situations—which is not true at all.

Thermodynamics is based on certain definitions and certain laws. One can assume the laws to be true since cases where they do not apply are not known (the axiomatic approach). One can alternatively derive the laws of Thermodynamics from Statistical Mechanics [3, 4]. In turn, one can derive Statistical Mechanics from the Theory of Information [5]. We shall discuss first the necessary definitions and then the laws, one at a time.

3.2 DEFINITIONS

A **system** is any part of the real world we choose to study. Everything else belongs to the **surroundings** of the system. **Extensive** properties depend on the size of the system; thus the value of the system property involves summation of each of its parts. Mass and volume are examples. **Intensive** properties are independent of the size of the system (i.e., of the amount of matter present). Color, odor, hardness, melting point, and boiling point are all examples. So also are the thermal expansivity and the isothermal compressibility (see more on them in Section 15.2). A ratio of two extensive properties is an intensive property; density, which is equal to mass divided by volume, is an example.

If the intensive properties of a system are uniform throughout, we describe the system as **homogeneous** and call it a **phase**. Any system may consist of one or more phases. If it consists of more than one phase, with differing phases characterized by different values for one or more intensive properties, we describe the system as **heterogeneous**. An **open phase** is one of variable material content. It can exchange material with other phases in the system or with the surroundings. Otherwise we have a **closed phase**. By analogy, we also have **open** and **closed systems**.

If we add together two components, say A and B, there are three typical scenarios:

1. A and B *like* each other and are **miscible**, hence they create one phase. Water and ethanol are an example;
2. A and B *tolerate* each other. There are small islands of phase A inside of phase B, or vice versa, depending on which component has a higher concentration than the other one. Typically it is then said that the two components are mutually **compatible**;
3. A and B *hate* each other. They form two separate phases; an example is water and gasoline (a mixture of hydrocarbons). One says that the components are **immiscible**.

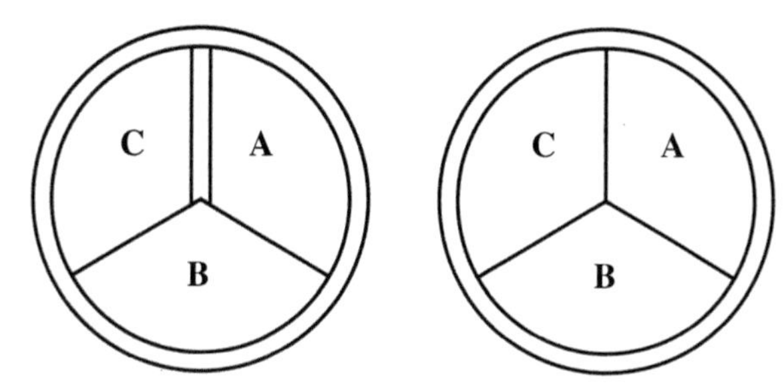

FIGURE 3.1 Illustration of the Zeroth Law of Thermodynamics. Double lines are adiabatic walls. Single lines are diathermic walls.

If we have a material consisting of more than two components, the same situations can be encountered. For example, if we mix hexane, octane, and decane, all three paraffinic hydrocarbons, we shall find they are miscible and form one phase. They can be separated by distillation, though, since they have different boiling temperatures—more on this in Section 3.11.

3.3 ZEROTH LAW OF THERMODYNAMICS

Consider a system consisting of materials A, B, and C as shown in Figure 3.1. The **adiabatic** wall (in Figure 3.1a) is a barrier such that materials on either side of it do not interact; specifically, as we will find out, no heat is gained or lost in an adiabatic process. The single wall, called **diathermic**, is such that it allows interaction; so—with time—materials on both sides of the wall can reach **equilibrium** at which point "nothing more happens". From such observations, the **Zeroth Law of Thermodynamics** is formulated: if System A is in equilibrium with System B, and System C is also in equilibrium with System B, then Systems A and C are in equilibrium as well.

This law was formulated by Sir Ralph Fowler and Edward A. Guggenheim [3, 4]. The property all three systems have in common is called **temperature** (T). Thus thermal equilibrium is transitive. This applies in both (a) and (b) of Figure 3.1. Systems not in thermal equilibrium are said to have different temperatures. The Law permits construction of a thermometer to measure T. Historically, since the First and Second Laws were already known from an earlier time, the new law was called the Zeroth Law. We shall discuss below why the concept of temperature needs to be understood before the concepts of energy and entropy.

3.4 FIRST LAW OF THERMODYNAMICS

The **First Law** can be stated in the form of an equation:

$$dU = w + q. \tag{3.1}$$

Here dU is the change in internal **energy** U of a system in a given process. w is **work** done on the system by its surroundings. q is **heat** flowing into the system from its surroundings. The process can be physical (such as melting) or a chemical reaction (such as polymerization). U is a **state function**, that is the value dU in any process depends only on the original and final state of the process. w and q are <u>not</u> state functions; their values for any process

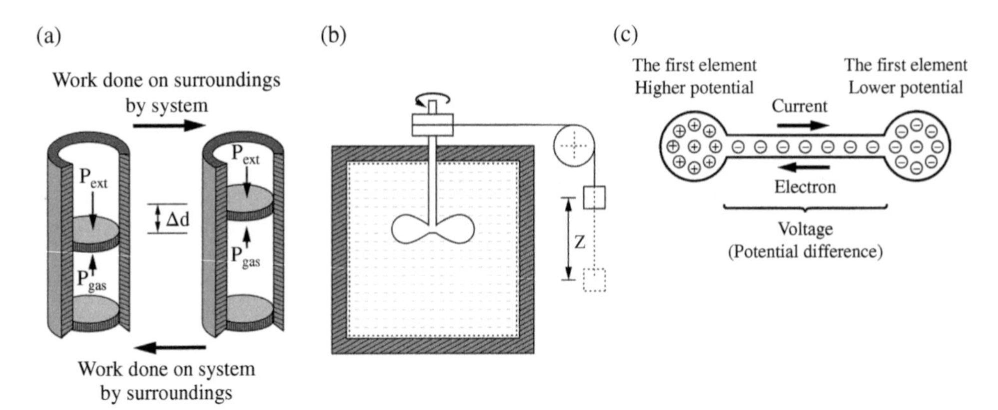

FIGURE 3.2 Three ways of performing work. (a) Heat added to the system causes expansion of the gas, and the piston goes up: work is done by the system. This example illustrates "the whole First Law" since heat and work are connected. (b) Work is done on the system: lowering a weight over a pulley causes stirring of the paddle, which increases the temperature of fluid in the vessel. Energy associated with ΔT is equal to that supplied by lowering the weight. (c) Work is done because of the voltage or potential difference ΔE which causes the current I to flow; note that we are saying the current flows from the positive side to the negative side—while in reality electrons are moving in the opposite direction. The amount of work done is proportional to the time interval Δt during which the current flows.

depend on the pathway (i.e., intermediate states) followed during the process. The First Law can be stated thus: the change in energy of a system during a given process is equal to the sum of work done on the system and heat flow to the system. By extension, negative values of w and q correspond to work done *by* the system and heat flow *from* the system to the surroundings, respectively. There are various ways of performing work on a system. Three such ways are illustrated in Figure 3.2. The respective equations are:

(a) $w = -\int P \cdot dV$

(b) $w = -\int mg \cdot dh$ (m=mass, g=local acceleration of free fall, h=height)

(c) $w = EI \cdot \Delta t$ (E=potential difference, I=electric current, t=time)

3.5 SECOND LAW OF THERMODYNAMICS

The **Second Law** can also be expressed by an equation:

$$S = k \cdot \ln \Pi. \tag{3.2}$$

Here S=entropy, k=the Boltzmann constant, Π=the number of states of the system (also called the sum over the states, from the name die Zustandssumme used by Max Planck). Consider an example: a 1-dimensional system of three atoms of two kinds, with two atoms represented by squares □ and the third by a round dot ●. Possible states of the system are:

Thus, $\Pi = 3$. We owe the clear understanding of entropy to Ludwig Boltzmann, Professor of Physics at the University of Vienna. Before and after Boltzmann much nonsense has been written on the subject of thermodynamics! You have seen the statement by Maxwell L. McGlashan as the motto of the present chapter. You see from the above example how simple the concept of the number of states is. If somebody tells you "entropy is related to disorder", ask them what "disorder" is; there is no definition.

Entropy can be more accurately thought of as *missing information*. To quantify the probabilities of possible states, we can in fact get to the answer by defining our ignorance of the matter via Eq. (3.2). The larger Π is, the less we know about the state of the system [5]. Such evaluation of information applies in general to the theory of decisions; there is a typical situation when we have to make a decision now, while the information we have is far from complete. As discussed by Myron Tribus in his now classical book [6], the theory of decisions helps us to make the best use of the information we have—what is based on evaluation of our ignorance or missing information. While the form of Eq. (3.2) is specified for thermodynamic entropy, ignorance in a broader sense is defined by a similar equation associated with probabilities [7–9]. Arrival of information, therefore, results in a decrease in entropy.

A student asked a question: if the concept of temperature is so important that the law dealing with it has to precede other laws, how is that importance reflected in subsequent laws?

To answer this question, consider an example. We have two containers with carbon dioxide, but in container A the carbon is isotope 12 while in container B it is carbon-14. We open a valve between the two containers. Now some molecules of CO_2 based on carbon-12 will go over to container B while some CO_2 molecules with carbon-14 will go into container A. Because ^{14}C emits β-radiation, one can measure with a Geiger-Müller counter or a proportional counter the number of radioactive molecules that went from container B into container A after, say, one hour. Clearly the number of states Π in the combined system A + B is much higher than in the case when the valve between A and B was closed. Thus, the combined system has a much higher entropy S. What has temperature T to do in our example? At a higher temperature the mixing of molecules between the two containers will be faster than at a low temperature. Thus, we see how T affects S.

3.6 THE SO-CALLED THIRD LAW OF THERMODYNAMICS

This is another sad story created by lack of understanding. Before Boltzmann, people used to calculate changes in entropy using the equation

$$\Delta S = \frac{q}{T}, \tag{3.3}$$

where as before q = heat and T = temperature. This equation works sometimes but not always. As just discussed, when we mix two isotopes of the same element, there is no heat effect; Eq. (3.3) tells us that $\Delta S = 0$, which is not true. Just as it was in our examples with the three atoms and with the mixing of two carbon isotopes of CO_2, any time we mix atoms of two or more kinds, entropy increases significantly.

Consider the case where T approaches zero ($T \to 0$). Equation (3.3) tells us $\Delta S \to \infty$, which is nonsense. In reality $\Delta S \to 0$. Trying to solve this problem with the insufficient definition of entropy, Walther Nernst announced in 1913 at the University of Berlin that $S(0) = 0$ (the so-called Nernst heat theorem), irrespective of what the $\Delta S = q/T$ equation says.

One can visit the lecture hall of Humboldt University at Bunsenstrasse 2 (named after Robert Bunsen, the inventor of a burner) where Nernst made his announcement. However, we do not need such attempts to fix the old equation at all. The Boltzmann equation (Eq. 3.2) which we use as the statement of the Second Law works always. What some people still call the Third Law (Nernst heat theorem, etc.) is not needed.

3.7 STILL MORE LAWS OF THERMODYNAMICS?

How many laws of thermodynamics do we need? There is no answer to this question. The Laws 0, 1, and 2 deal with equilibrium properties. The simplest way to describe these three laws is: Law 0 defines temperature, Law 1 defines energy, Law 2 defines entropy. As already said, when we measure an equilibrium property, nothing changes in the material.

However, as also stated already in Section 3.1, thermodynamics deals with transport properties as well. No law or laws of thermodynamics cover transport properties. This does not mean that we do not need such laws. It only means that an agreement between scientists and engineers on the contents of such laws has not been reached. Some people suggest that the basic law dealing with transport properties should consist of the so-called Onsager reciprocity relations that deal with fluxes. Incidentally, Lars Onsager received the Nobel Prize for his relations *30 years* after he formulated his equations. However, the British physicist Rolf Landsberg has proposed a vastly different formulation of a law covering transport properties (in terms of extensive and intensive variables) and there is no agreement.

3.8 THERMODYNAMIC POTENTIALS

We have considered in the previous sections energy U and entropy S; both are state functions. Such functions are also called **thermodynamic potentials**. There are a total of 16, and we shall need to mention a few more of them. The third one we shall discuss is the product of pressure and volume PV called **the Landau potential**, named so after the Ukrainian physicist Lev Landau who worked in Harkiv (the true name of the city, though due to the effects of Russian occupation it is sometimes written with a "K" as Kharkiv). With these three and with temperature, we can define more:

Enthalpy $$H = U + PV \tag{3.4}$$

Helmholtz function $$A = U - TS \tag{3.5}$$

Gibbs function $$G = H - TS \tag{3.6}$$

Some people [4] work also with

Massieu function $$J = \frac{A}{T} \tag{3.7}$$

Planck function $$Y = \frac{G}{T} \tag{3.8}$$

Be careful if somebody talks about "free energy" because this could be either A or G.

In order to solve some practical problems more easily, we need to define the so-called chemical potential and then to represent some of the thermodynamic potentials by differential equations. The term N represents number of particles. Thus, for a given component α in a multicomponent material, the **chemical potential** μ of that component is

$$\mu_\alpha = \left(\frac{\partial U}{\partial N_\alpha}\right)_{S,V,N_{\beta\neq\alpha}} = \left(\frac{\partial H}{\partial N_\alpha}\right)_{S,P,N_{\beta\neq\alpha}} = \left(\frac{\partial A}{\partial N_\alpha}\right)_{V,T,N_{\beta\neq\alpha}} = \left(\frac{\partial G}{\partial N_\alpha}\right)_{P,T,N_{\beta\neq\alpha}} \tag{3.9}$$

where partial differentiation is performed with numbers of particles for all components $\beta \neq \alpha$ held constant. We therefore have

$$dU = TdS - VdP + \sum_\alpha \mu_\alpha dN_\alpha \tag{3.10}$$

where the summation extends over all components in the material. Similarly

$$dH = TdS + VdP + \sum_\alpha \mu_\alpha dN_\alpha \tag{3.11}$$

$$dA = SdT - PdV + \sum_\alpha \mu_\alpha dN_\alpha \tag{3.12}$$

$$dG = SdT + VdP + \sum_\alpha \mu_\alpha dN_\alpha \tag{3.13}$$

Equations (3.10)–(3.13) are for open systems. If the system is closed, having no exchange of matter with the surroundings, then the last term in each of the equations is equal to zero.

Many relationships between thermodynamic quantities are derived from the above equations. By a way of illustration, we shall now consider Eq. (3.13). We get from it

$$\left(\frac{\partial G}{\partial T}\right)_{P,\,\text{all}\,N_\alpha} = -S \tag{3.14}$$

as well as

$$\left(\frac{\partial G}{\partial P}\right)_{T,\,\text{all}\,N_\alpha} = V \tag{3.15}$$

Now we invoke a rule from calculus: if we perform two differentiations, then the order of differentiations does not matter, the result is the same. Therefore

$$-\left(\frac{\partial S}{\partial P}\right)_{T,\,\text{all}\,N_\alpha} = \left(\frac{\partial V}{\partial T}\right)_{P,\,\text{all}\,N_\alpha} \tag{3.16}$$

We mentioned this relationship between P, V, and T in Section 3.1, but now we have an equation we can use.

3.9 THERMODYNAMIC STABILITY CRITERIA

One does not need to be an engineer or a scientist to know that a piece of clean shiny iron left alone will rust. The inverse process of leaving rusty iron and hoping that we get clean shiny iron just does not happen. Why? This is the time to discuss **thermodynamic stability criteria**. Here is the first criterion:

$$\text{In a natural process} \quad \Delta U < 0 \tag{3.17}$$

In other words, if we have a process—physical (such as freezing) or chemical (such as rusting when iron reacts with oxygen from air and iron oxides are formed)—and the process goes by itself, then the energy of the system undergoing the process goes down (decreases).

The stability criterion of Eq. (3.17) is based on the First Law. We also have a criterion based on the Second Law:

$$\text{In a natural process} \quad \Delta S > 0 \tag{3.18}$$

This is easy to understand. Go back to the story of two containers with CO_2 where we have $^{12}CO_2$ in one and $^{14}CO_2$ in the other one. Gas molecules move around when the valve between the containers gets opened; entropy increases. Once they are mixed, could we get $^{12}CO_2$ back in one container and $^{14}CO_2$ in the other one? No, this will not happen by itself.

Given the criteria (3.17) and (3.18), it is easy to derive from them further ones:

$$\text{In a natural process} \quad \Delta A < 0 \tag{3.19}$$

$$\text{In a natural process} \quad \Delta G < 0 \tag{3.20}$$

We now can answer the question concerning the rusting of iron. Apparently during rusting the Gibbs function decreases. The inverse process, de-rusting of rust, must have $\Delta G > 0$. If we have information about changes of the Gibbs function ΔG in a process, we can predict whether the process will or will not occur.

An important warning is in order here. Natural processes described by the stability criteria (3.17)–(3.20) will inevitably take place, and materials will go towards their equilibrium states if there is no perturbation from outside. How does a perturbation occur? See Figure 3.3. In the top of the Figure, the moving vehicle is in equilibrium with respect to the road surface, all four wheels on that surface. In the bottom part that equilibrium is lost and we see why. Of course *Figure 3.3 concerns human beings, but the situation with materials is practically the same*. When equilibrium exists, it can be perturbed by an outside agent. For materials one employs an agent such as change of temperature, application of a magnetic field, a change in pressure, and so on.

In Section 1.4 we have mentioned briefly that processing leads us to new materials with better properties. Clearly when we begin to make a better material, the starting materials are each in equilibrium—to be perturbed to achieve our objective. Should we, for instance, change both temperature and pressure? Well, that depends on the phases through which we intend to go towards a new material. Thermodynamics helps here since it provides us with the **Gibbs phase rule**:

$$f = c - p + 2 \tag{3.21}$$

FIGURE 3.3 Perturbation of equilibrium in a human system. *Source*: Adan, an illustrator of books by Willy Breinholst.

Here f is the number of degrees of freedom, that is the number of external parameters (temperature, pressure, for gases also volume) that can be varied without creation of new phases or disappearance of any of the existing ones; c is the number of components while p is the number of phases present. We shall see applications of Eq. (3.21) soon.

3.10 UNARY PHASE DIAGRAMS AND SUPERCRITICAL STATES

We shall now consider phase diagrams which show what phases exist in a given material at a given temperature T, pressure P, or volume V and composition. In this Section we consider **unary diagrams**, that is those with only one component. Our first example is magnesium.

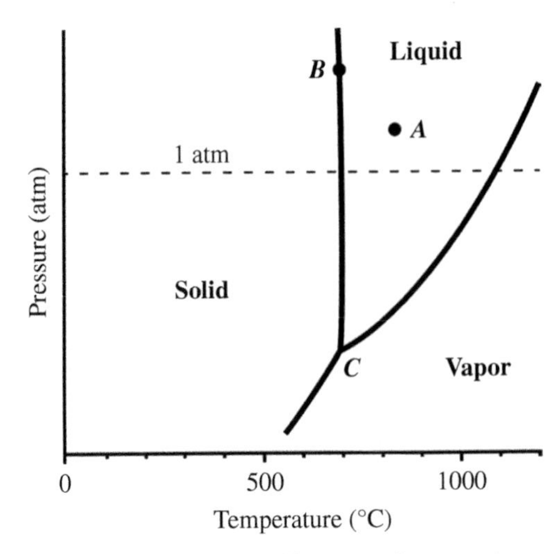

FIGURE 3.4 Phase diagram of magnesium.

Let us see what kind of information we can get from Figure 3.4. Consider an isotherm at 300°C, we move in the diagram vertically from the bottom upwards; the whole time we have only the solid phase. If we consider another isotherm, say, 1000°C, and do the same, we find at low pressures the vapor phase and at higher pressures liquid magnesium. We can also proceed differently, namely, going along an isobar, say, that for 1 atm, starting at 0°C; we shall have a solid, then a liquid and then vapor.

We can also focus on a point, such as point A in the liquid phase. Applying Eq. (3.21), the Gibbs phase rule, $f=2$; thus we can change P or T or both and stay in the liquid region. At point B, on the solid-liquid boundary, there is one degree of freedom. If P changes, T must also change in order stay on the boundary, where solid and liquid coexist. The point C, at which solid, liquid, and vapor coexist, is called the **triple point**. The Gibbs phase rule gives us $f=0$, that is T and P are fixed as there are no degrees of freedom. In other words, if we go with our material off the point C, we are going to lose one or two of the three phases. Actually the diagram itself tells us the same thing, but now we are also able to make the calculation even without looking at the diagram.

Consider now the unary phase diagram for carbon dioxide shown in Figure 3.5. Qualitatively the diagram for CO_2 is similar to that for Mg. In the *y*-shaped curve, the upper thick line, so to speak, is the solid-liquid boundary line along which solid and liquid coexist. Going isobarically from left to right across that line we shall have melting, and backwards we shall get freezing. We have also the right or lower side of the *y*-curve, consisting of the solid-vapor and liquid-vapor boundaries. If we follow the line upwards, we first pass through the coexistence of solid and gas (sublimation, well known for CO_2, is the process going across the line from left to right) and then coexistence of liquid and vapor (associated with the processes of vaporization and condensation). There is also a triple point where the two lines meet.

We shall use Figure 3.5 to explain a phenomenon not discussed before: the existence of the liquid-vapor supercritical state. If we follow the liquid-vapor line from the triple point upwards, we find the end of the line. That end is **the liquid-vapor critical point** characterized by T_{cr} and P_{cr}. For CO_2, the values of the parameters are indicated in the Figure.

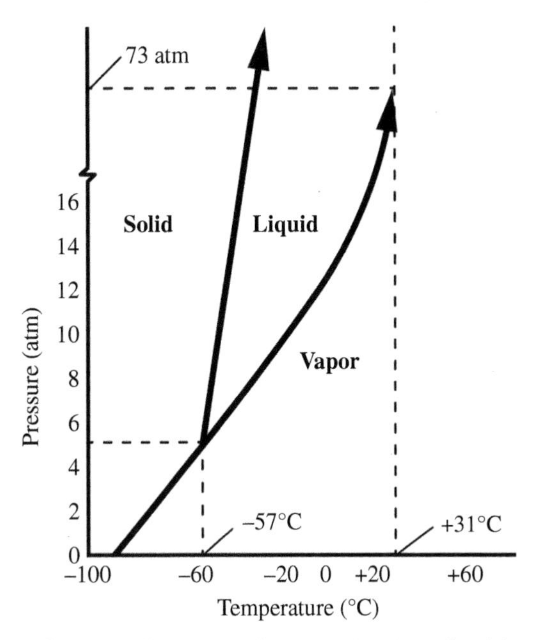

FIGURE 3.5 Phase diagram of carbon dioxide.

Above that temperature and that pressure we have the **supercritical fluid**. Now, any critical point of a one-component material including this one is defined by

$$\left(\frac{\partial^2 A}{\partial V^2}\right)_{T_{cr}} = \left(\frac{\partial^3 A}{\partial V^3}\right)_{T_{cr}} = 0 \tag{3.22}$$

How do we determine the derivatives in Eq. (3.22)? As said before, relations between thermodynamic quantities—called thermodynamic identities—are quite useful. From Eq. (3.12) we get for a closed system

$$\left(\frac{\partial A}{\partial V}\right)_T = -P \tag{3.23}$$

From the two equations above we obtain

$$\left(\frac{\partial P}{\partial V}\right)_{T_{cr}} = \left(\frac{\partial^2 P}{\partial V^2}\right)_{T_{cr}} = 0 \tag{3.24}$$

now in terms of quantities easy to visualize. Thus, there is a top point of the liquid-vapor coexistence line, that point has a name, and it can be defined by equations.

Is that all? Consider a point on the diagram in the liquid region, say, at 15 atm and 0°C. Imagine increasing the pressure keeping the temperature constant to a value above P_{cr}, say, 80 atm. Now keeping the pressure constant, we heat our material to, say, 40°C. We have supercritical fluid. Now we keep the temperature constant and decrease the pressure down to, say, 5 atm. Then along the 5 atm isobar we cool the material down to 0°C, the temperature at which we started. We are now in the gas phase. What have we seen along the way? Since we did not cross the liquid-gas boundary line, there was no vaporization of the liquid. The change was continuous.

Similarly, we can cover the same route *backwards*: starting at 0°C and 5 atm in the gas region and at the end "landing" also at 0°C but at 15 atm. Again we did not cross the boundary line which is the liquid-gas "border"; we went around it. There was no condensation; the transition from gas to liquid was also continuous.

One can ask: well, what is the use of these facts in working with materials? The semiconductor industry requires cleaning operations such as removal of photoresists, of thermal oxide layers, and of post-ash residues. Insufficient removal causes variations in electrical properties, hence havoc in functioning of semiconductor devices. There are also difficulties associated with deposition of metal lines on the devices. It turns out that these operations can be carried out more effectively through the use of carbon dioxide in its supercritical state [10]. One manipulates of course temperatures and pressures; actually small increases in pressure at low temperatures induce sizeable increases in fluid density.

Consider now the solid-liquid boundary line in Figure 3.5. There is no top end of that line marked in the diagram. Previously, we carefully noted "the liquid-vapor critical point". Does the solid-liquid coexistence line end upwards at **the solid-liquid critical point**? So far no such point has been found for any material. Therefore, there are people who say that such a point does not exist. We do not agree, however. Thermodynamics gives us equations such as Eq. (3.24) that apply equally to the solid-liquid critical point. We believe it is just a matter of time to obtain capacity to create pressures high enough to see that point.

In Section 2.3 we talked about the strange behavior of water caused by hydrogen bonds: water expands in the process of freezing to form ice. This is not the only anomaly of water. Let us have a look now at the unary phase diagram of water in Figure 3.6. The diagrams we have seen for Mg and CO_2 are "decent": when we increase the pressure, the melting temperature T_m goes up—as it does in other materials. However, for H_2O the melting temperature goes *down* when pressure increases. This phenomenon is responsible for the surface melting of ice that provides a lubricating film for ice skaters. It is also one more consequence of the hydrogen bonding.

3.11 LIQUID-VAPOR EQUILIBRIA

We have talked about systems containing one component; consider now multicomponent ones, beginning with their liquid-vapor equilibria. In Figure 3.7 we have a binary liquid-vapor zeotropic diagram, having neither extrema nor points where the liquid and vapor lines intersect. We shall call what we see in Figure 3.7 the **fish diagram**. Consider as an example the composition marked by the left vertical line called "line a". Below the fish the material is in the liquid phase. By heating the material, we eventually reach the bottom line of the fish (the bubble point curve, indicating the boiling temperature T_b of a given composition), and the first tiny amount of vapor phase appears. The vapor that forms will have the composition marked by the intersection of the composition line with the dew point curve (point A). The composition of remaining liquid will have more component 1 and less component 2, as indicated by following the isotherm line towards the right to where it intersects the bubble point curve (at point B). The process of getting a vapor from a liquid by heating is called **distillation**. As we see in the Figure, the vapor phase has less component 1 and more component 2. Therefore, formation of the vapor phase results in the composition of the remaining liquid moving to the right.

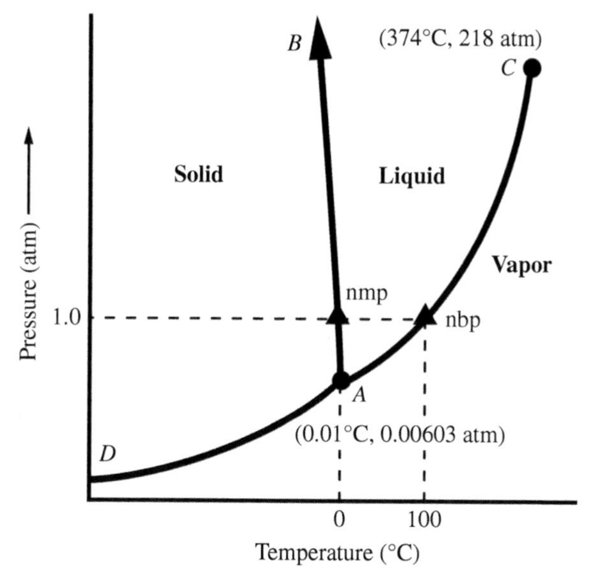

FIGURE 3.6 Phase diagram of H_2O, including the triple point A and the liquid-vapor critical point C. nbp=normal boiling point, at 1 atm. nmp=normal melting point, at 1 atm. Coordinates are shown in parentheses for T and P at the triple point at A and critical point C.

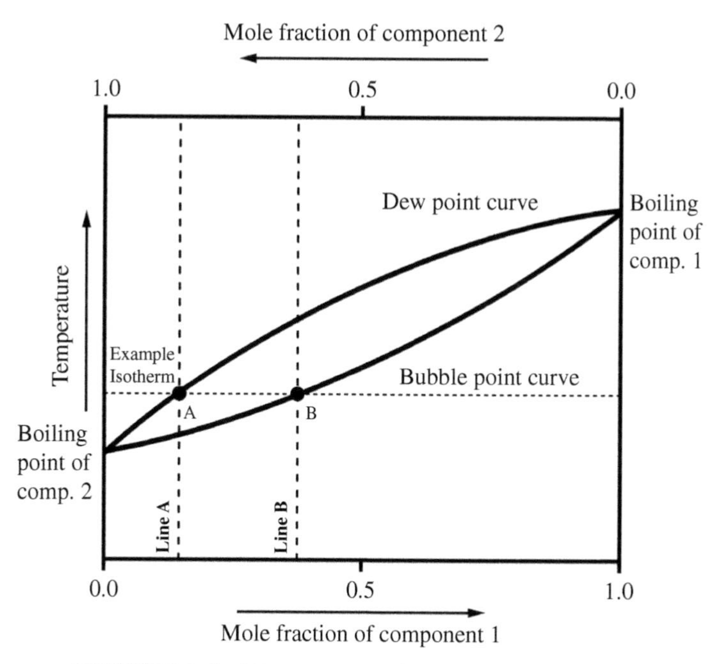

FIGURE 3.7 Binary zeotropic equilibrium diagram.

An important use of phase diagrams is in separation and purification of materials. A repetitive cycle of distillation is called rectification; the process is performed in rectification columns. A rectification column consists of several plates or trays such that the vapor can condense multiple times. Movement of vapor upwards and of liquid downwards is possible. Thus, the

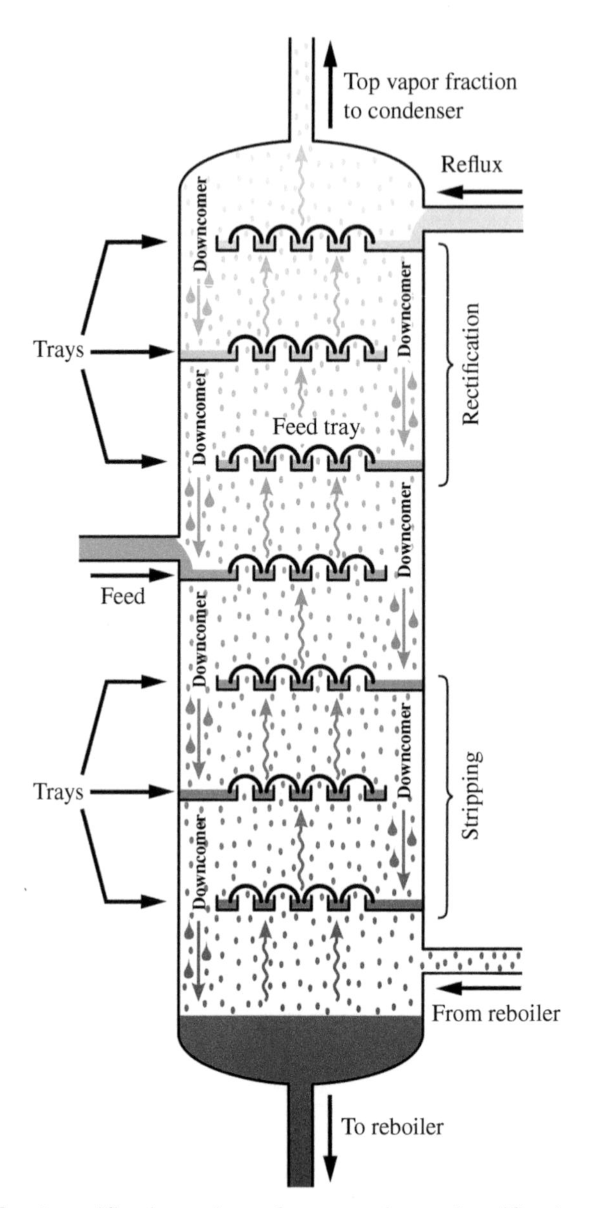

FIGURE 3.8 A rectification column for separation and purification of materials.

operation seen in Figure 3.8 of getting a vapor with a different composition than the liquid is repeated a number of times. At the top of the column we shall get much more component 2, in the liquid remaining at the bottom much more component 1 than in the original composition. In some locations one can see very large rectification columns from which a variety of materials including gasoline are obtained from petroleum; see also Section 8.3. A column is necessary for such procedures as a single distillation would not be sufficient.

Does the operation of separating two liquid components by distillation work always? Not necessarily. Nor do all binary liquid-vapor equilibrium diagrams look like fishes. Consider now the binary positive azeotropic diagram shown in Figure 3.9. In Figure 3.9

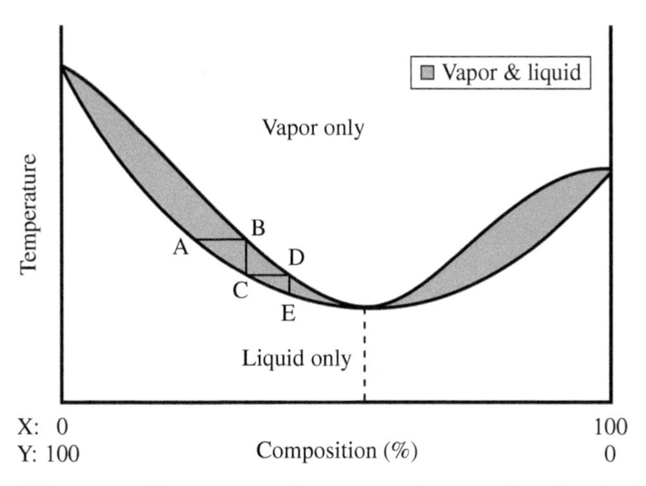

FIGURE 3.9 Binary positive azeotropic system. Azeotropic point is denoted by the dashed line.

we see that there is a point—other than the end points for pure components—such that liquid and vapor have the same composition. It is called the azeotropic point. This one is called a **positive azeotrope**. There exists the Raoult law—largely not true—according to which the vapor pressure line in a binary liquid system should be a straight line. Imagine now that for the same components X and Y in Figure 3.9, we would replace temperature T as the vertical-coordinate by pressure P. A material that has a high vapor pressure will have a low boiling point, and vice versa. Therefore, the same diagram in the P versus composition coordinates would look like a flip-flop of Figure 3.9; the vapor phase would be in the lower part of the diagram and the liquid phase in the upper part. In such a figure, the vapor pressure curve would be above the straight line connecting vapor pressures of the two components. Hence we identify positive deviations from the Raoult law and a positive azeotrope.

As you can now surmise easily, there are also negative azeotropes; one is shown in Figure 3.10. Additionally, if we represent the system shown in Figure 3.10 in the

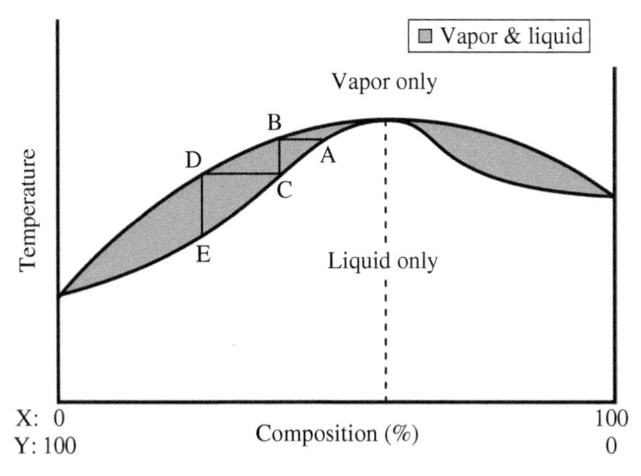

FIGURE 3.10 Binary negative azeotropic system. Azeotropic point is denoted by the dashed line.

vapor pressure versus composition coordinates, the curves will be below the straight line connecting the pure components, hence showing negative deviations from the Raoult law.

One azeotropic diagram is well known to manufacturers of ethyl alcohol. There exists a positive water + ethanol azeotrope—consisting of 95.63 wt.% C_2H_5OH—that has the boiling temperature $T_{bAz} = 78.2°C$. Ethanol has $T_b = 78.4°C$, only slightly higher than the azeotrope; that small difference is sufficient to make getting pure ethanol from its mixture with water by distillation or rectification *impossible*. If one obtains a so-called pure ethanol, it means that a third component called an **entraining agent** has been used.

How do we know whether two liquids will "behave decently" and form a fish type diagram or else form an azeotrope? A theory of azeotrope formation has been developed by Malesinski [11] already in the 1960s. His theory deals with binary as well as ternary and even quaternary azeotropes.

To deal with ternary systems, we need to know how to represent them in phase diagrams. In binary systems we could present the composition plus one more coordinate such as T or P. Ternary systems are represented by the Gibbs triangle, so named after J.W. Gibbs. There is no space in the two-dimensional Gibbs triangle for a coordinate such as temperature—unless one would create three-dimensional models. Thus, ternary phase diagrams pertain to an isotherm (T=constant) or an isobar (P=constant). Such a diagram is shown in Figure 3.11. Before we talk about the azeotropes shown there, we need to know how to use the triangle. The concentrations shown can be weight fractions, mole fractions, or still some other units of concentration. The points of the triangle show pure components. The sides show binary systems, and concentrations of the two components are easily determined by the scale on that side. How then do we evaluate ternary compositions corresponding to points inside the triangle? Consider any point inside the triangle (refer to insert of Figure 3.11). Draw three lines passing through that point that are parallel to each of the sides. Take any side, *it does not matter which*. The lines passing through the chosen

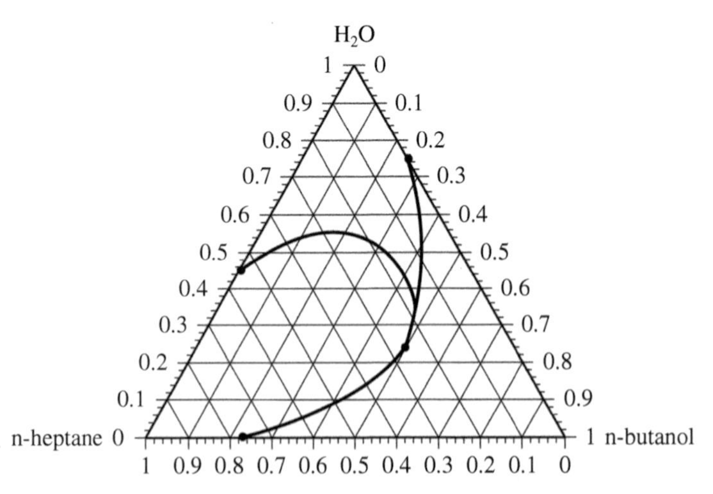

FIGURE 3.11 A ternary system with three binary azeotropes and a ternary azeotrope. *Source*: [12]. Reproduced with permission from Elsevier.

point cut that side into three segments. Call the end points of that side A and B. The segment adjacent to the point A represents the concentration of B (≈0.6), and vice versa (concentration of A ≈ 0.2). The segment in the middle (≈0.2) represents the concentration of the third component.

In Figure 3.11, the four points mark azeotropic compositions. There are three binary azeotropes and one ternary azeotrope. The latter is a saddle azeotrope, also called positive-negative azeotrope, since the ternary azeotropic boiling point is not below the binary azeotropes (as in a positive ternary azeotrope), nor above them (as in a ternary negative azeotrope) but in-between. The so-called valley lines connect the minima of boiling points and meet at the ternary azeotropic point.

There is a reason for displaying this particular ternary azeotropic system. Antimony and arsenic oxides often appear in Nature together. For example, senarmontite found in Algeria (and also in Canada to a lesser extent) contains some 20–30% antimony oxide (Sb_2O_3) and 7% arsenic oxide (As_2O_3). The usual procedure for separating Sb from As is roasting, a costly process of heating to temperatures exceeding 800°C. However, an alternative much less expensive procedure has been developed [12]. As_2O_3 reacts with normal (not branched) butanol as follows:

$$6n-C_4H_9OH+As_2O_3 \leftrightarrows 2(n-C_4H_9O)_3 As+3H_2O \qquad (3.25)$$

Butanol does not react with Sb_2O_3. This looks good, however reaction (3.25) is reversible; n-butanol is partly soluble in water, 9 g/100 mL. There is a solution: one adds n-heptane to the system and the reaction now proceeds to the right only:

$$6n-C_4H_9OH+As_2O_3 \xrightarrow{n-C_7H_{16}} 2(n-C_4H_9O)_3 As+3H_2O \qquad (3.26)$$

As a result of water produced by reaction (3.26), we now have As_2O_3 and Sb_2O_3 from the mineral plus the ternary liquid system: n-heptane + n-butanol + H_2O. As_2O_3 treated with a mixture of butyl alcohol and heptane is quantitatively dissolved in the organic phase producing tributyl arsenite. Sb_2O_3 quantitatively remains on the reactor bottom as sediment. Now back to Figure 3.11. The lowest boiling of the four azeotropes is the binary one of water + n-heptane, the parameters are $T_b = 79.2°C$ and the mole fraction $y_{water} = 0.451$. Pure n-heptane has at atmospheric pressure a normal boiling point $T_b = 98.4°C$. Thus, by the simple process of boiling, water generated as a result of the esterification reaction (3.26) is completely removed from the reaction zone as an azeotrope. The undissolved Sb_2O_3 is easily separated by filtration from tributyl arsenite. Subsequent hydrolysis of the latter produces As_2O_3 [12].

3.12 LIQUID-LIQUID EQUILIBRIA

Liquid-liquid equilibria are known to us from everyday life. Pouring water and gasoline into one container, we find that they form two phases. From the preceding Section we already have an idea how to deal with phase diagrams. An example of a **binary liquid-liquid equilibrium diagram** is shown in Figure 3.12. Inside of the miscibility gap there are two phases, outside only one phase. At the top of the gap we have a **liquid-liquid critical point**.

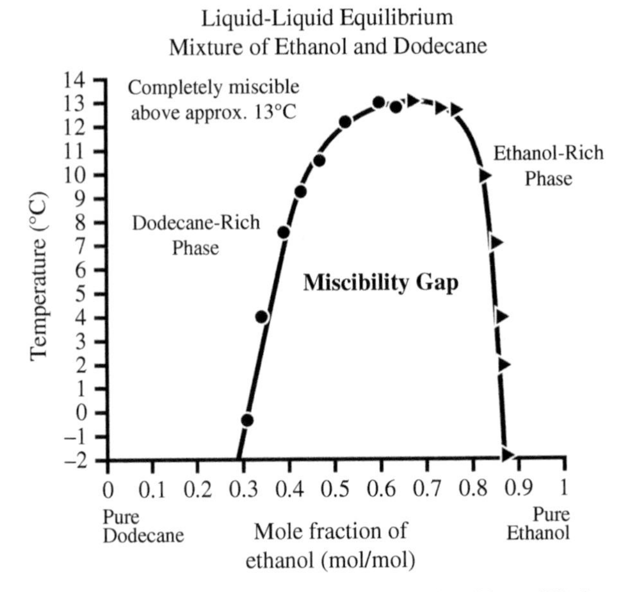

FIGURE 3.12 Binary dodecane + ethanol liquid-liquid equilibrium diagram.

Since this is a binary system, instead of Eqs. (3.22) and (3.24), working with mole fractions of one of the components y, we have

$$\left(\frac{\partial^2 G}{\partial y^2}\right) = \left(\frac{\partial^3 G}{\partial y^3}\right) = 0. \tag{3.27}$$

We already have also certain familiarity with the ternary phase diagrams and with concentrations represented on the Gibbs triangle; see now Figure 3.13.

Figure 3.13 shows us also a miscibility gap. Above that gap, closer to pure propionic acid, there is one phase. Inside the gap there are two phases. The tie lines (also called connodes) connect compositions of two phases in equilibrium. Obtaining such a diagram involves considerable work; a large number of ternary mixtures have to be prepared; when a sample "splits" into two phases, compositions of these phases have to be determined. Note that we do not need to display tie lines in binary diagrams such as in Figure 3.12; when we define a temperature, we get from the diagram compositions of the phases in equilibrium.

3.13 SOLID-LIQUID EQUILIBRIA

We see that the solid-liquid equilibrium diagram of Figure 3.14 is similar to that in Figure 3.7 (the liquid-vapor equilibrium diagram); only the pair of phases is different. A single phase, such as the α phase in Figure 3.14, consisting of two components fully miscible in the solid phase is known as a **solid solution**. The r.h.s. (right hand side of Figure 3.14 shows how one obtains a phase diagram. A material is cooled from the liquid state at a specified rate. However, the material will not undergo cooling the whole time at that rate,

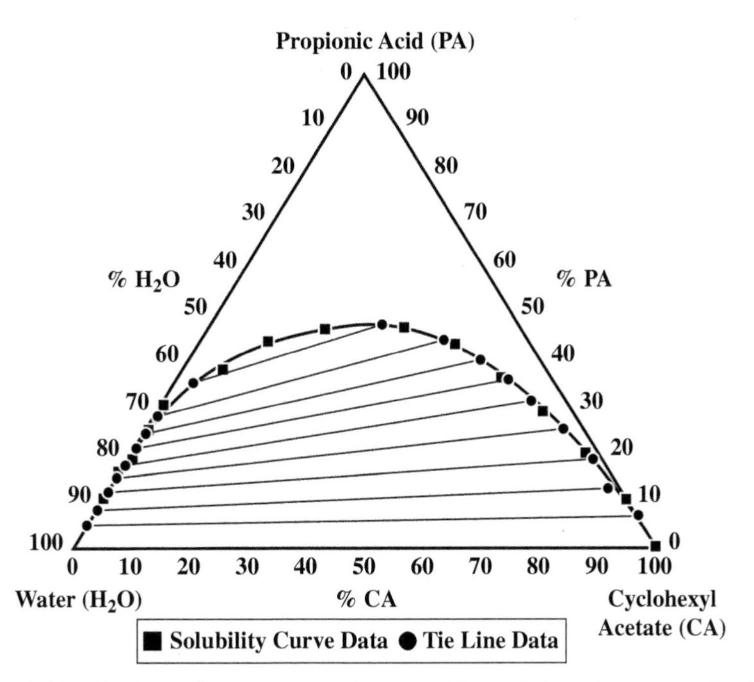

FIGURE 3.13 Isothermal water + propionic acid + cyclohexyl acetate liquid-liquid equilibrium diagram. *Source*: Modified from D. Özmen *et al.*, *Braz. J. Chem. Eng.* **2004**, *21* (4), 647. Reprinted with permission of D. Özmen. http://www.scielo.br/scielo.php?script=sci_abstract&pid=S0104-66322004000400014&lng=en&nrm=iso&tlng=en.

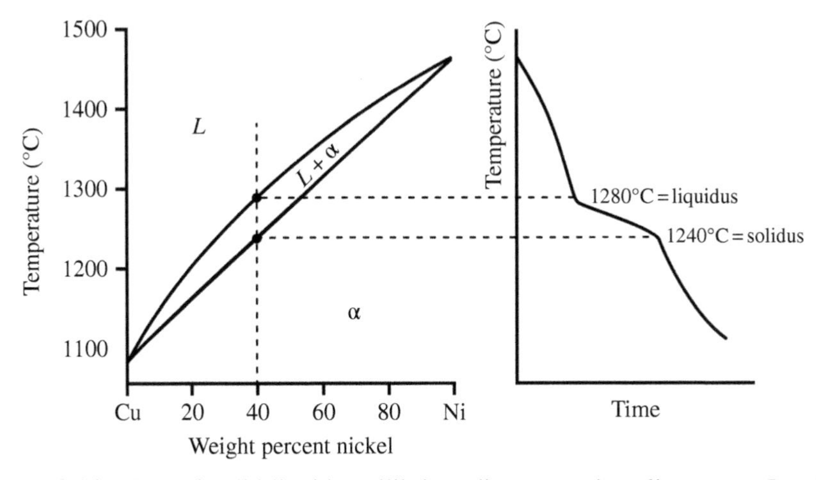

FIGURE 3.14 Cu + Ni solid-liquid equilibrium diagram and cooling curve. L refers to liquid phase; α corresponds to a solid phase.

and the temperature response of the material can be measured—in the case shown, probably by a thermocouple placed in the melt. Looking at the r.h.s. curve in Figure 3.14, we see first a relatively rapid drop in temperature down to the top side of the fish; inside the fish, the cooling rate is slower. Then there is another change of slope upon leaving the fish and entering into the solid state. Plotting such curves for the varied compositions, one obtains sufficient data to create the entire phase diagram. More commonly, given a phase diagram, one determines the composition of an alloy using differential scanning calorimetry (DSC), differential thermal analysis (DTA), or more recently a simultaneous thermal analysis to generate thermal data. The basis of these techniques has been well described by Menard [13], Lucas and her colleagues [14], and Gedde [15]. Variations exist such as photo DSC when DSC measurements provide us with results of ultraviolet or other irradiation of the sample [16, 17]. We shall return to determination of thermal properties in Chapter 15.

Having now seen a variety of phase diagrams, this is a good time to see how much information we can get out of them. Binary diagrams provide us with several kinds of information:

- the numbers of phases present
- the kinds of phases
- the compositions of phases coexisting in equilibrium
- the amounts of phases

The last item deserves more discussion.

How do we get the amounts of phases? Consider a middle region of Figure 3.14, drawn separately as Figure 3.15. First, think about the system which contains 33 wt.% Ni at 1250°C. In terms of phases present, it will be mostly liquid; we are quite close to the composition where there is liquid only. Consider similarly a system which contains 44% Ni; it will be mostly the solid solution called the α phase. Now go to the composition of 40% Ni indicated in the diagram. It will have less liquid than solid since it is closer to the solid phase. How much? Here we use the **lever rule** in terms of equilibrium concentrations. We can use either component, let us continue with Ni. Let us call x_{Ni} and x_{Cu} the respective weight fractions. The rule now gives us the percent fraction of liquid w_{L} as

$$w_{\mathrm{L}} = \frac{x_{\mathrm{Ni}}(\mathrm{solid}) - x_{\mathrm{Ni}}(\mathrm{overall})}{x_{\mathrm{Ni}}(\mathrm{solid}) - x_{\mathrm{Ni}}(\mathrm{liquid})} \times 100\% \tag{3.28}$$

In this particular case, $w_{\mathrm{L}} = (45-40)/(45-32)\cdot 100\% = 38.5\%$. Percent fraction of solid (w_{solid}) present at the specified temperature is of course $100\% - w_{\mathrm{L}}$.

We have seen in Section 3.11 that the fish type diagram is not the only one. Similarly with solid-liquid equilibria, there are binary systems with diagrams that look like the azeotropic ones for liquid-vapor systems. A binary system containing a point at which the composition of the solid is equal to that of the liquid is called **acrystallotropic**. Thus, the acrystallotropic solid-liquid equilibrium point corresponds to the azeotropic point in liquid-vapor equilibrium diagrams.

Furthermore, there are a variety of other types of solid-liquid binary equilibrium diagrams which do not have analogs in liquid-vapor equilibria. The most important ones are shown in Figure 3.16, not as complete diagrams but with illustrations of the characteristic central regions defining the unique behavior of each. We see in Figure 3.16 systems involving more than two phases; α, β, and γ are solid phases denoted with a parenthetical *s* while L refers to a liquid phase. The topmost diagram (Figure 3.16a) is eutectic, having miscibility of two solid phases, α and β, in fact a solid-solid equilibrium. The point at the

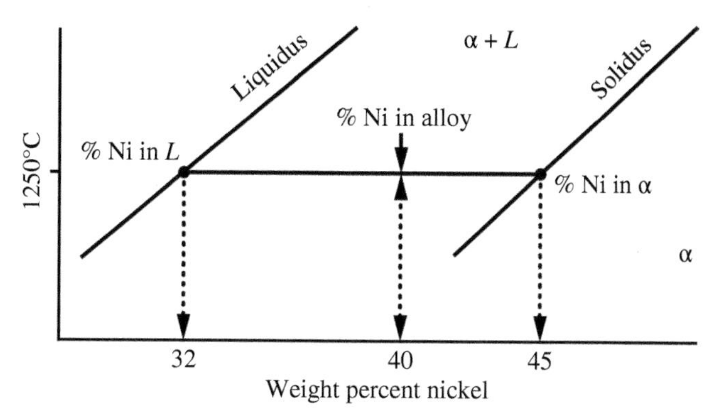

FIGURE 3.15 Expanded view of a middle part of the Cu + Ni solid-liquid equilibrium diagram.

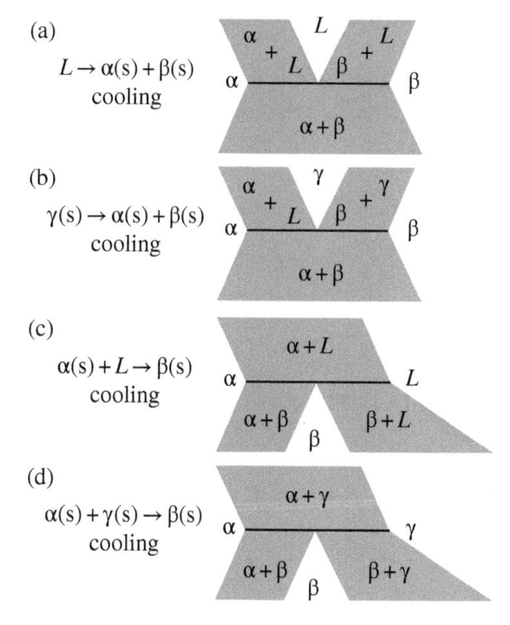

FIGURE 3.16 Several types of solid-liquid equilibrium diagrams (central parts without pure components). α, β, and γ phases are solids. From the top: (a) eutectic, (b) eutectoid, (c) peritectic, and (d) peritectoid. (In the figure, (s) = solid state, L = liquid state.)

vertex of the triangular part is called the eutectic point. Similarly in diagrams (b), (c), and (d), we have the eutectoid point, peritectic point, and peritectoid point. Written to the left of each diagram are the phase transitions that occur upon cooling the material through these transition points.

Let us examine in somewhat more detail the eutectic point. At that point three phases coexist: α, β, and liquid. Is there a practical use of binary materials systems in which eutectics are formed? Yes, we shall consider here a widely applied example. During winter in cold climates roads become icy and dangerous (remember brief discussion of friction in Section 1.6). There is a way out: sprinkling the road with the mineral halite (also known as

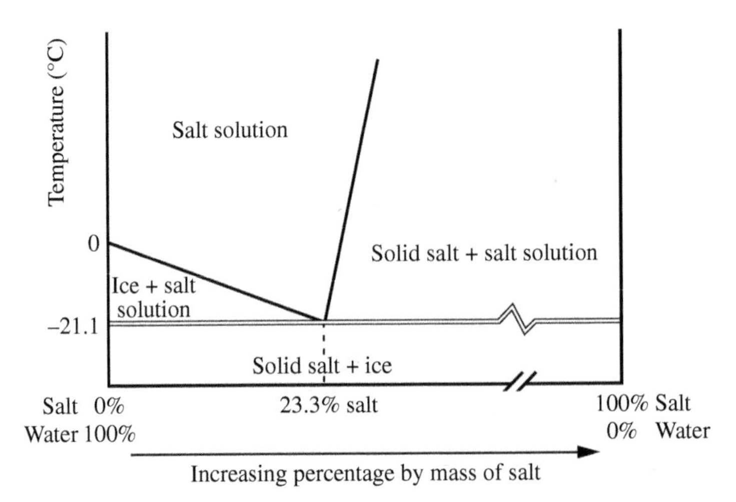

FIGURE 3.17 H_2O + NaCl solid-liquid equilibrium diagram showing eutectic point.

NaCl, a.k.a. the kitchen salt). H_2O and NaCl form a eutectic. As in any eutectic, the melting temperature is below those of the pure components; see Figure 3.17.

For the case that somebody might still have doubts whether the phase diagrams are important, we are going to tell now the story of the Mo + B disaster. It is believed to have happened in the aerospace industry. An engineer wanted to melt a pound of powdered boron into a single ingot. The melting point of boron is T_m(Be) = 2310 K, so he decided to use a new expensive molybdenum crucible; T_m(Mo) = 2895 K. The engineer filled the crucible with boron powder, put it into an induction furnace, evacuated the system, and started heating. Everything went well until ≈ 2200 K, when boron appeared to be melting, still more than 100 K below its melting point. At ≈ 2230 K the bottom of the crucible melted, hundreds of degrees below the melting point of molybdenum. The result of the experiment was Mo + B alloy all over the furnace, and on the floor too (the engineer was periodically looking inside). The whole story has quite a simple explanation: there exists Mo + B eutectic, containing mole fraction of boron $x_B < 0.1$; as you have guessed it, the eutectic temperature is $T_e = 2230$ K.

3.14 SELF-ASSESSMENT QUESTIONS

1. Define what thermodynamics is. Explain how we use it. Why will rust not by itself convert into iron and oxygen?
2. Formulate the First Law of Thermodynamics. Are the parameters in the equation representing that law similar to one another? How or why not?
3. The entropy S at absolute zero (which is 0 K = −273.15°C) is equal to zero. The Boltzmann formulation of the Second Law tells us that there is only one state. Why?
4. How can we perturb an equilibrium in human systems or in materials system?
5. Explain the unary phase diagram of water. Are other unary diagrams similar? Why?
6. Can one get from a liquid to a vapor state without boiling? Explain.
7. Draw and explain a binary solid solutions solid + liquid diagram. How are such diagrams determined?
8. Generally speaking, what kinds of information are available from phase diagrams?

REFERENCES

1. M.L. McGlashan, *J. Chem. Ed.* **1966**, *43*, 226.
2. S. Carnot, *Réflexions sur la puissance motrice du feu*, Bachelier: Paris **1824**.
3. R.H. Fowler & E.A. Guggenheim, *Statistical Thermodynamics—A Version of Statistical Mechanics for Students of Physics and Chemistry*, Cambridge University Press: Cambridge **1939**.
4. E.A. Guggenheim, *Thermodynamics—An Advanced Treatment for Chemists and Physicists*, 5th edition, North Holland Publishing Co.: Amsterdam **1967**.
5. A. Katz, *Principles of Statistical Mechanics—The Information Theory Approach*, Freeman: San Francisco **1967**.
6. M. Tribus, *Rational Descriptions, Decisions and Designs*, Pergamon Press: New York **1969**.
7. E.T. Jaynes, Information theory and statistical mechanics, in *Statistical Physics—1962 Brandeis Summer Institute Lectures in Theoretical Physics*, Vol. *3*, editor K. Ford, pp. 181–218, W.A. Benjamin, Inc.: New York **1963**.
8. S. Watanabe, Chapter 1, in *Knowing and Guessing—A Quantitative Study of Inference and Information*, John Wiley & Sons: New York **1969**.
9. C.E. Shannon, *Bell Syst. Techn. J.* **1948**, *27*, 379 and 623.
10. P.D. Matz & R.F. Reidy, *Solid State Phenomena* **2005**, 103–104, 315.
11. W. Malesinski, *Azeotropy and Other Theoretical Problems of Vapour-Liquid Equilibrium*, Wiley-Interscience: London **1965**.
12. W. Brostow, M. Gahutishvili, R. Gigauri, H.E. Hagg Lobland, S. Japaridze, & N. Lekishvili, *Chem. Eng. J.* **2010**, *159*, 24.
13. K.P. Menard, Thermal transitions and their measurement, Ch. 8, in *Performance of Plastics*, editor W. Brostow, Hanser: Munich/Cincinnati, OH **2000**.
14. E.F. Lucas, B.G. Soares, & E. Monteiro, *Caracterização de polimeros*, e-papers, Rio de Janeiro **2001**.
15. U.W. Gedde, *Polymer Physics*, Springer/Kluwer: Dordrecht/Boston **2001**.
16. K.P. Menard, W. Brostow, & N. Menard, *Chem. & Chem. Technol.* **2011**, *5*, 385.
17. W. Chonkaew, P. Dehkordi, K.P. Menard, W. Brostow, N. Menard, & O. Gencel, *Mater. Res. Innovat.* **2013**, *17*, 263.

4

CRYSTAL STRUCTURES

For a stone, when it is examined, will be found a mountain in miniature. The fineness of Nature's work is so great, that, into a single block, a foot or two in diameter, she can compress as many changes of form and structure, on a small scale, as she needs for her mountains on a large one; and, taking moss for forests, and grains of crystal for crags, the surface of a stone, in by far the plurality of instances, is more interesting than the surface of an ordinary hill; more fantastic in form and incomparably richer in colour—the last quality being, in fact, so noble in most stones of good birth (that is to say, fallen from the crystalline mountain ranges).

—John Ruskin, Modern Painters, Vol. 4, Containing part 5 of Mountain Beauty (1860), 311.

4.1 THE NATURE OF SOLID PHASES

We are familiar with the notion of the ideal gas, which serves as a standard for discussing the behavior of gaseous substances. Similarly, the ideal crystal serves as reference for discussing the nature of solid crystalline materials. The special case of liquid crystal materials is covered in Chapter 12. Comparing liquids and solids, the density of a solid differs only slightly from that of the corresponding liquid. Therefore the mean molecular separation must be comparable. There is long-range order in a crystalline solid but only relatively short-range order in a liquid or an amorphous solid. The structures of liquids and amorphous solids, however, are distinguished by the time scale of motion. Solid phase structures are generally classified as crystalline, polycrystalline (or semicrystalline), and amorphous (also called glassy). We will now explore the first two; the latter will be covered

Materials: Introduction and Applications, First Edition. Witold Brostow and Haley E. Hagg Lobland.

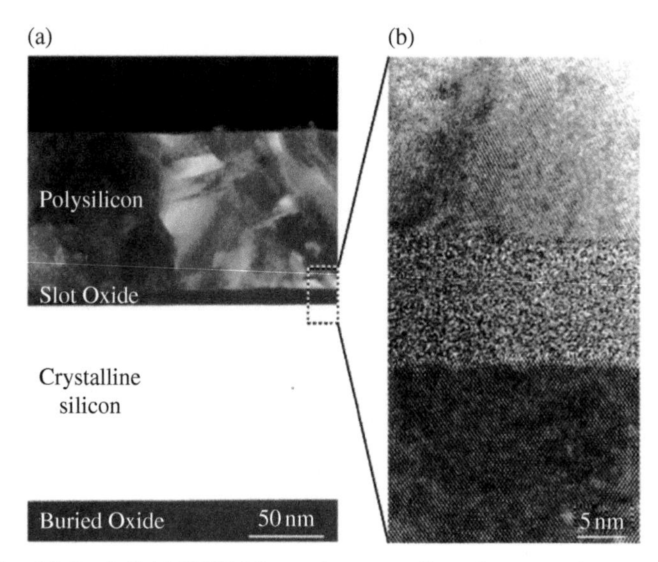

FIGURE 4.1 (a) Dark field STEM (scanning tunneling electron microscope) image of a fabricated material stack to be used as a slot waveguide for electrical injection. (b) Bright field TEM image of the slot region showing the polycrystalline, amorphous, and single crystalline layers (top to bottom). *Source*: Preston and Lipson [1]. Used with permission of Michal Lipson.

in Chapter 5. Their differences are illustrated by electron micrographs of real materials, shown in Figure 4.1 [1].

In the **ideal crystal** all atoms are arranged in a regular and definite geometry. Such a crystal does not exist, but it represents a useful approximation to real systems. Real crystals contain a certain number of defects. Some properties of crystals are practically independent of defects. These **structure-insensitive properties** include: basic volumetric properties (density, isobaric expansivity, isothermal or adiabatic compressibility) and thermal properties (enthalpy of fusion or of sublimation). On the other hand, **structure-sensitive properties** depend on the type and number of **defects** present in a given phase. Most mechanical properties are of this type. The colors of many crystals are due to impurity atoms (defects), and the number of impurities might be quite low. The Cullinan diamond mine in South Africa, not far from Pretoria, is the richest source of rare blue diamonds in the world. The reason for the blue color is the presence of an impurity, namely, boron atoms.

The same property may be either structure-sensitive or structure-insensitive depending on the *phase* involved. It cannot be overstated that a very small number of defects can change the value of a structure-sensitive property by several orders of magnitude. We have just mentioned blue diamonds. Therefore, defects in crystal structures are vastly important, and for that reason constitute a significant part of the present chapter.

Many solid materials contain several phases. As already stated in Chapter 3, we consider a **phase** as a portion of a system bounded by surfaces, with a distinctive and reproducible structure and composition. A phase can be distinguished from another phase if at the contacting surface there is a sharp (within a small number of molecular layers) change in composition or structure (or both).

PHASE TRANSITIONS

It is helpful to classify phase transitions [2]. In the Paul Ehrenfest classification scheme, an n-th *order phase transition* exhibits discontinuity in the n-th derivative of the Gibbs function G, but only change of slope in the $(n-1)$-th derivative. Accordingly, the liquid-vapor critical point discussed in Chapter 3 would be a second-order phase transition. Roy [3] also posits the presence of mixed cases. The mixed case and first- and second-order transitions are described schematically in Figure 4.2. The derivative function could be, for instance, the entropy S or volume V, since these can be expressed in terms of G; see again Eqs. (3.14) and (3.15). A second-order phase transition is characterized by a continuous first derivative of chemical potential, but discontinuous second derivative.

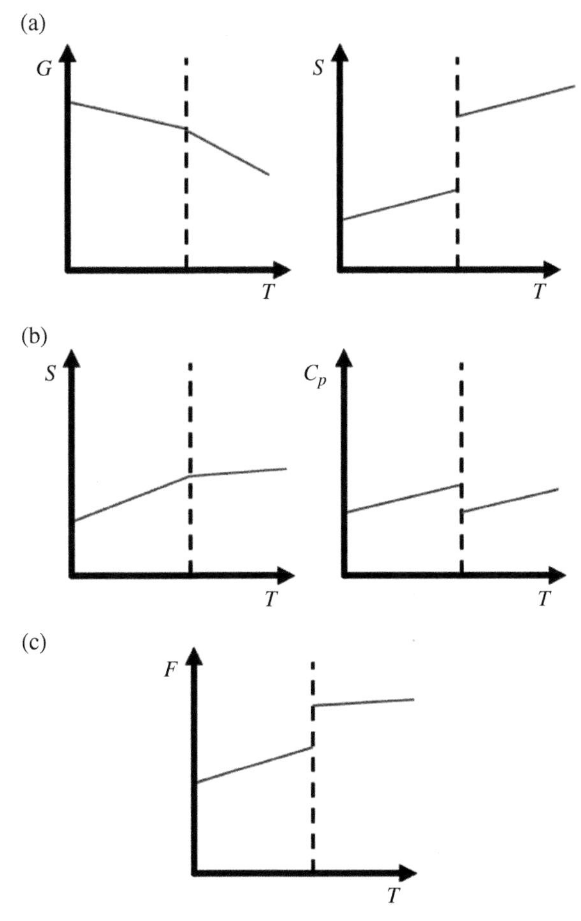

FIGURE 4.2 Examples of some types of phase transitions. (a) First order transition: discontinuity in the first derivative S and change of slope in Gibbs function G. (b) Second order transition: discontinuity in the second derivative heat capacity (C_p) and change of slope in the first derivative S. (c) Mixed transition: according to Rustum Roy [3], involving both change in slope and discontinuity of a function denoted F, such as entropy S, at the transition.

4.2 FORMATION OF SOLID PHASES

Before we work through the details of crystal structures, let us take a look at how solids are formed from liquids. Consider a liquid phase at a temperature not very far above the freezing point. In this state, the liquid molecules—which are moving rather slowly compared to what they do at temperatures well above freezing—have a tendency to form crystal-like temporary aggregates, or ordered microregions [4]. As the temperature drops below the melting point T_m (which is equivalently the freezing point), the aggregates are no longer temporary but become stable, even if the liquid phase persists as a result of **supercooling**. This stability is explained in two ways: first by the inequality (3.20), $\Delta G<0$ in a natural process, and second by the schematic shown in Figure 4.3. In Figure 4.3 we see that the Gibbs function change on fusion (G^f)

$$G^{f} = G(T)\ (\text{liquid}) - G(T)\ (\text{solid}) \tag{4.1}$$

is negative above the melting point, equal to zero at $T = T_m$, and persisting positive below the melting temperature. Therefore, although the Gibbs function is lower for the solid state at T^* below T_m, the transition to the solid state for any liquid is energetically unfavorable because G^f is positive. Since we know that solid does form, there must be another piece to the puzzle.

To understand the process of nucleating a solid, a little more mathematics is needed. First, we specify that G^f corresponds to a unit volume. Next, we specify that G^{nucl} is the Gibbs function change involved in formation of a unit volume of solid aggregate in a liquid phase. It is generally assumed that any such aggregate or solid nucleus is spherical. Thus, denoting the radius of the sphere by R, there is a negative contribution (or liberation of energy) to G^{nucl} equal to $-4\pi R^3 G^f/3$. There is also an energy cost in nucleation since it takes energy to create an interface between a newly formed solid aggregate and the pre-existing liquid. Denoting Γ as the surface energy required to form 1 cm^2 of free interface, the Gibbs function change in forming a spherical cluster of radius R is

$$G^{nucl} = 4\pi R^2 T - \frac{4}{3}\pi R^3 G^f. \tag{4.2}$$

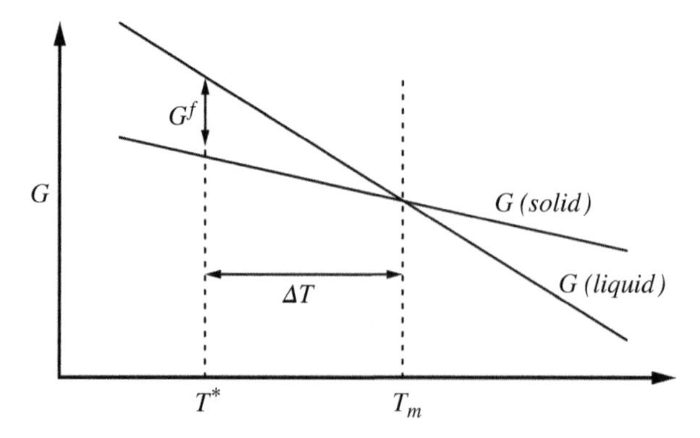

FIGURE 4.3 Schematic curves of Gibbs function G versus temperature for liquid and solid states of a crystalline material. G^f is the Gibbs function change on fusion.

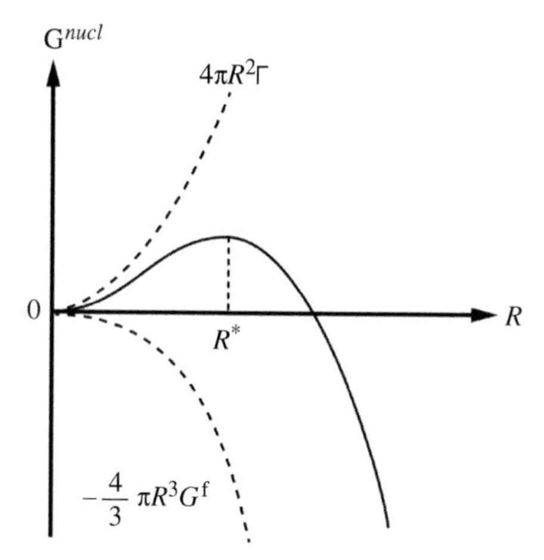

FIGURE 4.4 Gibbs function change G^{nucl} on formation of a solid nucleus in a supercooled liquid phase as a function of the nucleus radius R. Dashed lines represent the contributions featured on the right-hand side of Eq. (4.2); continuous line is the resulting G^{nucl} curve.

Figure 4.4 illustrates the behavior of each of the right-hand terms of this equation. As the nucleus grows from $R = 0$ to R^* (the critical radius), the Gibbs function increases. We remember Eq. (3.6); let us use it here and substitute Eq. (3.4) for H:

$$G = H - TS = U + PV - TS. \tag{4.3}$$

The nucleation process can be induced at this stage by furnishing energy to the system. When the nucleus reaches R^*, it is stable. Further addition of atoms to the nucleus is associated with a *decrease* of the Gibbs function, resulting in a natural process. The condition for this spontaneous growth is

$$\frac{dG^{\text{nucl}}\left(R^*\right)}{dR} = 0 = 8\pi R^*\Gamma - 4\pi R^{*2}G^{\text{f}}, \tag{4.4}$$

which yields

$$R^* = \frac{2\Gamma}{G^{\text{f}}}. \tag{4.5}$$

Substituting Eq. (4.5) into (4.2) gives a new expression for G^{nucl}:

$$G^{\text{nucl}}\left(R^*\right) = \frac{16\pi\Gamma^3}{3G^{\text{f}^2}}. \tag{4.6}$$

Drawing from our earlier discussion of the signs of $G^{\text{f}}(T)$, corresponding to Figure 4.3 and Eq. (4.1), and from the expression of Eq. (4.5), we find that

$$R^*\left(T_{\text{m}}, G^{\text{f}} = 0\right) = \infty \tag{4.7}$$

$$R^*\left(T > T_{\text{m}}, G^{\text{f}} < 0\right) < 0. \tag{4.8}$$

Since it is not possible to have either infinite or negative critical radius, we conclude that stable nucleation is only possible *below* the melting temperature. It is for this reason that supercooling is necessary. The nucleation process as we have just described it is known as **homogeneous nucleation**: it occurs spontaneously and randomly and requires large supercooling, typically by one or more hundreds of degrees. This is contrasted with **heterogeneous nucleation**, in which nucleation occurs at preferential nucleation sites. In practice, heterogeneous nucleation occurs most often, with the walls of the container or suspended foreign particles (sometimes introduced deliberately) serving as preferred sites for nucleation. As a consequence of the nucleation sites, which reduce the surface energy Γ, only a small extent of supercooling is needed.

The number of stable nuclei present in a given volume depends on the number of sites available for nuclei formation. The growth rate of the nuclei depends on the ability of atoms or molecules to diffuse towards a nucleus. Furthermore, the rate of particle movement is exponential. Overall then, the rate of nucleation of a solid phase from a liquid phase is proportional to the number of nucleation sites and the diffusion rate of particles. The reverse process of melting can also be considered from a thermodynamic point of view. That process involves absorption of energy by the crystal, which explains therefore the enthalpy of melting.

4.3 CRYSTAL STRUCTURES

The **crystal structure** of solids refers to the periodic arrangement of atoms in the crystal. We shall discuss in the next chapter diffractometry, which is an experimental method for determining crystal structures. Each such structure is further defined by a **lattice**: an infinite array of points in space, in which each point has identical surroundings. Each lattice point corresponds to either a single atom or to a group of atoms. In the latter case, each group contains the same number of atoms, of the same types, arranged in space in the same way around the lattice point. This is illustrated in Figure 4.5 by a fragment of a two-dimensional lattice. To present the concept, we simply use a fish to represent any atom or atom group. The figure shows a lattice consisting of points and an atom (or atom group)—represented by the fish. A group of atoms, such as this, associated with a lattice point is referred to as a **basis** or **motif**. In the lower part of the figure are two examples of the resulting array of "atoms" based on the two motifs shown. Although there are thousands of crystals, there are only fourteen ways identified to arrange lattice points in three-dimensional space so that each point has identical surroundings. The fourteen lattices, known as **Bravais lattices**, can each be represented by a unit cell. The **unit cell** is the smallest component of the crystal which, when stacked together with pure translational repetition, reproduces the whole crystal. In practice, the unit cell is a parallelepiped that represents the smallest repeating unit in a Bravais lattice. A **primitive** unit cell is one that contains only a single lattice point. Examples of primitive and non-primitive unit cells are shown on the lattice in Figure 4.5. It may appear that the primitive cells contain more than one lattice point, but bear in mind that the lattice points are shared by neighboring cells. Later in this section we will discuss the counting of lattice points and atoms in a unit cell.

The Bravais lattices are presented in detail in Figure 4.6. The artwork *Depth* by Maurits C. Escher shown in Figure 4.7 also provides an illustration of the monoclinic P lattice. A unit cell is described by the reference axes x, y, and z and by the angles α, β, and γ

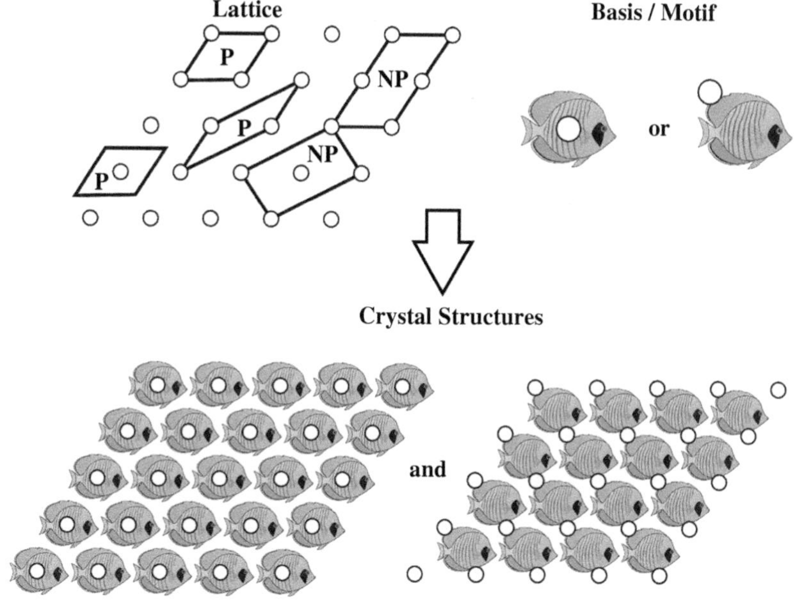

FIGURE 4.5 Two-dimensional example illustrating the definitions of lattice, basis, and crystal structure. Here the fish represents an atom or atom group; two motifs are shown, with the fish in different positions relative to the lattice point. A lattice times an atom group produces a structure. We see two crystal structures shown, based on the two motifs. Lattice points do not necessarily lie at the center of atoms, just as the fish does not have to be centered on the lattice points. Various unit cells are drawn on the lattice; P=primitive, NP=non-primitive.

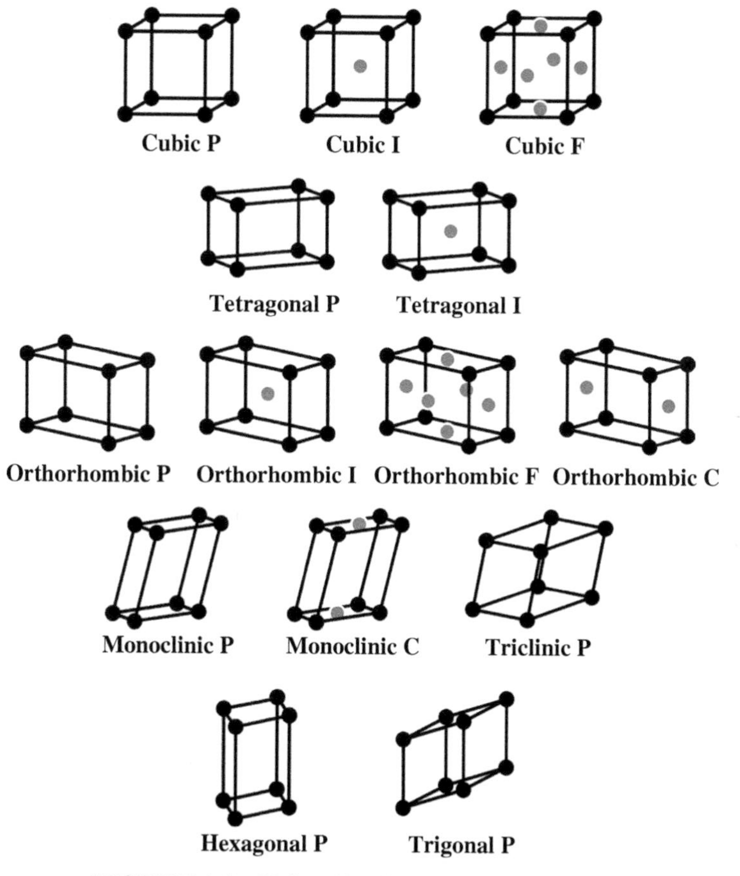

FIGURE 4.6 Unit cells of the 14 Bravais lattices.

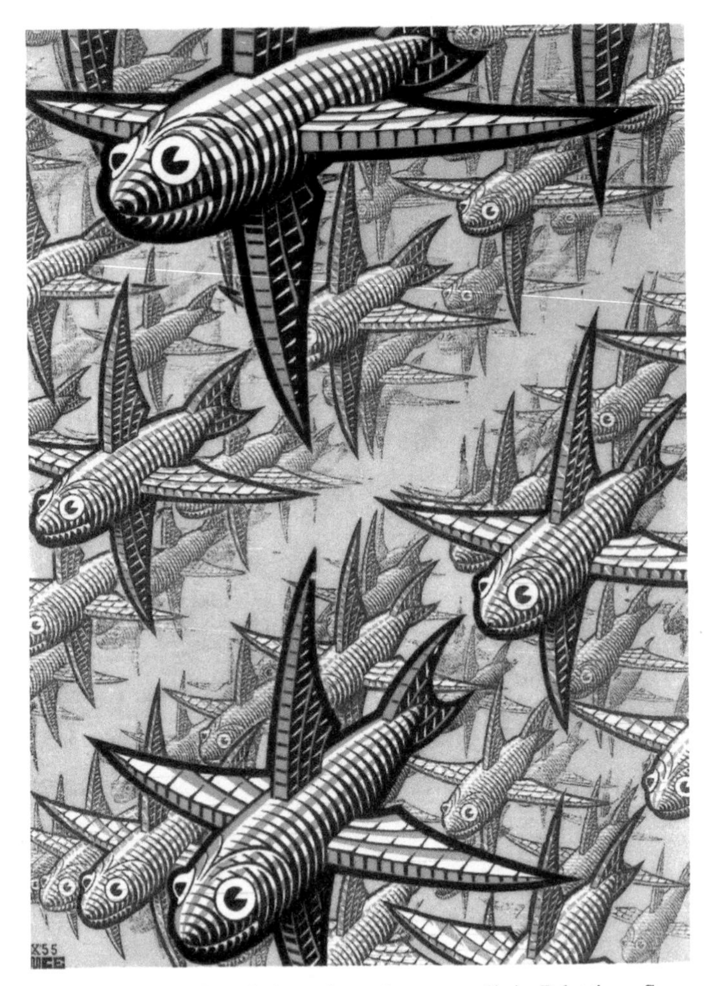

FIGURE 4.7 The artwork *Depth*, based on the monoclinic P lattice. *Source*: The M.C. Escher Company—The Netherlands. M.C. Escher's "Depth" © 2013 The M.C. Escher Company—The Netherlands. All rights reserved. www.mcescher.com.

between the axes. The **lattice parameters** a, b, and c refer to the lengths of the edges of the unit cell along the reference axes. **Axial ratios** are therefore the relative lengths of the crystallographic axes. Opposite faces of a unit cell are parallel, while the edge of a unit cell connects equivalent points. We can see in Figure 4.6 that the fourteen unit cells fall into seven categories. These categories are defined by their edge lengths (a, b, and c) and internal angles (α, β, and γ). The lattices are further distinguished by whether the unit cell is primitive or non-primitive, with the latter including body-centered, face-centered, and side-centered arrangements.

We have seen the simple cubic lattice in Figure 4.6; it is easy to visualize. If we look for examples of this lattice in nature, we find that among chemical elements only one has such a structure: Polonium (Po, atomic number 84). Po has 36 isotopes, more than any other element. Still further, its elastic anisotropy is greater than for any other solid. That is, it is about ten times easier to deform a Po crystal along the direction diagonal to the consolidated cubic cells than it is to deform the crystal in a direction perpendicular to any of its cubic faces. Po was discovered in 1898 by Marie Skłodowska-Curie and Pierre Curie in Paris. It was the first radioactive element discovered. It is named after Poland since Maria Skłodowska was born in Warsaw. In 1903 Marie & Pierre Curie received the Nobel Prize in Physics. Then in 1911 Marie received the Nobel Prize in Chemistry—she was the first woman to win a Nobel Prize and is the only woman to win the award in two different fields.

Counting the lattice points contained in a unit cell requires some explanation. We begin by considering 2D (two-dimensional) lattices. Shown in Figure 4.8 is a 2D lattice; all the parallelograms represent valid unit cells. We choose the square because it is the smallest and provides the most symmetry. In Figure 4.9 we see another 2D lattice; in this case the smallest unit cell does *not* reflect the symmetry within the structure. The larger unit cell, which is non-primitive, is needed because it reflects the symmetry of the lattice. If we consider the points as circles, we see that only a fraction of each point lies within the unit cell. Thus the points at each corner of the square lattice in Figure 4.8 count as ¼; likewise with the smaller unit cell in Figure 4.9. The 4 points times ¼ equal 1 lattice point; thus each of these cells is primitive. Because the larger cell in Figure 4.9 contains an additional point within the cell, it is not primitive. Looking at that non-primitive cell in Figure 4.9, we count the four corners as 1 plus the central point as 1 for a total of 2 lattice points in the unit cell.

We shall consider the structure of graphite as a final example for 2D lattices. Keep in mind that atoms are not the same as lattice points. Figure 4.10a shows a lattice with hexagonal symmetry on which is drawn the hexagonal unit cell, which is primitive. In Figure 4.10b we see the hexagonal structure of a single layer of graphite on which the hexagonal unit cell is placed in two different ways. In each of these unit cells, shown individually in Figure 4.10c and d, there is still 1 lattice point. However, there are 2 atoms in the motif. In Figure 4.10c the corner atoms contribute ¼ to the cell count (as with counting lattice points) and the atom within the cells counts as 1, making the total of atoms in the unit cell equal to 2. The two atoms are plainly seen in Figure 4.10d. In either case the two-atom motif is arranged in a definite way with regard to the lattice points such that translating the lattice—with its two-atom motif—in two dimensions results in the graphitic structure of Figure 4.10b. If there were to be atoms on the edge of a unit cell, they would contribute ½ to the unit cell count.

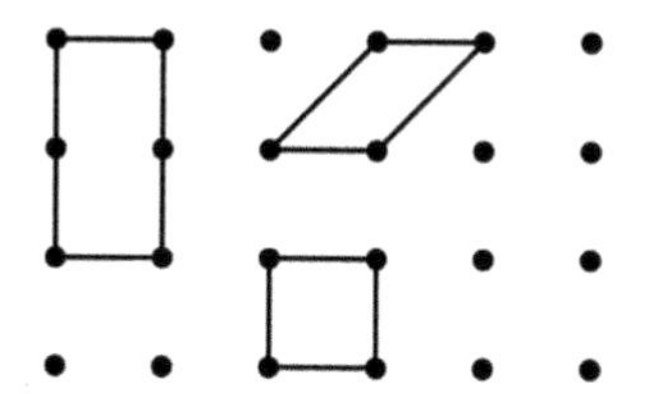

FIGURE 4.8 Three different unit cells are drawn on the lattice. The square cell is the simplest, contains the least number of points within its cell, and provides the symmetry of the lattice.

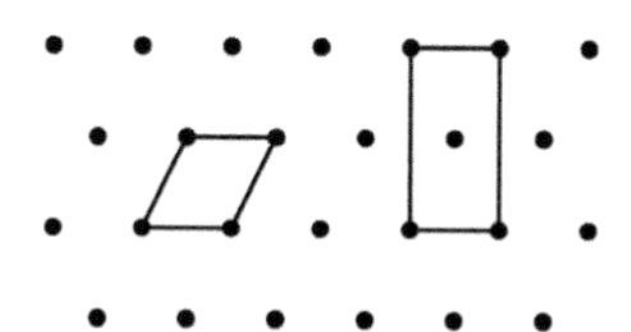

FIGURE 4.9 Both primitive and non-primitive unit cells are shown on the lattice. The non-primitive cell accurately represents the symmetry of the lattice.

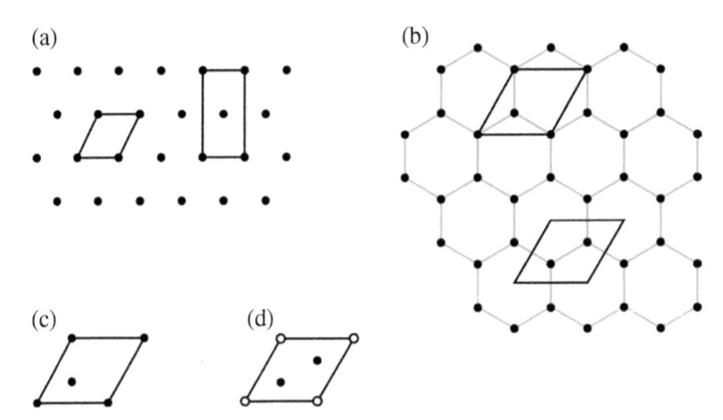

FIGURE 4.10 (a) A hexagonal lattice and the hexagonal unit cell. (b) Hexagonal structure of graphite on which atoms are drawn as open circles. The hexagonal unit cell (identical to that in (a)) is placed on the structure at two different positions, creating two motifs. (c) A motif with atoms at four corners and one interior atom. (d) A motif with two interior atoms in the hexagonal unit cell.

Counting the number of atoms in a three-dimensional lattice is similar, but the contributions to the cell count differ because the atoms are shared in three dimensions rather than just two. The guidelines for counting in three dimensions (applicable to atoms as well as lattice points) are summarized as follows:

- Vertex atom is shared by 8 cells; contribution is ⅛ atom per cell.
- Edge atom is shared by 4 cells; contribution is ¼ atom per cell.
- Face atom is shared by 2 cells; contribution is ½ atom per cell.
- Body or interior atom is unique to 1 cell; contribution is 1 atom per cell.

The most common types of unit cell are primitive (with one lattice point per unit cell), body-centered (with 2 lattice points per unit cell), and face-centered (with 4 lattice points per unit cell). For perspicuity, these three unit cells are shown in Figure 4.11.

Many common metals, such as Cu, Al, Ag, and Au, have face-centered cubic (FCC) structure. If we represent the atoms in the FCC cell as hard spheres, the atoms touch each other along the diagonal of each face. If r is the atom diameter and a is the unit cell dimension, then $a = 2r\sqrt{2}$. A similar calculation can be done to determine the unit cell dimension for the body-centered cubic (BCC) structure. Based on the counting guidelines given earlier, we can also count how many atoms are in a FCC or BCC unit cell. In FCC, the corner atoms are divided between eight neighboring unit cells, so ⅛ of each corner atom is in a cell. The face atoms are shared between only two unit cells. The total number of atoms in an FCC unit cell is therefore $(8 \times 1/8)$ corners $+ (6 \times 1/2)$ face $= 4$ atoms. For BCC, there is one interior atom besides the eight corner atoms; thus there are 2 atoms in a BCC unit cell.

Another way to define the unit cell is by the three *lattice translation vectors* **a**, **b**, and **c**. The lattice parameters a, b, and c are the lengths of the respective vectors. The three lattice vectors define a coordinate system called the *crystal system*; the seven crystal systems defining the 14 Bravais lattices were described in Figure 4.6. Given a unit cell then, one also has to specify positions of all atoms in it. This is normally done in terms of the three lattice translation vectors, with one corner of the unit cell chosen as the origin of the crystal system. For instance, an atom in the center of a cubic cell has coordinates ½, ½, and ½. Crystallographers

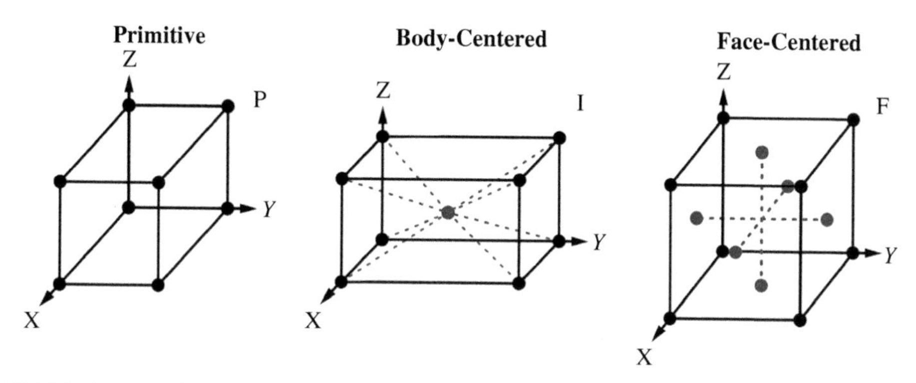

FIGURE 4.11 Unit cells: Simple (or primitive), Body-centered, and Face-centered. Coordinate axes are labeled *x*, *y*, and *z*. Labels are defined as: P = primitive, I = body-centered, F = face- or side-centered.

also need a method to describe the angular position of any given crystal face or plane with respect to the lattice vectors. A plane that makes intercepts in *x*, *y*, and *z* units on the three crystallographic axes can be defined in terms of the so-called **Miller indices**, obtained by carrying out the following operations: determine the intercepts of the plane in terms of unit cell dimensions; take the reciprocals; clear fractions and reduce to lowest (integer) terms; present the indices enclosed in parentheses and without commas or spaces between the numbers. To illustrate the procedure, we shall consider two examples with a cubic lattice.

Corresponding to our two examples, a plane on the other side of the origin would have negative indices. For example, the plane at −1/2, −3/4, and −1/2 (which is parallel to the plane in Example 4.1) has negative indices written as $(\overline{323})$. All planes (in a given crystalline material) with the same set of absolute values of indices, in this case 3, 2, and 3, will have

---

EXAMPLES: CALCULATING MILLER INDICES

Example 4.1 If the *x*-, *y*-, and *z*-intercepts of a plane are ½, ¾, ½, then

- Take reciprocals: 2, 4/3, 2
- Clear fractions: 6, 4, 6
- Reduce to lowest terms: 3, 2, 3

Thus, the Miller indices of the plane are (323). Any plane that is parallel to this particular plane and on the same side of the origin has exactly the same indices: for example, the plane cutting the axes at 1, 3/2, and 1.

Example 4.2 In Figure 4.12, the shaded plane in the cubic lattice is parallel to the *z*-axis. Thus, the intercepts are 1, 1, ∞. To determine the Miller indices:

- Take reciprocals: 1, 1, 0
- Clear fractions: (none)
- Reduce: (already in lowest form)

Miller indices of the plane are (110). Thus, we see how Miller indices allow us to avoid dealing with the value of infinity.

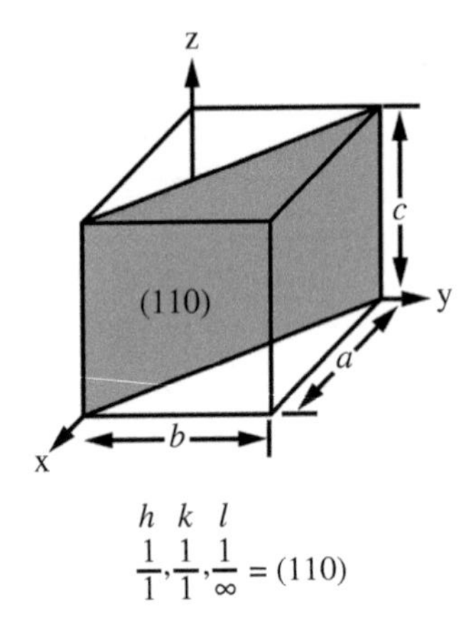

FIGURE 4.12 The (110) crystal plane (shaded)—shown to illustrate the process of determining Miller indices—corresponding to Example 4.2 in the text.

identical arrangements of points, and therefore identical physical properties. This is very important because, in general, the properties of a crystal are different in different directions, that is along different planes. A generic Miller index is denoted (*hkl*) and applies to all crystal systems but hexagonal, in which a fourth axis is considered. In cubic systems, the planes (011), (101), and (110) are parallel to the *x*-, *y*-, and *z*-axes, respectively. Each crystal face intersects two of the crystallographic axes.

The presence of a regular lattice in the structure of crystals has profound effects on the properties of crystalline materials. Professor G.C. Amstutz of the Mineralogical Institute at the University of Heidelberg in Germany said that "Matter's latticed waves are spaced at intervals corresponding to the frets on a harp or guitar with analogous sequences of overtones arising from each fundamental. The science of musical harmony is in these terms practically identical with the science of symmetry in crystals" [5]. Indeed, a software program at http://www.voicesync.org generates resonant three octave compound tones from a list of more than 3000 minerals. The tones are generated using powder x-ray diffraction parameters (corresponding to crystal lattice parameters).

Now let us return to the atoms that make up the crystal. In the simplest approximation, we treat atoms or molecules or ions in a three-dimensional space as points. The next simplest approximation consists in representing them by spheres. In 1611, Johannes Kepler asserted that there is no way of packing equivalent spheres at a greater density than that of a face-centered cubic arrangement. This is referred to generally as the problem of dense packing. In 1953 it was shown that the maximum number of touching nearest–neighbor spheres is twelve [6]. This apparently put an end to many attempts of forcing in a thirteenth sphere somehow. Then in August 1998, Prof. Thomas Hales, at the University of Michigan, announced a computer-based solution as a proof of the Kepler Conjecture (although the 250 manuscript pages and 3 gigabytes of computer files had to be rigorously checked by the scientific community to ensure validity of this proof). Thus, close packing—which amounts to having 12 nearest neighbors—is considered as the most efficient way for equal sized spheres to be packed in three dimensions. In two dimensions, each sphere touches six other spheres. Note that lower density packing is possible; not all materials have close-packed structures. The number of atoms adjacent to any particular atom (i.e., the number of neighbors) is called the *coordination number*. This number is an indication of how tightly the atoms are packed. The simple cubic lattice has a coordination number of 6; BCC has a coordination number of 8; FCC has the theoretical maximum of 12.

The close packing of spheres is realized in the FCC and hexagonal lattices; these are illustrated in Figure 4.13 (in 2D) and in Figure 4.14 (in 3D). In Figure 4.14, spheres are represented by circles. The first layer is denoted A; the second layer is placed on top of the first. There are two alternative sets of hollows or indentations in which the atoms of the third layer can be arranged. If the third layer lies in indentations that place the atoms directly in line (eclipsed) with the first layer, the layering is denoted ABA. This yields the hexagonal close-packed (HCP) structure, shown schematically in Figure 4.14. If instead the third layer lies in the alternative indentations leaving it staggered with respect to both previous layers, we have ABC layering, which is referred to as cubic close–packing (CCP) and is the face-centered cubic (FCC) structure, illustrated in Figure 4.14. In close-packed structures about 74% of the space is occupied. Most of the empty space in between layers is located at octahedral and tetrahedral interstitial sites, shown schematically in Figure 4.13.

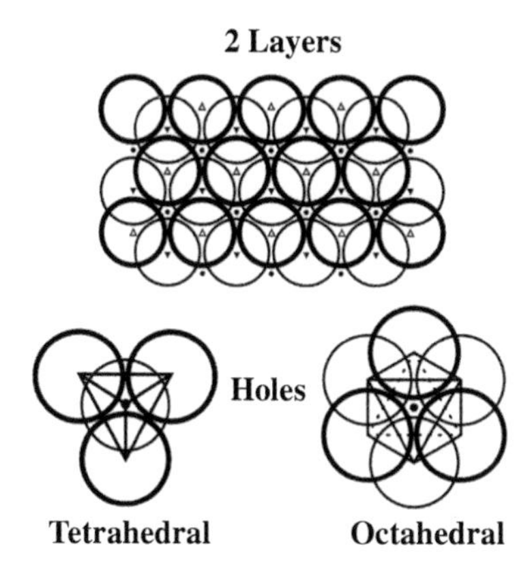

FIGURE 4.13 Close-packing of two layers of atoms, in two dimensions. The holes are referred to as interstitial sites. Tetrahedral holes have four nearest sphere neighbors, distributed between layers. Octahedral holes have six nearest neighbors, also distributed between layers (not in a single plane).

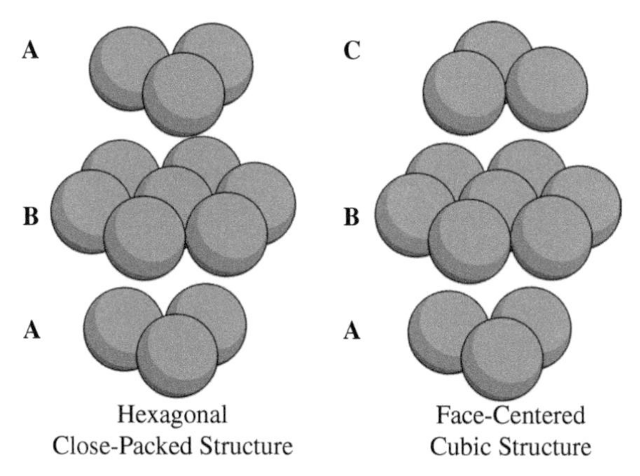

FIGURE 4.14 Three-dimensional hard sphere representation of atomic stacking in close-packed structures.

LATTICE ENERGY

Given a crystal structure and an intermolecular potential $u(R)$, one can calculate the lattice energy of a crystal. Let us consider a phase consisting of N atoms of one kind only with each of the atoms having z nearest neighbors. The simplest possible assumption is that each atom rests at the minimum of the potential well of interaction with each of its neighbors, that is, at the point R_m shown in Figure 2.5. Then the configurational energy U^c of the crystal is

$$U^c = \frac{Nzu_m}{2} \tag{4.9}$$

where u_m is the minimum of the interaction potential versus distance $u(R)$ curve in Figure 2.5, that is at R_m. The factor ½ has been included in order to not count each interaction twice. In reality, atoms do vibrate around their lattice positions, and this produces the so-called acoustic contribution U^{ac} to the lattice energy. However, we can make the assumption that the configurational and acoustic contributions are independent of one another and therefore write the lattice energy as

$$U^{latt} = U^c + U^{ac}. \tag{4.10}$$

In Chapter 1 we stressed the fact that structure plus interactions determine properties of a material. Crystal phases illustrate well the basic rule outlined in Section 1.4, that if we know any two vertices of the MSE triangle we can determine the third. Consider first that three quarters of all metals crystallize in only three structures: hexagonal close packed, face-centered cubic, and body-centered cubic. Second, the bonding (a.k.a. interactions) in metals is (not surprisingly) predominantly metallic. Thus, metals are typically excellent electrical conductors, malleable and ductile to some degree, lustrous, and have moderate to high melting points.

An important feature of crystals is the possibility of **polymorphism**. Two or more crystals are polymorphs if they have the same chemical composition while consisting of different arrangements of identical structural units. Carbon, which may crystallize as diamond or as graphite, is a good example. The diamond crystal lattice is shown in Figure 4.15; from the point of view of the Bravais classification, this is the FCC lattice with two atoms per lattice point. Gallium arsenide has a similar structure but with a different motif. Another material with polymorphic crystal forms is the element tin: above 286 K, the stable form is β-tin or "white" tin, which is shiny and metallic; below 286 K, the non-metallic α-tin or "gray" tin is stable. The former has a tetragonal structure, the latter a cubic one. In practice, the entire transition does not occur all at once at exactly 286 K; the transition time is slow, and the temperature can be lowered to some extent with the white tin remaining in a metastable state. An interesting anecdote about the effects of the transition between these two forms involves the army of the famed Napoleon Bonaparte, a French military and political leader [7]. When Napoleon's army invaded Moscow in the winter of 1812, the temperature dropped so low for so long that the tin buttons of the soldiers' uniforms crumbled, at which the soldiers

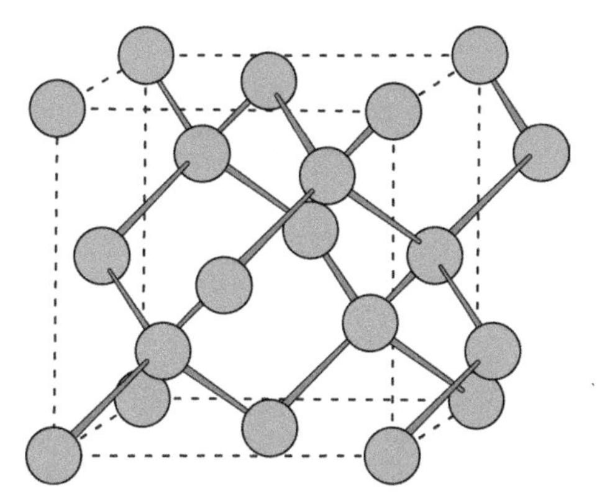

FIGURE 4.15 The diamond crystal lattice.

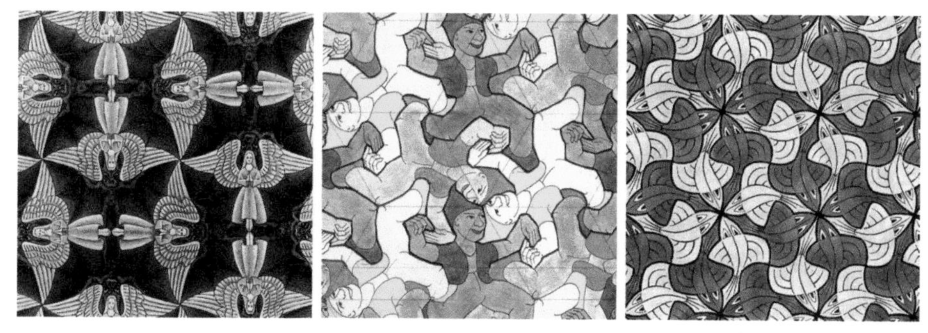

FIGURE 4.16 Crystalline symmetry: from the left, 2-fold, 3-fold, and 4-fold. *Source*: The M.C. Escher Company—The Netherlands. M.C. Escher's "Symmetry Drawing E45", "Symmetry Drawing E21", and "Symmetry Drawing E119" © 2013 The M.C. Escher Company—The Netherlands. All rights reserved. www.mcescher.com.

lost morale, believing it was caused by the wrath of God. In reality, the metallic white tin underwent a polymorphic transition to the nonmetallic gray tin.

Crystal structures also represent the symmetry found in nature. Symmetry can be considered from a mathematical or aesthetic standpoint, while it has implications also for behavior of chemical species. Artistic versions of 2-fold, 3-fold, and 4-fold symmetry in crystals by Maurits C. Escher—whose brother was a crystallographer—are shown in Figure 4.16. It should be noted that X-ray diffraction (XRD) patterns are the primary means of identifying the symmetry of crystalline materials; thus XRD is the key tool of crystallographers. Materials with 5-fold symmetry were discovered in 1980s [8] and will be discussed in the next Chapter. Such symmetry is not defined by any of the space groups in Figure 4.6 and therefore does not allow for regular lattice translations. Such materials are called quasi-crystals, and we shall consider them in Section 5.1.

Apart from all else, a crystal is indeed something to admire. However, a perfect crystal with every atom in place simply does not exist. Thus, the next section is devoted to structural defects of crystals.

4.4 DEFECTS IN CRYSTALS

Colin Humphreys, Goldsmiths' Professor of Materials Science at Cambridge University, said [9] that "Crystals are like people; it is the defects in them which tend to make them interesting!" Most crystal properties are determined by the defects present. Since this may seem unusual, consider a simple example: suppose we want to tear a piece of fabric in half. It may be quite difficult to initiate a tear. However, once a start has been made, perhaps by an incision with scissors, continuation of the tear is relatively easy. Similarly in crystals, we find that a defect or a dislocation might be difficult to produce, but once it exists it is easy to propagate. The defect, like the incision, might be small in size, but the property of the whole—in our example the resistance of fabric to tearing—is changed considerably. In general it is true that very small defects, of microscopic or even atomic size, are capable of producing large changes in properties of materials.

When solid phase nuclei grow as described in Section 4.2, they "swallow" more and more molecules from the liquid phase. In the last stages of the freezing process there are only a few layers of liquid molecules between the solid **grains**. A liquid molecule between two crystalline grains will interact with both of them. Thus, liquid layers that are the last to freeze form a transition zone between grains; the specific structure of this zone, known as a **grain boundary**, depends on the nature and the orientation of the adjacent grains. As a consequence of the growing process, crystals exhibit several kinds of **structure defects**. They are classified into groups, differing in dimensionality and size:

- Zero-dimensional imperfections or **point defects**.
- One-dimensional imperfections or **line defects**.
- Two-dimensional imperfections include **planar defects** and **interfacial defects**.
- Three-dimensional imperfections or **bulk defects**.

In a pure material, point defects come in two types. **Vacancies** are empty lattice sites. **Interstitials** are atoms placed between lattice sites. The cases of missing or extra electrons are dealt with separately in our discussions of electrical conductivity. We can also have impurities or foreign atoms in a material. **Substitutional impurities** sit on the lattice sites; **interstitial impurities** sit between the sites. As might be expected, foreign atoms found as interstitials are typically small compared to the size of the matrix atoms. Likewise, foreign atoms that are larger in size can typically only be accommodated as substitutional impurities. Vacancies and interstitials are represented in a simple diagram in Figure 4.17. As a result of point defects, atoms are displaced from perfect lattice sites. The resultant distortion produces a displacement field around the dislocation, creating a stress field that introduces compression and tension in the material.

Vacancies are present in all crystalline materials. Although they increase entropy, they decrease thermodynamic stability because of other energy effects (e.g., higher interfacial energy). Vacancies affect the rate of diffusion of atoms or ions through a solid. Moreover they can be used to tune electrical properties in a solid, for example in ceramics.

An interstitial defect, namely, an extra atom or ion inserted in between lattice sites, introduces compression and distortion. This effect can actually strengthen metallic materials; such a defect provides resistance to the movement of larger dislocations, thereby making permanent deformation difficult. Substitutional defects, replacing an atom or ion with a different one, may have similar effects. The doping of P or B into the normal lattice sites of Si is an example of substitutional defects.

How can one determine the number of defects in a crystal? Crystal defects exist in thermodynamic equilibrium, and the Gibbs function of the imperfect crystal is lower than that for the perfect crystal. Consider the change of the Gibbs function $\Delta G = \Delta H - T\Delta S$ upon formation of the defects. For simplicity let us assume that all defects are identical, such as one kind of impurity. On this basis one obtains

$$n_{\mathrm{d}} = Ne^{\left(\frac{-E_{\mathrm{d}}}{kT}\right)}. \tag{4.11}$$

Here n_{d} = number of defects; N = number of lattice sites per cm^3 or mole; E_{d} = energy of defect formation; as before, T = absolute temperature and k = Boltzmann constant. We have mentioned the information theory approach to thermodynamics in Chapter 3. Following that approach, we would find the following: probabilities are exponential because the formula for entropy (the Second Law of Thermodynamics) is logarithmic.

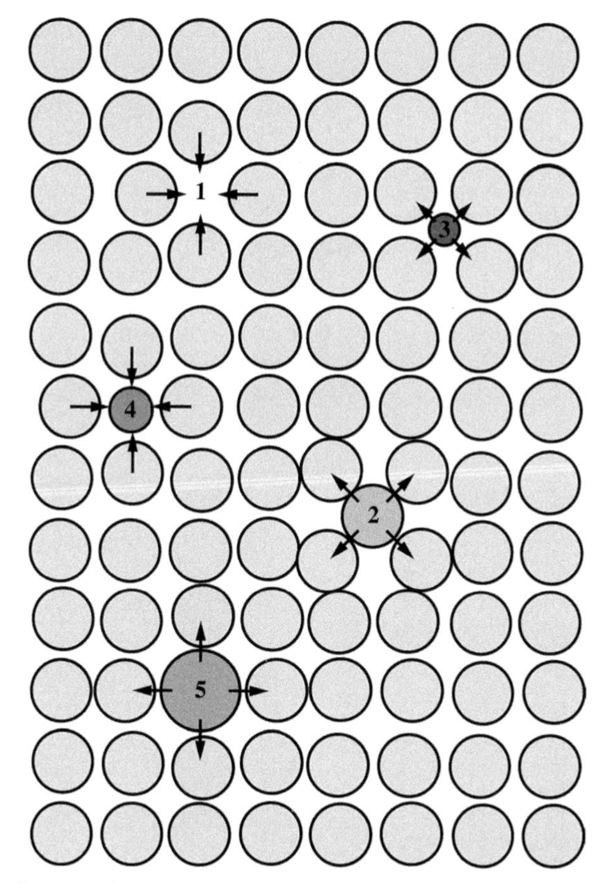

FIGURE 4.17 Schematic representation of point defects in a crystal: (1) vacancy; (2) self-interstitial (same atom type as matrix atoms); (3) interstitial impurity; (4) and (5) substitutional impurities. Arrows indicate the local stresses imposed by the point defects. *Source*: Adapted from http://phelafel.technion.ac.il/~korens/defects.pdf. Reprinted with permission from L.V. Zhigilei, University of Virginia.

In ionic crystals, we can also identify *pairs* of point defects. Owing to the requirement of charge neutrality in ionic crystals, a cation vacancy must be coupled with either an anion vacancy or a cation interstitial. The reverse pairing applies to an anion vacancy. The defect pair cation vacancy + anion vacancy is referred to as a **Schottky imperfection**. The defect pair of vacancy + interstitial is known as a **Frenkel defect**. Further kinds of point defects may be produced in ionic lattices by foreign ions that carry different charges than the lattice ions.

Crystal defects of higher dimensions can be considered as collections of point defects. Line defects are of two types: **edge dislocations** and **screw dislocations**, both pictured in Figure 4.18. Mixed dislocations involve superposition of edge and screw components. Furthermore, dislocations can move around; the movement is referred to as **slip**.

Dislocations are characterized by the so called Burgers vector, named after Dutch physicist Jan Burgers. We compare a perfect crystal lattice (Figure 4.18a) with crystals containing a dislocation (Figure 4.18b and 4.18c). Consider Figure 4.18b, which shows an edge dislocation. Start at the point M, go down 3 atoms to point N, then left 5 atoms to

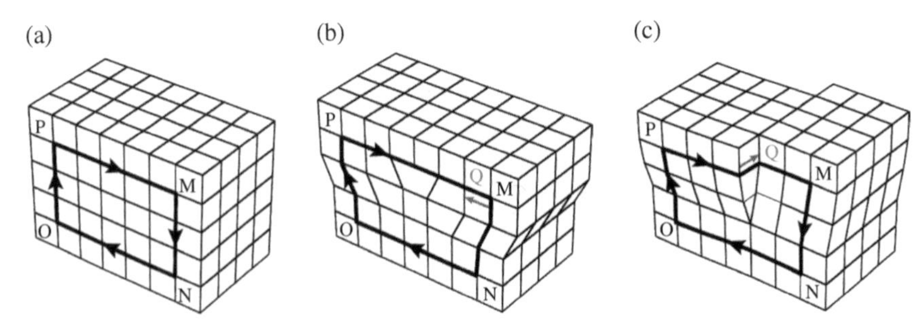

FIGURE 4.18 Representations of line defects, shown with a corresponding perfect lattice (a) for comparison: (b) edge dislocation and (c) screw dislocation. Burgers vectors are shown (shaded arrows) for each type of dislocation.

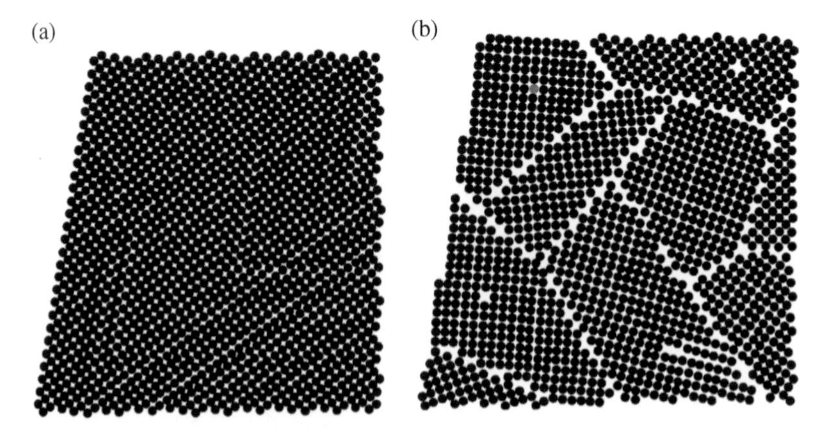

FIGURE 4.19 Two-dimensional representation of (a) a perfect single crystal and (b) a poly-crystal with many defects and grain boundaries. In (b) one can see defects such as extra atoms (of the same and different (gray colored) types), missing atoms, and a missing half row of atoms. *Source*: Modified after http://phelafel.technion.ac.il/~korens/defects.pdf. Reprinted with permission of Prof. Helmut Föll at the University of Kiel.

point O. In a perfect crystal one would return to the point M by going up 3 and right 5 atoms. However, because there is an edge dislocation present, going from point P right to point M, one has to take an extra step; this extra step is the **Burgers vector**.

There is a similar situation in Figure 4.18c, which represents a screw dislocation. If we start at point M, move downward to N then left to O and back up to P, we find that to get back to M we need to traverse 6 rather than the expected 5 atoms. Because of the screw dislocation, we need to take an extra step. As with the edge dislocation that extra step is the Burgers vector.

Most two-dimensional defects involve interfaces, thus it behooves us to consider what kind of interfaces we have in crystalline materials. The following list distinguishes the relevant types of interfaces:

1. Interfaces between solids and gases are called **free surfaces**.
2. Interfaces between regions where there is a change in the electronic structure but no change in the periodicity of atom arrangement are known as **domain boundaries** (discussed in Chapter 18).
3. *Grain boundaries* were discussed earlier in this section.
4. Interfaces between different phases are called **interphase boundaries**; this involves a change in structure and chemical composition across the interface.

Consider the perfect single crystal illustrated in Figure 4.19a. The only interfaces are at the edges, the free surfaces in contact with the surrounding air. But we know that nothing is perfect; the structure of a real material is more like that of the poly-crystal in Figure 4.19b. There are many defects; apart from missing and extra atoms, there is an interrupted row of atoms and there are grains separated by boundaries.

Grain boundaries are frequent. **Twin boundaries** are a special case of tilt boundaries in which the grain structure on one side of the interface mirrors exactly the structure on the other side. An example is shown in Figure 4.20. During twinning formation caused by deformation of HCP tin, there is sound emission; this phenomenon is called **crying tin**. **Shape-memory alloys** are twinned. Deformation causes them to untwin, and upon heating the alloy returns to the original twin configuration, thus restoring the original shape. We shall discuss shape-memory materials in Section 12.6. Twin boundaries are preferentially formed in crystals with low symmetry, such as HCP or monoclinic.

Stacking faults are a kind of planar defect not involving an interface. As can be expected, they involve a defect in the stacking sequence of crystalline layers. We described, for instance, in Section 4.3 the layers for FCC and HCP structures. A stacking fault is essentially an agglomeration of vacancies and additionally is bordered by an edge dislocation.

Lastly, there are **bulk defects** in solid crystalline materials. These may be introduced during the production or fabrication of materials. **Inclusions** are unwanted second-phase particles introduced during production; small portions of clay appearing in ordinary glassware are an example. The most important kinds of bulk defects arising in fabrication are:

- **Casting defects**: examples are shrinkage cavities from volume contraction and gas holes from evolution of gases during solidification.
- **Working or forging defects.**
- **Welding or joining defects.**

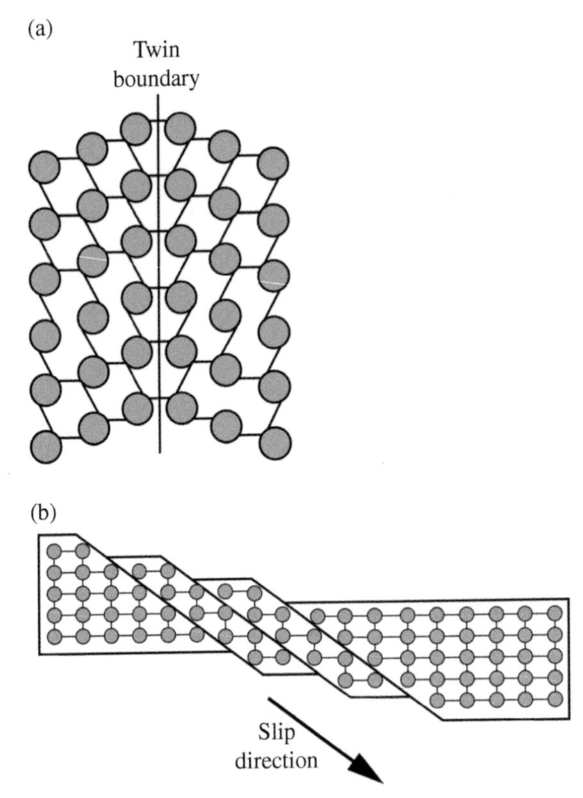

FIGURE 4.20 (a) Schematic representation of a twin boundary. (b) Schematic of twin (or tilt) boundaries from slip due to deformation.

TABLE 4.1 Crystalline Defects, Corresponding Affected Properties, and the Practical Applications

Type of Defect	Properties Affected	Application or Use
Impurity atoms—Substitutional	electrical conductivity, mechanical strength	Semiconductor diodes and transistors
Impurity atoms—Interstitial	magnetic coercivity, dielectric strength, optical transparency	Coloring of glasses and plastics
Vacancies	solid-state diffusion, mechanical creep	Annealing
Dislocations, Twin boundaries	plastic (permanent) deformation, ductility	Strain hardening
Grain boundaries	dislocation movement, electrical resistance, magnetic coercivity, optical transparency	Mechanical hardening, strong permanent magnets
Stacking faults	mechanical strength, electrical resistance	Mechanical strengthening, conductivity modulation

Defects of the latter two categories are typically **cracks**, which we were familiar with before studying MSE.

It bears repeating that defects have a significant effect on material properties. Those effects are not always detrimental. Table 4.1 summarizes the effects of several defect types

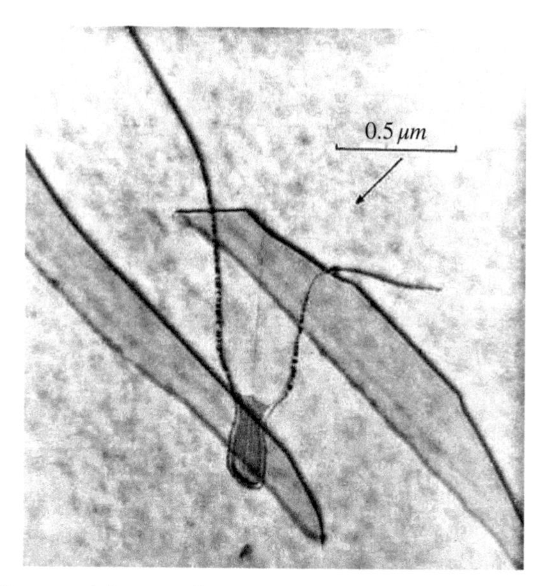

FIGURE 4.21 Micrograph image of a crystal lattice defect. *Source*: Föll [10]. Reproduced with permission of Prof. Helmut Föll at the University of Kiel.

and indicates features or applications that use the defects to advantage. The properties and applications listed are discussed further in Part 3 of this textbook, from which the importance of defects will be increasingly evident. Figure 4.21 shows a real crystal lattice defect. In the words of Helmut Föll, "It is remarkable because:

- It contains all 5 kinds of one-dimensional defects (dislocations) that occur in FCC crystals.
- It contains a two-dimensional defect (stacking fault); two more stacking faults are seen in the background.
- It is partially decorated with three-dimensional defects (precipitates).
- It came into being by the agglomeration of zero-dimensional defects: Si interstitials.
- It was formed during the processing of Si integrated circuits
- It is a killer defect, short-circuiting a transistor.
- It can only be identified by transmission electron microscopy, one of the most important methods for the investigation of defects" [10].

4.5 SELF-ASSESSMENT QUESTIONS

1. Very small crystals floating in a liquid will not grow naturally. Why?
2. Consider a body centered cubic lattice (BCC). How many atoms are inside the unit cell?
3. Describe how point defects affect properties of crystals.
4. In an ionic lattice a cation "walks away". What do you expect will happen with the cation removed from or moved within the structure?

REFERENCES

1. K. Preston & M. Lipson, *Optics Express* **2009**, *17*, 1527.
2. P. Ehrenfest, *Proc. Kon. Akad. Amsterdam* **1933**, *36*, 153.
3. R. Roy, A syncretist classification of phase transitions, in *Phase Transitions—1973* (H. K. Henisch and R. Roy, Eds.), pp. 13–28, **1973**, Pergamon: New York.
4. M. Pokorny & H. U. Astrom, *Phys. & Chem. Liquids* **1972**, *3*, 115.
5. G.C. Amstutz, Introduction, in *Sacred Geometry: Philosophy and Practice* (R. Lawlor, Ed.), p. 4, **1982**, Thames & Hudson: New York.
6. K. Schutte & B.L. van der Waerden, *Math. Ann.* **1953**, *125*, 325.
7. B. Weintraug, Tin Disease and Ernst Julius Cohen (1869–1944), Chemistry in Israel, *Bull. Isr. Chem. Soc.* **2002**, April (Issue 9), 31–32.
8. D. Shechtman, I. Blech, D. Gratias, & J.W. Cahn, *Phys. Rev. Letters* **1984**, *53* (20), 1951.
9. J.J. Hren, J. Goldstein, & D.C. Joy (eds.), *Introduction to Analytical Electron Microscopy*, Plenum Press: New York **1979**. Also quoted in Chapter 2 of *Structure-Property Relations in Nonferrous Metals* by A.M. Russell & K.L. Lee, **2005**, John Wiley & Sons, Inc.: Hoboken, NJ.
10. H. Föll, *Defects in Crystals*, http://www.tf.uni-kiel.de/matwis/amat/def_en/index.html, accessed April 30, **2015**.

5

NON-CRYSTALLINE AND POROUS STRUCTURES

Jakichże przyczyn kołowrót mógł wyrzucić kostkę takiego skutku?
—Helena to Ludmir in the play "Pan Jowialski" by Aleksander Count Fredro; translated from the Polish: *What carousel of causes could have tossed out such an outcome?*

5.1 QUASICRYSTALS

Quasicrystals were mentioned in the previous chapter; here we shall add to the discussion. But first refer back to the image in Figure 1.3. We said that one sometimes has to play the detective when trying to understand a material's behavior. In the motto above, Helena similarly stresses the causality to Ludmir—asking for a reason for the situation the young man has just described to her. Is causality easy to follow? Furthermore, if you see an effect you cannot assign to a known cause, are you in error or are you seeing something really new?

In 1982 Dan Shechtman of Technion (National Israel Institute of Technology in Haifa) was on his sabbatical working at the US National Bureau of Standards (NBS, now the National Institute of Standards and Technology or NIST) in Gaithersburg, Maryland. He discovered in an Al + Mg alloy diffraction patterns resembling those of a single crystal. However, the structure was inconsistent with lattice translations we have seen in the previous Chapter. Shechtman presented his results first to his colleagues at NBS. Not only was he ridiculed; he was expelled from the research group to which he belonged. Back in Haifa, Shechtman enlisted the cooperation of Ilan Blech. They described their results in a paper submitted to the *Journal of Applied Physics*—it was rejected. Then they added two widely respected colleagues—and the four of them published in 1984 the first paper on

Materials: Introduction and Applications, First Edition. Witold Brostow and Haley E. Hagg Lobland.

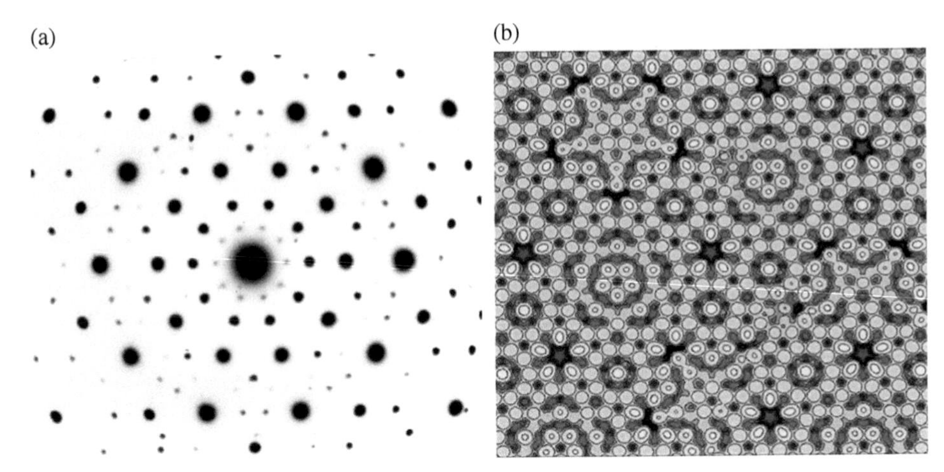

FIGURE 5.1 (a) Original electron diffraction pattern from the icosahedral phase. *Source*: Shechtman *et al.* [1]. Reproduced with permission of American Physical Society. Image courtesy of D. Shechtman. (b) Image of the potential energy surface (PES) of an adsorbed atom on a 5-fold icosahedral quasicrystal surface. Source: Reprinted from P.A. Thiel, B. Unal, C.J. Jenks, A.I. Goldman, P.C. Canfield, T.A. Lograsso, J.W. Evans, M. Quiquandon, D. Gratias & M.A. Van Hove, A distinctive feature of the surface structure of quasicrystals: intrinsic and extrinsic heterogeneity. *Israel J. Chem.* **2011**, *51*, 1326–1339, Copyright (2011) with permission from John Wiley & Sons.

quasicrystals—this time in *Physical Review Letters* [1]. The earlier paper by Shechtman and Blech was published a year later in yet another journal [2].

Shechtman received the 2011 Nobel Prize in Chemistry for this discovery. He says quasicrystals can be used for strengthening stainless steel; among the applications are electric shavers. Among non-crystalline materials, quasicrystals are the closest to crystals. What are quasicrystals? Look at Figure 5.1. **Quasicrystals** have structures which are ordered but not periodic. A quasicrystalline pattern can *fill the space fully*—an essential requirement for any structure. As already noted, the pattern does *not* have translational symmetry; recall the procedure of using replicas of a crystal unit cell to create a whole crystal. Bear in mind that crystals can have only 2-, 3-, 4-, and 6-fold rotational symmetries; refer to Figure 4.16 which shows three of these four symmetries. Quasicrystals can have, for instance, 5-fold symmetries. Shechtman and his colleagues describe quasicrystals as having structures intermediate between crystals and liquids [1]. In Sections 5.3 and 5.4, we shall talk about two ways of characterizing structures of *all* phases.

5.2 MINERALOIDS

There is a widely held opinion that minerals have to be crystalline. Indeed, according to the Merriam-Webster Dictionary, a mineral is a solid homogeneous crystalline chemical element or compound that results from the inorganic processes of Nature. However, **opals** are not crystalline while they are one of the most precious gemstones used in jewelry. Therefore, proponents of terminological purity (a mineral is a crystal…) invented new terms: opals are **mineraloids**, or else they are **colloidal crystals**. There seems to be no generally approved definition of a mineraloid, except for: looks like a mineral but is not

FIGURE 5.2 An opal, with the play of colors visible. *Source*: Dpulitzer, https://commons.wikimedia.org/wiki/File:Coober_Pedy_Opal_Doublet.jpg, Coober Pedy crystal opal from Dead Horse Gully Field with novaculite backing (doublet). Used under CC-BY-SA-3.0.

crystalline. The British and World English Dictionary says that a mineraloid differs from a mineral in some respect, especially in not having a crystalline structure.

The chemical structure of opals is $SiO_2 \cdot nH_2O$; n is a variable. The phase structure consists of small colloidal silica spheres (such as with 10 μm diameter) arranged in certain local repeating unit patterns. There are in fact two kinds of patterns: those formed by those spherical particles and those formed by voids between them. Herein lies the difference between crystals and mineraloids since, as we know, in crystals the repeating units are atoms or molecules.

An example of an opal is shown in Figure 5.2. Different colors may be seen when viewed from different directions, a phenomenon called **play of colors**. We shall return to opals in Section 16.3.

5.3 DIFFRACTOMETRY

Thus far, we have not discussed the methods of obtaining experimental information on structures, except for a brief mention. The key technique is **diffractometry**. When a beam of radiation or particles, most often *X-rays*, *electrons*, or *neutrons*, "attacks" a material, a scattering pattern is obtained [3]. The pattern is a source of information on the material structure; it can be obtained for crystals, non-crystalline solids, liquids, and also dense gases. A schematic of an X-ray diffraction machine is shown in Figure 5.3. As seen there, angle of incidence θ is the half of the scattering angle.

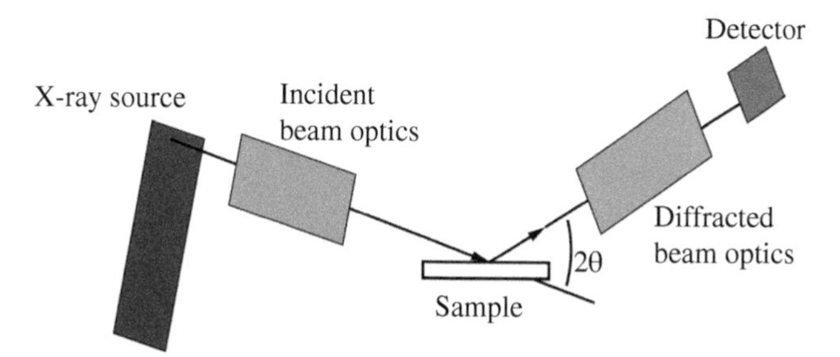

FIGURE 5.3 A schematic of the X-ray diffraction principle for a material in reflection geometry; 2θ is the scattering angle.

To understand diffractometry better, we need to invoke here the **wave-particle duality**; all particles exhibit both wave and particle properties. More accurately, in 1924 Prince Louis-Victor-Pierre-Raymond de Broglie stated in his doctoral dissertation [4]:

$$\lambda = \frac{h}{p} \tag{5.1}$$

Here λ = wavelength, h = Planck's constant, p = momentum = mass·velocity. de Broglie received his PhD in 1924 not because the committee understood or agreed with what he had written but because he was a prince.... He published his results in a journal a year later [5].

Using Eq. (5.1), we find that electrons and neutrons have wavelengths smaller than a nanometer. Electrons do not penetrate as deeply into matter as X-rays, hence electron diffraction reveals structure *near the surface*. Neutrons do penetrate easily and have an advantage that they possess an intrinsic magnetic moment that causes them to interact in distinctive ways with atoms having different alignments of their magnetic moments. In actual experiments the sample is gradually rotated and the pattern of diffraction recorded as a function of the orientation. In the case of electrons going into the material, the results depend on the electron density as a function of the location in the material. This is in fact what we need, information in terms of the positional coordinates in the material. To obtain such information from the scattering patterns, one performs a mathematical operation called the Fourier transform (after the French mathematician Joseph Fourier, the University in Grenoble bears his name). We thus get away from the wavelength (or frequency) and scattering angles.

Specifically for X-rays going into a crystal, William Lawrence Bragg formulated what we now call Bragg's law [6]:

$$2d \sin\theta = n\lambda \tag{5.2}$$

Here d is the interplanar distance (i.e., lattice spacing), and n is an integer. As stated earlier, θ is the angle of the incident wave. Thus the path length difference between incident and scattered waves is an integer number of wavelengths.

5.4 THE BINARY RADIAL DISTRIBUTION FUNCTION

The **binary radial distribution function** $g(R)$ is also called the **pair correlation function**. Here R is the distance between two particles such as atoms, molecules, ions, polymer chain segments, or colloid particles. There can be "mixed" particles such as an atom and an ion. $g(R)$ is the probability of finding a particle at a distance R from a reference (central) particle. In practice, we take a narrow interval of distances $R + dR$. To better understand the concept of $g(R)$, see Figure 5.4.

If one takes dR as somewhat larger than the particle diameter, one can form shells from the particles surrounding the central (black) one. This is shown in Figure 5.5. We see that the first shell (dark gray particles) has all particles at approximately the same distance from the central one. However, at bottom left we see that light gray particles are at almost the same distance as one of the white particles. In other words: with increasing R, the shells are

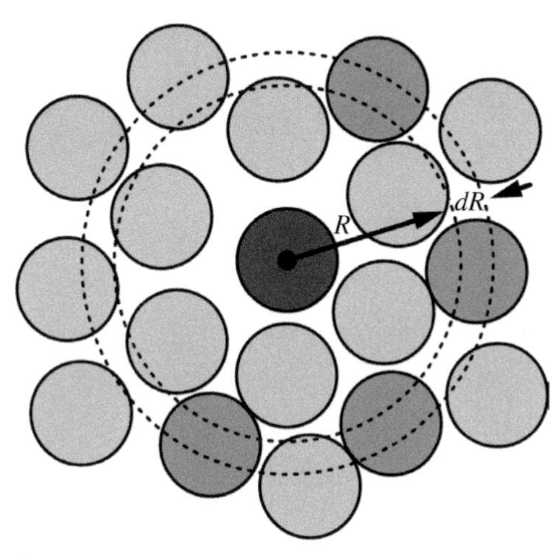

FIGURE 5.4 The scheme presents the concept of radial distribution. It illustrates the method for obtaining the radial distribution function $g(R)$ for a system of atoms, as described in the text.

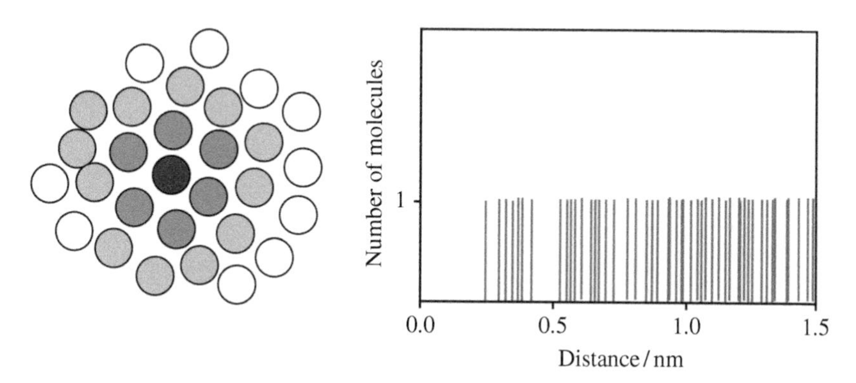

FIGURE 5.5 An illustration of shells surrounding an atom and the corresponding histogram as a function of distance R. Distances correspond roughly to gaseous Ar. *Source*: R.K. Thomas, *Liquids and Solutions: 2nd Year Michaelmas Term: Introduction*, http://rkt.chem.ox.ac.uk/lectures/liqsolns/liquids.html. Reprinted with permission from R.K. Thomas, Oxford University.

less and less "organized" or more and more diffuse. This is reflected in the histogram shown beneath the atomic arrangement. Initially we see a cluster of atoms at $R < 0.5$ nm, corresponding to the first shell of atoms surrounding the reference atom. As we move further away from the central atom, it becomes increasingly difficult to distinguish a pattern for any subsequent coordination shells.

How does one determine $g(R)$? As noted above, the diffractometry results subjected to the Fourier transform give us the materials structure, which can be expressed as $g(R)$. See some results in Figure 5.6 for argon. While these results are per unit volume in nm^3, we have to take into account that the volume of each shell is proportional to R^3. A correction can be taken for this, and normalized $g(R)$ functions for Ar are shown in Figure 5.7.

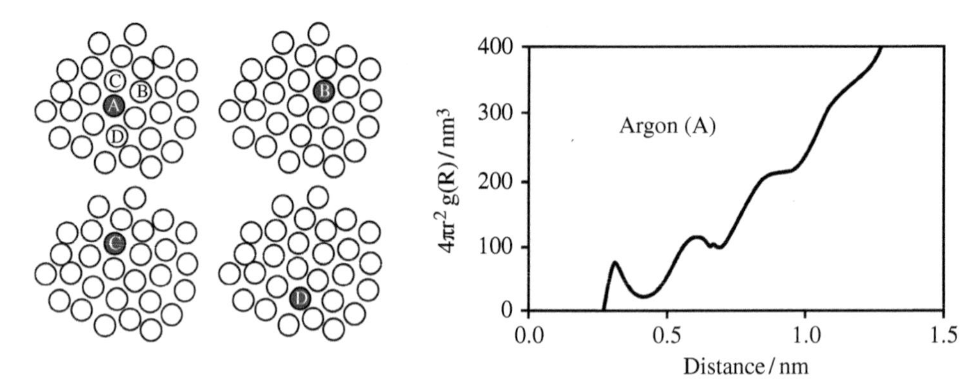

FIGURE 5.6 Construction of the radial distribution function $g(R)$, called in this figure $n(r)$, for Ar. Shown is the function based on atom A as the reference. (y-axis label is $4\pi r^2$ $g(R)$ / nm^3.) *Source*: R.K. Thomas, *Liquids and Solutions: 2nd Year Michaelmas Term: Introduction*, http://rkt.chem.ox.ac.uk/lectures/liqsolns/liquids.html. Reprinted with permission from R.K. Thomas, Oxford University.

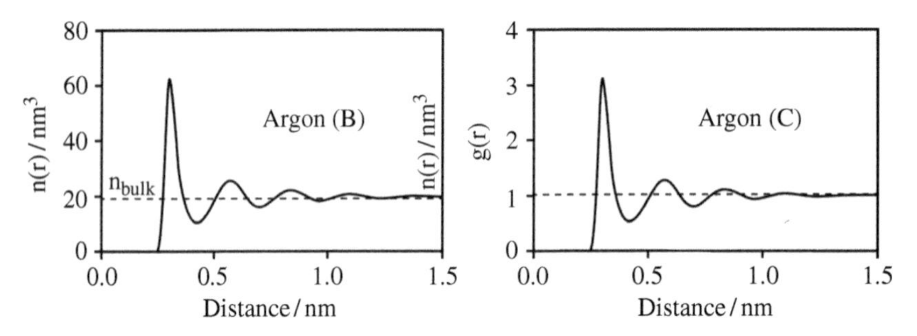

FIGURE 5.7 The normalized radial distribution function $g(R)$ for two argon samples; B and C notations refer to atoms specified in Figure 5.6. *Source*: R.K. Thomas, *Liquids and Solutions: 2nd Year Michaelmas Term: Introduction*, http://rkt.chem.ox.ac.uk/lectures/liqsolns/liquids.html. Reprinted with permission from R.K. Thomas, Oxford University.

The normalization also is performed with respect to the bulk density n_{bulk} such that at $R = \infty$ we have $g(R) = 1$.

The $g(R)$ diagram obtained from diffractometry results is the average over the whole sample. In our representation this corresponds to each atom in turn considered as the central atom (identified by dark shading and in Figure 5.6 named by letters A–D) before performing the averaging. The radial distribution function method can be applied to any material. Staying with argon, let us look at its $g(R)$ diagrams in the solid, liquid, and gaseous states. In Figure 5.7 we show the diagrams for two solids which were labeled B and C in Figure 5.6. The distinctive pattern for solid state Ar is immediately evident. In Figure 5.8 we repeat the same diagram for the solid, but we also include a $g(R)$ diagram for a liquid as well as for two Ar gas materials at two temperatures. Though the peaks are wide and the spacing between them varies, the peaks show higher densities than in the liquid or gaseous phases. We can also appreciate the difference between Ar in the gas state at 90 K versus 300 K. We see what the effect of raising the temperature (and thus also increasing the volume of the material) is on $g(R)$.

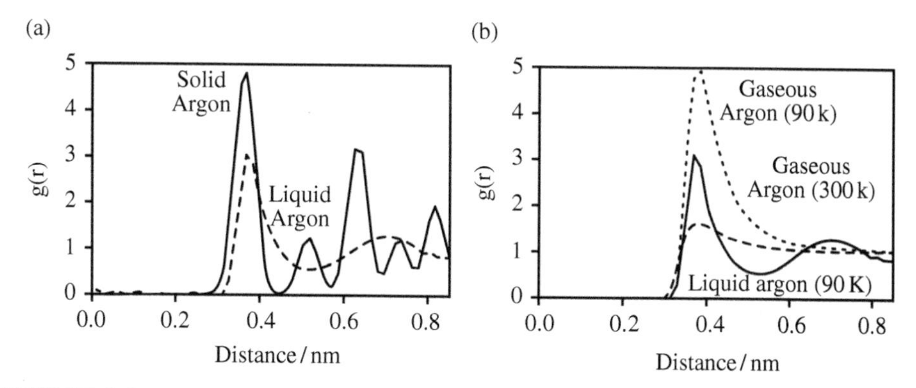

FIGURE 5.8 (a) $g(R)$ functions for solid and liquid Ar. (b) $g(R)$ functions for liquid and gaseous Ar. *Source*: R.K. Thomas, *Liquids and Solutions: 2nd Year Michaelmas Term: Introduction*, http://rkt.chem.ox.ac.uk/lectures/liqsolns/liquids.html. Reprinted with permission from R.K. Thomas, Oxford University.

We stated above that the $g(R)$ approach works for any material. How about crystals? Along any crystallographic direction, the $g(R)$ function looks like a series of very narrow peaks centered at specific distances. Thus the $g(R)$ approach provides exactly the same information as crystallography does—expressed in a different form.

5.5 VORONOI POLYHEDRA

At the end of Section 5.1 we said that there are two methods for describing structure of any material type. We shall now describe the second method. It is based on a construction of the Ukrainian mathematician Hryhoriy Voronoi. Born in the village of Zhuravka near Kyiv in 1868, he created his seminal papers in the first decade of the 20th century while working in Warsaw [7–9]. Mathematics then was already truly international; he published his papers in French (writing his first name as Georges) in a German journal [6, 7] while working in a predominantly Polish environment.

To understand what Voronoi created, let us first consider the principle in two dimensions; see Figure 5.9. As in the $g(R)$ approach, the particle can be an atom, a molecule, an ion, or a polymer chain segment. Also similarly, we pick a central particle. We then draw lines connecting that central particle with centers of its neighbors. On each such line we draw at

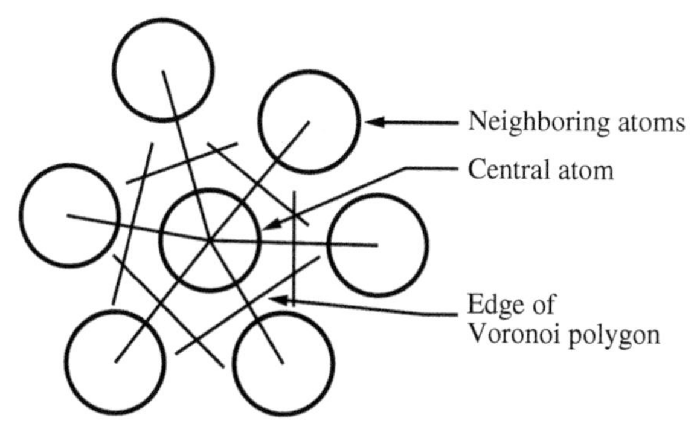

FIGURE 5.9 The Voronoi polygon for a particle.

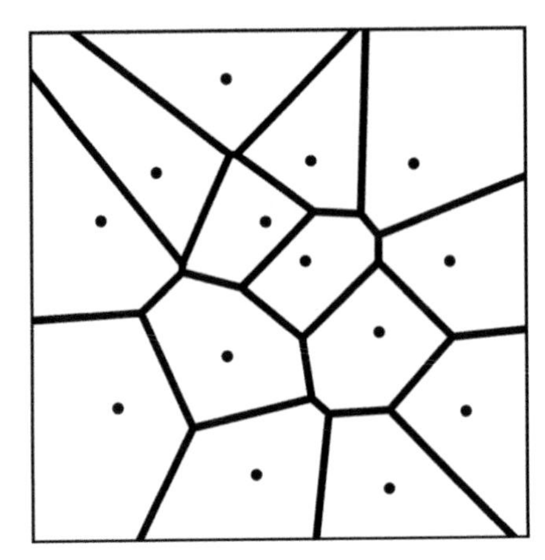

FIGURE 5.10 An example of Voronoi tessellation in 2 dimensions. Particles are shown as dots with polygon defined by thick lines. *Source*: Modified after [9]. Reprinted with permission from *Journal of Materials Education*.

the midpoint a line perpendicular to the connecting line. The perpendicular lines form a polygon. The particle in the center now becomes the "owner" of that polygon. The same can be done for other particles in the system. An example of a result is shown in Figure 5.10.

The procedure for creating a set of Voronoi polygons is called **Voronoi tessellation**. It can be performed also in 3 dimensions; instead of polygons we then have **Voronoi polyhedra**. Instead of bisecting lines we have in 3D dissecting faces. The Voronoi polyhedra have a number of properties important for us:

- The entire volume is divided between points.
- There is only one way of constructing the Voronoi diagram, thus one set of Voronoi polyhedra for a given set of points.
- This method can also be used to represent a crystalline structure, which would merely result in the unit cell.
- Main applications of the Voronoi method in MSE are for liquids and amorphous solids.

VORONOI TESSELLATION BEYOND MSE

Since the Voronoi tessellation is a general mathematical approach, it has a large variety of other applications:

- In astronomy to describe results of fragmentation of celestial bodies such as of two meteorites after a collision
- Mapping out territories of wild animals
- Computing weighted averages for average rainfall for a region
- Optimizing circuit board layout
- Dispatching taxi-cabs [10]
- Biology: creation of plant growth patterns
- Foam structures; see Section 5.8

Knowledge of the sizes and shapes of the Voronoi polyhedra, average number of faces per polyhedron in a given material, average surface area of a face, etc., enables visualization and handling of these phases.

In Figure 5.9 we started by connecting a central particle with centers of its neighbors. If we fill the whole space (2D or 3D) with such connecting lines, we perform the **Delaunay tessellation** [11]. We thus obtain the **Delaunay diagram**, also a unique representation of a system of points; see an example in Figure 5.11. Voronoi polygons (in 2D) or Voronoi polyhedra (in 3D) and so-called Delaunay simplices are mathematical **duals**. The process is shown schematically in Figure 5.12.

How can one describe crystals using the Voronoi or the Delaunay tessellations? In crystals the Delaunay figures are exactly the unit cells. The Voronoi polyhedra are the duals. We get exactly the same information from traditional crystallography as from the Voronoi-Delaunay approach. We shall return to the Voronoi and Delaunay diagrams in Section 5.5—so as to describe quantitatively the difference between amorphous solids and liquids.

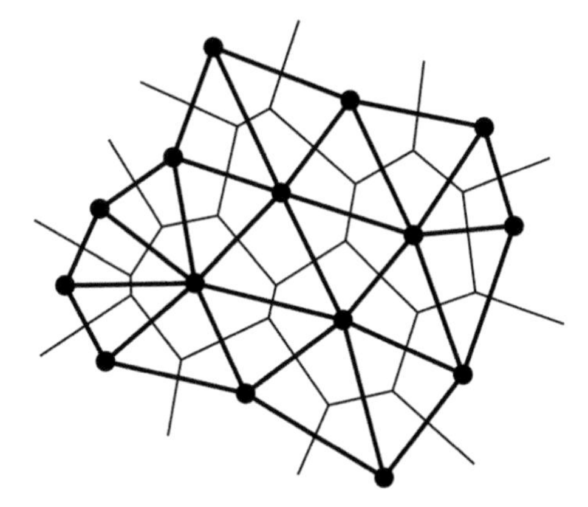

FIGURE 5.11 Delaunay tessellation (black) and Voronoi (gray) diagram for a set of points (particles, as black dots). *Source*: [12]. Reprinted with permission from *Journal of Materials Education*.

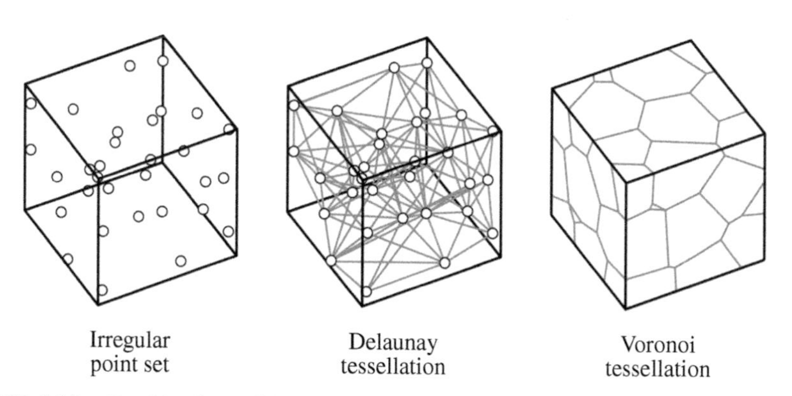

FIGURE 5.12 Partitioning of 3D space (a cube in the present case) containing a randomly placed set of particles (dots) into Voronoi polyhedra. *Source*: [12]. Reprinted with permission from *Journal of Materials Education*.

Needless to say, the radial distribution function and the Voronoi tessellation approaches are not mutually exclusive. Barbara Bertoncelj and her colleagues [13] have applied both approaches to **bulk molding compounds** which commonly consist of a polymer, discontinuous glass fibers, and a mineral filler. The fiber distribution in the composite can be analyzed by scanning electron microscopy and on that basis Voronoi diagrams as well as $g(R)$ curves created. From the results, the percentage of fiber yielding the most homogeneous distribution is easily determined—10–15 wt.% of fibers in the named example [13]. In the analysis, that concentration yielded also the highest level of reinforcement as determined by dynamic mechanical analysis and flexural testing.

5.6 THE GLASS TRANSITION

We have discussed the melting and freezing transitions in crystals in Chapter 3. Both occur at the same single temperature usually represented by T_m where m stands for melting. In Figure 3.14 we have also seen freezing for a certain composition of the binary Cu + Ni system. Now we shall examine the related thermal transitions for an **amorphous solid**, or equivalently **glassy solid** (we shall use these terms interchangeably in reference to non-crystalline materials). When an amorphous solid is heated, one of two things happens. Either the solid *melts*—in the sense that it changes from a solid into a liquid—or in the case of crosslinked polymers the solid undergoes the **glass-to-rubber transition**. That is, it acquires elastomeric properties, which we shall discuss later in Chapter 9. Qualitatively, we can deduce from the name that the material is going from a harder glass-like state to a more flexible rubber-like state.

One usually represents the transition of a glass to either a melt or a rubber by the **glass transition temperature** T_g, an analog of T_m. Here lies a problem! As we know, the melting or freezing transitions occur at a single temperature defined by the material composition. The glass transition, on the other hand, occurs over a temperature interval. In most cases people do not have enough patience to determine that range. However, in 1958 André J. Kovacs at the Louis Pasteur University in Strasbourg investigated carefully and patiently a polymer called poly(vinyl acetate) cooling it down and determining its specific volume as it varied with temperature [14]. The results are presented in Figure 5.13.

We see in Figure 5.13 two curves pertaining to two cooling rates. For the case when the whole process took only 0.02 hours (h), Kovacs assumed that the liquid cooling curve—the upper region of the curve—can be approximated by a straight line; the same assumption was applied to the line representing the glassy state—the lower portion of the curve, approaching the y-axis. Intersection of these two lines gives the point A, whose x-coordinate gives us the glass transition temperature T_g. Looking at the curve of experimental data points, we see that there is indeed a region of temperatures—switching from the top straight line to the bottom one—over which the glass transition takes place. We further see that the point A used to locate the T_g is *not* on the experimental curve. But this is only a half of the story. Look now at the other curve—the reason we are praising Kovacs for his patience—when his experiment lasted 100 hours. Assuming two straight lines again, we arrive at point C instead of A; dropping down to the x-axis we identify a different glass transition temperature T_g'. Thus, the location of Tg depends on the cooling rate. In phase transitions of the first order, such as melting (see the end of Section 4.1), the transition temperature (e.g., T_m) does not depend on the heating rate. The glass transition therefore does not meet the criteria for a first order transition, and the exact nature of the transition is still debated by some [12]—while in the next Section we provide a quantitative way of defining the difference. Nobody has beaten André Kovacs since 1958 in patient and accurate determination of such a diagram.

Location of the glass transition region by changes of the specific volume with temperature is just one of the options. Another option based on calorimetry is explained in Section 15.3; melting a glassy material to obtain a liquid requires energy from outside and the glass transition region can be located this way. In Section 14.5 we discuss the dynamic mechanical analysis (DMA) technique and the fact that it provides more than one way of locating the glass transition region.

A distinct question is: *what causes the glass transition to take place?* One answer is already clear from the above discussion: cooling a liquid, cooling a flexible non-glassy elastomer, or else heating a glassy material. These are not the only options, however. We note in Section 14.5 how increasing the frequency in DMA shifts the glass transition region to higher temperatures. Briefly, at low temperatures there is little free volume, therefore low mobility of atoms, molecules, and polymer chain segments; this corresponds to high frequencies when there is not enough *time* for these entities to relax. Conversely, low frequencies in DMA correspond to high temperatures; at low frequencies the same entities have time to relax, while at high temperatures they have space (free volume) to relax.

There is also an interesting option discovered by João F. Mano [15, 16]. Chitosan membranes have been studied in water + ethanol solvents—varying the mixed solvent composition and using DMA. Swelling and thus plasticization of chitosan were expected to be affected by the solvent composition. Glassy behavior was found below 25 vol.% of water—seen in the storage modulus E' versus water concentration diagram. E' falls vertically at that concentration [16]; this is similar to the diagram in Figure 14.18—only there the

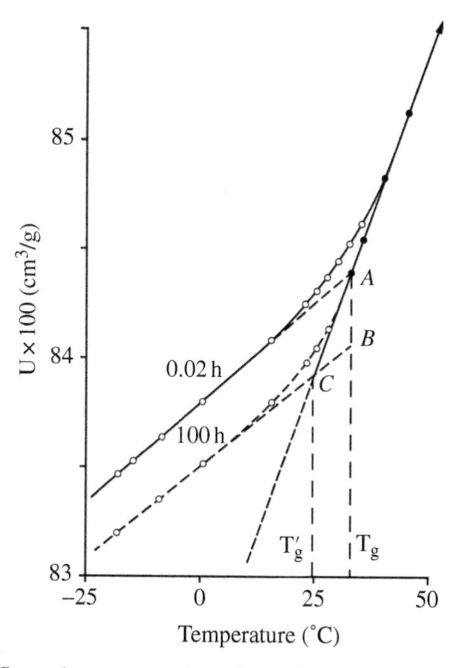

FIGURE 5.13 Specific volume as a function of temperature for poly(vinyl acetate) determined at two cooling rates and including the glass transition. *Source*: Kovacs [14]. Reprinted with permission of John Wiley & Sons.

horizontal coordinate is the temperature. These findings are explained in terms of the mobility of the segments in the chitosan network. That mobility is low—unless water molecules can "push their way" inside and provide more free volume. Thus, hydration—or lack of it—has to be added to the list of factors causing the occurrence of the glass transition.

The fact that free volume determines molecular mobility can be used to advantage. When we wish to predict behavior of a material at longer times and a given temperature, we can perform measurements, for instance, of mechanical properties as a function of time, at higher temperatures. Given more free volume, the same phenomena will take place sooner at a higher T. Measurements of a given property performed at several temperatures can be combined into a single curve versus time by shifting. This means using the **time-temperature correspondence principle** and relies on the calculation of temperature shift factors [17]. There is a similar **time-stress correspondence principle** [17]. From the discussion above one expects a **time-frequency correspondence principle**—which in fact works well too. As further expected from our discussion above, the **time-humidity correspondence principle** has been formulated by Mano [15]. Note that for these principles the terms "superposition" and "equivalence" are also in use.

In this Section we need to discuss still one more aspect. In Chapter 3 we saw the melting temperature T_m as a function of composition, along with similar diagrams for the boiling temperature T_b, in binary systems. What does the dependence of T_g on composition in binary systems look like? In fact the possibilities vary widely [18–20], as shown in Figure 5.14. When are such diagrams useful? For instance, when we are dealing with encapsulation of drugs by polymers [19].

T_G COMPOSITION DEPENDENCE IN BINARY SYSTEMS

We see in Figure 5.14 thick (black) lines going through the points. These lines have been calculated using the following equation [18]:

$$T_g = \varphi_1 T_{g,1} + (1-\varphi_1)T_{g,2} + \varphi_1(1-\varphi_1)\left[a_0 + a_1(2\varphi_1 - 1) + a_2(2\varphi_1 - 1)^2\right] \quad (5.3)$$

Here indexes 1 and 2 pertain to the components, and φ_1 is the mass fraction of the low-T_g component. Other lines pertain to other equations which do not fare so well in comparison with experimental results. Equation (5.2) has been referred to as the BCKV equation [20].

The quadratic polynomial, centered around $2\varphi_1 - 1 = 0$, is used to represent deviations from linearity. The nature of that of deviation is described mainly by parameter a_0, which reflects mostly differences between the strength of hetero- (intercomponent) and homo- (intracomponent) interactions. Composition-dependent energetic contributions from hetero-contacts, entropic effects, and structural heterogeneities affect the magnitude and sign of the higher-order parameters a_1 and a_2. Similarly, the number and magnitude of parameters needed for representation of an experimental $T_g(\varphi)$ pattern give quantitative measures of a system's complexity. One finds that irregular deviations from the linear mixing rule are closely tied to the compositional dependence of the total free volume and the free volume distribution around pertinent chain segments. Furthermore, it is expected that all such "asymmetric" entropic or enthalpic contributions will be reflected in the volumes and numbers of faces of the composite polyhedra of structural particles.

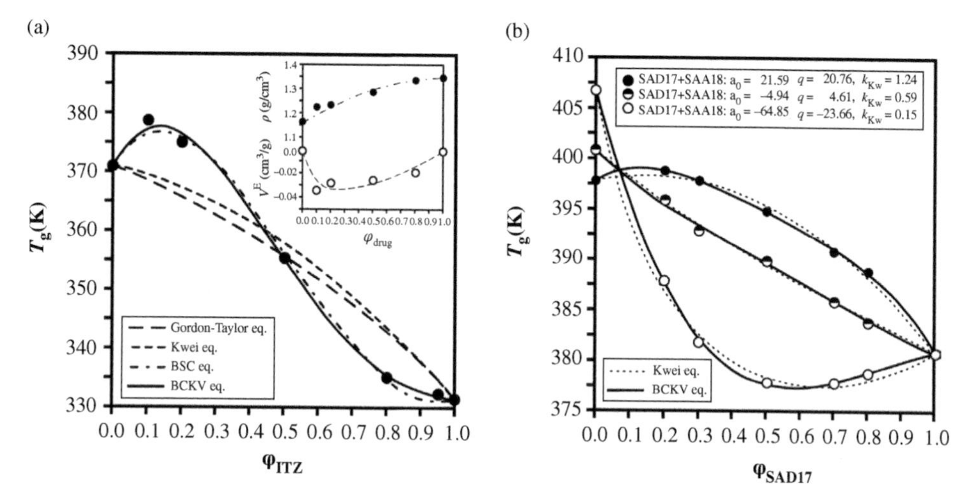

FIGURE 5.14 (a) Models describing (or attempting to describe) T_g versus composition dependence of itraconazole (ITZ, an HIV drug compound)+PLS-630 (a copolymer) blends. Inset: compositional variation of density ρ and excess mixing volume V^E (per gram of mass) for the same mixture. *Source*: Babu *et al.* [19], inset after data from Babu *et al.* [19]. Reproduced with permission of Wiley. (b) Dependence of T_g on composition in certain binary systems: poly(styrene-*co*-*N*,*N*-dimethylacrylamide) with 17 mol.% of *N*,*N*-dimethylacrylamide (SAD17)+poly(styrene-*co*-acrylic acid) with 18, 27, or 32 mol.% acrylic acid. Thick lines are fits to Eq. (5.2) (the BCKV equation). Source: Kalogeras [20]. Reprinted with permission from Elsevier.

5.7 GLASSES AND LIQUIDS

Since the transition from a glass to a liquid is continuous, is there a way to pinpoint differences between these two states? Actually, there is a way [21]—related to our discussion in Section 5.4. The Delaunay simplices in 3D are tetrahedrons. The perfect regular tetrahedron has the minimum circumradius. The applicable question then is how far from regularity are Delaunay tetrahedrons in glasses and liquids? One can introduce a quantity called **tetrahedricity** [22], defined as

$$\tau = \frac{\sum_{i>j}\left(l_i - l_j\right)^2}{\left(15\langle l\rangle^2\right)} \tag{5.4}$$

Here l_i is the length of the i-th simplex edge while $\langle l\rangle$ represents the average edge length value. Equation (5.4) has an important advantage: when the Delaunay tetrahedron is perfect, $\tau=0$. Thus, low τ values go with small distortions from the regular shape.

Analysis of a model material [21] shows that in the *liquid state* low density (high tetrahedricity) atomic configurations form a percolative cluster with significant free volume, hence with mobility of atoms and relative ease in atomic rearrangements. **Percolation** is the capability to walk from one edge of the material to the opposite side through the bulk of the material (not along edges). Percolative clusters exist also in the *glassy state*, but they are, in contrast, formed by "frozen" high density configurations with very low τ values [21]. A distinction between the liquid and glassy states is therefore defined on the basis of concepts first elucidated by Voronoi. This is one piece of a challenging puzzle; further analysis of the distinctions between the liquid and glassy states—since both exhibit randomness—is found in Ref. [21].

There exists an ideal gas model—while no such gases exist. We have noted in Chapter 4 that there are no ideal crystals but one can compare real crystals to the ideal model. There is also an **ideal amorphous solid** (IAS) model developed by Stachurski and his colleagues [23, 24]. While, for instance, in Figure 5.3 we see short-range order, there is no such order, nor medium-range order, in the IAS. With respect to elucidating the nature of the glassy state, this is the last piece of the puzzle we shall consider here.

An ideal amorphous solid is a model of an artificial amorphous material in which there are no imperfections in the random packing of the atoms. This requires accepting the notion that there can be perfection in randomness. Is it possible? The answer is yes. A set of random data can be fitted to Gaussian distribution function. If the fit is perfect then it can be said that the data set is perfectly random. By analogy, if the coordinates of the atomic positions in the IAS form such a data set, and there are no exceptions to the randomness, then it is a random packing of atoms. We have defined thus a necessary requirement for an amorphous (glassy) solid.

Such random packing can be achieved by starting with a cluster containing one inner sphere and a number, say k, of outer spheres touching the inner sphere. A necessary starting point is that the outer spheres must be placed at random on the surface of the inner sphere. Then joining the centers of any three adjacent outer spheres will form irregular triangles. All such three-adjacent sphere sites are places where the next layer of spheres can be added, until all sites are filled in, and the original cluster now has three layers. This process of addition of spheres onto irregular sites formed by three-adjacent spheres can be continued ad infinitum, thus growing the IAS model to any required size.

The IAS model need not be mono-atomic. To create a metallic glassy alloy of composition $Zr_{42}Ti_{12}Cu_{14}Ni_{10}Be_{22}$, the spheres are chosen with radii equal to the atomic radii of the elements, and the addition of a specific type of sphere must be in proportion to the composition. More details are described in Refs. [23–25]. The IAS, as just described, appears as a good model for the atomic arrangement in metallic glasses. To test this, it can be evaluated as a predictor of certain properties of a specific metallic glass. One such test is the prediction of X-ray scattering patterns through use of the Debye equation, which can be written as:

$$I(q) = Nf^2\left[1 + \frac{1}{v}\sum_j \left[g(r_j) - 1\right]\frac{\sin(qr_j)}{qr_j}\right] \tag{5.5}$$

where I is the scattered intensity, N is the number of atoms, f is the atomic scattering factor of the mono-atomic glass, or an average of atomic scattering factors if it is multi-atomic, and v is unit volume occupied by the N atoms. Subscripts i and j are indices for the atoms, and $g(R)$ was discussed in Section 5.3. The variable q, the scattering vector, is equal to 2 $\sin(\theta)/\lambda$, with λ as the wavelength of the radiation and θ as half the scattering angle. An idea about the capabilities of the IAS model is provided by Figure 5.15 [24, 25]. Figure 5.15a shows an outer surface view of a two phase hard sphere packing model of a Ge + As + Se model containing 103 atoms. In Figure 5.15b we see revealed the embedded tetrahedral $GeSe_2$ network (in blue).

Since the positions of all atoms in the IAS are known and the properties of individual elements are known, calculations of other physical properties, such as magnetic

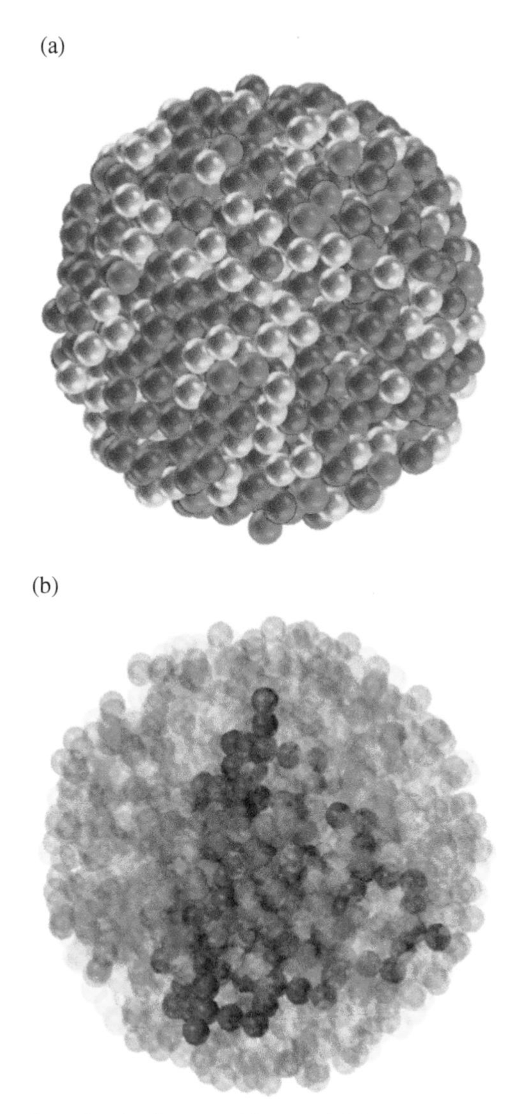

FIGURE 5.15 Two phase hard sphere packing model for the Ge–As–Se glasses comprising 103 atoms. (a) An outer surface view, and (b) the embedded tetrahedral $GeSe_2$ network is revealed (blue). *Source*: [24]. Reprinted with permission from Elsevier.

susceptibility or dielectric constant, can be carried out. One can then compare properties of a given amorphous material with the IAS of the same composition—as done in Figure 5.15 for scattering intensity. This in turn facilitates development of new non-crystalline materials with improved properties.

5.8 AGING OF GLASSES

Over time, glasses undergo **aging**, a process also called **devitrification**; the crystalline regions formed in the glass during this process are called *stones*. As discussed by Struik [26], aging occurs because a non-equilibrium state is frozen-in during the cooling of a material from a point above to a point below the glass transition region. Above the region, retardation or relaxation times are short enough to enable material properties such as volume and entropy to follow the changes of temperature. However, below the glass transition region, the retardation times become too long. The entire region is typically characterized by a single temperature T_g. According to Struik [26], to a first approximation the state existing at T_g is frozen-in so that the material properties at $T<T_g$ deviate from those expected of an equilibrium state at T.

The consequence of this situation is *aging*: a slow structural process that induces changes in many materials properties. Its essential mechanism is material **densification**. Volume and entropy were already mentioned; there are also changes in mechanical properties, an increase in viscosity, and decrease in creep and stress relaxation rates (see Chapter 14 on these processes).

Can we counteract aging? One way to do it is of course to raise the temperature to above the glass transition region—but this might not be appropriate for given service conditions. An example of de-aging known from everyday life is **ironing** of textiles. As a consequence of aging, clothing loses its "fresh" look and becomes wrinkled. To reverse the process, one first puts water into the garment; water goes into the pores causing *expansion*, and thus counteracts densification which took place. Using an iron provides *heating*, hence more expansion. *Steam* leaving the pores also contributes to expansion. The iron also mechanically preserves the new expanded or equivalently de-aged configurations of polymeric chains in the textile. In a variety of cases, reversal of the aging process, as in this example, is desired, and this is not only for materials but also for humans. Reversal of human aging, a process involving the shortening of telomeres [27], is achieved by a mechanism of telomerase activation [28, 29]. De-aging is thus beneficial to practitioners of MSE.

5.9 POROUS MATERIALS AND FOAMS

First of all, what are the differences between porous materials and foams? In the literature authors talk about one category or the other; there seem to be no generally agreed upon definitions. There can be liquid + gas phase foams, such as **soap bubbles**, with air inside the bubbles. **Froth** on a glass of beer belongs to the same category, but the gas phase is largely carbon dioxide. Liquid + gas foams are unstable, usually burst within seconds, and can be colorful. There are also solid + gas foams, and Kraynik [30] says that both classes of foams have similar geometric structures.

Foams are created deliberately; as in soap bubbles and beer froth, foams consist of a gas (often but not necessarily air) and comparatively thin walls between the gaseous bubbles. If one would insist on defining a difference between foams and porous materials, we could find that difference in the walls. **Porous materials** can have any amount of solid between the bubbles or pores, not necessarily thin walls only. Materials with pores can be created deliberately, or else they form "by themselves". Moreover, one does not apply the term "porous materials" to liquid + gas systems.

Among foams, we can distinguish two classes: **open cell** and **closed cell**. In the former the gas can move between cells (bubbles, pores) and can also be exchanged with the environment. An example is a bathroom sponge; either air or water can move easily inside the sponge. Materials from the latter category, closed cell foams, are required for certain applications. In a sleeping mat, for instance, open cells would result in squeezing much air out of the mat, with the usefulness of the mat most assuredly reduced. Sport shoes require some open and some closed cells for "springiness". Whether the cells are open or closed, they are not all of the same size. We thus see that the *distribution* of cell sizes is important.

To characterize cells and their size distributions, we go back to Section 5.4 of the present chapter. Given that information, it is now understood easily that cells in a foam can be characterized precisely by the Voronoi diagram for a given foam. This approach has been pioneered by Shulmeister and coworkers [31] and Windle and coworkers [32, 33], and then used by others. The latter group has dealt with the Voronoi cells' irregularity, dealing with the effects of mechanical forces by using the finite elements approach (wherein a material is decomposed into a set of small elements). An interesting conclusion reached was that at low strains the more irregular structures have a higher tangential modulus while at higher strains the situation is reversed [32, 33]. Tangential modulus is the slope of the stress versus strain curve at a predefined value of stress or strain. Thus, the Young's modulus is a special case of the tangential modulus and pertains to low values of both stress and strain (below the proportionality limit, see Section 14.1). It seems that at lower strains the irregular cells resist deformation better, possibly becoming more regular. At high strains cell buckling becomes important and then the irregular cells become more vulnerable.

A particular methodology for defining the structures of foams in terms of the Voronoi diagrams has been formulated by Kraynik and coworkers [30, 34, 35]. One creates the Voronoi diagram for a system of densely packed random spheres; see Figure 5.16 on the left. The material is not necessarily in thermodynamic equilibrium, so a process of relaxation is performed leading to a minimum of the Helmholtz function; refer back to the thermodynamic stability criterion (3.19). In the middle of Figure 5.16 we have a monodisperse system, begun with spheres of equal sizes and arrived at by relaxation. On the right side of the figure we see a structure representing spheres of various sizes; as expected, small cells have fewer faces while large polyhedra have more faces. In the monodisperse system, the probability is high of polyhedral cells having 14 faces. In the random Voronoi structure, the probability distribution is similar, only with the peak shifted to a slightly higher number of faces. However, as polydispersity increases, the probability of finding a cell with 14 faces decreases.

An important feature in such analyses is the ratio of the sample size to the cell size. Tekoglu, Gibson, Pardoen, and Onck [36] studied four modes of deformation of foams: compression,

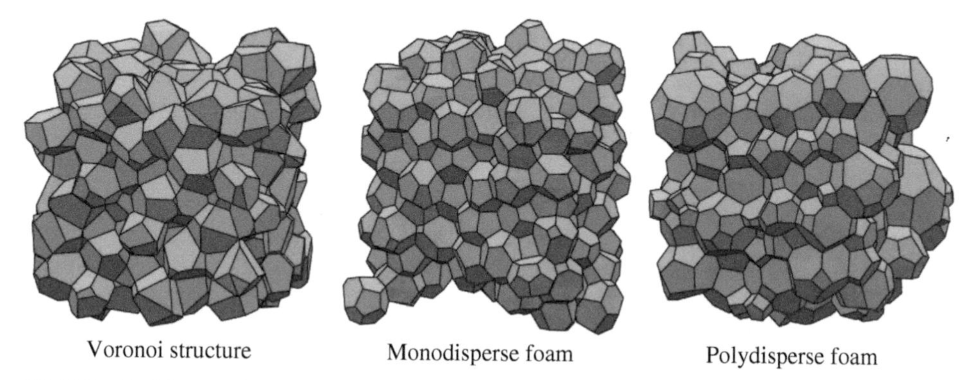

FIGURE 5.16 Models of monodisperse and polydisperse foams; after Kraynik [30]. The Voronoi structure was relaxed to produce the monodisperse foam. These disordered structures contain 216 cells in the representative volume (unit cell). *Source*: Reprinted from A.M. Kraynik, Foam structure: from soap froth to solid foams. *MRS Bull.* April 2003, 276–278, © 2003 with permission from Cambridge University Press.

shear, indentation, and bending. They assumed that the foams have random structures, and they used Voronoi tessellations to characterize those structures as well as finite elements to deal with force imposition. They worked in two dimensions—easier to visualize as we already know—dealing with both open and closed cell foams. Using then the finite element method, they characterized the deformations. They confronted their predictions with experimental results for aluminum foams but declared also that "the tendencies are shared by a large family of foams made of different materials" [36].

In the same study [36] the authors determined that effective Young's modulus of Voronoi samples decreases under compression and bending while it increases under shear and indentation. Apparently, under *compression* there is an increase of the volume fraction of weak boundary layers; the respective cells are more compliant than the cells in the bulk. Localized deformation bands are formed. The smaller the sample size is, the easier the band formation. The strain localization in *bending* occurs through similar mechanisms as under compression. The difference is that in bending the tensile and compressive strains increase when moving away from the neutral axis; therefore weak boundary layers present at the stress-free edges have a larger contribution to the overall mechanical response in comparison to the cells in the bulk. The reduction in effective mechanical properties is therefore larger in bending than in compression.

In *indentation*, the deformation is mainly confined to a zone under the indenter; localized bands that connect the two edges of the indentation are formed [36]. The depth of the deformation is comparable to the width of the indenter. Possibly not only the cells under the indenter offer resistance but also cells around the indented area provide "lateral support" against deformation. In *shear*, rotational zones of one up to two average cell sizes form at the boundaries—where the samples are bonded to rigid loading platens—and act as strong boundary layers with a larger resistance to deformation. The volume fraction of these strong layers increases with decreasing size of the material specimen. Thus, in contrast to the weakening effect of boundary layers under compression and bending, the boundary layers that form in shear exert a strengthening effect and lead to an increase in the modulus [36].

Consider more generally the difference between foams and solid materials in their modes of transmitting energy coming from outside sources (no limitation on the type or rate of force application). A solid element will undergo yielding and flow (in metals plastic flow) as a whole in response to an incident stress (when the elastic limit is exceeded). By contrast, in a foam the energy from an "attack" is frequently absorbed locally by plastic deformation. Individual foam bubbles are crushed, being capable of undergoing enormous amounts of local strain and thereby absorbing large amounts of energy, but outside of that location the bulk of the foam has a good chance to remain intact. The same principle is used in the design of cars to enable the controlled crushing of the engine cavity in a collision.

We now return to porous materials. While we have attempted early in this Section to point out a difference between porous materials and foams, an industrial company called Barat Ceramics provides its own definitions [37]: porous ceramics have a maximum porosity of approximately 30 vol.% while foam ceramics have open porosity of ≈80 vol.%.

There are interesting cases where one not only determines porosity but more importantly varies it in a controlled manner. Bouville and his colleagues [38] have developed a way to create macroporous crystallographically textured ceramics by a two-step operation of ice templating and template grain growth (TGG). In **ice templating** a temperature gradient across a suspension causes ice crystals of the liquid to nucleate and grow, rejecting the solid particles and forming a porous template. Then as the ice crystals are removed by sublimation the pore spaces remain, surrounded by densely packed walls of the primary particles from the original suspension. This structure serves as a seed for grain growth, with the aligned platelets forming a template for that growth, hence the name TGG. Just such a combined technique was successfully applied to alumina and to sodium potassium niobate (a piezoelectric, see Section 17.6). Thus, structure control is achieved at two different length scales [38].

A useful category of porous ceramics is **aerogels**, in which we have a gas (for instance, air) instead of a liquid component in the gel. The liquid is typically removed by supercritical drying, a "gentle" process compared to relatively "brutal" ordinary evaporation of the liquid; the latter yields a **xerogel** that can undergo material collapse as shrinkage is very high. Aerogels have very low density and low thermal conductivity—advantages in certain applications. These materials have been modeled by Quintanilla, Reidy, Gorman, and Mueller [39] using a Gaussian approach. They use an intersection of two independently constructed Gaussian random fields. A key aspect of the approach is the chord length probability density function. If a straight line is drawn through a random material, the chords are defined as the segments entirely in one phase, with both end points on an interface. One thus obtains two cord length probability density functions, one for each phase. The model has been confronted with experimental results for aerogels based on tetraethoxysilane (TEOS) and satisfactory agreement of predicted and experimental morphologies found [39].

Of course, we can have foams or porous materials in any category of materials involving solids. For metals, Banhart [40] makes several distinctions as follows:

- "Cellular metals: the most general term, referring to a metallic body in which any kind of gaseous voids are dispersed. The metallic phase divides space into closed cells which contain the gaseous phase.
- Porous metals: a special type of cellular metal restricted to a certain type of voids. Pores are usually round and isolated from each other.
- (Solid) metal foams: a special class of cellular metals that originate from liquid-metal foams and, therefore, have a restricted morphology. The cells are closed, round, or polyhedral and are separated from each other by thin films.
- Metal sponges: a morphology of a cellular metal, usually with interconnected voids" [40].

We see here, as earlier, the association of thin walls with foam structures. As discussed by Badiche and colleagues [41], metal foams—including those they studied made from nickel—have a variety of applications such as for: thermal insulations, electromagnetic shielding, and energy absorption (e.g., during car crashes). Consider by contrast soft polymeric foams used, for instance, for repetitive absorption of spilled petroleum [42, 43].

While one usually thinks about ceramic, metallic, or polymeric foams as unrelated entities, Colombo and Hellman [44] have created a method of making ceramic foams starting with polymeric ones. Their polymeric foams—that contain inorganic substituents—are subjected to high temperature pyrolysis in an inert atmosphere. After pyrolysis, only the inorganic components remain; ceramic foams thus obtained exhibit outstanding mechanical properties as well as excellent thermal stability. Colombo and Hellman subjected their foams repetitively to 1200°C in air and so demonstrated their durability. The Colombo-Hellman method not only produces outstanding ceramic foams but is also relatively simple to implement [44].

5.10 SELF-ASSESSMENT QUESTIONS

1. Explain qualitatively the binary radial distribution function, describing what it is used for and how it is so used.
2. Describe the construction of Delaunay and Voronoi tessellations. What types of materials can be represented this way?
3. Compare and contrast quasicrystals to crystals and amorphous solids.
4. Does a glass transition temperature exist in real materials? What do the values of that temperature listed in various publications or reports really tell us?
5. Describe ironing a coat as a polymer de-aging process.
6. Foams can be used in ways that their solid counterparts cannot. Metal foams are used to protect automobile passengers in collisions. On the other hand, solid metal used around the wagon wheels and bicycle wheels of times past gave the passengers a rough ride. Explain these behaviors. What can you say about energy transmission in these situations?

REFERENCES

1. D. Shechtman, I. Blech, D. Gratias & J.W. Cahn, *Phys. Rev. Letters* **1984**, *53*, 1951.
2. D. Shechtman & I. Blech, *Metall. & Mater. Trans. A* **1985**, *16*, 1005.
3. G.H. Michler & F.J. Balta-Calleja, *Nano- and Micromechanics of Polymers: Structure Modification and Improvement of Properties*, Hanser: Munich/Cincinnati, OH **2012**.
4. L.-V. de Broglie, *Recherches sur la théorie des quanta*, Thesis, Paris **1924**.
5. L.-V. de Broglie, *Ann. de Physique* **1925**, *3*, 22.
6. W.L. Bragg, *Proc. Cambridge Philos. Soc.* **1913**, *17*, 43.
7. G. Voronoi, *Z. reine & angew. Math.* **1908**, *134*, 198.
8. G. Voronoi, *Z. reine & angew. Math.* **1909**, *136*, 67.
9. W. Brostow & V.M. Castaño, *J. Mater. Ed.* **1999**, *21*, 297.
10. http://cgm.cs.mcgill.ca/~godfried/teaching/projects.pr.98/tesson/taxi/taxivoro.html, accessed September 10, **2012**.
11. B. Delaunay, *Izvestia Akademii Nauk SSSR—Otdelenie Matematicheskih i Estestvennyh Nauk* **1934**, *7*, 793.
12. I.M. Kalogeras & H.E. Hagg Lobland, *J. Mater. Ed.* **2012**, *34*, 69.
13. B. Bertoncelj, K. Vojislavljevic, J. Rihtarsic, G. Trefalt, M. Huskic, E. Zagar & B. Malic, *Express Polymer Letters* **2016**, *10*, 493.
14. A.J. Kovacs, *J. Polym. Sci. A* **1958**, *30*, 131.
15. J.F. Mano, *Macromol. Biosci.* **2008**, *8*, 69.
16. S.G. Caridade, R.M.P. da Silva, R.L. Reis & J.F. Mano, *Carbohydrate Polymers* **2009**, *75*, 651.
17. W. Brostow, *Pure & Appl. Chem.* **2009**, *81*, 417.
18. W. Brostow, R. Chiu, I.M. Kalogeras & A. Vassilikou-Dova, *Mater. Letters* **2008**, *62*, 3152.

19. R.J. Babu, W. Brostow, O. Fasina, I.M. Kalogeras, S. Sathigari & A. Vassilikou-Dova, *Polym. Eng. & Sci.* **2011**, *51*, 1456.
20. I.M. Kalogeras, *Thermochim. Acta* **2010**, *509*, 135.
21. N.N. Medvedev, A. Geiger & W. Brostow, *J. Chem. Phys.* **1990**, *93*, 8337.
22. N.N. Medvedev & Yu.I. Naberukhin, *J. Non-Cryst. Solids* **1987**, *94*, 402.
23. L.Th. To, D. Daley & Z.H. Stachurski, *Solid State Sciences* **2006**, *8*, 868.
24. R. Feng, Z.H. Stachurski, M. Rodriquez, P. Kluth, L.L. Araujo, M.C. Ridgway & D. Bulla, *J. Non-Cryst. Solids* **2014**, *383*, 21.
25. Z.H. Stachurski, *Fundamentals of Amorphous Solids: Structure and Properties*, Wiley-VCH: Weinheim **2015**.
26. L.C.E. Struik, Physical aging: influence of the deformation behavior of amorphous polymers, Chapter 11, in *Failure of Plastics*, ed. W. Brostow, Hanser: Munich/Vienna/New York **1992**.
27. J. Hooper, The Man Who Would Stop Time, *Popular Science*, August **2011**, p. 50.
28. B.B. de Jesus, E. Vera, K. Schneeberger, A.M. Tejera, E. Ayuso, F. Bosch & M.A. Blasco, *EMBO Mol. Med.* **2012**, *4*, 1.
29. J.W. Anderson, Compound and method for increasing telomere length, WO Patent App. PCT/US2012/023901, August **2012**, http://www.google.com/patents/WOA1?cl=en, accessed March 24, 2016.
30. A.M. Kraynik, *MRS Bull.* April **2003**, 276.
31. V. Shulmeister, M.W.D. Van der Burg, E. Van der Giessen & R. Marissen, *Mech. Mater.* **1998**, *30*, 125.
32. H.X. Zhu, J.R. Hobdell & A.H. Windle, *J. Mech. & Phys. Solids* **2001**, *49*, 857.
33. H.X. Zhu, S.M. Thorpe & A.H. Windle, *Internat. J. Solids & Structures* **2006**, *43*, 1061.
34. A.M. Kraynik, D.A. Reinelt & F. van Swol, *Phys. Rev. E* **2003**, *67*, 031403.
35. A.M. Kraynik, D.A. Reinelt & F. van Swol, *Phys. Rev. Letters* **2004**, *93*, 208301.
36. C. Tekoglu, L.J. Gibson, T. Pardoen & P.R. Onck, *Progr. Mater. Sci.* **2011**, *56*, 109.
37. Barat Ceramics: Advanced Materials Solutions, Porous and Foam Ceramics, http://barat-ceramics.com/www/barat-en/werkstoffe/detail.htm?recordid=12458254F18, accessed March 1, **2016**.
38. F. Bouville, E. Portuguez, Y. Chang, G.L. Messing, A.J. Stevenson, E. Maire, L. Courtois & S. Deville, *J. Am. Ceram. Soc.* **2014**, *97*, 1736.
39. J. Quintanilla, R.F. Reidy, B.P. Gorman & D.W. Mueller, *J. Appl. Phys.* **2003**, *93*, 4584.
40. J. Banhart, *J. Metals* **2000**, *52* (12), 22.
41. X. Badiche, S. Forest, T. Guibert, Y. Bienvenu, J.-D. Bartout, P. Ienny, M. Croset & H. Bernet, *Mater. Sci. & Eng. A* **2000**, *289*, 276.
42. A.M. Atta, W. Brostow, H.E. Hagg Lobland, A.-R.M. Hasan & J.M. Perez, *RSC Advances* **2013**, *3*, 25849.
43. A.M. Atta, W. Brostow, H.E. Hagg Lobland, A.-R.M. Hasan & J.M. Perez, *Mater. Res. Innovat.* **2015**, *19*, 459.
44. P. Colombo & J.R. Hellman, *Mater. Res. Innovat.* **2002**, *6*, 260.

PART 2

MATERIALS

6

METALS

Life should be like the precious metals, weigh much in little bulk.
—Seneca, Roman philosopher, mid-1st century AD.

6.1 HISTORY AND COMPOSITION

What are metals? Depending on the source, there are varied definitions, but metals have been known to mankind from the earliest days. A broad description is most fitting. In general, metals are opaque, lustrous elements that are good conductors of heat and electricity. They are typically malleable and ductile as well as more dense than other elements. As illustrated by the periodic table in Figure 6.1, the majority of elements are classified as metals. With the exception of mercury, metals are solid at room temperature. The conductivity of metals is largely due to a tendency to have between 1 and 3 (valence) electrons in the outer shell and to lose the valence electrons easily. Metals form oxides (compounds with oxygen) that are basic (i.e., alkaline). In general they have low electronegativities (refer to Section 2.3).

Having provided a broad description of metal properties, we could now proceed to describe the details of metallurgy and metallic systems necessary for a proper understanding of metals. However, it is useful to first know where metals are found in nature and in what form they are found. Metals are an integral part of the Earth's crust and are found in almost all rocks and soils. Yet, metals are rarely found in nature in their pure metallic state; primarily gold, silver, mercury, and copper have been found as (more or less) pure metals (the **native state**), though not in great abundance. Instead, metals exist in compounds with other elements (metallic and non-metallic) forming **minerals**. Minerals are inorganic solids with regular chemical compositions and structures (and are also the source of some

Materials: Introduction and Applications, First Edition. Witold Brostow and Haley E. Hagg Lobland.

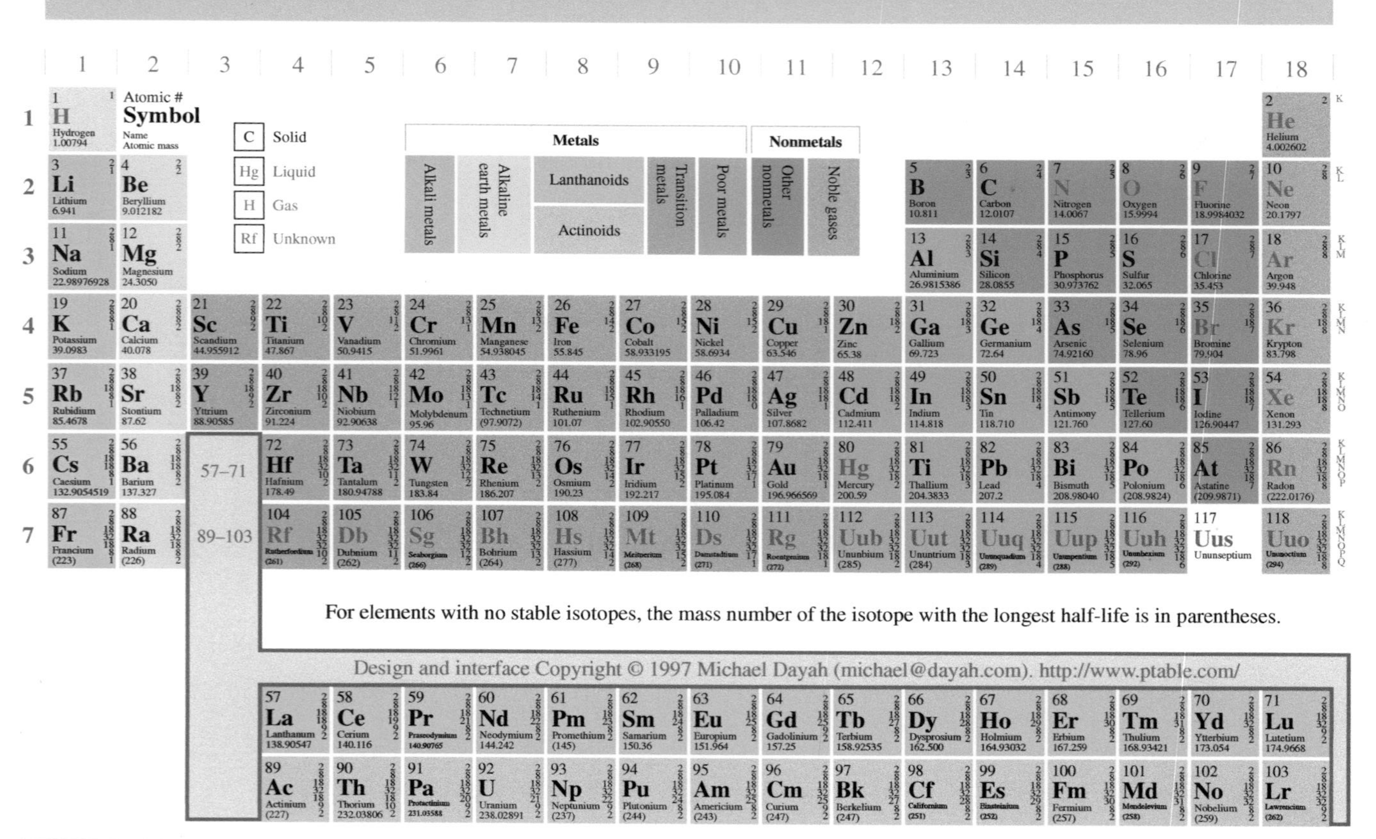

FIGURE 6.1 Periodic table of chemical elements. *Source*: Reprinted with permission of Dayah, M. (1997, October 1). *Dynamic Periodic Table*. Retrieved July 19, 2013, from Ptable: http://www.ptable.com.

ceramics—refer to Chapter 7). A given metal element can be a component in many different naturally occurring minerals. The metals contained in minerals are also responsible for many of the beautiful colors of gemstones. For example the copper in chalcopyrite produces a bright yellow, while the copper-phosphate compound in turquoise gives it a blue color. The rocks found in our earth are composites of different minerals. The term **ore** refers to mineral deposits with a high concentration of metal such that the metal can be mined for a profit. **Smelting** is the term used to describe the process of heating ores to get out the pure metal. The most common metal compounds found in nature [1] are:

1. **Oxides**, which are the source of aluminum (bauxite), iron (magnetite and hematite), tin (cassiterite), chromium (chromite), manganese (pyrolusite), nickel (laterite), and uranium (uraninite).
2. **Carbonates** of sodium, calcium (limestone), and magnesium (magnesite).
3. **Sulfides**, which host metals such as copper (chalcopyrite), zinc (sphalerite), nickel (pentlandite), lead (galena), mercury (cinnabar), and iron (pyrite).
4. **Sulfates**, such as that of calcium (gypsum).
5. **Phosphates**, in which vanadium is found.
6. **Silicates** (silicon oxides) of zirconium and aluminum.
7. **Halides** (compounds with chlorine, fluorine, bromine, and iodine) of sodium (sodium chloride is common table salt) and potassium.

Radioactive astatine, also a halogen (sitting underneath iodine in the periodic table), does not provide a natural source of metals in the form of halide salts owing to its short half-life. The most stable isotopes of astatine have half-lives less than a minute, while astatine itself is essentially unavailable in nature. It is estimated that there is less than 30 g of it in the Earth's crust at any time, while only minute amounts have been prepared via bombardment of bismuth-209 with alpha particles in a cyclotron.

A very basic timeline of the history of metals extends back to 6000 BC [2]. The seven so-called "Metals of Antiquity" are gold, copper, silver, lead, tin, iron (smelted), and mercury. All of these except lead and tin can be found in the native (pure, naturally occurring) state (though native iron from meteors is somewhat rare). Gold artifacts of jewelry and ornamentation are abundant, while smelted copper was used for tools, weapons, and other implements. Bronze was created by adding tin to copper, yielding a stronger alloy. In the fourth chapter of the Biblical book of Genesis an early descendent of Cain is described as "an instructor of every craftsman in bronze and iron (Genesis 4:22, NKJV)." Lead, as Galena, was known by the ancient Egyptians and used as an eye paint [2]; it was later used by the Romans for containers and conduits. With the burgeoning establishment of iron smelting in Egypt, the Bronze Age gave way to the Iron Age, which extended into the 1st century AD. During the following millennium, developments in processing made high quality iron and steel rather widely available. By the mid-18th century high-quality steel was available inexpensively by commercial production. A new ore-reducing process developed in the 19th century made aluminum available cheaply. That discovery facilitated the use of aluminum alloys for aircraft. By the mid-20th century, specialty alloys were being created for improved performance in demanding applications, such as in turbines or internal combustion engines. By the mid to late 20th century, alloys (of iron, cobalt, and titanium) were routinely used to replace or mend human body parts (see Chapter 11 for more on Biomaterials). So-called superalloys developed for jet engines made space travel possible during the latter part of the 20th century. Refer to the history by Cramb [2] for more details on the history of other metals (arsenic, antimony, zinc, and so on).

Now in the 21st century, the main areas of research and development of metals are in fact quite varied. On the one hand, plastics and composites are replacing metals in many applications. However, for applications in which only metals suffice, we will likely see some improvements in the alloys used. As ores are depleted, we may also see more impetus on recycling of metals. Steel remains the most commonly used metal of the 21st century.

Now a few words about structure. Most solid metals have crystalline structures (see Chapter 4), but we shall discuss glassy metals in Section 6.6. A crystalline metal of a particular composition may occur in various lattice structures and form different phases depending on temperature and pressure, as we shall see in subsequent sections. **Alloys** have been mentioned; they are mixtures of two or more metals. The crystal structures of alloys therefore accommodate atoms of different sizes. As discussed in Chapter 2, the so-called "cloud of electrons" associated with metallic bonding is responsible for the conductivity of metals. The combining of different metal elements to form alloys can be used as a means to modify the electronic properties of metals and also their thermal, mechanical, and tribological properties. In Chapter 4 we discussed slip associated with defect movement. Many metals crystallize in the FCC or BCC lattices, which having many slip systems allow for greater plastic deformation (see Chapter 14) in those materials. The HCP lattice has fewer slip systems; but in general there is a higher probability of defect movement through slip in crystalline metals than in crystalline ceramics, as we shall see in the next Chapter. The ductility of metals is in part a result of this structural aspect.

6.2 METHODS OF METALLURGY

Metallurgy has been defined by DiBenedetto [3] as "the art and science of utilizing metallic elements." One may speculate whether the fact that in this definition art precedes science is accidental. It certainly was an art in the past, see Figure 6.2. Metallurgy begins with the materials furnished by Nature. Given an ore, the main steps of the metallurgical process are the following:

1. Preliminary concentration of the metal, possibly including roasting
2. Reduction of metal oxides
3. Refining and removing of impurities
4. Sizing and shaping
5. Cold working into a final product
6. Heat treatment of metal phases

Since a native ore contains also many other compounds and substances besides the metal of interest, **preliminary concentration** of the metal is necessary. Different techniques are used in this step as needed. If the components differ in density, a gravity separation might be applied. To chemically separate a metal from the rest of the ore, there are various methods of **leaching**. Some ores may be obtained in soluble form. For example, copper ore can be acquired as copper and copper sulfates and then leached in sulfuric acid (H_2SO_4).

FIGURE 6.2 *Blacksmith Apprentice, Williamsburg, VA*, watercolor by Raymond H. Pahler, 1986. *Source*: Reprinted with permission of R.H. Pahler.

Very pure copper can then be obtained by electrolysis in an electrolytic cell—in which case later stages such as reduction (discussed next) are omitted. **Selective leaching** removes an element from a solid solution alloy (we discuss solid solutions in Section 3.13). **Dezincification of brass** is the best known example. Sometimes **roasting** is needed to change the chemical nature of the metal or the impurities. The mineral galena (PbS) is typically roasted since it cannot be easily reduced to metallic lead by the standard process with carbon or carbon monoxide. The essential reaction in roasting PbS is:

$$2PbS + 3O_2 \rightarrow 2PbO + 2SO_2 \quad (6.1)$$

Once an ore has been concentrated, and if the metal component is in the form of an oxide, the next step is typically **reduction**. We can continue the example with galena to illustrate this process. The lead oxide (PbO) produced by roasting can be rapidly reduced to metallic Pb around 1300 K. Reactions with carbon dioxide or with coke are possible:

$$PbO + CO \rightarrow Pb + CO_2 \quad (6.2)$$

$$2PbO + C \rightarrow 2Pb + CO_2 \quad (6.3)$$

Galena can also be added into the smelt, participating in the reduction process as

$$2PbO + PbS \rightarrow 3Pb + SO_2 \quad (6.4)$$

The establishment of a reduction regime requires attention to the phases involved and any changes in thermodynamic functions accompanying the reduction. Consider **iron reduction**: a **blast furnace** is filled with the ore, coke, limestone, and air; see a schematic of the furnace in Figure 6.3. Limestone is removed from the earth by blasting with explosives. It is then crushed and screened to a size between 0.5 and 1.5 inch to become blast furnace flux. This flux can be pure high calcium limestone, dolomitic limestone containing

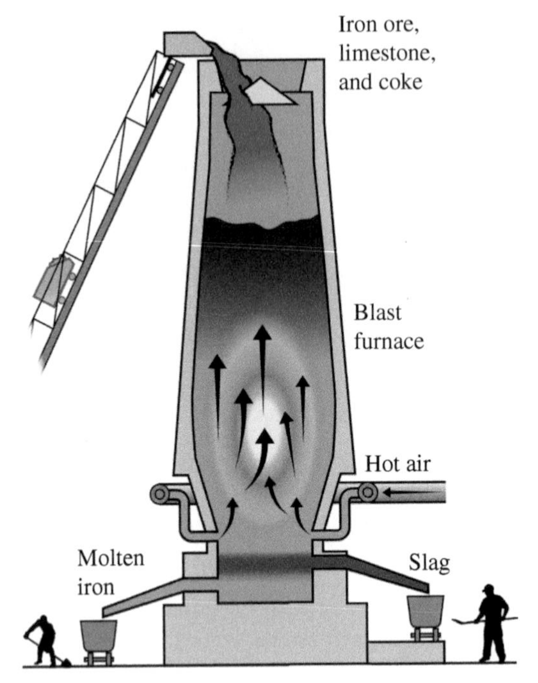

FIGURE 6.3 Diagram of a blast furnace.

magnesia, or a blend of the two types of limestone. Analogous to the reactions (6.1) and (6.2) with PbO, at least two possible reactions are imaginable:

$$Fe_2O_3 + 3CO \rightarrow 2Fe + 3CO_2 \quad (6.5)$$

$$Fe_2O_3 + 3C \rightarrow 2Fe + 3CO \quad (6.6)$$

In principle, decomposition of Fe_2O_3 into elements is possible:

$$Fe_2O_3 \rightarrow 2Fe + \tfrac{3}{2}O_2 \quad (6.7)$$

But we know from experience that iron rusts easily, so we have good reason to believe that the reverse process (decomposition) is not a natural one; recall Section 3.9. Considering then reactions (6.5) and (6.6), we note that the reduction reaction (6.6) occurs in the solid state only. Solid phase reactions are usually much slower than gas phase reactions owing to slower diffusion of molecules, atoms, and ions in solids. From the point of view of kinetics then, reaction (6.5) would predominate. A similar argument can be used for lead reduction (6.2) over (6.3).

From the thermodynamics perspective discussed in Section 3.9, it is convenient to consider the Gibbs function changes ΔG expected from the reduction reactions. Since C, CO, and CO_2 are involved, other reactions also possible are:

$$3C + 3O_2 \rightarrow 3CO_2 \quad (6.8)$$

$$3C + \tfrac{3}{2}O_2 \rightarrow 3CO \quad (6.9)$$

$$C + CO_2 \rightarrow 2CO \quad (6.10)$$

Based on data collected by DiBenedetto for these reactions [3], let us consider the dependence of ΔG on temperature. Comparing the (gaseous and solid state) reactions at 500 K and at 1500 K, we have

		ΔG / kJ mol^{-1}	
		500 K	1500 K
(6.7)	$Fe_2O_3 \rightarrow 2Fe + \frac{3}{2}O_2$	+687.8	+440.6
(6.8)	$3C + 3O_2 \rightarrow 3CO_2$	−1186.2	−1186.2
(6.9)	$3CO \rightarrow 3C + \frac{3}{2}O_2$	+465.7	+731.8
(6.5)	$Fe_2O_3 + 3CO \rightarrow 2Fe + 3CO_2$	−32.7	−13.8
(6.7)	$Fe_2O_3 \rightarrow 2Fe + \frac{3}{2}O_2$	+687.8	+440.6
(6.9)	$3C + \frac{3}{2}O_2 \rightarrow 3CO$	−465.7	−731.8
(6.6)	$Fe_2O_3 + 3C \rightarrow 2Fe + 3CO$	+222.1	−291.2

ΔG for many reactions can be found in handbooks of chemistry, physics, or metallurgy. From the values above we can draw several conclusions. First, our intimation regarding reaction (6.7) was correct: ΔG values are positive, even if decreasing a little at higher temperature. Recall the inequality (3.20) and the fact that the "derusting" process is not a natural one. Second, reaction (6.5) of iron oxide with CO gas is a natural process at both temperatures. Reaction (6.6) of iron oxide with coke—here represented as a sum of reactions (6.7) and (6.9), with (6.9) involving a gaseous substrate—is not a natural process at 500 K. However, the process of reaction (6.6) *becomes natural* at 1500 K and is even thermodynamically favored over reaction (6.5) at 1500 K. At this point, we need to consider the *phases* involved in the melting and separation procedure.

From the diagram it can be seen that the starting materials pass vertically from the top of the furnace through a range of temperatures before being separated at the bottom. On *lower levels*, reaction (6.5) proceeds directly. It is thermodynamically feasible, and the necessary CO is furnished from reactions (6.9) and (6.10). On *higher levels*, reaction (6.10) proceeds to a much lesser extent, but CO comes from lower levels feeding the reduction process (6.5). On the other hand, the solid phase process of reaction (6.6), even if thermodynamically natural at high temperatures, transpires to little or no degree. We need also to remember the limestone which descends in the blast furnace and decomposes as follows:

$$CaCO_3 \rightarrow CaO + CO_2 \tag{6.11}$$

When iron ore contains sulfur in the form of FeS, then CaO formed in the last reaction is used to decompose the sulfide:

$$FeS + CaO + C \rightarrow CaS + FeO + CO \tag{6.12}$$

CaS so formed becomes part of the **slag**. The slag contains also silica (SiO_2), alumina (Al_2O_3), magnesia (MgO), or calcia (CaO). Liquid slag then trickles through the coke bed to the bottom of the furnace; it floats there on top of the liquid iron since it has *lower density*.

Together with molten iron and slag, the blast furnace process produces also hot "dirty gases." These gases exit the top of the blast furnace and proceed through gas cleaning equipment; particulate matter is removed and also the gas is cooled. This gas, with a significant energy value, is burned as a fuel in the "hot blast stoves" which are used to preheat the air entering the blast furnace to become "hot blast". Any of the gas not burned in the stoves is sent to the boiler house and is used to generate steam. Steam turns a turbo blower that generates the compressed air known as "cold blast" that comes to the stoves. The production of iron, important on its own account, has thus illustrated factors essential in metal reduction processes.

For those metals that are not sufficiently pure after reduction, the third stage of the metallurgical process is **refining and removing impurities**. We already talked about sulfur. Not surprisingly, many different techniques are used to accomplish this. A wide difference in oxidation potential between the metal and impurity can provide a means for separation and removal. For instance, blasting oxygen or air through an ore containing Si produces SiO; the latter can be removed as a slag. Electrolysis, mentioned earlier in connection with copper, is feasible in some cases. **Zone melting**, a popular technique, operates on the principle that because of differences in solubilities, impurities concentrate in the moving liquid phase rather than the solid phase. The method consists of moving a heating unit slowly from one end of a long metal block to the other, rejecting impurities at the end. The efficiency is improved by repeating the process several times or by using several heating units. Metals of very high purity can be obtained by zone melting.

With a metal sufficiently pure, the next step is to give it the desired size and shape. There are now many metal-fabricating techniques; the main ones are casting, sintering, forging, rolling, and extrusion.

Casting entails pouring molten metal into a prepared mold cavity. The cavity may be the shape of the final product, or it may be some other shape referred to as an *ingot* that will then be subjected to further processing by one of the other previously mentioned methods of fabrication. There are multiple forms of casting:

1. **Centrifugal casting**, wherein the mold is rotated so that centrifugal force assists metal flow.
2. **Die casting**, wherein pressure is applied to force the molten metal into the die cavity (which is usually made of metal).
3. **Investment casting**, a multi-step process facilitating the casting of complex parts by preparing a ceramic investment (mold) from a wax pattern that can then be melted out of the ceramic shell before the molten metal is cast into it.
4. **Permanent mold casting**, wherein a metal mold is prepared and re-used many times with liquid metal being fed into the mold by gravity only.
5. **Sand casting**, wherein the mold is made from sand or loam and must be destroyed to recover the casting.

Sintering requires metal (or metals) in powder form. The powdered metals are mixed at desired concentrations, pressed into a die, and ultimately sintered in a furnace. Consequently the products of sintering may exhibit properties different than alloys prepared from molten (liquid) metal phases. The distinguishing mark of sintered metals is the capacity to easily control their porosity. Bearings, for example are sintered metal products. Interestingly bronze metal bearings soaked beforehand in oil become "self-lubricating" for a long time, even for their entire working life.

Now we come to steps five and six in the metallurgical process, which involve hardening of the material. Whether the method is rolling, elongation in tension, wire drawing, etc., the mechanism of hardening is essentially that of producing new structural defects in the metal. The density of defects increases significantly even after a short time of cold working. Moreover, these are *fine defects* (line dislocations, point defects, twins, etc.) in contrast to *coarse defects*, 3D ones that might arise during casting; recall Section 4.4. The strength or hardness of a metal is approximately proportional to the density of fine defects. Therefore *each defect* is an important contributor to the mechanical resistance of the material.

Said another way, **cold working** entails imposing a stress on the material; the ensuing **work hardening** or **strain hardening** of the material is comprised of the multiplication of defects and dislocations in response to the imposed stress. The dislocations subsequently move to accommodate the strain. We shall consider this issue again from the point of view of mechanical behavior in Section 14.1. The driving force for the movement of dislocations is the tendency to minimize the energy U of the system, inequality (3.17) at work again. A minimum energy configuration, once attained, will persist. Kuhlmann-Wilsdorf developed a detailed quantitative theory of work hardening in which a relationship between the dislocation density and the flow stress was obtained. The Kuhlmann-Wilsdorf theory has been discussed in several papers [4] along with experimental evidence that supports it.

Cold working is typically followed by heat treatment. The former process usually incurs some internal stresses that are then reduced by the latter. The main types of heat treatments are annealing, quench hardening, tempering, solution treatment, and precipitation hardening.

Annealing of metal products consists of three stages: recovery, recrystallization, and grain growth. Consider a block of metal that during cold working has been rolled down from 2 cm thickness to 1 cm and then cut into smaller pieces or samples. Now imagine that each sample is annealed, that is heated, for a defined period of time such as one hour at selected various temperatures and then the hardness of each piece measured at room temperature—to see to what extent the annealing has changed the hardness. The higher is the annealing temperature, the lower is the hardness. Hardness as a property will be discussed in Chapter 14. The resultant curve of hardness versus annealing temperature is represented qualitatively in Figure 6.4.

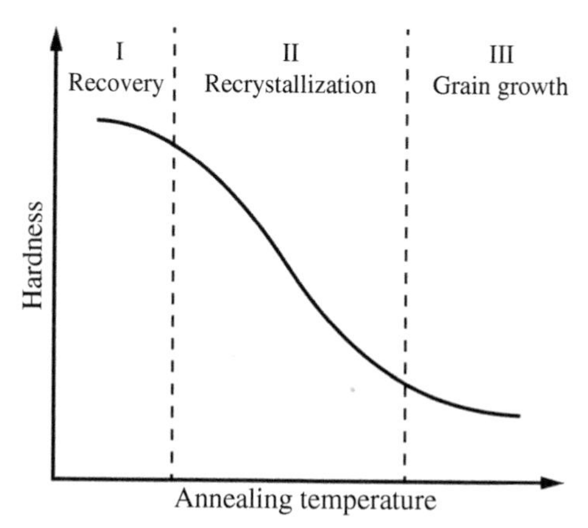

FIGURE 6.4 Metal hardness versus annealing temperature (a qualitative diagram).

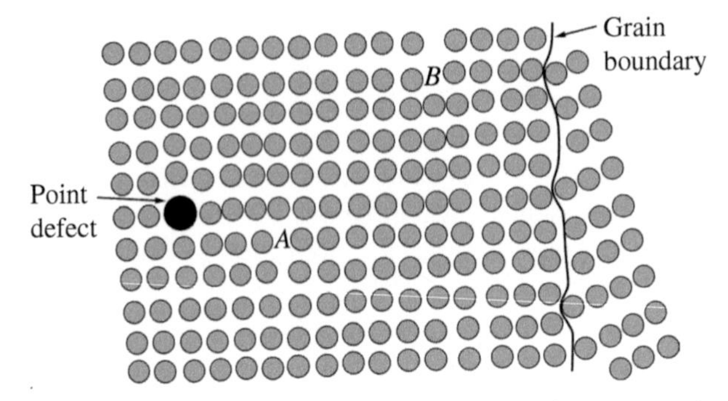

FIGURE 6.5 A dislocation at point A, for consideration of two scenarios. What happens if the dislocation moves to the left? To the right?

During the **recovery** stage, there is some movement of defects within existing grains (grains and grain boundaries in crystalline materials were discussed in Section 4.4) that leads to a lower strain energy. There are no changes in grain structure, and the density of defects changes very little (possibly some defects with attractive forces between them move together). During recovery there may be a slight hardening which accompanies a rearrangement of dislocations into a more stable configuration than existed in the cold-worked state. Rearrangement of point defects (such as vacancies) may actually result in strengthening during recovery; see Figure 6.5.

It is sometimes useful in MSE to give inanimate atoms the properties of humans—in this case the capacity to make decisions and undertake actions in consequence. This way we understand materials better! In Figure 6.5 there is a dislocation at point A, in fact similar to that seen in Figure 4.16a. If the dislocation at point A "decides" to move to the left, it is stopped by the point defect. If it moves to the right, it interacts with the dislocation at B and further to the right it is blocked by the grain boundary. These inhibitions of atomic movement translate to the macroscale as a lack of permanent deformation. Thus we see how the story of defects is connected to strengthening metals.

Following the recovery stage is **recrystallization**, in which small defect-free grains are nucleated. During subsequent **grain growth** the recently nucleated grains consume the pre-existing defects, thereby eliminating a large percentage of defects. The remaining defects ride on the surfaces of growing grains and eventually "settle" at the new grain boundaries. Such effects appear also in submicrocrystals of copper [5]. The driving force for grain growth is the surface energy of the grain boundaries. The outcome of grain growth is a decrease in the yield strength of the material since the yield stress is inversely proportional to the mean grain diameter. As we have already seen in Figure 6.3, the lower defect density associated with grain growth manifests in lower hardness. However, ductility increases. The advantage of hot working over cold-working is that the material may be subjected to much larger deformations without over-hardening or cracking the material. On the other hand, there are obvious energy costs associated with hot working. Hence we see how **economics** interferes in making materials.

To have an idea how several metals compare in hardness to a variety of other materials, we now look at Table 6.1. As discussed in Chapter 14, there are several measures of hardness. In fact, two are involved in Table 6.1. However, the ordering of materials based on hardness is not affected.

TABLE 6.1 Hardness of Some Widely Used Materials

Material	Brinell or Vickers* Hardness
Pure aluminum	15
Pure copper	35
Mild steel	120
304 stainless steel	250
Hardened tool steel	650/700
Hard chromium plate	1000
Sand	1000*
Chromium carbide	1200*
Tungsten carbide	1400*
Titanium carbide	2400*
Diamond	8000*

Quench hardening is a multistage treatment consisting of heating, soaking (i.e., holding at a specified temperature), and then quenching, where **quenching** means cooling at a specified rate. This technique is used for medium and high carbon steels and for tool steels to obtain high hardness.

Tempering is used to reduce brittleness of hardened steels. The procedure entails heating to specified temperature and soaking for a defined length of time, followed by slow cooling. The mechanisms of tempering and quench hardening are similar to that of annealing; however the distinct time and temperature specifications for each technique allow for unique modification of properties.

In **solution treatment**, an alloy is heated to its *single phase* region on the phase diagram. The system is then quenched fairly rapidly in order to retain that phase. Figure 6.6 shows a phase diagram of the Pb + Sn system on which such a treatment is possible. Consider a point in Figure 6.6, for instance, containing 10 wt.% tin at 25°C. We heat this composition up to 150°C where we have only the α phase. We have just completed the first stage of the solution treatment. The second stage will be cooling rapidly back to 25°C. Formation of large β phase (mostly Sn) particles will now be quite difficult, due to a very

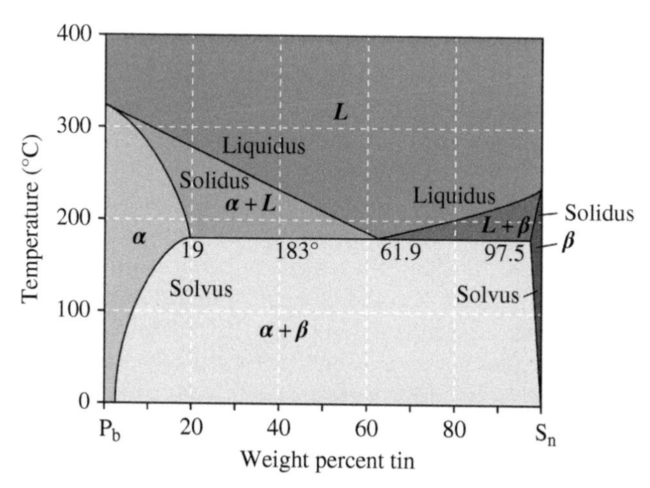

FIGURE 6.6 The Pb + Sn solid-liquid equilibrium diagram. Solvus, Solidus, and Liquidus refer to the boundary lines.

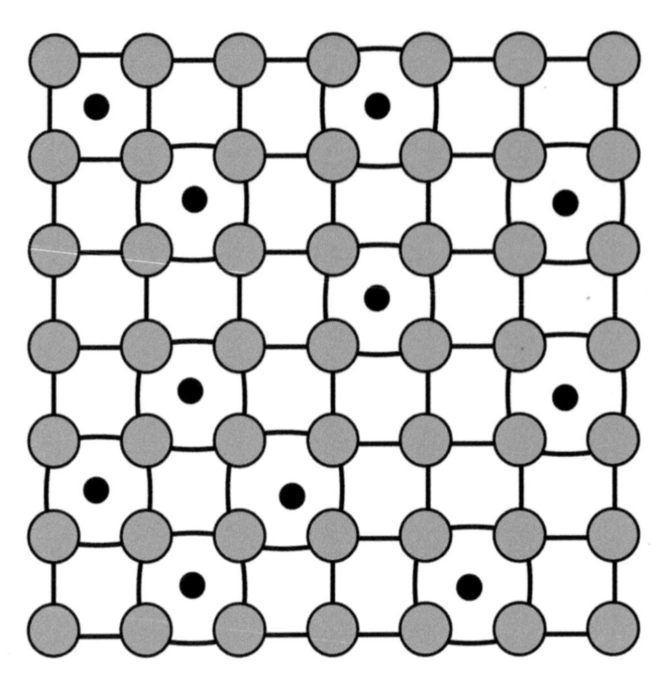

FIGURE 6.7 Interstitial (black) impurity atoms in a lattice of gray atoms.

slow rate of diffusion of molecules in the solid. However, we are in a two phase α+β region; very small particles of the β phase will participate. Now, the covalent radius of a Pb atom is 180 pm (picometers), for Sn the respective value is 145 pm. Thus, there is a possibility that some atoms of tin will appear in interstitial positions; see Figure 6.7.

While in Figure 6.7 the difference in size between the two kinds of atoms is larger than between Pb and Sn, we get an idea of the strain it puts on the crystalline lattice. What happens if the minority (impurity) atoms are larger than those of the main constituent? This situation is shown in Figure 6.8; now we have impurity atoms in substitutional positions. In both cases, the following rule (not a new one) works: Each impurity particle represents a defect in the main phase, and thus increases mechanical strength of the alloy.

Consider a block of metal that has undergone the solution treatment just described. If it is afterwards heated to a higher temperature but still within a two phase region (that is below the eutectic point), the diffusion rate in the solid phase will increase considerably, and this will help many nuclei of the β phase to precipitate. Holding the alloy at a temperature such an elevated temperature is called **aging** the alloy. The heat treatment consisting of solution treatment followed by aging is referred to as **precipitation hardening**, since the precipitation of larger β particles hardens the alloy. Thus, solution hardening (i.e., a solution treatment including precipitation) is related to the loss of mobility of dislocations due to their interaction with the strain fields that exist around the precipitated or "solute" atoms. In a continuous solution, the peak strength does not occur at the 50/50 composition because, for example, an oversized atom in a solid solution causes a greater misfit strain than an undersized atom. The higher strain field around the solute atom results in lower mobility and higher strength. There are quite a few alloys that are suitable for precipitation hardening; they include Al+Cu, Al+Mg, Ni+Ti, and Cu+Be. The last of these alloys is non-sparking when it strikes steel; it is therefore used for wrenches in the petroleum industry. The two-phase alloy of α/β brass (Cu+40% Zn) is stronger than either of the pure metal constituents.

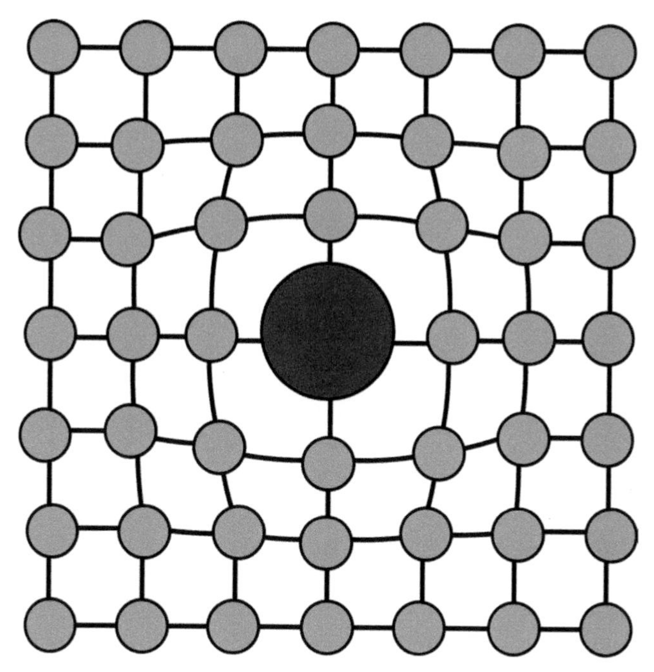

FIGURE 6.8 Metal strengthening by a substitutional impurity atom.

An outcome of all these hardening procedures is grain refinement. Crystals are strengthened by grain boundaries which block dislocations, creating pile-ups. Thus, a steel that has more grain boundaries per unit volume blocks the movement of dislocations better than a steel with fewer such boundaries. In Section 5.4 we noted that the Voronoi polyhedra can be used for representation of any materials, crystalline, or otherwise. Oluwole and Akinkunmi [6] have represented structures of carbon steels by the polyhedra. See in Figure 6.9 an example of a steel containing 0.33% carbon; on the left we have a micrograph and on the right the Voronoi diagram.

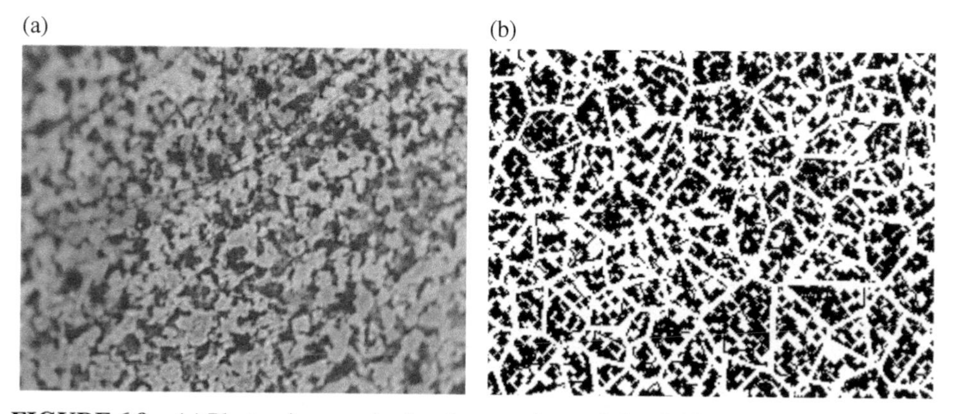

FIGURE 6.9 (a) Photomicrograph of carbon steel containing 0.33% carbon (Micrograph) X200. (b) Voronoi tessellation model of the steel structure at left. *Source*: Oluwole and Akinkunmi [6]. Used under CC BY.

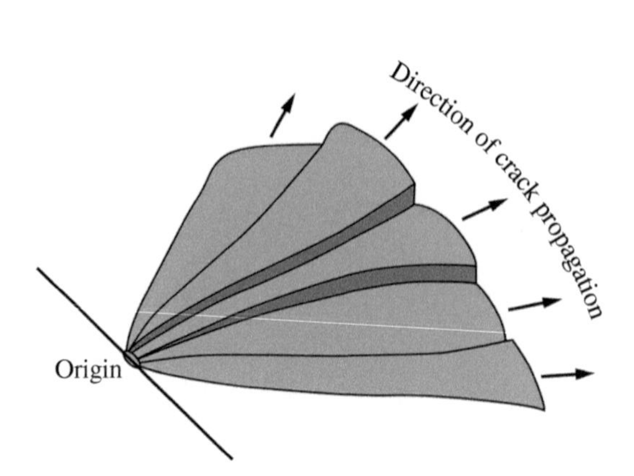

FIGURE 6.10 The Chevron pattern of crack propagation.

The importance of defects in determining the strength of metals has been discussed already in reference to the preceding figures. Considering again Figure 6.5 as an illustration, if the dislocation at point A moves to the left, it is stopped by the point defect. If it moves to the right, it interacts with the dislocation at B and further to the right it is blocked by the grain boundary.

Voids in the microstructure lead to fracture in metals. Fracture surfaces from brittle failures leave behind distinctive crack structures. A Chevron pattern is frequently observed on brittle crack surfaces, as shown in Figure 6.10, which indicates the origin of the crack, showing how the pattern is formed from one starting point. There is further discussion of crack propagation in Section 6.7, where we address the topic as it relates to amorphous versus crystalline metal structures.

The traditional metallurgical processes of casting and cold and hot working are still prevalent. In order to reduce energy costs associated with heating and machining, there is some shift towards processes that permit the formation of finished shapes directly from liquids and powders (**near net shape manufacturing**). Of course, energy is required also in the preparation of metal powders. There are reasons to use both techniques, and a systems approach (discussed in Chapter 19) to metallurgical processing may be the best route to determining optimum processing methods.

6.3 ALLOYS

We have referred several times to alloys up to this point. Before moving on, we shall draw attention to their particular features. We already established that a metal is any of a category of electropositive elements that usually have a shiny surface, are generally good conductors of heat and electricity, and can be melted or fused, hammered into thin sheets, or drawn into wires. Typical metals forms salts with nonmetals, basic oxides with oxygen, and alloys with one another. An alloy is a metal composed of more than one element. Engineering alloys include the cast-irons and steels, aluminum alloys, magnesium alloys, titanium alloys, nickel alloys, zinc alloys, and copper alloys. For example, **brass** is an alloy of copper and zinc; **bronze** is copper and tin; **rose gold** is copper and gold; **solder** is often lead and tin, plus perhaps some additives. Titanium alloys are frequently used as biomaterials; magnesium alloys—boasting lower density than aluminum—are used in aircraft, cars, and bicycles. Alloys are prepared to attain properties more desirable than those of the individual metal comprising them. In this aspect, they are similar to composites.

A category of alloys with good future are **high-strength low-alloy** (HSLA) steels. These micro-alloyed steels are important structural materials and contain small amounts of alloying elements, such as niobium, titanium, vanadium, and aluminum, which enhance the strength through the formation of stable carbides, nitrides, or carbonitrides. As expected, hardness is enhanced. Such steels contain less than 0.1% of the alloying additives, used individually or in combination. Yield strength improvements of two or three times that of plain carbon-manganese steel can be attained. High-strength low-alloy steels seem to be gradually replacing traditional ones in the manufacturing of industrial parts that are typically produced by more expensive processes [7].

There is also a category known as **shape-memory alloys**. Common alloys of this type are copper-aluminum-nickel and nickel-titanium materials. After being distorted, shape-memory alloys can go back to their original shape by heating them to a critical temperature. Muscle wire, an alloy of this type, can expand and contract when an electrical current is applied. Shape-memory alloys are discussed in more detail along with other smart materials in Chapter 12.

6.4 PHASE DIAGRAMS OF METAL SYSTEMS

While this topic deserves a section, we have already seen several such diagrams. In Chapter 3 we have seen the unary diagram of magnesium (Figure 3.4) and the binary Cu + Ni diagram (Figure 3.13), and we have learned how such diagrams are obtained. We have learned how to use the lever rule (Figure 3.14). We have learned about eutectic, eutectoid, peritectic, and peritectoid diagrams (Figure 3.15). We have also seen above the eutectic Pb + Sn diagram in Figure 6.6. The most important things about phase diagrams have been already said. It is worth repeating that the diagrams provide information that enables us to prepare materials for specific applications and also allows us to see how to apply effective strengthening procedures.

6.5 FERROUS METALS: IRON AND STEEL

The **iron + carbon phase diagram** is considered to be the most important of all. It is shown in Figure 6.11. The phase diagram Fe + C is quite complicated. Even pure iron has three solid phases. Iron metallurgists need to know this diagram by heart. For us in this one-semester course it suffices to have the diagram as a reference. If you are told that you have austenite steel at your disposal, you can look it up in the diagram. The reason that Fe is always accompanied by C is that Fe is soft. A knife made of pure iron would not survive long even if used for bread cutting only. Also the manufacturing process of getting Fe from iron ores involves carbon—as we have seen in Section 6.2. Thus, carbon serves here two different purposes.

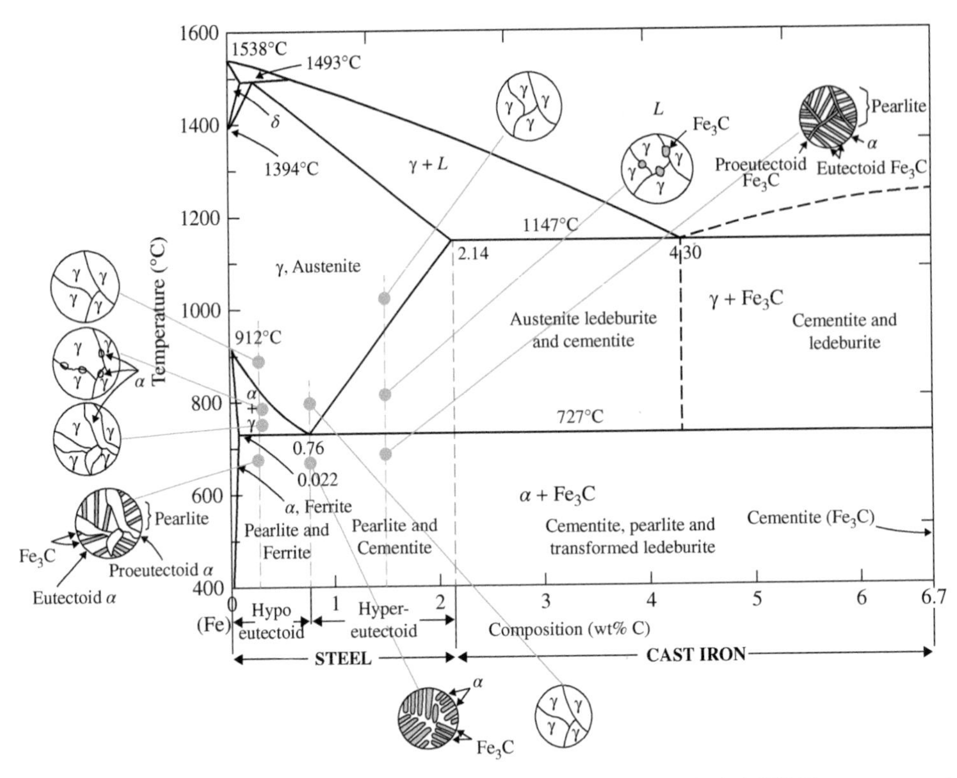

FIGURE 6.11 Iron + carbon phase diagram. *Source*: http://sekolah007.blogspot.com/2013/04/diagram-fe-fe3c.html. Public Domain.

Carbon steel is typically distinguished by the carbon content. In order of increasing carbon content (from approximately 0.07% to 1.6%), steels are referred to as **dead mild steels**, **mild steels**, **medium carbon steels**, **high carbon steels**, and **carbon tool steels**; see also Figure 6.13. Alloy steels typically contain (individually or in some combination) Mn, Ni, Cr, Mo, W, Si, S, or P. Each type of alloy steel has its own defining set of properties; mainly they are considered as either structural or stainless steels. **Cast iron** materials, in contrast to steels, have higher carbon content (usually between 2.5 and 3.8% C).

The presence of alloying elements—several of which have been named in the previous paragraph—has interesting consequences. One of these is that metal carbides are formed [8–10]. Structures at power plants include heat resistant steels. During service, carbide crystallites in such steels slowly change their sizes. On this basis Baltusnikas, Levinskas, and Lukosiute have devised a method to monitor the carbide crystallite size changes under service conditions as a function of time [8–10] and thus evaluate steel aging.

Owing to their iron content, ferrous materials are magnetic. They also tend to corrode easily. Alloying with Ni significantly reduces the magnetism of steel, while addition of Cr improves corrosion resistance. A useful comparison of ferrous metals is provided in Figure 6.12.

Non-Ferrous Metal Types and Features			
Name	**Composition**	**Properties**	**Uses**
Aluminum	Pure metal	Greyish-white, soft, malleable, conductive to heat and electricity. It is corrosion resistant. It can be welded but this is difficult. Needs special process.	Aircraft, boats, window frames saucepans, packaging and insulation, pistons and cranks.
Aluminium alloys- (Duraluminium)	Aluminium 4% copper + 1% manganese	Ductile, malleable, work hardens.	Aircraft and vehicle parts.
Copper	Pure metal	Red, touch, ductile, high electrical conductor, corrosion resistant. Can work hard or cold. Needs frequent annealing.	Electrical wire, cables and conductors, water and central heating pipes and cylinders. Printed circuit boards, roofs.
Brass	65% copper + 35% zinc	Very corrosive, yellow in color, tarnishes very easily. Harder than copper, good electrical conductor	Castings, ornaments, valves, forgings
Lead	Pure metal	The heaviest common metal. Soft, malleable, bright and shiny when new, but quickly oxidizes to dull grey. Resistant to corrosion.	Protection against X-Ray machines. Paints, roof coverings, flashings, hiding things from Superman.
Zinc	Pure metal	A layer of oxide protects it from corrosion, bluish-white, easily worked.	Makes brass, coatings for steel galvanizes corrugated iron roofing, tanks, buckets, rust-proof paints.
Tin	Pure metal	White and soft, corrosion resistant.	Tinplate, making bronze.
Gilding metal	85% copper + 15% zinc	Corrosion resistant, golden color, enamels well.	Beaten metalwork, jewelry.

FIGURE 6.12 Comparison of ferrous metals.

6.6 NON-FERROUS METALLIC ENGINEERING MATERIALS

Non-ferrous metallic materials, namely, metals other than iron and alloys not containing iron, are produced in smaller quantities than steels and irons; but their importance exceeds their proportion in production. Non-ferrous metals are not magnetic; they also tend to be more resistant to corrosion than are ferrous materials. The characteristic features of non-ferrous metals can guide the materials scientist in the design and development of materials and products. Figure 6.13 provides a useful comparison of some common non-ferrous metals and alloys.

Aluminum and aluminum alloys are known for their light weight (because of low density) as well as high thermal and electrical conductivities. The **coherent oxide** that forms on the surface provides good corrosion resistance to aluminum. Aluminum is commonly alloyed with various combinations of Cu, Mg, Mn, Si, and Zn and is frequently wrought or cast to produce desired products.

Ferrous Metal Types and Features			
Name	**Composition**	**Properties**	**Uses**
Mild Steel	0.15 to 0.30% carbon.	Tough, High tensile strength, ductile. Because of low carbon content it can not be hardened and tempered. It must be case hardened.	Girders, plates, nuts & bolts, general purpose.
High Speed Steel	Medium carbon, tungsten, chromium and vanadium.	Can be hardened and tempered. Can be brittle. Retains hardness at high temperatures.	Cutting tools for lathes.
Stainless Steel	18% chromium, and 8% nickel added.	Corrosion resistant.	Kitchen draining boards, pipes, cutlery, aircraft.
High Tensile Steel	Low carbon steel, nickel, and chromium	Very strong and tough.	Gears, shafts, engine parts
High Carbon Steel	0.70 to 1.40% carbon.	The hardest of the carbon steels. Less ductile, tough and malleable.	Chisels, hammers, drills, files, lathe tools, taps and dies.
Medium Carbon Steels	0.30 to 0.70% carbon.	Stronger and harder than mild steels. Less ductile, tough and malleable.	Metal ropes, wire, garden tools, springs.
Cast Iron	Remelted pig iron with smal amounts of scrap steel.	Hard, brittle, strong, cheap, self-lubricating. White cast iron, grey cast iron, malleable cast iron	Heavy crushing machinery, car cylinder blocks, vices, machine tool parts, brake drums, machine handle and gear wheels, plumbing fitments.

FIGURE 6.13 Comparison of non-ferrous metals.

Like aluminum, **copper** has high electrical and thermal conductivities. It also has very good corrosion resistance and is easily formable. Copper is used for wires, cables, sheets, tubes, and pipes. Familiar alloys include brasses (containing zinc) and bronzes (with up to 20% Sn); there are also Cu+Ni alloys (used in applications ranging from coins to turbine blades, bolts, and nails). The latter are especially resistant to corrosion.

Zinc, **lead**, and **tin** materials are also important non-ferrous metals. Zinc is used for galvanizing iron and steel. Pb+Sn solder is commonly used for joining together metal workpieces.

We need to discuss also **coated metals** and **clad metals**. **Tinplate**, used for food containers, is a coated metal in which a mild steel sheet is coated with a very thin layer of tin. **Galvanized iron**, another example of coated metal, consists of mild steel with a thin Zn coating to prevent rust. Clad metals are coated on both sides, thus possessing a sandwich structure (see Section 10.5 about composites with sandwich structures). Oil storage tanks make use of clad metals in order to prevent rusting on both the inside and the outside of the tank. To avoid the expense of a solid stainless steel tank, a thick layer of less-expensive ordinary steel—prone to rusting—is rolled with two thin layers of stainless steel.

Before we move to amorphous metals, we need to contrast ordinary metals and metal alloys with **single crystal metals** (no grain boundaries "by definition"). Turbine blades in jet engines are usually metallic. A number of groups work in this area, advocating—for instance—**nickel-based superalloys** [11, 12]. We stress that defects reinforce metallic phases. It is thus no surprise that, in a single crystal turbine blade, the absence of grain boundaries gives a decrease in yield strength. On the other hand, at the high service temperatures of turbine blades, the absence of grain boundaries also *decreases creep*—an important factor in this application. Thus we see how engineers are faced with difficult choices.

The key method of making single crystals was reported in 1918 by the Polish chemist **Jan Czochralski** [13]. A precisely oriented rod-mounted seed crystal is dipped into the molten material. The seed crystal's rod is slowly pulled upwards and rotated simultaneously. By precisely controlling the temperature gradients, rate of pulling, and speed of rotation, it is possible to extract a large, single-crystal cylindrical ingot from the melt. The mechanism of the process is based on surface tension which we discussed in Section 2.5. Occurrence of unwanted instabilities in the melt can be avoided by investigating and visualizing the temperature and velocity fields during the crystal growth process. It is normally performed in an inert atmosphere, such as argon, in a nonreactive chamber, such as quartz. While Czochralski developed his technique working with metals, it is now particularly important for silicon and other semiconductor materials. We shall talk about semiconductors and their applications in Chapter 17.

6.7 STRUCTURES OF METALS IN RELATION TO PROPERTIES

Properties depend on sizes and also shapes of constituent structural units; this is true for all materials, not only for metals. Here is a short survey for metals. We shall discuss in Section 6.8 the case of a dendritic microstructure in a glassy metal—and its effects on properties.

We recall a discussion in the beginning of Chapter 4: mechanical properties are structure sensitive, that is, strongly dependent on the presence and amount of defects; recall again Figure 6.5. This is generally true, but not the whole story for metals; their behaviors in response to radiation are also structure sensitive. Demkowicz and coworkers demonstrate that *interfaces* are important in determining the radiation response of metallic composites [14–17]. While pure metals such as copper or niobium have a certain response to radiation, if Cu and Nb are joined into layered composites, the interfaces where these two metals are bonded are especially good traps for radiation-induced point defects. Furthermore, the interfaces accelerate the recombination of these defects. The result is much higher resistance to radiation than in either of pure metals [18].

One can make crystalline **metal foams**. Thus, **aluminum foams** have densities between 0.08 and 0.85 g/cm^3 and high strength/weight ratios. Both open cell and closed cell structures are made. Aluminum foams are used in cars where they provide very high energy absorption, much higher than expected on the basis of their weight. We shall see in Figure 10.6 other structures with much empty space inside, only there we have sandwiches with different material on top and bottom surfaces than in the middle, while in Al foams we have aluminum only.

There are ballistic devices known as explosively formed penetrators (EFPs) whose forming and deformation processes result in a variety of microstructures, while those structures are in fact difficult or expensive to test and model. Expense arises from the need to reproduce the explosive conditions used to form the projectile (penetrator). Difficulty is in overcoming the insurmountable number of parameters involved in microscopic modeling of such materials.

How does an explosively formed penetrator work? Essentially there are three components: a *sleeve* or tube containing a relatively thin, concave *metal liner* inserted at the front with *highly explosive material* packed behind it. For instance, the liner may be a thin copper disc (with concave curvature). Upon device detonation, forwards propulsion of the disc is accompanied by a simultaneous transformation of the projectile into a penetrator having a shape more like a bullet or missile (video demonstrations of EFPs in action are available online [18]). Explosively formed penetrators so formed are capable of penetrating through thick layers of steel and other materials. Lawrence Murr and coworkers, among others, have conducted extensive studies of the deformation processes and microstructures involved in ballistic penetrator formation, in hypervelocity impact, and in penetration phenomena [19, 20]. Very high strains and strain rates are conditions unique to the deformation processes associated with explosively formed projectiles. Deformation can be extreme; features include dynamic recrystallization and grain boundary sliding [21]. For testing purposes, shock loading of metals and alloys is used to impose severe plastic deformation. Murr and Esquivel have used a simple model to explain differences in microstructure resultant from plane shock and spherical shock in connection with stacking fault energy and the shock wave geometry [22]. Understanding of such microstructures and the processes of extreme deformation has important applications not only for ballistics and weapons technology but also, for instance, in friction stir welding, a procedure for joining dissimilar metals and alloys that could be applied in automotive and aircraft manufacture [21].

Nanostructures have sizes not larger than 100 nm. If one thinks that nanostructures have been utilized only now in the 21st century, one finds that this is not the case. Mayan pigments in the pyramid in Chichen Itza, Yucatan, Mexico, have been studied [23]. It turns out that they contain nanoparticles of iron and chrome. Applications of nanometals are not limited to strictly metallic systems. Textile industry produces waste water containing dyes such as indigo carmine, harmful to the respiratory track if inhaled and an irritant to the skin and eyes. Ways of cleaning textile waste water have been developed. They involve using nanoparticles of Fe, Cu, and Ni in combination with a composite based on corn (maize) stems [23].

Let us provide at least one more example. A **plasmon** is a quantum of plasma oscillation, a quasiparticle; see again Eq. (5.1) and its discussion. **Plasmonic nanoparticles** are particles whose electron density can couple (interact) with electromagnetic radiation of wavelengths that are far larger than the particle. This is a consequence of the nature of the dielectric-metal interface between the medium and the particles, different than in a pure metal where there is a maximum limit on what size wavelength can be effectively coupled based on the material size. What differentiates these particles from normal surface plasmons is that plasmonic nanoparticles also exhibit unusual scattering, absorbance, and coupling properties based on their geometries and relative positions. There is ongoing research of applications for metal nanoparticles in solar cells, spectroscopy, signal enhancement for imaging, and cancer treatment.

6.8 GLASSY METALS AND LIQUID METALS

The interest in non-crystalline or **glassy metals** (also called metallic glasses) is growing. They resist corrosion better than crystalline metals because they are free of grain boundary defects and second-phase particles [24, 25]. Glassy metals can resist localized corrosion such as pitting or crevice attack—phenomena "popular" in crystalline metals. Glassy metals are also stronger than crystalline metals, even those with similar composition; see Figure 6.14.

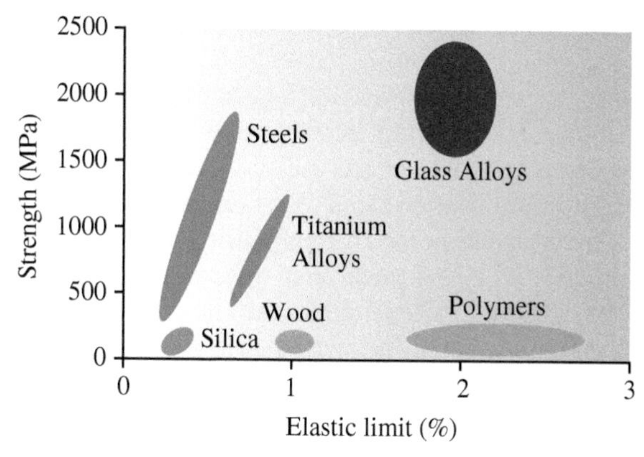

FIGURE 6.14 Strength of several classes of materials versus elastic limit (refer to Chapter 14 for mechanical properties). *Source*: After [26]. Reprinted with permission from William Johnson.

The history of glassy metals followed an interesting progression through the first few decades after its inception at the end of 1950s. In the earlier days, one was able to make only very thin layers of glassy alloy on conductive substrates, and these were only achievable at very high cooling rates between 10^5 and 10^6 K/s. By the 1970s metallic glass was formed via cooling rates as low as 10^3 K/s with rods of these materials developed having diameter on the order of millimeters. This trend continued, and by the early 1980s, centimeter ingots of vitrified materials were being developed at cooling rates well below 100 K/s. In 1993 William L. Johnson and coworkers at the California Institute of Technology (Caltech) succeeded in creation of amorphous rods having diameter of several centimeters by rates of 1 K/s [27, 28]. On the other hand, the same group successfully applies very high cooling rates such as 1 million K/s [29]; rapid thermoplastic forming of the undercooled liquid into complex net shapes is then possible under rheological conditions typically used in molding of plastics.

Let us turn to more fundamental questions as we attempt to understand this phenomenon: among the different alloys in the liquid state, which are the most likely to form glasses upon cooling? What characteristics are relevant? As it turns out, there are only a few such characteristics to look for. To begin, let us define the *reduced glass transition temperature* t_g by

$$t_g = \frac{T_g}{T_L} \tag{6.13}$$

where T_g is the glass transition temperature (in degrees K) and T_L is the alloy melting temperature—more properly called the liquidus temperature (see again Figure 3.13). The ratio of these two temperatures forms a convenient measure of the competition for thermodynamic stability between the liquid and crystalline phases. Alloys that have both low liquidus temperatures and high reduced glass transition temperatures (in the range of ≈0.6) are glass forming.

One further characteristic affecting glass formation in alloys is liquid/liquid phase separation. A number of experimental methods (including small angle neutron scattering, field-ion microscopy, and others) have revealed this phenomenon to be of importance. Crystal growth is preceded by a decomposition, which produces two liquids that appear to provide critical nucleation sites. Recall Figure 3.12. This phase separation is considered the rate-limiting step in crystal growth [27].

As mentioned at the opening of this Section (and as Figure 6.14 further demonstrates), glassy metals have a number of physical advantages over their crystalline counterparts. Why, in this situation, is the manufacture and use of crystalline metals so much larger than that of glassy ones? There are two problems. First, metals have the natural tendency to crystallize. To make glassy metals, we have to apply various *tricks to prevent crystallization*. Prevention of crystallization of a *pure* metal is so difficult that nobody has yet invented a good way of doing this. Therefore, glassy metals thus far are always alloys. The elements present are not necessarily all metals, in fact nonmetals help to prevent crystallization. There is also a second problem to be discussed shortly.

VISCOSITY AND METALLIC GLASS FORMING

Rheological behavior is important in identifying metals likely to form glasses. In particular, if the viscosity η of the alloy is relatively high at temperatures well above the glass transition, then glass alloys are more easily realized. A Vogel-Fulcher law can also be employed as

$$\eta = \eta_0 e^{\left[DT_0/(T-T_0)\right]} \tag{6.14}$$

where $\eta_0 = hA_0/v$ and h, A_0, and v denote Planck's constant, Avogadro's number, and the molar volume of the liquid, respectively. The parameter D is called the "fragility index," and T_0 the Vogel-Fulcher temperature. We see from Equation (6.14) that the viscosity diverges as we approach T_0. The fragile glass forming liquids are contrasted with strong glass forming liquids—as discussed by Saiter, Saiter, Grenet, and coworkers [30, 31]. Glass-forming alloys typically have a fragility index $D > 10$, unlike elemental metallic liquids where $D \approx 1$. Reduced fluidity of alloy melts is thus a characteristic of the "strong" glass-formers when they are at elevated temperatures, near the melting temperature. Crystal nucleation is therefore suppressed and more favorable conditions exist for glass formation.

Regarding methodology, let us make a list of popular techniques for making glassy metals:

Splat quenching was invented at the California Institute of Technology in 1960 [32]. The technique consists of propelling at high velocities small molten globules against a thermally conductive metal substrate. Alloys so obtained include: 75% Au + 25% Si; Pd + Si; Fe + P + C; Cu + Zr. One can ask naturally: if the technique has been known for so long, why don't we have it applied much more widely? What happened was that glassy metals were considered for quite a while to be just laboratory curiosities.

Vacuum evaporation: ingredients are vaporized by heating and then deposited on a substrate cooled to 77 K or below (liquid nitrogen is used as the cooling agent). Materials so obtained include: Mg + Cu; Mg + Sb; Bi + Ge.

Electrolytic deposition: similar to the popular nickel plating. Substrate is placed in a liquid solution (electrolyte). An electrical potential is applied between a conducting area on the substrate and a counter electrode (usually platinum). A chemical reduction-oxidation process takes place, resulting in the formation of a layer of material on the substrate. The technique is used to obtain alloys of P with either Ni, Fe, Co, or Pd.

Sputtering or plasma jet deposition: target of desired composition and a substrate are placed in a plasma; we recall saying in Section 1.5 that applications of the plasmatic state will grow. In a plasma jet deposition system we have ionized gas maintained by an electric field. Gas ions accelerated towards a target become deposited on a substrate.

Melt spinning: amorphous metal wires have been produced by sputtering molten metal onto a spinning metal disk. The rapid cooling, on the order of millions of degrees per second, is too fast for crystals to form and the material is "locked in" a glassy state.

Blow molding: this technique is similar to the polymer blow molding which will be discussed in Chapter 9. It allows obtaining shapes not achievable with crystalline metals [33].

A second problem with glassy metals is that they are prone to fatigue. Significant progress in this area was made in 2009 when a combined group from Caltech, the US Department of Energy Lawrence Berkeley National Laboratory, and University of California at Berkeley developed a metallic glass alloy named DH3, made from five elements—roughly a third zirconium, a third titanium, and the remainder niobium, copper, and beryllium [34]. In bulk samples of DH3 the researchers induced a second phase in the form of narrow pathways of crystalline metal permeating the metallic glass in dendritic (treelike) patterns; its growth was carefully controlled by processing a partially molten liquid-solid mixture. The resulting **dendritic phase** acts as a local arrest point to any crack that begins to propagate in the glass. The goal is to match the mechanical and microstructural scales of the material: the microstructural scale is the space of the dendritic branches, while the mechanical scale is the length of a crack that breaks the material. Properties of the DH3 alloy are such that it is not only stronger than many structural metal alloys but has a fatigue limit more than 30% higher than ultra-high-strength steel and aluminum + lithium alloys; see Figure 6.14. The DH3 microstructure is shown in Figure 6.15.

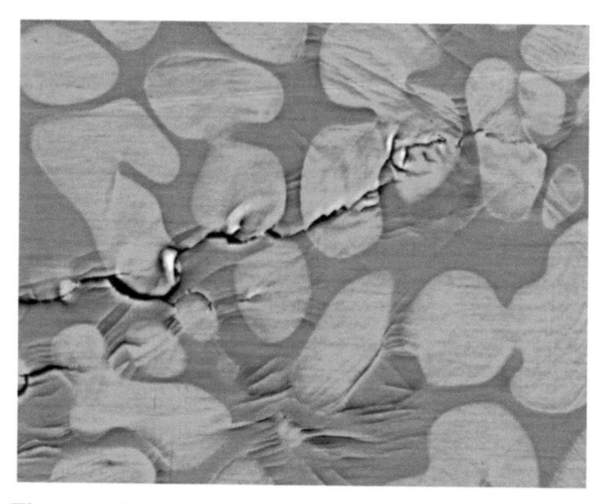

FIGURE 6.15 The two-phase structure of DH3, a metallic glass alloy of five different metals with glassy matrix and a crystalline dendritic phase; when the spaces between the dendritic branches are the right width, they can stop cracks before they propagate. *Source*: Launay *et al.* [34]. Reprinted from *Proceedings of the National Academy of Science*.

Since the development of DH3, the Caltech group has made even more progress in making composites with glassy metal matrices [26]. Their materials exhibit an unusual combination of high strength, high toughness, and very good processability and are able to be fabricated into cellular structures of egg-box topology. Under compressive loading, the egg-box panels are capable of undergoing extensive *plastic collapse* at very high plateau stresses—enabling absorption of large amounts of mechanical energy. In terms of specific mechanical energy absorbed, such panels far outperform panels of similar topology made of aluminum or fiber-reinforced polymer composites, and even surpass steel structures of highly buckling-resistant topologies. Johnson's group at Caltech has also shown that catastrophic shear failure can be suppressed by the introduction of secondary microstructures within the glass matrix. Multiple shear bands are also created with this technique, allowing a dramatic increase in the global plasticity of the materials [35]. Apparently extensive plastic shielding ahead of an opening crack is important [36]; limited plastic yielding by shear-band sliding in the presence of a flaw takes place. Palladium turns out to be a useful constituent in that regard. In response to stress, the material bends rather than cracks. Higher strength than that of steel has been achieved [37].

Clearly glassy metals will have more and more applications. Current examples include:

- $Ti_{40}Cu_{36}Pd_{14}Zr_{10}$, believed to be noncarcinogenic, about three times stronger than titanium, and with elastic modulus closely matched to that of bones. The alloy has high wear resistance, is claimed not to produce abrasion powder, and does not undergo shrinkage on solidification. A surface structure can be generated that is biologically attachable by surface modification using laser pulses, allowing better joining with bone.
- $Mg_{60}Zn_{35}Ca_{5}$, rapidly cooled to achieve amorphous structure, is being investigated as biomaterial for implantation into bones as screws, pins, or plates to fix fractures. Unlike traditional steel or titanium, this material dissolves in organisms at a rate of roughly 1 mm/month and is replaced with bone tissue. The dissolution speed can be adjusted by varying the content of zinc.

Ferromagnetic metallic glasses also seem to have a bright future. Their low magnetization loss is useful in high efficiency transformers. Additionally, electronic article surveillance often uses metallic glasses, largely because of their magnetic properties.

We have discussed above crystalline aluminum foams. Glassy metals can also be made in the form of **foams**. An example is a $Pd_{43}Ni_{10}Cu_{27}P_{20}$ alloy [38]. To create a foam, the alloy is mixed with hydrated B_2O_3 which releases gas at elevated temperature and/or low pressure. Glassy metallic foam biocompatibility appears achievable with air bubbles made so as to match metal density to tissue density and metal elasticity to tissue elasticity. We shall discuss biomaterials in Chapter 11, including the requirement of biocompatibility.

Thanks to their exceedingly high elastic strain limits, glassy metals display enhanced energy storage density in comparison with other materials. This property has been utilized regularly in the world of sports, such that amorphous metals are featured in a variety of sporting goods. On the golf course in particular, irons and drivers made from glassy metals show superior performance in driving golf balls thanks to their elastic energy storage [39].

Except for fatigue, glassy metals tend to have much better mechanical properties than crystalline metals have. Owing to the irregular structure, cracks cannot propagate in straight lines as they do in crystals—see Figure 6.16. For the same reason, glassy metals have higher corrosion resistance than crystalline metals have. Irregular structures do not allow oxygen to penetrate easily, and certainly not in straight lines, thus the chemical reactions of

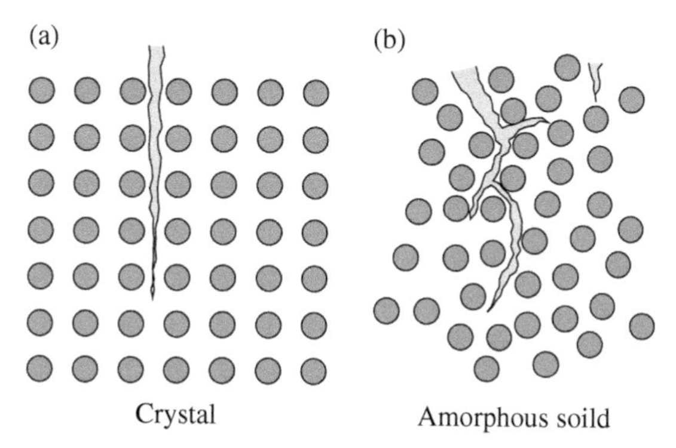

FIGURE 6.16 Crack propagation through (a) crystalline and (b) amorphous metals.

corrosion are inhibited. In crystalline metals corrosion propagates relatively easily along grain boundaries. Thus, glassy metals are not prone to pitting or crevice corrosion "popular" in crystalline metals.

Heterogeneous composites with glassy metal matrices and metal or ceramic fillers can be made. An example is $Zr_{57}Nb_5Al_{10}Cu_{15.4}Ni_{12.6}$ reinforced with W, WC, Ta, and SiC [40]. Such composites have been tested in compression and tension. Compressive strain to failure increased by more than 300% compared with the unreinforced alloy while the energy to break in tensile testing increased by more than 50%.

We know that noble gases are monoatomic and that they solidify at low temperatures. However, very small solid precipitates can be made from those materials even at the room temperature by implanting them in metals by ion beam irradiation [41]. Interestingly, even in this area glassy metals are better than crystalline ones. In crystals the required beam energies are between 50 and 300 keV while in glassy metals 4 keV are sufficient [42].

The promise provided by glassy metals has also led to the development and use of special techniques such as determination of heat capacity and degree of undercooling by electrostatic levitation (ESL) [43–45] and electromagnetic levitation (EML) [46]. Alternating current modulation calorimetry was used in the TEMPUS (Tiegelfreies ElektroMagnetisches Prozessieren Unter Schwerelosigkeit) electromagnetic levitation facility on board the Space Shuttle Columbia during the International Microgravity Laboratory Mission 2 [47].

Synthesizing glassy metallic nanostructures with control over morphology, surface chemistry, and length scale is clearly important—but difficult. The techniques used are often lengthy, require expensive precursors/stabilizers, and limit the control over nanoscale morphology and chemistry. Mukherjee and his colleagues [48] have proposed tuning metallic-glass nanostructures to a wide range of morphologies; the surface is enriched with a catalytic noble metal. By combining thermoplastic nanofabrication together with electrochemical processing, hierarchical metallic nanostructures with large electrochemical surface area and high catalytic activity are synthesized. The authors claim [48] that, due to the versatility in processing and independent control over multiple length scales, the approach may serve as a tool-box for fabricating complex hierarchical nanostructures for a wide range of applications.

There is also a related issue: how to choose the components for a glassy metallic material? Johnson, Goddard III, and their collaborators used molecular dynamics simulations for copper containing binary and ternary alloys to answer that question [49]. The **molecular dynamics computer simulation** technique was created by Berni Alder and T.E. Wainwright in 1959 [50], and reviews of its use exist [51, 52]. One solves Newton's equations of motion as a function of time for particles such as atoms, ions, or polymer chain segments. The work for copper and its alloys by Johnson, Goddard III, and their collaborators [49] has shown that the key parameter is the **atomic size ratio** λ. For large size ratios of $0.95<\lambda<1.0$, crystallization occurs; with moderate size ratios of $0.60<\lambda<0.95$, a glass phase forms; and for small size ratios of $\lambda<0.60$, the alloy phase separates into pure phases which crystallize.

We have talked above about behavior of crystalline metals under radiation. Behavior of glassy metals is also interesting. Baumer and Demkowicz [53] have performed molecular dynamics simulations of glassy metals under irradiation. They have demonstrated that heavy ion or neutron irradiation can cause amorphous metal alloys to undergo spontaneous anisotropic plastic deformation. Thus, shape changes take place without application of external mechanical forces.

6.9 SELF-ASSESSMENT QUESTIONS

1. Can we get metallic iron out of Fe_2O_3? If yes, how?
2. Why is α/β brass (Cu + 40% Zn) stronger than either of its pure metal constituents?
3. Discuss the role of carbon in ferrous metals and alloys.
4. What is the purpose of annealing and cold working in crystalline metals? Describe in detail how the processes affect the materials' structure.
5. Compare glassy and crystalline metals in terms of resistance to cracks and resistance to corrosion.
6. Why are glassy metals not more widely used?
7. How can we predict whether a given pair of components has the capability to form a glassy metal?

REFERENCES

1. Eurometaux, In What Form Are Metals Found? *Eurometaux: European Association of Metals*, http://www.eurometaux.eu/MetalsToday/MetalsFAQs/Formmetalsarefound.aspx, accessed August 1, **2012**.
2. A.W. Cramb, *A Short History of Metals*, http://neon.mems.cmu.edu/cramb/Processing/history.html, accessed September 19, **2013**.
3. A.T. DiBenedetto, *The Structure and Properties of Materials*, McGraw-Hill: New York **1967**.
4. D. Kuhlmann-Wilsdorf, in *Work Hardening*, editors J.P. Hirth & J. Weertman, Gordon and Breach: New York **1968**, p. 27; D. Kuhlmann-Wilsdorf, *Metal. Trans.* **1970**, *1*, 3173; D. Kuhlmann-Wilsdorf, in *Advances in Understanding of Monotonic and Cyclic Workhardening*, editors A.W. Thompson & R. Pelloux, AIME **1976**; D. Kuhlmann-Wilsdorf, *Proceedings of the Centenary Celebration of the Indian Association for the Cultivation of Science*, Calcutta **1977**.
5. N.A. Koneva, E.V. Kozlov, N.A. Popova & M.V. Fedorischeva, *Mater. Sci. Forum* **2010**, *633–634*, 605.
6. O.O. Oluwole & A.L. Akinkunmi, *J. Min. & Mater. Char. & Eng.* **2011**, *10*, 309.

7. D.A. Skobir, *Mater. & Tech.* **2011**, *45*, 295.
8. A. Baltusnikas, R. Levinskas & I. Lukosiute, *Mater. Sci. Medziagotyra* **2007**, *13*, 286.
9. A. Baltusnikas, R. Levinskas & I. Lukosiute, *Mater. Sci. Medziagotyra* **2008**, *14*, 210.
10. A. Baltusnikas, I. Lukosiute & R. Levinskas, *Mater. Sci. Medziagotyra* **2010**, *16*, 320.
11. J. Belan, *Mater. Sci. Medziagotyra* **2008**, *14*, 315.
12. J.Y. Hwang, R. Banerjee, J. Tiley, R. Srinivasan, G.B. Visvanathan & H.L. Fraser, *Metal. & Mater. Trans. A* **2009**, *40*, 24.
13. J. Czochralski, *Z. phys. Chem.* **1918**, *92*, 219.
14. M.J. Demkowicz, R.G. Hoagland & J.P. Hirth, *Phys. Rev. Lett.* **2008**, *100*, 136102.
15. M.J. Demkowicz & R.G. Hoagland, *Internat. J. Appl. Mech.* **2009**, *1*, 421.
16. K. Kolluri & M.J. Demkowicz, *Phys. Rev. B* **2010**, *82*, 193404.
17. L. Zhang & M.J. Demkowicz, *Appl. Phys. Lett.* **2013**, *103*, 061604.
18. W. Han, M.J. Demkowicz, N.A. Mara, E. Fu, S. Sinha, A.D. Rollett, Y. Wang, J.S. Carpenter, I.J. Beyerlein & A. Misra, *Adv. Mater.* **2013**, *25*, 6975.
19. L.E Murr, E. Ferreyra, S. Pappu, E.P. Garcia, J.C. Sanchez, W. Huang, J.M. Rivas, C. Kennedy, A. Ayala & C.-S. Niou, *Mater. Character.* **1996**, *37*, 245.
20. E.V. Esquivel & L.E. Murr, *Mater. Sci. & Tech.* **2006**, *22* (4), 438.
21. L.E. Murr, *Mater. Tech.* **2007**, *22*, 193.
22. S. Pappu & L.E. Murr, Microstructural and computer simulation studies on some EFP materials, in *Shock Compression of Condensed Matter*, editors M.D. Furnish, L.C. Chhabildes & R.S. Hixon, American Institute of Physics: New York **2000**, pp. 1141–1144.
23. G. Lopez Teltes, R.A. Morales Luckie, O.F. Olea-Mejia, V. Sanchez-Mendieta, J. Trujillo Reyes, V. Varela Guerrero & A.R. Vilchis Nestor, *Nanoestructuras Metalicas*, Editorial Reverte: Barcelona/Bogota/Buenos Aires/Caracas/Mexico **2013**.
24. N.R. Sorensen & R.B. Diegle, Corrosion of amorphous metals, in *ASM Handbook, Vol. 13—Corrosion*, ASM International: Metals Park, OH **1987**, pp. 864–870.
25. J.M. Greneche, N. Randianantoandro, Y. Labaye, M. Tamine, H. Guérault, M. Stephan & M. Henry, *Mater. Sci. Medziagotyra* **1999**, *5*, 3.
26. Prof. William L. Johnson Research Interests, http://www.its.caltech.edu/~matsci/wlj/wlj_research.html, accessed April 30, **2015**.
27. W.L. Johnson, *Mater. Res. Soc. Bull.* **1999**, *24*, 42.
28. W.L. Johnson, *J. Metals* **2002**, *54* (3), 40.
29. W.L. Johnson, G. Kaltenboek, M.D. Demetriou, J.P. Schramm, X. Liu, K. Samwer, C.P. Kim & D.C. Hofmann, *Science* **2011**, *332*, 828.
30. J.M. Saiter, E. Dargent, M. Cattan, C. Cabot & J. Grenet, *Polymer* **2003**, *44*, 3995.
31. A. Saiter, J.M. Saiter & J. Grenet, *Eur. Polymer J.* **2006**, *42*, 213.
32. W. Klement, R.H. Willens & P. Duvez, *Nature*, **1960**, *187*, 869.
33. J. Schroers, *Adv. Mater.* **2010**, *22*, 1566.
34. M.E. Launay, D.C. Hofmann, W.L. Johnson & R.O. Ritchie, *Proc. Natl. Acad. Sci. USA* **2009**, *106*, 4986.
35. J.P. Schramm, D.C. Hofmann, M.D. Demetriou & W.L. Johnson, *Appl. Phys. Letters* **2010**, *97*, 241910.
36. C.C. Hayes, C.P. Kim & W.L. Johnson, *Phys. Rev. Letters* **2000**, *84*, 2901.

37. M.D. Demetriou, M.E. Launey, G. Garrett, J.P. Schramm, D.C. Hofmann, W.L. Johnson & R.O. Ritchie, *Nature Mater.* **2011**, *10*, 123.
38. J. Schroers, C. Veazey & W.L. Johnson, *Appl. Phys. Letters* **2003**, *82*, 370.
39. R.D. Conner, R.B. Dandliker, V. Scruggs & W.L. Johnson, *Internat. J. Impact Eng.* **2000**, *24*, 435.
40. R.D. Conner, H. Choi-Yim & W.L. Johnson, *J. Mater. Res.* **1999**, *14*, 3292.
41. A. vom Felde, J. Fink, T.H. Muller-Heinzerling, J. Pfluger, B. Scheerer, G. Linker & D. Kaletta, *Phys. Rev. Lett.* **1984**, *53*, 922.
42. T. Miyauchi, H. Kato & E. Abe, *Mater. Res. Lett.* **2014**, *2*, 94.
43. M.B. Robinson, D. Li, J.R. Rogers, R.W. Hyers, L. Savage & T.J. Rathz, *Appl. Phys. Lett.* **2000**, *77*, 3266.
44. T.J. Rathz, M.B. Robinson, R.W. Hyers & J.R. Rogers, *J. Mater. Sci. Lett.* **2002**, *21*, 301.
45. S. Mukherjee, Z. Zhou, W.L. Johnson & W.-K. Rhim, *J. Non-Cryst. Solids* **2004**, *337*, 21.
46. M. Leonhardt, W. Loser & H.-G. Lindenkreuz, *Acta Mater.* **1999**, *47*, 2961.
47. Team TEMPUS, Containerless processing in space: recent results, in *Materials and Fluids Under Low Gravity*, editors L. Ratke, H. Walter & B. Feuerbacher, Springer: Berlin **1996**, p. 233.
48. S. Mukherjee, R.C. Sekol, M. Carmo, E.I. Altman, A.D. Taylor & J. Schroers, *Adv. Funct. Mater.* **2013**, *23*, 2708.
49. H.-J. Lee, T. Cagin, W.L. Johnson & W.A. Goddard III, *J. Chem. Phys.* **2003**, *119*, 9858.
50. B.J. Alder & T.E. Wainwright, *J. Chem. Phys.* **1959**, *31*, 459.
51. S. Fossey, Chapter 4, in *Performance of Plastics*, editor W. Brostow, Hanser: Munich/ Cincinnati, OH **2000**.
52. W. Brostow & R. Simoes, *J. Mater. Ed.* **2005**, *27*, 851.
53. R.E. Baumer & M.J. Demkowicz, *Mater. Res. Lett.* **2014**, *4*, 221.

7

CERAMICS

Diamonds are a girl's best friend.

—Marilyn Monroe as Lorelei Lee in the 1953 film "Gentlemen prefer blondes" directed by Howard Hawks.

7.1 CLASSIFICATION OF CERAMIC MATERIALS

Ceramics can be defined as inorganic and non-metallic materials with a great variety of electrical and thermal properties. Depending on the chemical composition and bonding as well as on structure, entirely different electrical conductivity can be expected: from metallic-type conductors (like TiN or TiB_2) via semiconductors (Si, SiC, GaAs, CdTe) to electrical insulators (SiO_2, Al_2O). Approximate electric neutrality is achieved by either a combination of cations and anions or simple elemental materials such as carbon.

Note here that **organic materials** are defined as those containing carbon in covalent compounds, not elemental carbon. By deduction, **inorganic materials** do not contain carbon in covalent compounds. These definitions are widely accepted but not quite accurate. *Carbides* are compounds but are not classified as organic materials. We should perhaps use the following definition: organic materials are those that contain carbon in compounds other than carbides. There is also the special case of structures with covalent bonds between carbon atoms: these include graphite and carbon nanotubes or fullerenes, with 2D configurations, and also diamond, a special case having pure covalent bonds in a 3D structure.

Ceramics are represented by many compounds, such as oxides, silicates, and aluminosilicates, which are the most common and predominant components of the Earth's crust. Man-made ceramic materials have a long history, as we will see in this chapter. However, such compounds as carbides, borides, and nitrides, as well as carbon nanotubes, are

Materials: Introduction and Applications, First Edition. Witold Brostow and Haley E. Hagg Lobland.

relatively new with less than 100 years of history; note that these compounds do not exist in the Earth's crust (with exception of carbon nanotubes) due to the oxygen-rich atmosphere. Therefore, the word ceramics is used most commonly in reference to silicon and aluminum oxides, often modified by other elements.

The word ceramic derives its name from the Greek *keramos*, meaning "pottery", which is derived from an earlier Indo-European root meaning "heat" or "to burn". Thus, conventional ceramics are a very old class of materials. Many thousand years ago figurines were made from clay, either by itself or mixed with other materials, and then hardened in fire. Later ceramics were fired with **glazing** to create smooth waterproof decorative surface coatings. Ceramics include domestic, industrial, and building products as well as art objects. For instance, Chinese porcelain vases, roughly 600 years old, from the Ming dynasty are famous.

While traditional ceramics are still important and recognized in society, newer so-called advanced or technical ceramics are being used as space shuttle tiles, artificial bones and teeth, electronic components, optic fibers, and cutting tools, to name a few. Ceramics are often classified according to their functions and properties into groups such as glasses, clays, refractories, abrasives, cements, and pure oxides. Few of these groups are mutually exclusive. For instance, alumina Al_2O_3 is a pure oxide and also a refractory material as well as an important abrasive. Considering our ongoing characterization of materials based on their structure, interactions, and properties, we divide ceramic materials into four classes: (1) crystalline; (2) non-crystalline, including inorganic glasses; (3) glass-bonded; and (4) cements. We shall soon have a look at each of these categories.

Let us consider further the complex character of chemical bonds within a compound. When the covalent-to-ionic ratio is significantly high, like in carbides, extremely high melting point and high hardness can be expected. However, increasing ionic character causes very different properties like essential solubility in water, typical for clays containing kaolinite, due to platelet structure and complex configuration of chemical bonds. Therefore, in order to predict or explain the hardness, strength, and inertness of ceramic materials, chemical bonds are of significant importance.

7.2 HISTORY OF CERAMICS

Very old man-made ceramics have been uncovered in former Czechoslovakia by archaeologists. Owing to the firing process, it is difficult to ascertain a good estimate on how old such materials are, but these are estimated to be many thousands of years old. These ancient ceramics which were typically in the form of slabs, balls, and animal and human figurines were composed of animal fat and bone mixed with bone ash and clay-like material. Although the purpose of the ceramics is unknown, it seems that their use was not functional. Pottery making of functional vessels flourished through many centuries, while the manufacture of glass appears to have been related to pottery making.

It is easy to dismiss ceramics in the larger scheme of human history, but their impact on humanity is significant. Modern iron and steel and non-ferrous metal production requires refractory ceramics for furnaces, troughs, and ladles. Likewise, other industries also depend on refractories, including chemical, petroleum, and energy conversion operations as well as those involving glass and other ceramics. The construction industry uses many ceramics: brick, cement, tile, and glass, to name a few. Brick, which has long been in existence, will not burn, melt, dent, peel, warp, rot, rust, or be eaten by termites. Tile is durable, hygienic, and also aesthetic.

The success of Thomas Edison's incandescent light bulb (in 1879) was facilitated by the specific properties of glass. Furthermore, ceramics are essential to the electronics industry: insulators, semi-conductors, superconductors, and magnets contain or are composed entirely of ceramics. Consider also the importance of ceramic spark plugs, invented in 1860 to ignite fuel for internal combustion engines; more about the plugs is in the next Section.

The telecommunications industry utilizes optic fibers: these are high-quality transparent silica (glass) fibers that improve the speed and reliability of information transmission. The use of optic fibers also reduces the need for copper mining (to create copper wires). Ceramics are used as filters and catalytic converters to decrease pollution and capture toxic materials; they are used as neutron barriers to encapsulate nuclear waste. Ceramics exist also for use in oil spill containment. An additional role of ceramics in the present day is as thermal barriers, such as the thermal barrier tiles used to protect spacecraft from temperatures around 1,600°C.

Thus, from dinnerware to semi-conductors, ceramics have and continue to function in various applications. Now we shall examine more closely the structure and composition of ceramics.

7.3 CRYSTALLINE CERAMICS

Crystalline ceramics typically contain a combination of metals and nonmetals. The stronger the metal, the more ionic (e.g., in MgO) and less covalent (e.g., in SiO_2) is the bonding character. An ionic crystal lattice is shown in Figure 7.1. We know from earlier discussion that the cations (colored black) come from metal atoms which have lost one or more electrons. The anions (light gray) come from nonmetal ions which have captured one or more electrons. The bonds in Figure 7.1 are ionic, but covalent bonds in crystalline ceramics are possible also.

Crystalline ceramics are single compounds (such as in Figure 7.1, e.g., NaCl) or mixtures of two or more compounds (e.g., $MgO + Al_2O_3$). Many of these compounds exist

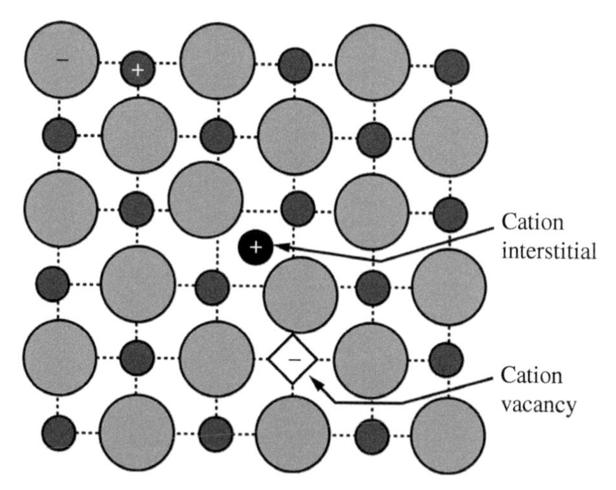

FIGURE 7.1 Illustration of an ionic lattice with a cation vacancy and a cation interstitial—shown here as an example of a metal + nonmetal ceramic.

in some form in nature as **minerals**: inorganic, naturally formed crystalline solids. Good bond stability results in the characteristic high mechanical strength of ceramics. High melting temperatures are typical of most ceramics, also the result of highly stable chemical bonds. Materials used because their melting temperatures T_m are high are called **refractory**. The melting temperatures of refractory ceramics are listed in Table 7.1 and compared to the melting temperatures of corresponding metals. We see that *tungsten* (known in Great

TABLE 7.1 Melting Temperatures (T_m) of Selected Ceramics and Refractory Materials

Material Name	Formula	T_m / K
Hafnium carbide	HfC	4420
Tantalum carbide	TaC	4120
Zirconium carbide	ZrC	3790
Niobium carbide	NbC	3770
Tungsten[a]	W	3640
Hafnium boride	HfB_2	3520
Titanium carbide	TiC	3390
Thoria	ThO_2	3380
Zirconium boride	ZrB_2	3330
Tantalum boride	TaB_2	3273
Tantalum[a]	Ta	3269
Titanium boride	TiB_2	3253
Tungsten carbide	WC	~3120
Magnesia	MgO	3071
Zirconia	ZrO_2	3043
Boron nitride	BN	~3000 (sublimes)
Molybdenum[a]	Mo	2895
Beryllia	BeO	2843
Silicon carbide	SiC	2773
Zircon	$ZrO{\cdot}SiO_2$	2768
Niobium[a]	Nb	2741
Boron carbide	B_4C	2723
Samarium oxide	Sm_2O_3	2600
Europium oxide	Eu_2O_3	2510
Alumina	Al_2O_3	2320
Chromium oxide	Cr_2O_3	2260
Torsterite	$2MgO{\cdot}SiO_2$	2100
Chromium[a]	Cr	2100
Mullite	$3Al_2O_3{\cdot}2SiO_2$	2080
Titanium[a]	Ti	2070
Platinum	Pt	2045
Silica (cristobalite)	SiO_2	1988
Titania	TiO_2	1878
Palladium[a]	Pd	1827
Iron[a]	Fe	1812
Cobalt[a]	Co	1763
Nickel[a]	Ni	1728
Beryllium[a]	Be	1550

[a]Denotes a metal.

Britain as wolfram) and *tantalum* metals have T_m in the range of highly refractory ceramics. Hafnium carbide possesses the highest melting point among known materials.

The presence of multiple elements in ceramics leads to their many varied structures. According to principles discussed earlier in Chapter 4, the crystal structure depends on the relative atomic size of the component elements as well as the bond type and strength. Fundamentally then, the structure depends on electronic configurations. Rock salt crystallizes in the simple cubic form, illustrated in Figure 4.6, with alternating lattice points filled by sodium and chloride ions. Zinc blende (ZnS) as well as cadmium sulfide (CdS) and aluminum phosphide (AlP) crystallizes in the close packed cubic structure shown in Figure 7.2. The radius of the Zn^{2+} ion is 0.074 nm while that of the S^{2-} ion is 0.170 nm.

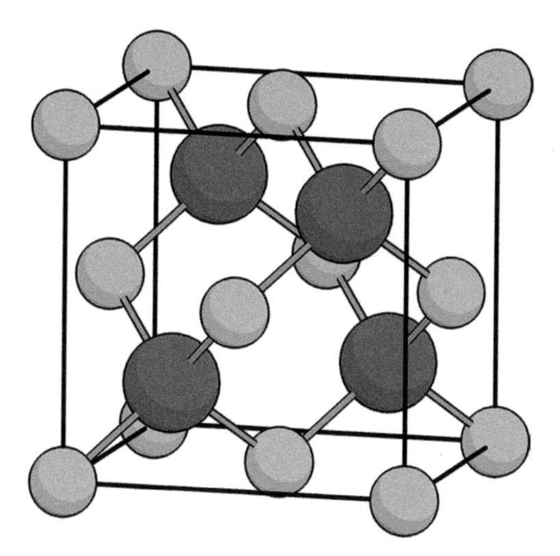

FIGURE 7.2 The zinc blende structure.

Electric neutrality can be also obtained in chemical compounds due to the combining of anions originating from one element (i.e., oxygen) with cations of two different elements having various numbers of valence electrons; these are often **binary oxides**. **Spinels**, which have important magnetic properties (see Chapter 18), are a group of oxides with the general formula AB_2O_4, where A and B are different metals (or the same metal in different oxidation states, as Fe^{2+} and Fe^{3+}). Some examples are magnesium aluminate spinel $MgAl_2O_4$ and cuprospinel $CuFe_2O_4$. The structure of spinels is essentially cubic, a combination of rock salt and zinc-blende structures. The oxygen ions form a face-centered cubic lattice while the A^{2+} ions occupy tetrahedral sites, and the B^{3+} ions sit in the octahedral sites. The unit cell contains 32 oxygen ions, 8 A^{2+} ions, and 16 B^{3+} ions; the spinel cell is illustrated in Figure 7.3. There is also the **inverse spinel structure** that contains the trivalent ions in the tetrahedral sites and an equal division of trivalent and divalent ions in the octahedral sites. Interestingly, the spinel structure may often contain vacancies as a regular part of the crystal.

The **perovskite** structure is another that is common among ionic ceramics; it is defined by the structure of calcium titanium oxide $CaTiO_3$. The mineral was discovered in the Ural Mountains by Gustav Rose in 1839 and is named after Russian mineralogist Lev Perovski. Examples of other perovskites are barium titanate ($BaTiO_3$) and magnesium silicate ($MgSiO_3$) as well as many other oxides with the general formula ABO_3. The ideally cubic structure of perovskite is illustrated in Figure 7.4; however stringent requirements for the relative ion size result in orthorhombic and tetragonal phases as common variants.

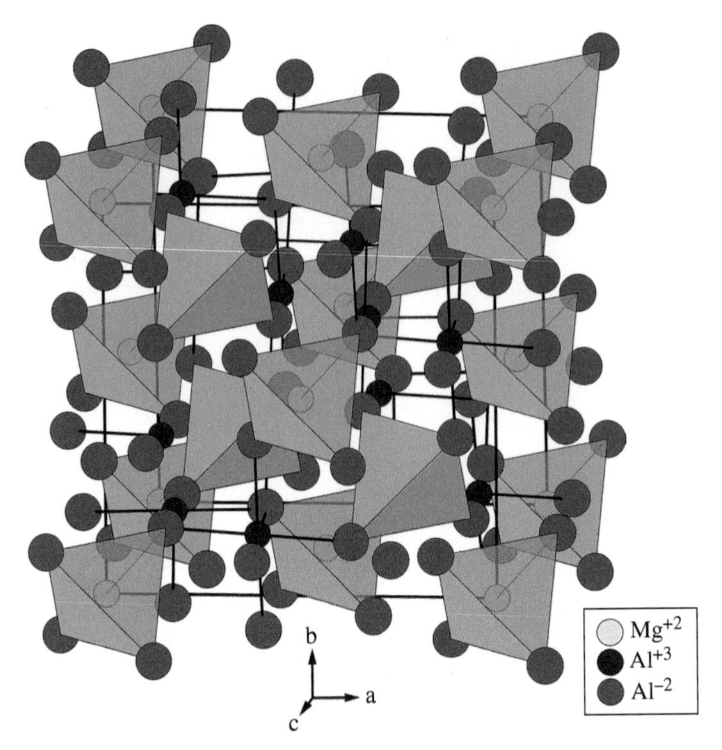

FIGURE 7.3 A spinel structure. *Source*: After Materialscientist, http://commons.wikimedia.org/wiki/File:Spinel_structure_2.jpg. Used under CC-BY-SA-3.0 http://creativecommons.org/licenses/by-sa/3.0/.

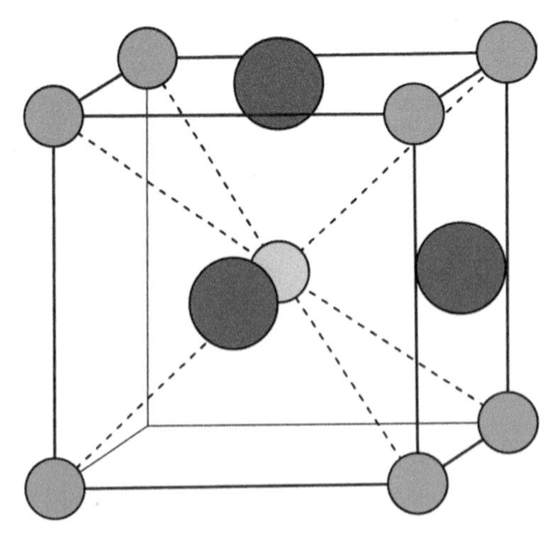

FIGURE 7.4 Perovskite cell structure.

On the other hand, many ceramic crystal structures accommodate atoms or ions widely differing in diameters, which affects the coordination number. For example, the observed cation/anion size ratio for rock salt structures varies between 0.732 and 0.414, depending on the ions present. For wurtzite (ZnS) (having a different structure than zinc-blende), the cation/anion size ratio is 0.25. For comparison, recall glassy metals; we know from Section 6.7 that a good atomic size ratio for them is between 0.60 and 0.95.

It is not always simple to grow good crystals of ceramic materials. For instance, Roy and White [1] who were studying a large number of phases in the Ti + O system concluded after eight years of extensive research that "by utilizing a *variety* of methods which provide controlled oxygen fugacities varying from roughly 10^2 to 10^{-30} one can prepare all the higher titanium oxides with at least the *stoichiometry* controlled." Consequently there exist a variety of **fabrication processes** as well as a sizeable body of literature addressing specific growth processes by different methods [2–5]. Since many of the polycrystalline ceramics are formed from powders, there are so-called particulate forming processes, divided into several main categories: powder pressing, injection molding, ceramic extrusion, slip casting, gel casting, and tape casting. Each of these is a method to obtain the basic shape of the final ceramic product.

Powder pressing consists in filling a die with powder, compressing it, and then releasing the part. Depending on the conditions during compression, powder pressing is specified as uniaxial cold pressing (CP), hot pressing (HP), cold isostatic pressing (CIP), or hot isostatic pressing (HIP). Both HP and HIP are the methods of simultaneous shaping and sintering. **Injection molding** involves feeding the material via a rotating screw into a mold cavity. This method is similar to injection molding of plastics; it is used mainly for small ceramic products. The **extrusion** method is used typically to create products that are long and with uniform cross-section, like alumina pipes. **Slip casting** is used to form shapes from a slurry; this process is used for making fine china, sinks, and thermal insulation parts, among others. **Gel casting**, like slip casting, involves shape forming from a slurry. However, gel casting incorporates an organic monomer in the slurry, which polymerizes after the slurry has been poured into a mold and thereby allows forming of large, complex parts such as turbine rotors. For both slip and gel casting, the liquid from the slurry gets removed to produce a consolidated part. **Tape casting** is used to form thin sheets (e.g., graphite foil) from a slurry. The solvent evaporates from a thin layer of slurry spread over a flat surface; the resultant ceramic sheet can be removed from the supporting surface. Among other applications, tape casting is used to manufacture multilayer ceramics such as those used for capacitors and dielectric insulators. After the shape has been formed by any of the above methods, the resultant so-called **green ceramic body** must be dried and then fired or sintered.

Temperature and humidity control during the drying and firing processes is essential to achieve the desired shrinkage without cracking and to minimize defects. Drying too quickly can result in irregular shrinkage, spallation, and other defects. **Sintering** is the process of heating the green ceramic to a high temperature that is yet below the melting point to allow the separate particles of the material to diffuse to neighboring powder particles. A reduction in surface energy by the decreasing vapor-solid interfaces drives this solid-state diffusion. Porosity is thereby reduced to an extent determined by the initial porosity and the sintering time and temperature. Pressure may be applied to reduce the sintering time (and also the porosity). In the case of hot pressing, sintering occurs simultaneously with the pressing. Finally, it should be noted that the above descriptions provide a basic understanding of how crystalline ceramic products are formed. Variations of the particulate forming processes exist while additives or catalysts may also be incorporated with the ceramic powder to obtain the desired results.

Crystalline ceramics are found in a variety of applications. We already talked above about *refractory ceramics* that can withstand very high temperatures and are non-reactive or exhibit outstanding corrosion resistance. These properties are largely dependent on composition and porosity. It is worth pointing out again that refractory ceramics are used to make kilns for firing other ceramics and for equipment in which to melt and form metals. Some ceramics are used as *abrasives*, for instance, in polishing paper, cutting tools, and even toothpaste. *Advanced ceramics* (also known as *engineering ceramics*) will be discussed at length in Section 7.9. These ceramics are prepared to meet the rigorous demands for applications in optic fibers, microelectromechanical systems (MEMS), piezoelectrics, etc.

Let us now consider some examples of crystalline ceramics from the point of view of applications. Silicon carbide (SiC) is a highly refractory oxidation resistant construction ceramic that can withstand temperatures above the melting point of steel. It is used as a coating on metals, carbon-carbon composites, and other ceramics to provide protection at very high temperatures—this is in addition to its use for heating elements in furnaces.

Silicon nitride (Si_3N_4) has properties similar to SiC but with lower oxidation resistance and melting temperature (it sublimes at $\approx 2180\,K$). It is used in turbine engines, permitting higher operating temperatures and better fuel efficiencies with low apparent density compared to traditional metal alloys. See in Figure 7.5 the use of ceramic materials inside a turbine engine.

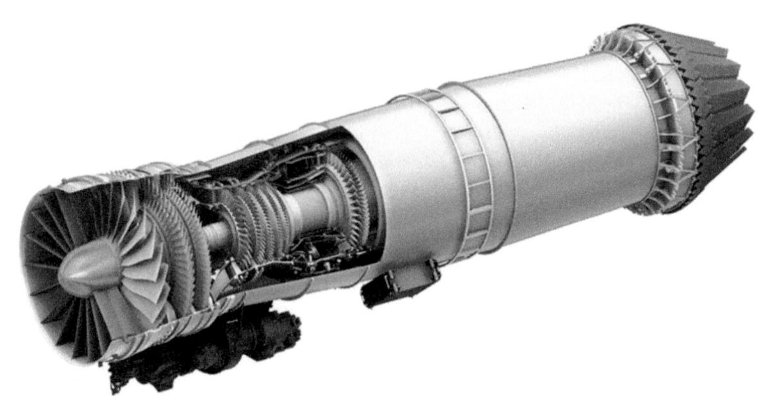

FIGURE 7.5 Cutaway of a turbine engine. *Source*: F136 concept, http://www.defenseindustrydaily.com/the-f136-engine-more-lives-than-disco-03070/. Used under CC0 1.0, Public Domain.

Titanium dioxide (TiO_2) is a crucial component in catalysis due to its photocatalytic and antibacterial properties. However, it has also wide application in electronics; for example $BaTiO_3$ has ferroelectric, photorefractive, and piezoelectric properties. TiO_2 is also used extensively as a white pigment to make paints and even to whiten milk. It is used also in sunscreen lotions to provide protection against ultraviolet rays.

Zirconia (ZrO_2) is used in many devices where friction resistance is of significant importance, e.g., milling balls or mortars. It is also used to make oxygen gas sensors for automotive applications and to measure dissolved oxygen in molten steels. Zirconia monocrystals (cubic lattice) with remarkable refraction and good abrasion resistance are used as gemstones for jewelry since their appearance is similar to that of diamond.

Alumina (Al_2O_3) is used as a low dielectric substrate for the electronic packaging that houses silicon chips. One application of this is insulators of spark plugs. Alumina is used in other applications as well, including as a catalyst in chemical reactions; as a component in decorative paints (especially in the automobile industry); and as an abrasive.

7.4 NETWORK CERAMICS: SILICATES AND SIALONS

A large and important class among ceramics are **silicates**. The basic unit of silicates is the silicon-oxygen tetrahedron, which has O atoms in the four corners and a Si atom in the center. Infinite three-dimensional networks may be formed from such tetrahedra. The most common arrangements are shown in Figure 7.6. Owing to their skeletal framework, such network ceramics are sometimes referred to as **skeletal ceramics**.

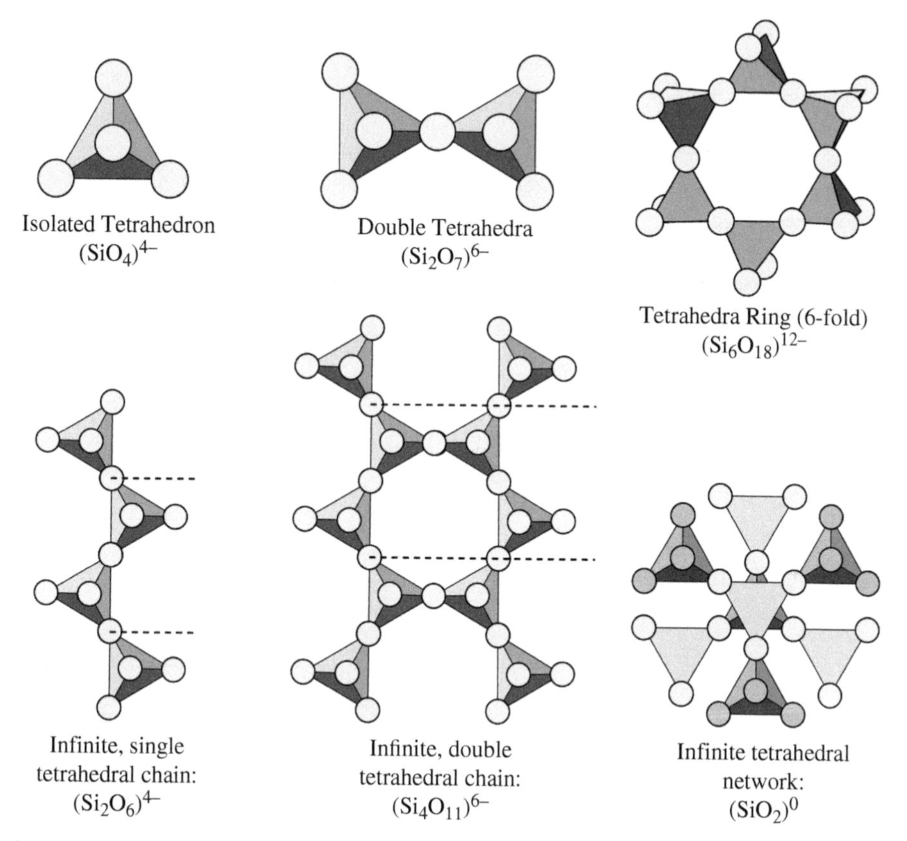

FIGURE 7.6 Silicate structures. *Source*: Modified after http://geologycafe.com/images/Silicates.jpg. Public Domain.

Simple structural elements shown in Figure 7.6 do not specify a particular crystalline form since the sequences of the relative spatial orientations of the tetrahedra may differ. For example, there are three polymorphs of silica SiO_2, namely, tridymite, quartz, and crystobalite, all consisting of rings of six silicate ions but with different extended structures and different properties. Like SiO_2, B_2O_3 and Al_2O_3 are network formers, consisting therefore of extended structures. Silicon-oxygen tetrahedra can also be connected in quasi-linear chains as for the mineral **asbestos** $3MgO{\cdot}2SiO_2{\cdot}2H_2O$. Additional schematics of the arrangements of SiO_4 tetrahedra in silicates are depicted in a handbook by Martienssen and Warlimont [6].

Silicates typically derive from soil, rock, clay, and sand; and most are complex, formed from substitutions of various metal ions into the network. **Feldspars** are alkali alumina silicates; they are rock-forming materials and often found in nature, for example as: potash feldspar

$K_2O{\cdot}Al_2O_3{\cdot}6SiO_2$, soda feldspar $Na_2O{\cdot}Al_2O_3{\cdot}6SiO_2$, and lime feldspar $CaO{\cdot}Al_2O_3{\cdot}2SiO_2$. Several silicates are known as gems, such as **garnets** and **olivines**, while **zeolites** are used as water softeners and as molecular sieves. In addition to these, there are **layered silicates**, which are based on the unit $(Si_2O_5)^{2-}$. Three oxygens in each tetrahedron are shared, for example in the clay **kaolinite** $Al_2Si_2O_5(OH)_4$. **Mica** $KAl_2(AlSi_3O_{10})(OH)_2$ and **talc** $H_2Mg_3(SiO_3)_4$ also have sheet structures (i.e., layered). Some of these exist as crystalline materials but are functional as glass-bonded ceramics.

It should be emphasized that SiO_2 is a glass-forming compound. This is why fine china pottery which contains high concentrations of silica might even seem semi-transparent. Window glass made of amorphous silica becomes frosted glass after long-term partial crystallization. Silicate glasses and glass-bonded silicates will be discussed in Sections 7.6 and 7.7. The crystalline silicates are ubiquitous, found in a wide range of applications from cosmetics to food additives, pigments to light bulbs, jewelry to piezoelectrics.

We have seen methods of characterizing the structures of crystals in Chapter 4 and of non-crystals in Chapter 5. Silicate glasses have complicated structures—though they are based on elements seen in Figure 7.6. Raines and his colleagues have developed a powerful approach based on diffractometry in 2D called **ankylography** [7]. Using diffraction data from a soft X-ray laser, ankylography can provide a 3D image of a test object from a single 2D diffraction pattern. One uses a so-called **Ewald sphere**, a geometric construct in diffractometry that involves the relationship between the wave vector of the incident and diffracted beams and the diffraction angle for a given reflection. 3D image creation proceeds in ankylography by means of an iterative algorithm; apparently *a priori* information is not needed. In Figure 7.7 we show structure determination for a sodium silicate glass sample.

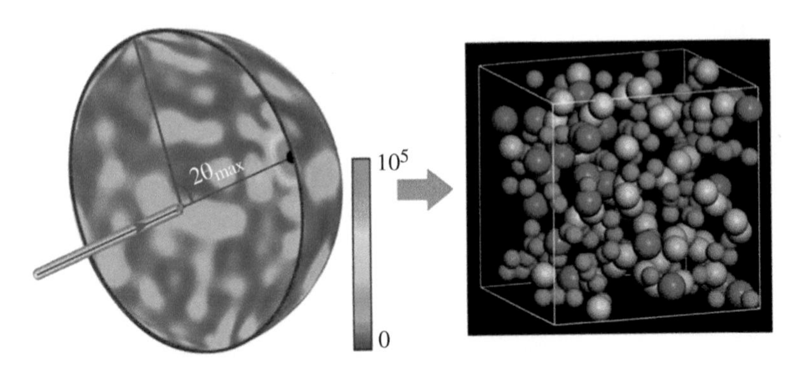

FIGURE 7.7 Determination of a structure of a sodium silicate glass by ankylography based on a simulated 2D spherical diffraction pattern alone. Left: the Ewald sphere; Right: 3D structure with red, purple, and yellow spheres representing, respectively, the positions of O, Na, and Si atoms. The accuracy is 0.2 nm. *Source*: Reprinted by permission from Macmillan Publishers Ltd: [7], copyright (2010).

In some ways similar to silicates, **sialons** are composed of silicon, aluminum, oxygen, and nitrogen, hence the name from the symbols Si—Al—O—N. They are essentially alloys of *silicon nitride*, Si_3N_4. Silicon nitride itself is a good engineering material having high strength, low wear, thermal shock resistance, oxidation and corrosion resistance, low friction, and a relatively high decomposition temperature (sublimation at $\approx$2180 K). A significant drawback however is the difficulty of fabricating silicon nitride: because of the covalent bonding, diffusion in a pure phase is slow, and therefore the material cannot be sintered to the maximum theoretical density by firing. Instead, shaping must be accomplished by less convenient methods such as hot pressing. Silicon nitride commonly exists in two different forms denoted α and β. Sialons are isostructural with either the α and β forms of Si_3N_4 or with *silicon oxynitride* (Si_2N_2O). The latter is built up of SiN_3O tetrahedra while parallel sheets of silicon-nitrogen atoms are joined by Si—O—Si bonds.

7.5 CARBON

Carbon as an element does not easily fall within a traditional classification as metal, ceramic, or polymer. It exists in several different crystalline polymorphs and in amorphous forms as well. We consider it here since graphite is sometimes considered as a ceramic and because the crystal structure of diamond (given in Figure 4.13) is similar to that of the ceramic known as zinc blende, which we have seen in Figure 7.2. To recap from Section 4.3, polymorphism occurs when two or more crystals have the same chemical composition but consist of different structural arrangements.

Diamond is a metastable form of carbon with each atom linked tetrahedrally to four other atoms. In the Bravais classification, diamond has a face-centered cubic lattice, with two atoms per lattice point; its structure—marked by cubic symmetry—is shown in Figure 4.13. Diamond has long been recognized as the hardest of all known materials. Though this status may be challenged, there still seems to be no material entirely competitive with diamond. Owing to its structure and strong covalent bonds, diamond also has very high strength, relatively high temperature stability, insolubility, and unusually high thermal conductivity for a non-metallic element. As we know from experience, diamond is optically transparent in the visible and infrared regions of the electromagnetic spectrum. Large diamonds dug from the earth are sold as gemstones; see again the motto of this Chapter. However, industrial diamonds are synthesized in hot presses—at extremely high temperatures and pressures—for use in cutting tools.

Philip Ball provides a detailed and fascinating account of diamonds including discussion of their global market and the difficult quest for a method to produce synthetic diamonds [8]. The properties of diamond would be advantageous in a wide range of applications, however the expense of diamonds, even synthetic ones, makes extensive use cost prohibitive. The preparation of diamond thin films by chemical vapor deposition (CVD) seeks to exploit diamond's properties at somewhat lower cost. Diamond films prepared in this manner are polycrystalline without the long-range crystalline regularity of natural diamond. Nevertheless, the mechanical, electrical, and optical properties of the thin films are approaching those of bulk diamond, rendering the films potentially advantageous in certain applications. Owing to diamond's conductivity, the films can be used for thermal management in electronic and opto-electronic devices. Since diamond films are optically transparent, they can be applied as mechanical reinforcement to lenses or windows without loss of optical clarity. Paul May has elaborated on existing and potential applications of diamond thin films [9]—while we might ponder whether diamonds would lose some of their grand appeal if they become a regular component in a host of familiar applications and consumer products.

By contrast with diamond, the carbon polymorph known as **graphite** is much softer. It is the stable form of carbon at temperatures up to 3000 K (although sublimation may occur at lower temperatures under some conditions) and pressures up to 10^4 atm. Also crystalline, graphite has a sheet-like structure consisting of atoms covalently bonded to three co-planar atoms in a hexagonal arrangement. Adjacent layers are held together by weaker bonds of van der Waals and metallic type. The graphite structure is illustrated in Figure 4.10b. Electric conductivity is high—due to the partly metallic character—in the direction parallel to the hexagonal layers. Due to the anisotropy in structure and the difference in bond strengths within and between layers, graphite sheets can easily slide past each other, a feature that gives rise to the familiar lubricating properties of graphite. Graphite is used in pencil tips, high temperature crucibles, dry cells, electrodes, as a lubricant, and more.

Graphene, a 2-dimensional form of carbon, has garnered much attention since 2004, when Andre Geim, Konstantin Novoselov, and coworkers isolated individual planes of graphene using adhesive tape [10]. Graphene consists of a monolayer or one-atom-thick planar sheet of sp^2-bonded carbon atoms arranged in a honeycomb lattice, as shown schematically in Figure 7.8. Thus, graphene is the basic structural element of carbon allotropes such as graphite, charcoal, nanotubes, and fullerenes. Unlike these other forms of carbon, however, graphene possesses unique properties owing to its one-atom thickness. The electronic properties of graphene are particularly noteworthy and are the subject of much research. There are ideas for potential applications of graphene in various areas. For example, it has been demonstrated that addition of graphene to an epoxy results in higher wear resistance and lower friction [11].

FIGURE 7.8 Structure of graphene. *Source*: AlexanderAIUS, https://commons.wikimedia.org/wiki/File:Graphen.jpg. Used under CC-BY-SA-3.0.

The form of **buckminsterfullerene**, a hollow sphere of sixty carbon atoms (denoted C_{60}), is shown in Figure 7.9. Named after Buckminster Fuller, the inventor of geodesic domes, this molecule is also known as a **buckyball** because of its soccer-ball shape. In general, the term *fullerene* refers to carbon molecules in the form of a hollow sphere, ellipsoid, or tube. Buckyballs are considered crystalline in the solid state since they pack together in a face-centered cubic array. An insulator in the pure state, fullerenes can be rendered conductive or semi-conductive by incorporating certain dopants (refer to Chapter 17 for more on doping and semiconductors).

Carbon nanotubes (CNTs) are cylindrical fullerenes, or basically rolled up graphene layers. Carbon nanotubes may be single-walled or multi-walled and have open or closed ends. They are known for exceptional tensile strength in the longitudinal direction. The electrical, thermal, and chemical properties of carbon nanotubes are also of interest to scientists developing new composites such as polymer matrix+CNTs [12, 13] or else alumina+CNTs [14]. See Figure 7.10 for structures of CNTs.

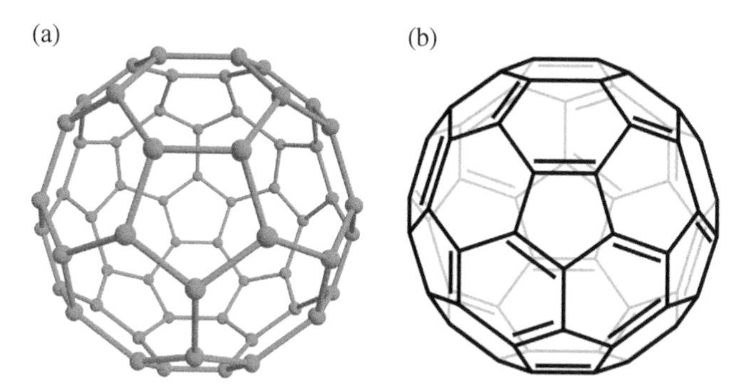

FIGURE 7.9 The structure of buckminsterfullerene, C_{60}, as a ball and stick model at left (a) and showing bond structure at right (b). *Source*: (a) Mstroeck and Bryn C, https://commons.wikimedia.org/wiki/File:C60a.png. Used under CC-BY-SA-3.0. (b) Benjah-bmm27, https://en.wikipedia.org/wiki/File:Buckminsterfullerene-2D-skeletal.png. Public Domain.

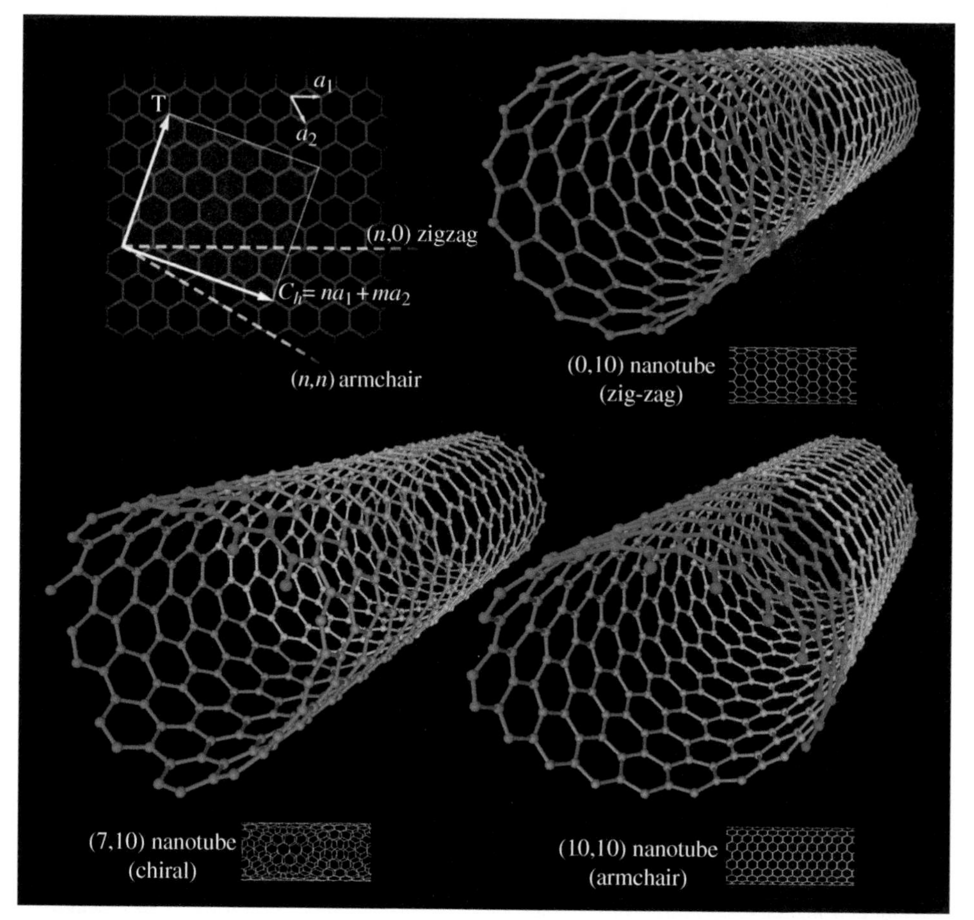

FIGURE 7.10 Types of carbon nanotubes. *Source*: Mstroeck, https://commons.wikimedia.org/wiki/File:Types_of_Carbon_Nanotubes.png. Used under CC-BY-SA-3.0.

Amorphous forms of carbon appear in **coal** and **carbon black**. Carbon black is used in printing ink, in automobile tires, and as an additive polymer blends and composites. We also recall from our Chapter on metals that carbon is an important component of **carbon steel** alloys.

A very rare allotrope of carbon known as **lonsdaleite** (named after Dame Kathleen Lonsdale, a British crystallographer and Professor at University College London), also called hexagonal diamond, has been identified as being even harder than diamond. Pictured in Figure 7.11, it has a hexagonal structure in a diamond-like network. It is formed in nature from graphite in meteorites; the high heat and stress from the meteorite's impact with the earth converts the graphite to the structure of lonsdaleite. Under identical loading conditions, lonsdaleite has been demonstrated to have higher impact strength than diamond. Given its rarity however, it is not likely to replace diamond. The hardness of the wurtzite structure of boron nitride (w-BN) may approach that of lonsdaleite and of diamond. However, procedures for its synthesis are not yet established enough to posit w-BN as a replacement for diamond in applications demanding high hardness.

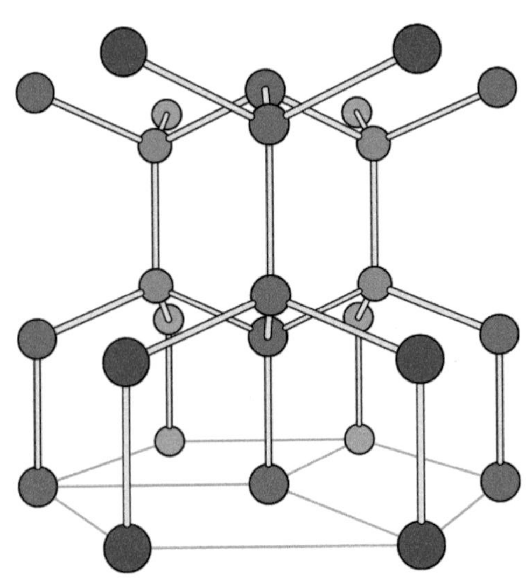

FIGURE 7.11 The structure of lonsdaleite. *Source*: Modified after Materialscientist, https://commons.wikimedia.org/wiki/File:Lonsdaleite.png. Used under CC-BY-SA-3.0.

Cubic boron nitride (c-BN) is second in hardness only to diamond and has a similar covalent structure. Cubic boron nitride exhibits high abrasion resistance and thermal conductivity compared to conventional abrasives such as silicon carbide and aluminum oxide. The thermal stability of c-BN and its ability to maintain sharp cutting edges during the machining of ferrous materials make it the product of choice in advanced grinding systems.

Hexagonal boron nitride (h-BN) is also known as "white graphite" and has a (hexagonal) crystal structure similar to that of graphite. This crystal structure imparts excellent lubricating properties. Therefore, h-BN is an important compound in lubrication engineering. Unlike graphite, h-BN can be used during operation in air at high temperatures, up to 1000°C. Further high temperature stability is exhibited in vacuum up to 1400°C and in inert gas up to 1800°C. Hexagonal boron nitride is superior to graphite in terms of chemical inertness, manifested by its non-wettability: h-BN is not wetted by glasses, salts, and (most) metals, therefore it provides strong resistance to chemical attack. Another advantage of h-BN is that complex shapes can be machined from a hot pressed structure. Extensive studies have been done in order to find the most suitable sintering additives for h-BN [15]. It was shown by changes in wettability that spinels (e.g., YAG [yttrium aluminum garnet] or $MgAl_2O_4$) can improve densification, however further research is needed to identify agents that can lead to entire densification of h-BN.

7.6 GLASSY CERAMICS

The importance of **glasses** is evident by their widespread use—in both critical and mundane applications. To understand glasses, we first need to consider the thermodynamics and stability of such materials. It was explained in Chapter 3 how the stability of materials including liquids and solids is determined by the Gibbs function G. Glassy materials are essentially liquid phases that have failed to crystallize upon cooling, consequently glasses are *metastable*. We looked also at the effect of concentration on the glassy state, for instance, the change in the glass transition temperature T_g in binary mixtures (Figure 5.13). We also recall a miscibility gap in a binary liquid mixture in Figure 3.11—with easy to imagine consequences of cooling a two-phase liquid into the glassy state.

As mentioned in Chapter 5, it is possible for glasses to undergo slow crystallization due to the metastability of the glassy state. That process is called **devitrification**, the partly crystallized glass is called devitrified, and the crystalline regions in the glass are called *stones*. The presence of stones is usually attributed to structural strains, such as those that might be introduced during forming, and they persist if annealing at high temperature is not sufficiently long. Devitrification also occurs in familiar glass products such as bottles; the stones are translucent, very weak, and brittle. One rich French wine connoisseur bought a bottle of 120 years old burgundy. He kept it for two years in his wine cellar, a prized possession, annoying his friends. Given their continuing comments, he invited them one day to drink that burgundy. He went to his cellar—finding there his burgundy on the floor together with stones from the bottle.... On the other hand, devitrification is introduced deliberately in some special cases to create a very fine-grained ceramic that is free of porosity.

The *nature* of the glassy state was discussed in Chapter 5. Ceramic glasses we consider in this Chapter are random three-dimensional networks. Many ceramic glasses consist primarily of oxides. Consequently, the oxide components of a glass can be distinguished based on the *functions* they perform in the network. The classifications are:

1. *Glass formers*: the main components of glassy materials forming the basis of the random three-dimensional network. Oxides include SiO_2, B_2O, GeO_2, P_2O_5, V_2O_5, and As_2O_3.
2. *Intermediates*: components with properties near to but not quite the same as those of glass formers. They are unable to form the glass network alone, but they can link with the network once it exists—hence we can think of them as "me too" components after the glass formers. In this category are Al_2O_3, Sb_2O_3, ZrO_2, TiO_2, BeO, and ZnO.
3. *Modifiers*: components with no glass-forming tendencies, whose main role is to decrease the melting point of multicomponent mixtures containing glass formers. These metal oxides fill voids in the network and modify density, viscosity, color, electrical, and other properties of glasses. Included here are MgO, Li_2O, BaO, CaO, SrO, Na_2O, and K_2O.

Glass forming methods are varied and include some of those used to fabricate crystalline ceramics and glassy metals. Typical glass forming processes are pressing, blowing, drawing, and fiber forming. Each involves melting of the mixture and rapid, non-equilibrium solidification. A now classic technique of Sarjeant and Roy [16] relies on the splat quenching method (described in Section 6.8) to obtain pure oxide glasses of V_2O_5, TeO_2, MoO_3, and WO_3.

Glasses (and other ceramics) can also be prepared from a gel by the **sol-gel** technique. A sol is a colloidal suspension of particles having diameters less than 0.1 μm. Via chemical reaction of organometallic compounds, an amorphous semi-solid network structure forms. Liquid is lost in this process, and a non-crystalline solid gel forms. The gel is then heated to densify the material and (if desired) drive out any residual organic components. In either case, the temperatures required for densification of a gel are much lower than what is required by other processing methods. Gel glasses appear to have more uniform microstructures than glasses produced by the fusion of mixtures of oxides.

Applications of non-crystalline ceramics are well known from everyday life: windows, bottles, mirrors. Common glass is typically the type called **lime glass**: 70–75% SiO_2, 12–18% Na_2O, 0–1% K, 5–14% CaO, 0.5–2.5% Al_2O_3, 0–4% MgO. Composition matters [17]; in soda lime silicate glasses, replacement of sodium with calcium results in significant structural changes. Cormack and Du [17] have demonstrated such changes by performing **molecular dynamics** computer simulations explained briefly in Section 6.8.

The first known **mirrors** were made in the 3rd century BC and were typically round and with a metal back. The first glass mirrors were invented in the 1st century AD by Romans. Later the use of mirrors had its ups and downs, being considered by some as tools of the devil; even in 17th century Russia the use of mirrors was considered a sin by the Russian Orthodox Church. Before that in the 15th century, the Venetian island of Murano became the center of glass and mirror making. There was on the island "The Council of Ten" with the task of protecting their glass and mirror making technique secrets at any price; the glassmakers were well paid but kept in isolation. The cost of one Venetian mirror then was comparable to the cost of a large naval ship. See an example of a Venetian mirror in Figure 7.12.

Given the costs of a Venetian mirror, Jean-Baptiste Colbert, France's powerful Controller General of Finance under King Louis XIV in the 17th century, lured some mirror makers from Murano to France and created for them the *Manufacture royale de glaces de miroirs*. Thus aristocratic ladies in France were now able to have mirrors made in their country. The Council of Ten reacted; mirror makers from Venice were poisoned. It seems none of the French ladies looking into their mirrors were seeing the faces of the Murano mirror makers who died making those mirrors. But now the technology was out, not to be contained again. Construction of the Hall of Mirrors (*Galerie des Glaces*) in the royal Palace of Versailles began in 1678; see a photo from that gallery in Figure 7.13. A French company called Saint-Gobain, based on the *Manufacture royale*, is still in business making new mirrors and maintaining old ones. Long after creation of the Hall of Mirrors, the French Prime Minister Georges Clemenceau chose the Hall to sign the Treaty of Versailles that ended World War I on June 28, 1919. US President Woodrow Wilson made important contributions to wording of the treaty—as a consequence there is Woodrow Wilson Square in Warsaw because of Wilson's insistence on a statement in the treaty granting Poland direct access to the Baltic Sea.

FIGURE 7.12 A Venetian mirror. *Source*: After http://www.invitinghome.com/Mirrors/venetian-mirror-594.htm. Reprinted with permission from http://www.invitinghome.com/.

FIGURE 7.13 A section of the Hall of Mirrors. *Source*: Edwinb, https://commons.wikimedia.org/wiki/File:Hall_of_Mirrors.JPG. Public Domain.

7.7 GLASS-BONDED CERAMICS

Glass-bonded ceramics consist of crystalline phases held within a glassy matrix. **Fired clay** (sometimes called fine china) materials are among the most important in this category of ceramics. Clays were mentioned in our earlier discussion of silicates (Section 7.4), and **natural clays** are sheet silicate structures, believed to be the weathered remains of various types of rocks, especially of the feldspar minerals in those rocks. Granite and basalt, among others, wear down to form clay deposits. Clays are noted for their strong intralayer bonding but weak interlayer bonding. Polar molecules such as water may be attached easily to the surfaces of the layers, resulting in plasticity, swelling, and easy slipping. These properties are lost when the clay is dried; however the moistening process is completely reversible. In contrast to the network silicates discussed in Section 7.4, the clays are **phyllosilicates**, or sheet silicates; a classification scheme is given in Figure 7.14. The four main groups of clay minerals are **kaolinites**, **smectites**, **vermiculites**, and **micas**.

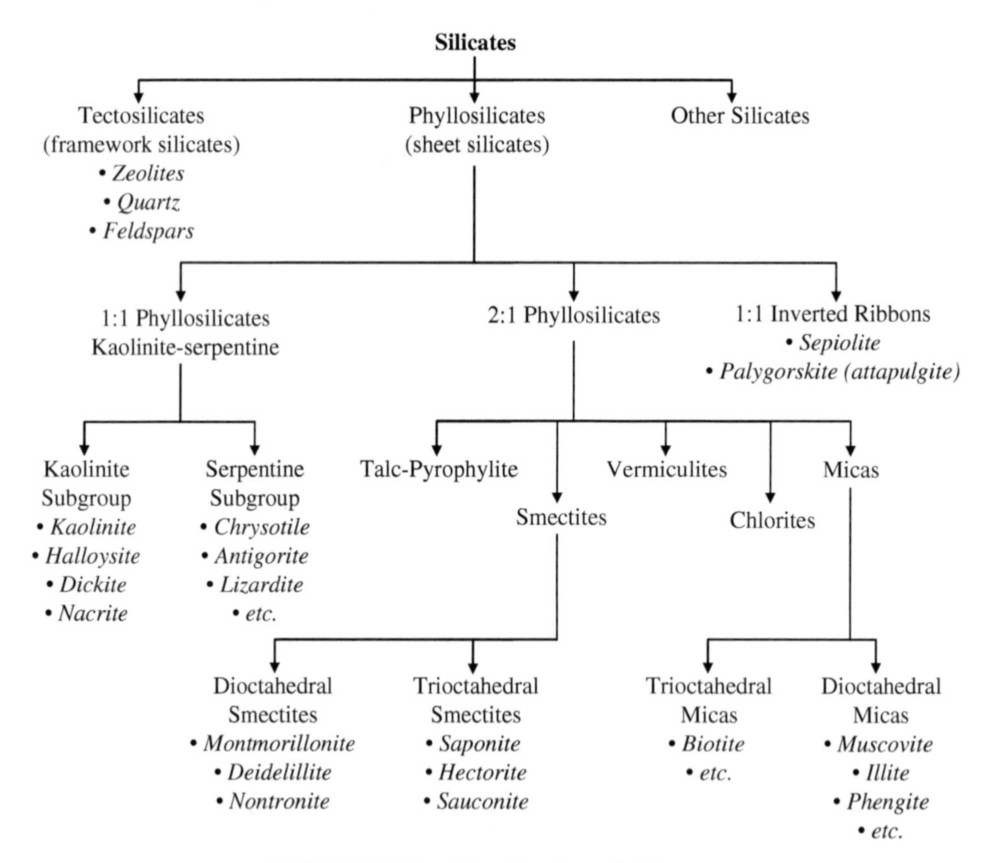

FIGURE 7.14 Classification of silicates.

Kaolinite has a double-layered structure in which a $(-Si_2O_5^{2-}-)$ layer is ionically bound to an $(-Al_2(OH)_4^{2+}-)_x$ layer. This structure is shown schematically in Figure 7.15. Each double-layered sheet has a dipole moment; therefore in addition to the always present dispersion forces, there are electrostatic attractions in kaolinite. The spacing between sheets in kaolinite is 0.72 nm. In montmorillonite the intersheet spacing is larger by a factor

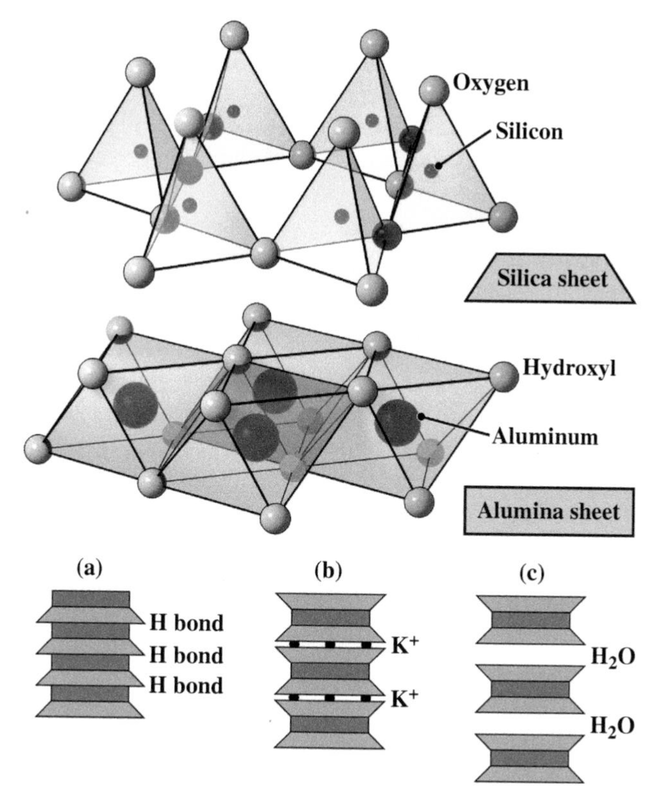

FIGURE 7.15 The top part of figure shows the basic structural units of clay minerals and the silica and alumina layers formed by them. There is the sheet structure of silicon tetrahedron arranged in a hexagonal network and the sheet structure of aluminum–hydroxyl octahedron. The lower part shows schematically the structures of (a) kaolinite, (b) illite, and (c) montmorillonite, based on the sheets defined in the upper part of the figure.

of two, namely, 1.5 nm. This facilitates greater water absorption, which in turn leads to production of a so-called interstitial solution.

Each sheet in **montmorillonite** consists of three layers; the middle layer of $(AlO(OH)—)_x$ is sandwiched between two layers of $(—Si_2O_5^{2-}—)$, so there is no dipole moment. **Muscovite** has sheets similar to those in montmorillonite, with the inter-sheet spacing of ≈1 nm. Potassium ions are between the sheets, and they weakly hold the sheets together, as shown in Figure 7.15 for illite, which is structurally similar to muscovite and differs mainly in its concentration of the component elements. Muscovite is the most common of the mica group of minerals. **Mica** is familiar for its easy cleavage into "atomically smooth" surfaces. During cleavage the ionic potassium bonds break and the ions adhere about equally to each face. Molecular formulas for most clays are difficult to define. For instance, the molecular formula for montmorillonite is usually given as $(Na, Ca)_{0.33}(Al, Mg)_2[Si_4O_{10}](OH)_2{\cdot}nH_2O$. For bentonite, no molecular formula is even specified.

Fired clay ceramics are made from a combination of clay, flint, and feldspar. We have noted some of the mineral components contained in clays. Clays also contain—in smaller quantities than Al_2O_3, SiO_2, and H_2O—silicates involving CaO, Fe_2O_3, FeO, Na_2O, and

MgO. **Flint** is mainly SiO_2 in the crystalline form of quartz; as such it is an inexpensive refractory component. The **feldspar** used in fired clay ceramics is typically a mixture of the three types of feldspar mentioned in Section 7.4. Feldspar is a low-melting component which forms a glass during firing and binds together the refractory crystalline components. It is for this reason fired clay materials are called **glass-bonded** ceramics. Their structures are necessarily more complex than those of glasses.

Fired clay ceramics are often classified on the basis of their properties after firing, in particular on porosity. Usually **porosity** is defined as a percentage of unit volume "filled" by voids or pores. **Apparent porosity** is measured by the amount of water that can be absorbed by the unit volume; thus voids that are sealed off from the outside surfaces are not included. **Total porosity** includes both apparent porosity and internal pores. **Bricks** are considered as fired clay ceramics. The technology of making them is at least 5,000 years old since the first bricks (noted in historical records) were made around 3,000 BC in Mesopotamia from clay. An oven type chamber called a **kiln** is used to generate very high temperatures, typically in the range between 870°C and 1100°C. The precise temperature required is defined by the type of clay used. Kilns can be either periodic or continuous.

The main classes of glass-bonded (a.k.a. fired clay) ceramics are:

- **Earthenware**, such as bricks, wall tiles, ceramic filters, and porous drainage pipes. Fired at 1100–1300 K. Apparent porosity 6–16%.
- **Fine china**, mostly tableware. Fired at 1400–1500 K. Apparent porosity often below 1%.
- **Stoneware**, used for tableware, tiles including roofing tiles, and drainage pipes. Fired above 1500 K. Apparent porosity < 3%, often ≈ 1%. Total porosity much higher. When low apparent porosity needed, *glazing* already defined above is applied.
- **Porcelain**, used for fine tableware and parts of scientific equipment. Fired above 1600 K, negligible apparent porosity, glazing also used. *It is difficult to distinguish porcelain from fine china.*

The origins of porcelain date back to 200 BC. Centuries later, translucent porcelain was manufactured in China and sold around the world—at exorbitant prices, comparable to those of gold and silver. In fact, porcelain was also called white gold. Outside of China, the materials from which porcelain was made and the method of its manufacture were a mystery.

Only early in the 18th century did the situation change. August the Strong, King of Poland and of Saxony, employed an alchemist named Johann Friedrich Böttcher (or Böttger). By definition, **alchemists** were working on procedures of conversion of ordinary metals into gold (we know they never succeeded and we know why). Böttcher took his duties quite seriously working on *Goldmachentinktur*, a gold making **tincture** (a solution of a low volatility substance). He almost missed a byproduct which was porcelain, but his superior Ehrenfried Walther von Tschirnhaus forced him to work on it. There is also an opinion that von Tschirnhaus invented the process while Böttcher only perfected the details. On March 28, 1709, Böttcher reported to the King that the technology of making porcelain was available. August the Strong made Böttcher the head of the first porcelain manufacture in Europe, namely, in Meissen in Saxony. Products of what is now Staatliche Porzellan-Manufaktur Meissen GmbH are well known and appreciated around the world, and marked by the characteristic logo of two blue crossed swords (Figure 7.16).

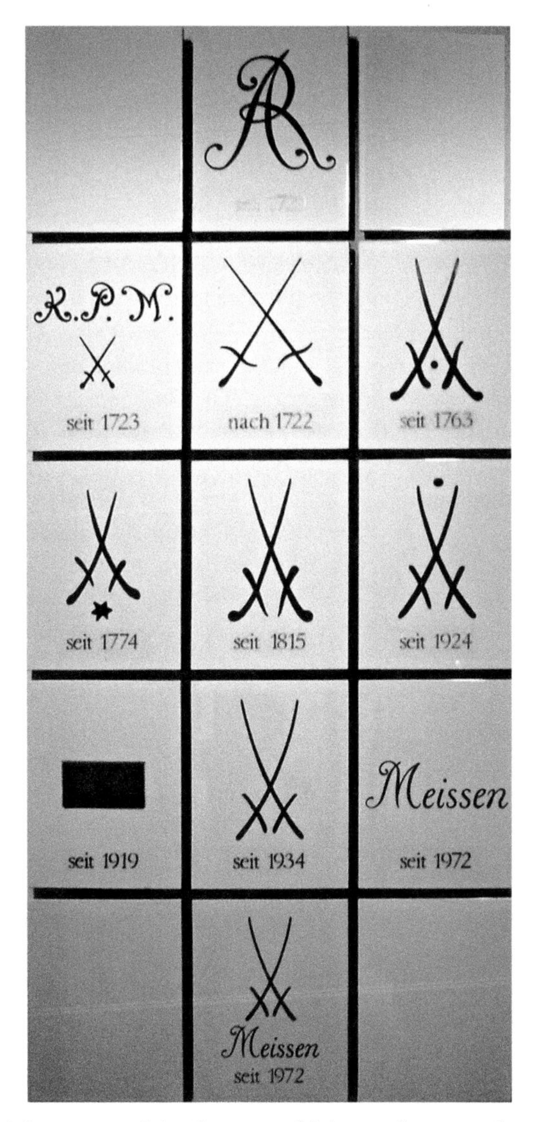

FIGURE 7.16 Meissen porcelain tiles, on which are shown trademark symbols (shown here in black) of Meissener Porzellan Manufaktur. *Source*: Ingersoll, https://commons.wikimedia.org/wiki/File:Meissen-Porcelain-Sign-2.JPG. Public Domain.

7.8 CEMENTS

The last major group of ceramics to be discussed are the cements. **Cements**, which are familiar from practical use, are typically composed of compounds of lime CaO, alumina Al_2O_3, and silica SiO_2. After being mixed with water to form a paste, cements set and harden, consuming a significant amount of water in the process. **Nonhydraulic** ones set in air, and **hydraulic** cements set and harden under water. The most common cements are:

- **Lime mortar**, nonhydraulic, known for over 5,000 years. $CaCO_3$ is *calcined* (medieval terminology that means heated) to $\approx$1300 K to produce *quicklime* = CaO (finely divided hence reactive). Then CaO is *slaked* (mixed with water) to form powdered *slaked lime* = $Ca(OH)_2$. The latter is mixed with sand and more water to produce lime mortar. The mortar sets when the excess water dries out and hardens due to the action of atmospheric CO_2 by the carbonization reaction:

$$Ca(OH)_2 + CO_2 \rightarrow CaCO_3 + H_2O \tag{7.1}$$

- **Plaster of Paris**, formed by heating gypsum $CaSO_4 \cdot 2H_2O$ to $\approx$450 K. When mixed with water and left for some time, the plaster sets to form paste.
- **Keenan's cement**, obtained from gypsum calcined at $\approx$820 K. At this temperature the hydration water is removed from the mineral. After mixing with water (now a physical mixture, not hydrated water) and hardening, Keenan's cement is practically insoluble in water and much stronger than the plaster of Paris. This cement is used in the production of imitation marbles and also in wall construction.
- **Portland cement**, hydraulic, the most important of all and the principal component of concrete (a composite we shall discuss in Chapter 11). Looks in appearance like limestone from Portland Hill, England, and is a mixture of: tricalcium silicate $3CaO \cdot SiO_2$; dicalcium silicate $2CaO \cdot SiO_2$; tricalcium aluminate $3CaO \cdot Al_2O$; and tetracalcium aluminoferrite $4CaO \cdot Al_2O_3 \cdot Fe_2O_3$. Proportions of these minerals vary, which obviously affect the final mechanical properties.

Portland cement is used in construction of houses, roads, bridges, and more. Given its importance, we discuss it now in further detail. The raw materials are obtained from nature, usually as quarried rock. A mixture of materials containing lime, usually $CaCO_3$ in the form of limestone or chalk, and silica and alumina (usually from clay) are ground and fired, during which carbon dioxide is released. Iron ore may be added as necessary. The mixture is calcined at ~1700 K whereupon it is important that no free lime remains as that would hydrate and disintegrate the hardened cement. The resultant product, called **clinker**, is ground to a fine powder to which a small amount of gypsum is added to prevent flash setting, which can happen in rapid hydration of tricalcium aluminate. As with other cements, the powder is mixed with water to form a paste and is thereby ready for use. Other cement options are now being explored, such as the use of magnesium oxide and magnesium sulfate rather than limestone [18]. An alternative route pioneered by Davidovits [19] consists in making geopolymers, that is man-made rock geosynthesis leading to high strength cement.

The *setting* of the cement paste takes between 30 minutes and 10 hours, with subsequent strengthening or *hardening* typically measured up to 28 days. Because the reactions involve thermal effects, they can be measured by microcalorimetry—which can be applied to long term processes. Foundations of this technique were developed most notably by Swietoslawski [20]. A microcalorimetric study of cement hardening by Zielenkiewicz and coworkers [21, 22] has found measurable thermal effects 2.5 years after the start of the process.

There is a natural question that arises: why is cement such a strong material? Let us first look at the so-called silicate garden [23] in Figure 7.17. In Figure 7.17a, we observe the rapid growth of a crystal in solution. The mode of growth by dissolution and then precipitation is shown schematically in Figure 7.17b.

As seen in the schematic diagram in the bottom part of Figure 7.17, a shell of insoluble silicate is formed around the crystal—permeable however to water from outside. With high osmotic pressure created by water inside, at some point that shell ruptures, which results in formation of fine hollow silicate tubes. These tubes interlock—giving mechanical strength.

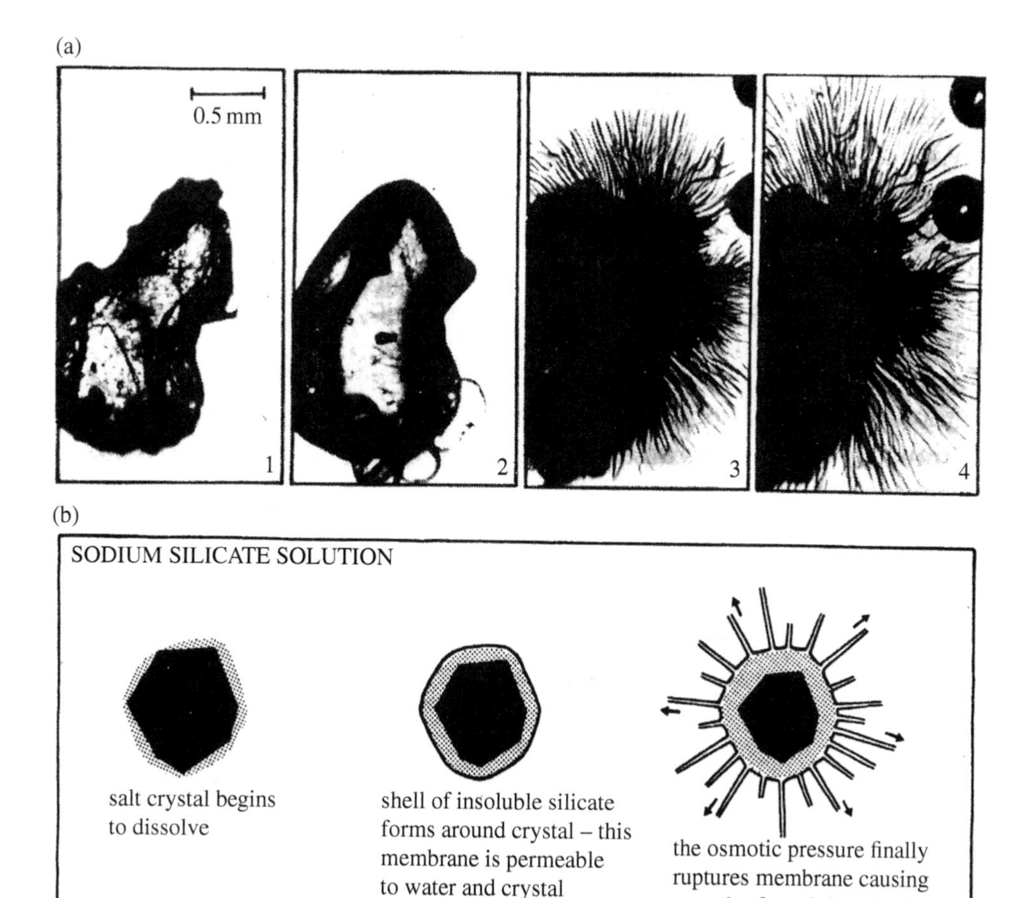

FIGURE 7.17 The growth of a "silicate garden". (a) Growth from a crystal of $Co(NO_3)_2$ immersed in dilute sodium silicate solution; sequence of 1–4 occurs in approximately 30 seconds. (b) Schematic diagram of growth of a "silicate garden". *Source*: (a) Reprinted by Permission from Macmillan Publishers Ltd: from [23], copyright (1976). (b) Image courtesy of D.D. Double.

Given Figure 7.17, we can better understand the electron micrographs of cement growth in Figure 7.18. The **interlocking** is clearly visible in Figure 7.18. While the nature of the process based on interlocking in silicate gardens is quite similar to that of cement hydration and setting, there is a very large difference in the time scale. Silicate garden growth including membrane rupture takes place in 30 seconds. We already know that cement hydration takes years.

7.9 ADVANCED AND ENGINEERING CERAMICS

Modern ceramics are designed to function in increasingly sophisticated systems and products. Many high performance applications demand superior mechanical properties combined with wear resistance, corrosion resistance, and thermal stability. So-called **engineering or advanced ceramics** are fabricated to meet such specifications. Of course

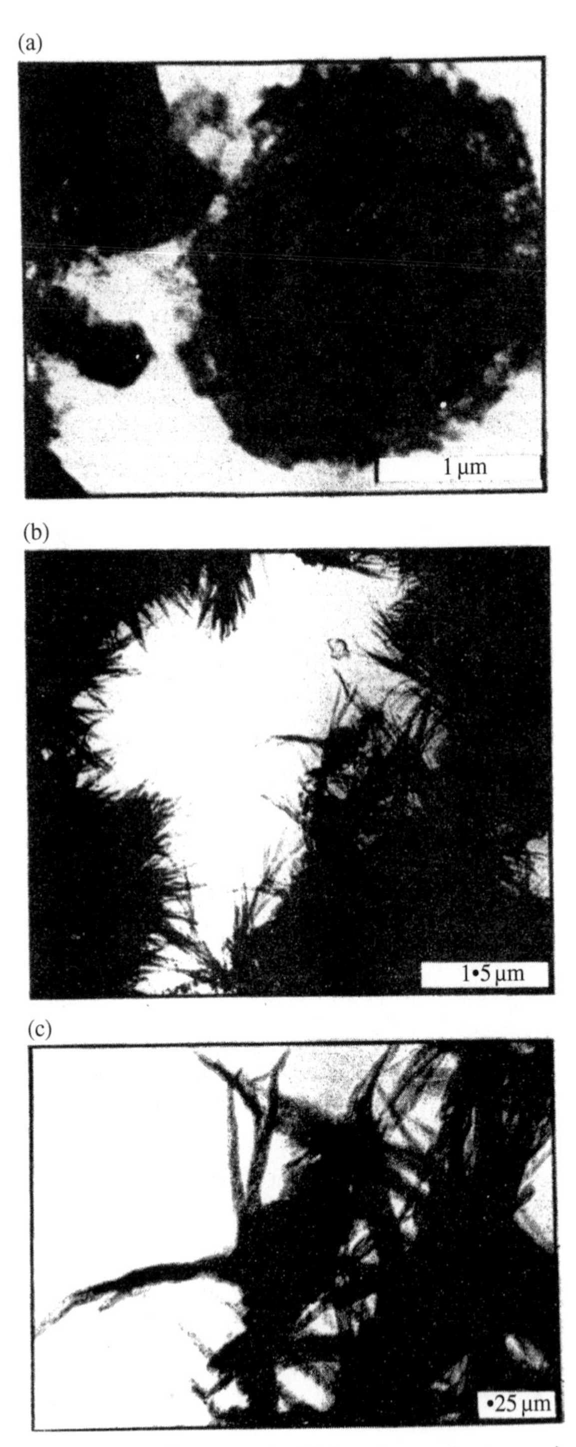

FIGURE 7.18 Cement hydration seen in high voltage electron microscopy (HVEM). Cement is shown (a) after 3 hours, showing the initial gel coating around a cement grain and small angular crystals of Portlandite (at left); (b) after more than 1 day, showing fibrillar development of gel around the cement grains, which is illustrated in more detail in (c). *Source*: Reprinted by Permission from Macmillan Publishers Ltd: from [23], copyright (1976).

refractory ceramics (mentioned earlier) can also withstand high temperatures; but more than this one property is required for ceramics to be used as engine components, cutting tools, electronic capacitors, piezoelectrics, catalysts, or catalyst supports.

Engineering ceramics can be divided into four general categories: (1) structural ceramics, (2) electrical and electronic ceramics, (3) ceramics coatings, (4) chemical processing and environmental ceramics. Examples of applications in each category are given in Table 7.2.

TABLE 7.2 Types of Engineering Ceramics

Type	Applications
Structural	Industrial wear parts, bioceramics, cutting tools, engine components, armor
Electronic	Capacitors, insulators, substrates, integrated circuits packages, piezoelectrics, magnets, superconductors
Coatings	Applied on: engine components, cutting tools, industrial wear parts
Chemical processing	Filters, membranes, catalysts, catalyst supports

A given material compound may be used in differing categories of applications, each taking advantage of characteristic properties of the ceramic. For instance, titanium nitride is a representative ceramic for electronics as well as for coatings for the reasons that it is characterized by good thermal and electrical conductivity as well as strong chemical inertness and excellent abrasion resistance. TiN can be used successfully in chemical sensors as well [24].

Common structural ceramics are the carbides, such as WC, SiC, and TiC. The last of these can also be used as an electrochemical sensor [25]. There may be other advanced ceramics compounds, such as borides, characterized by metal-like electrical conductivity that can be used similarly in sensing applications.

Due to more advanced technologies and tougher requirements imposed on materials, new technological methods have also to be developed for miniaturization or for higher purity. The development of techniques for chemically prepared powders, for example by the sol-gel method described in Section 7.6, really fueled progress on engineering ceramics. The availability of increasingly pure and ultra-fine powders offers more control over the composition, microstructure, and crystal structure as compared to minerals-based ceramics. Chemically prepared ceramic powders are typically the oxides, nitrides, carbides, and borides of various metal species.

While the market for advanced ceramics is expanding, there are still challenges to be hurdled. To lower manufacturing expenses, new forming methods are needed. Under research investigation (with some commercial use) are methods such as gel casting, freeze casting, injection molding, and rapid prototyping. Such methods could reduce machining costs, which for ceramics can be as much as 50% of the manufacturing cost.

Nanoporous silica is an example of increasing interest in nanoporous materials. Rimsza and Du [26] have performed both molecular dynamics and experimental work, varying the porosity between 30 and 70%. Structures were characterized by bond angle distributions and by binary radial distribution functions (see again Section 5.3). Surface area/volume ratios were obtained from both MD simulations and experiments and a reasonable agreement found. Similar comparisons were performed for mechanical properties. This seems to be a nice example of inspiration for creation of new materials based on computer simulations.

Ceramic matrix composites (CMCs) are a relatively new type of material (see Chapter 10 for more on CMCs). Significant improvements in CMCs are expected during the next decades. Major fields for such materials are in heat engines and as thermal shields for plasma reactor technologies.

Some fuel cells use metal agents as catalysts. Solid oxide fuel cells (SOFCs), as their name implies, utilize ceramics, in this case as the electrolyte. Many so-called smart materials are also composed of ceramics; a few of these are discussed further in Chapters 12 and 17. On the horizon is the possibility of viable applications for ceramic high-temperature superconductors (see Chapter 18). Meanwhile a variety of bioceramics—ceramic materials designed for use in the biological environment, and sometimes but not always incorporating biological materials with their structure—are already being used for bone repair and replacement and for dentistry (refer to Chapter 11 for more on biomaterials).

Throughout this Chapter we have seen that brittleness is a concern with ceramics. In general, there are two approaches taken to increasing the fracture toughness and structural stability of ceramics [27]. The first of these relies on the **flaw control**. It endeavors to utilize and modify processing conditions to achieve the end goal of toughening (or reducing brittle failure). It entails investigation of, for instance, powder fabrication—to control chemical homogeneity and particle size. Other lines of research are on powder consolidation and packing—to create green bodies (see Section 7.3) with consistently high density—and on sintering control. A second approach, which is to control microstructures so they will have greater ability to retard crack propagation, might be characterized as **flaw tolerance**. Five major mechanisms of ceramic toughening are listed along with representative examples in Table 7.3. It is evident from Table 7.3 that the mechanisms of transformation toughening and microcracking are important particularly in zirconia ceramics.

TABLE 7.3 Ceramic Toughening Mechanisms

Mechanism	Highest Toughness ($MPa\,m^{1/2}$)	Example Materials
Transformation	≈20	ZrO_2 (MgO)
		HfO_2
Microcracking	≈10	Al_2O_3/ZrO_2
		Si_3N_4/SiC
		SiC/TiB_2
Metal dispersion	≈25	Al_2O_3/Al
		Al_2O_3/Ni
		WC/Co
Whiskers/platelets	≈15	Si_3N_4/SiC
		Si_3N_4/Si_3N_4
		Al_2O_3/SiC
Fibers	≥30	CAS[a]/SiC
		LAS[b]/SiC
		Al_2O_3/SiC
		SiC/SiC
		SiC/C
		Al_2O_3/Al_2O_3

Source: Adapted from [27]. Reprinted with permission from Elsevier.

[a] Calcium aluminum silicate glass ceramic.

[b] Lithium aluminum silicate glass ceramic.

ZIRCONIA AND TOUGHENING MECHANISMS

We shall now give further consideration to zirconia (ZrO_2), also called zirconium oxide, as an example of an advanced ceramic. Zirconia—not to be confused with zircon ($ZrSiO_4$)—is a white crystalline material with properties very similar to those of metals. At room temperature and pressure, the structure of pure zirconia is monoclinic (m) (refer to Section 4.3 regarding lattice structures). As temperature increases, the structure transforms to tetragonal (t) by $\approx$1440 K and then to a cubic (c) fluorite structure around 2640 K followed by melting around 2990 K. Importantly, these transformations are **martensitic**. That is, they occur by coordinated shifts in lattice positions without diffusion or transport of atoms. Additionally, the transformations occur over a temperature range rather than at a specific temperature, and they elicit shape deformation (and therefore bulk volume changes). Stability of the c and t structures can be favored by the addition of lower valence oxides such as CaO, MgO, La_2O_3, and Y_2O_3 into the zirconia material. The c and t phases are more symmetric and less strained than the m phase. The amount of dopant added to stabilize the c and t phases is significant; for example, as much as 8 mol% of dopant Y_2O_3 is added to fully stabilize the cubic phase. Oxygen vacancies are created in the process to retain charge neutrality.

The study of doped zirconia ceramics has led to the defining of three classes of transformation-toughened zirconia. These are outlined, according to Kelly and Denry [27] in Table 7.4. Central to the classifications is the possibility of phase transformations, the metastability of the c and t phases, and the utilization of dispersion or doping to stabilize a particular phase and thereby exploit its properties. Dispersion-toughened ceramics consist of zirconia particles dispersed in another matrix such as alumina or mullite. In this case particle size plays a critical role in limiting the t $\rightarrow$ m transformation. On the other hand, partially stabilized zirconia (PSZ) results from the addition of dopants to a ZrO_2 matrix. The phenomenon of martensitic transformation in zirconia can have a limited effect compared to that in metals, since little plastic deformation can occur due to covalent and ionic bonds. In the absence of plastic deformation, there is still one stress lowering mechanism available, namely, a phase transformation to a structure with larger unit cells.

Moreover, part of the strengthening mechanism consists in having dispersed precipitates of t-ZrO_2 in a matrix of c-ZrO_2. As indicated in Table 7.4, these materials are typically labeled by the dopant with the abbreviation PSZ appended, e.g., Y-PSZ. The microstructure of this class of materials is complex, and Mg-doped zirconia is one of the toughest among transformation-toughened ceramics. The third class, tetragonal zirconia polycrystals, are noted for their fine grain size. Kelly and Denry [27] delve further into the mechanisms behind the toughening associated with the structures of this class.

Transforming ceramics such as zirconia are contrasted with linearly elastic brittle materials, for which the highest strength and highest toughness occur in the same material. The associated mechanisms and mathematical descriptors are elaborated by Kelly and Denry [27] and elsewhere. The outcome, the so-called strength-toughness "disconnect," is a continuum of materials with high-strength lower-toughness materials at one end and high-toughness lower-strength materials at the other. The former are sensitive to processing flaws while the latter are tolerant of flaws and damage. For example, Mg-PSZ, in contrast to Y-TZP (refer to Table 7.4), can withstand twice the indentation load yet with no loss in strength. Zirconia ceramics are considered for biomaterials, aeronautic components, and certain high temperature applications, among others.

TABLE 7.4 Forms of Transformation Toughened Zirconia

1. Zirconia (dispersed phase) toughened ceramics: ZTA (alumina), ZTM (mullite), In-Ceram Zirconia (dental material, VITA Zahnfabrik)
2. Partially stabilized zirconia (PSZ): Ca-PSZ, Mg-PSZ, Y-PSZ, Denzir-M (dental material, Dentronic AB)
3. Tetragonal zirconia polycrystals (TZP): Y-TZP, Ce-TZP, and dental materials—DC Zirkon (DCS Precident, Schreuder & Co.), Cercon (DENTSPLY Prosthetics), Lava (3M ESPE), In-Ceram YZ (VITA Zahnfabrik)

Source: Adapted from [27]. Reprinted with permission from Elsevier.

A relevant question in our consideration of advanced engineering ceramics is "From where do we acquire the fundamental knowledge to design advanced ceramics?" The design process can be based on trial-and-error, on past observations, or on newly developed technology. In other instances, nature provides the inspiration. For example, abalone shells may be considered as a kind of ceramic, a kind that is extraordinarily tough. Scientists hope that through reverse engineering (refer to Figure 1.1, go downwards from Properties) of abalone shells, they will learn the secrets of the complex microarchitecture that makes these shells so crack- and shatter-resistant. In turn, they hope to use any insight gained for the purpose of creating ceramics with similar properties. An abalone shell consists mainly of multisided calcium carbonate "tablets" packed closely in layers [28]. The tablets are adhered to one another and cushioned by a rubbery polymer. Another way to think of it is as a brick-and-mortar type structure [29]. In the event of microcrack formation, shattering is unlikely because the crack is propagated along tortuous paths that essentially diffuse the crack. The polymer also absorbs some of the damage and aids in preventing fracture. To mimic the natural design of nacre (i.e., abalone shell), one has to develop a method to produce similar ceramic "bricks" of a specified shape and (small) size and then to identify a polymer that both adheres to the selected ceramic material and possesses suitable mechanical and thermal properties. Multiple research groups have tried to create nacre-like materials; in consequence a review article exists to discuss the variety of experimental methods [30]. Interestingly, nacre-like structures have been used in bulk materials, coatings, and free-standing films with different results. This is owing to the fact that the dimensions of the material strongly affect the properties. Nature continues to hold the answers to many of our questions and the solutions to many of our problems.

7.10 GENERAL PROPERTIES OF CERAMICS

The defect structure of ceramics plays an important role in their properties, not only mechanical properties but also tribological, electrical, optical, etc. For a detailed discussion of defects in crystal structures, refer back to Chapter 4. Crystalline ceramics can have impurities and vacancies in the crystal lattice, which significantly affect their performance. However, diffusion is difficult in ionic (and covalent) solids, thus the *rate* of diffusion is an important factor. The diffusion rate—of defects, atoms, and electrons—in ceramics affects electrical conductivity. Ceramics can therefore be insulating, conducting, or semi-conducting.

Considering ceramics as a class of materials, it is important to keep in mind several distinctive features and properties. The combination of great stiffness and refractoriness generates also an essential limitation for some applications. Ceramics are most often noted for their tendency towards *brittle fracture*. Under tensile stresses, both crystalline and non-crystalline ceramics typically fracture before plastic deformation occurs (see Chapter 14). After crack formation in crystals, cracks may propagate through crystalline grains or along grain boundaries. Ceramics respond to a variety of stresses most often by limited elastic deformation and then cracking. Plastic deformation is difficult because of the difficulty of slip (of defects) and the presence of very few slip systems (contrasted with metals). The result is that ceramics are generally *brittle*. For non-crystalline ceramics, viscous flow may allow some amount of plastic deformation.

Ceramics are the hardest known materials. We shall see in Chapter 14 a comparison of metals and ceramics on Mohs scale of hardness. The phenomenon of creep, which is observed often for metals and polymers, is manifested rarely in ceramics. It usually occurs only as a result of very high stresses at elevated temperatures.

We note in various locations improved understanding provided by computer simulations techniques, such as Monte Carlo or molecular dynamics. **Quantum mechanical simulations** can also be useful. Thus, Jiang and Srinivasan [31] explained mechanisms of unexpected strain stiffening in some ceramics. They elucidate two different mechanisms. The extraordinary stiffening of Fe_3C is a result of the strain-induced reversible "cross-linking" between weakly coupled edge- and corner-sharing Fe_6C slabs. A strong, three-dimensional covalently bonded network is formed; it resists large shear deformation. On the other hand, in Al_3BC_3 stiffening occurs because of strong repulsion between Al and B; a compressed Al—B bond unsettles the existing covalent bond network. Jiang and Srinivasan say [31]: "These discoveries challenge the conventional wisdom that large shear modulus is a reliable predictor of hardness and strength of materials, and provide new lessons for materials selection and design".

Finally, as a class of materials, ceramics are frequently valued for their typically high compressive strength. It is this property that encourages the use of ceramics as structural components (for example in building materials). Improvements in other properties have broadened the spectrum of applications for ceramics. Technical ceramics used in various high performance applications have been discussed in greater detail in the previous Section.

7.11 SELF-ASSESSMENT QUESTIONS

1. Discuss reasons for using ceramic materials.
2. Describe the structures, bonding, and composition that characterize the major classes of ceramics.
3. Discuss spatial structures of various forms of carbon and effects of those structures on properties.
4. Describe the functions of the main classes of oxide components of glassy ceramics.
5. Explain the process of cement hydration in terms of strength of the resulting structure.
6. Describe the structure and properties of abalone shells.

REFERENCES

1. R. Roy & W.B. White, *J. Cryst. Growth* **1972**, *13/14*, 78.
2. L.J. Gauckler, Th. Graule, & F. Baader, *Mater. Chem. & Phys.* **1999**, *61*, 78.
3. M. Esfehanian, R. Oberacker, T. Fett, & M.J. Hoffmann, *J. Am. Ceram. Soc.* **2008**, *91*, 3803.
4. N. Garmendia, I. Santacruz, R. Moreno, & I. Obieta, *J. Eur. Ceram. Soc.* **2009**, *29*, 1939.
5. A. Sanson, P. Pinascoa, & E. Roncari, *J. Eur. Ceram. Soc.* **2008**, *28*, 1221.
6. W. Martienssen & H. Warlimont, eds., *Springer Handbook of Condensed Matter and Materials Data*, Vol. *1*, Springer: New York/Heidelberg **2005**, p. 433.
7. K.S. Raines, S. Salha, R.L. Sandberg, H. Jiang, J.A. Rodríguez, B.P. Fahimian, H.C. Kapteyn, J. Du, & J. Miao, *Nature* **2010**, *463*, 214.
8. P. Ball, *Made to Measure: New Materials for the 21st Century*, Princeton University Press: Princeton, NJ **1997**, pp. 313–343.
9. P.W. May, *Phil. Trans. Royal. Soc. London A* **2000**, *358*, 473.
10. K.S. Novoselov, A.K. Geim, S.V. Morozov, D. Jiang, Y. Zhang, S.V. Dubonos, I.V. Grigorieva, & A.A. Firsov, *Science* **2004**, *306*, 666.
11. R. Shah, T. Datashvili, T. Cai, J. Wahrmund, B. Menard, K.P. Menard, W. Brostow, & J.M. Perez, *Mater. Res. Innovat.* **2015**, *19*, 97.
12. F.H. Gojny, J. Nastalczyk, Z. Roslaniec, & K. Schulte, *Chem. Phys. Letters* **2003**, *370*, 820.
13. F.J. Carrión, C. Espejo, J. Sanes, & M.D. Bermudez, *Compos. Sci. & Tech.* **2010**, *70*, 2160.
14. J.-W. An, D.-H. You, & D.S. Lim, *Wear* **2003**, *255*, 677.
15. M. Ziemnicka & L. Stobierski, *Proceedings of the 10th International Conference of the European Ceramic Society—CD ROM Edition*, ed. European Ceramic Society, Göller Verlag GmbH: Baden-Baden **2008**, p. 1077.
16. P.T. Sarjeant & R. Roy, *J. Am. Ceram. Soc.* **1967**, *50*, 500.
17. A.N. Cormack & J. Du, *J. Non-Cryst. Solids* **2001**, *293–295*, 283.
18. D. Bradley, "Storing carbon dioxide in cement: Green concrete", *Technology Review* 2010, *113*, 56.
19. J. Davidovits, *J. Mater. Ed.* **1994**, *16*, 91.
20. W. Swietoslawski, *Microcalorimetry*, Reinhold: New York **1947**.
21. E. Margas, A. Tabaka, & W. Zielenkiewicz, *Bull. Acad. Polon. Sci. Sér. Chim.* **1972**, *20*, 329.
22. W. Zielenkiewicz & E. Margas, *Bull. Acad. Polon. Sci. Sér. Chim.* **1973**, *21*, 251.
23. D.D. Double & A. Hellawell, *Nature* **1976**, *261*, 486.
24. B. Baś, R. Piech, M. Ziemnicka, W. Reczyński, & M. Robótka, *Electroanalysis* **2009**, *21*, 1773.
25. B. Baś, R. Piech, E. Niewiara, M. Ziemnicka, L. Stobierski, & W.W. Kubiak, *Electroanalysis* **2008**, *20*, 1655.
26. J.M. Rimsza & J. Du, *J. Am. Ceram. Soc.* **2014**, *97*, 772.
27. J.R. Kelly & I. Denry, *Dental Materials* **2008**, *24*, 289.

28. J.C. Diop, "R&D 2002: Nano Ceramics", *Technology Review*, December **2002**/January 2003, https://www.technologyreview.com/s/401725/rd-2002-nano-ceramics/, accessed March 28, 2016.
29. K. Bourzac, "Ceramics That Won't Shatter", *Technology Review* at www.TechnologyReview.com, originally published December 4, **2008**, http://www.technologyreview.com/news/411301/ceramics-that-wont-shatter/, accessed February 9, 2016.
30. I. Corni, T.J. Harvey, J.A. Wharton, K.R. Stokes, F.C. Walsh, & R.J.K. Wood, *Bioinspir. Biomim.* **2012**, *7*, 1.
31. C. Jiang & S.G. Srinivasan, *Nature* **2013**, *496*, 339.

8

ORGANIC RAW MATERIALS

The world will not run out of geologic or biologic resources; they will merely become more expensive.

—W.T. Lippincott [1] in an editorial entitled "The materials revolution: Can we handle it?"

8.1 INTRODUCTION

A class of materials often overlooked in traditional classification schemes is organic raw materials. This is the situation in spite of its indispensability to economies worldwide and its preciousness as the starting point for innumerable products. Here "organic" refers to the hydrocarbon composition.

A goal of the present chapter is to acquaint the reader with the chemical compounds that make up this class of materials. To make that knowledge usable, we delve into the acquisition methods, processing, and applications of organic raw materials while also describing basic properties and structures of the various members of this class.

In Part 1, called Foundations, we discussed distinctions—thermodynamic and others—among the states of matter. Organic raw materials are found in nature as solids, liquids, and gases. Among those existing in the liquid state, we find both mixtures and solutions. Organic liquids are more complicated than condensed gases while polymer solutions are notably more complex than liquid hydrocarbons.

We have to distinguish between liquid mixtures and solutions. In a **liquid mixture** all components have "equal rights", that is to say, we have two or more liquids simply co-existing together. In a **liquid solution** there is one or more components designated as **solvent** (if more than one we have a mixed solvent) and also one or more

Materials: Introduction and Applications, First Edition. Witold Brostow and Haley E. Hagg Lobland.

components designated as **solute**(s). If we have aqueous sodium chloride (aq. NaCl), that is a solution of sodium chloride in water, then clearly the former is the solute and water the solvent. If we have 50% water and 50% ethanol, and we decide to call it a solution, then it becomes a matter of personal preference which will be designated the solvent.

8.2 NATURAL GAS

Natural gas extracted and collected from within the Earth's crust is utilized for two main purposes: chemical synthesis and combustion. Natural gas is not a single chemical compound but refers to a mixture of hydrocarbon gases including primarily nitrogen (N_2), methane (CH_4), and ethane (C_2H_6) (hydrocarbon chemical structures are shown in Figure 8.1). The composition of natural gas depends on its origin (i.e., geographic location) and is therefore varied from one source to another.

For some synthesis processes, natural gas can be used directly. More often, however, it must first be separated into its constituents. Separation and purification processes depend on phase equilibria of the gaseous components. Hence an understanding of thermodynamics and phase diagrams (presented in Chapter 3) is key in the design of these industrial processes.

A commonly used method of separation is the liquefaction of natural gas by cooling it to temperatures around 100 K and then manipulating both temperature and pressure for extraction of each constituent gas. By this technique non-combustible gases such as helium and nitrogen are removed while combustible components are separated and reserved for further use. The liquid–vapor and liquid-liquid phase equilibria play importantly in the separation processes. Also associated with separation and purification are

Compound	Chemical Formula	Structural Formula	Line Formula
Methane	CH_4	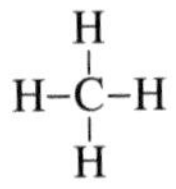 H H–C–H H	CH_4
Ethane	C_2H_6	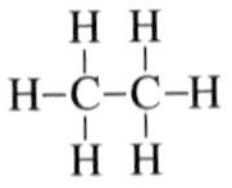 H H H–C–C–H H H	H_3C-CH_3
Propane	C_3H_8	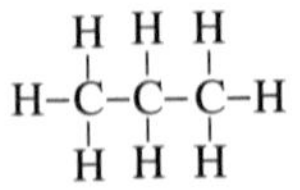 H H H H–C–C–C–H H H H	$H_3C-CH_2-CH_3$
Butane	C_4H_{10}	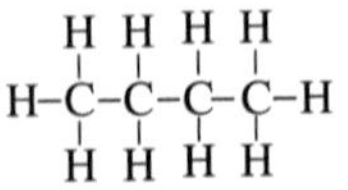 H H H H H–C–C–C–C–H H H H H	$H_3C-CH_2-CH_2-CH_3$

FIGURE 8.1 Structure and formula of basic hydrocarbons.

TABLE 8.1 Enthalpies of Combustion (H^{comb}) of Different Materials (Approximate Values)

	Atomic Percentage			
Material	H	C	O	$H^{comb}/(kJ\ g^{-1})$
Turf (dry)	44	40	16	15.9
Lignite	37	49	14	15.9
Wood	46	32	21	18.8
Methanol	66.7	16.7	16.7	22.7
Anthracite	27	70	2	30.5
Graphite	0	100	0	32.8
Benzene	50	50	0	41.8
Petroleum	63	36	0.4	47.3
Natural gas	76	20	0.4	48.5
Methane	80	20	0	55.5
Hydrogen	100	0	0	119.9

rectification/distillation and extraction processes, which we shall examine with more detail in the next section of this chapter (see also Section 3.12). When a gas component possessing some desired properties has been separated and purified, it then enters as a substrate into a chemical reaction (for synthesis of other chemical products) or it is used for combustion.

Applications of gas combustion do not rely solely on natural gas; gases obtained by processing other raw materials are also used. As already mentioned, composition of natural gas is varied from one source to another. Consequently, a substitute natural gas is often obtained by coal gasification, a procedure in which coke (carbon solids) is usually produced simultaneously. We shall talk more about coal in Section 8.4. Composition of substitute gas depends primarily on the temperature regime of the process and can be determined by calculations from heat and mass balances. Enthalpy of combustion of natural gas is listed in Table 8.1 (and shall be discussed further in subsequent sections).

The H^{comb} values tell an important part of the story but not the whole story. According to the Table, one should use hydrogen whenever possible. However, hydrogen mixed with oxygen explodes very easily.

Calculation procedures for combustion of a gas are based on the species involved, on energy conservation principles, and on the tendency towards an equilibrium state [2]. Coca-Cola, or fruit juice, or even mineral water are not good combustibles because, technologically speaking, their main components are nothing but the ashes of burnt hydrogen; and ashes are (naturally) the worst combustibles imaginable.

While natural gas has been used by humans for a long time, there is a technology of obtaining natural gas used first in 1947 but which became widely used only in the 21st century. **Hydraulic fracturing** or **fracking** is the process of drilling and injecting fluid into the ground at a high pressure in order to fracture shale rocks and release natural gas inside. **Shale** is a fine-grained sedimentary rock composed of mud, a mix of flakes of clay minerals and very small fragments of other minerals, quartz and calcite in particular. The ratio of clay to other minerals is variable. Well stimulation is only conducted once in the life of the well. After hydraulic pressure is removed from the well, small grains of proppant (sand or aluminum oxide) hold these fractures open, and the rock achieves dynamic equilibrium.

The fluid is typically water but it contains some additives and sand. Water with additives is more effective; see Section 13.5 on drag reduction. Where there is no water available through pipelines, it has to be delivered by tanker trucks to the site.

Those who oppose the process of fracking point out that methane concentrations are 17 times higher in drinking-water wells near fracturing sites than in normal wells. Clearly the factor of 17 is not universal. Opponents also claim that contaminants other than methane appear in potable water as a result of fracking. However, more objective analysis tells us that risks can be foreseen and contained while advantages are very large [3].

Organic liquids are generally subdivided into two groups: those of low molecular weight, and polymers. The raw materials petroleum and coal tar are important members of the former group. We shall discuss petroleum in the next Section and coal tar in the following one.

8.3 PETROLEUM

As entire volumes are dedicated to petroleum, we shall limit our discussion to what is of practical use to MSE. We shall use the terms petroleum and oil so that both have the same meaning.

As with natural gas, the composition of petroleum depends on the source. Likewise, the principal uses are chemical synthesis and combustion; and whatever the use, separation and purification processes are typically based on knowledge of phase equilibria. An important difference between natural gas and petroleum is that the components of natural gas are mainly aliphatic (regardless of source) while for petroleum the ratio of aliphatic to aromatic hydrocarbons varies significantly with the origin. Final commercial products, such as gasoline, reflect that variation. Nevertheless, the compositions of different gasolines are described such that their enthalpies of combustion can be easily calculated from tables of combustion data.

In spite of its importance as fuel, petroleum should not be neglected in its central role as a substrate for chemical synthesis. For example, olefins are one class of compounds synthesized from petroleum, and some 70% of petroleum-derived olefins are used in the manufacture of important aliphatic polymers: namely polyethylene, polypropylene, polyvinyl chloride, polyacrylonitrile, and polybutadiene. The specifics of the synthetic processes belong to the realm of chemistry and are treated for instance in the classical book by Waddams [4].

Petroleum consists of a mixture of hydrocarbons of various molecular weights and degrees of chemical saturation. To create useful materials, the constituents are separated by distillation, that is rectification. Thus, the **rectification column** is the leading tool used in both industry and the laboratory. Such columns were mentioned in Chapter 3; it is relevant also here, thus a schematic of the device is given in Figure 8.2. Recall that we have both liquid and vapor at each plate in the column.

Separation of petroleum components then requires efficient distillation of a multicomponent mixture, a problem that in the simplest case may be reduced to that of knowing binary boiling point isobars for all pairs of components involved. There are typically many of such pairs, and when homologs are part of the mixture they may be approached as series of systems. For instance, in ternary systems analyses may be conducted with two components fixed and the third being a variable member of a homologous series. Likewise, in the

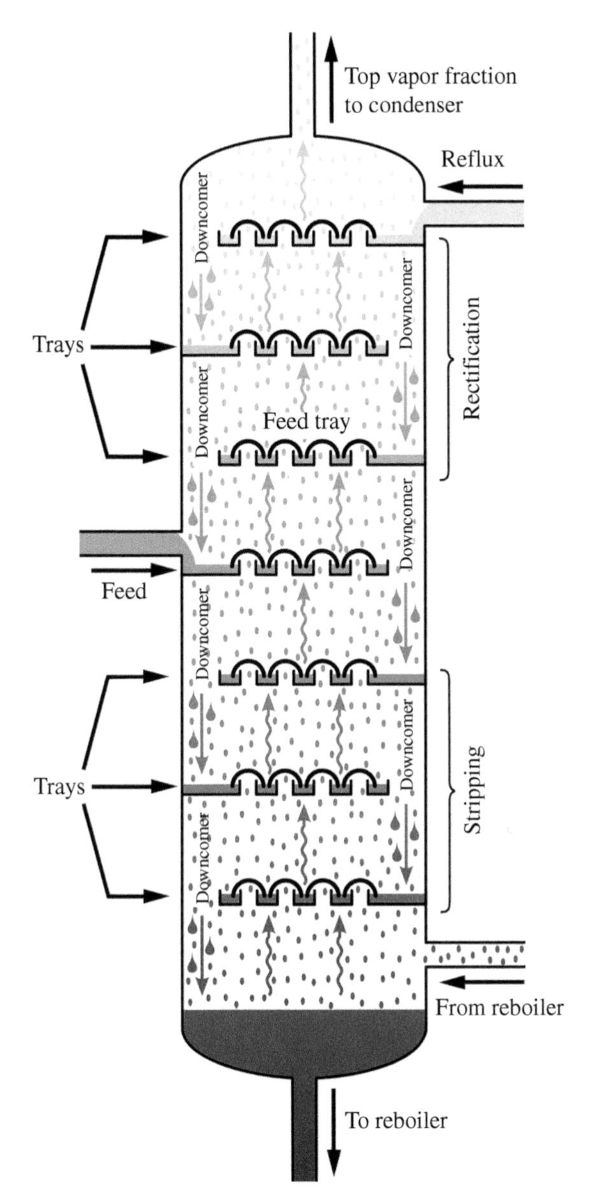

FIGURE 8.2 Schematic diagram of a rectification column used for separation and purification of materials, such as for separating different hydrocarbons in petroleum.

case of boiling isobars of petroleum mixtures, a series of curves are examined: one for each member of a given homolog series with a common second component. The series approach has been advocated by several. Swietoslawski discussed various possible types of zeotropic, azeotropic, and polyazeotropic behavior, where polyazeotropic mixtures are those in which numerous multicomponent azeotropes are formed during distillation [5]. Although they apply to any kind of liquid system, they were developed particularly to address multicomponent organic raw materials such as coal tar and petroleum.

PETROLEUM AND MULTICOMPONENT MIXTURES

It would be advantageous to predict the boiling point or vapor pressure curves of these multicomponent mixtures, but this is generally very difficult. If the chemical potentials of components are known as functions of temperature, pressure, and composition, then the boiling point or vapor pressure curves in mixtures can be predicted. The chemical potential was defined by Eq. (3.9); alternatively the same information could be represented in terms of activities, activity coefficients, or relations between these coefficients. Since all of these quantities are thermodynamic, there exist various theoretical and semi-theoretical schemes for predicting them, some of them reliable for predicting activities in hydrocarbon mixtures. An accurate method of predicting volumetric and thermal properties of hydrocarbon, fluorocarbon, or alcohol mixtures has been developed [6, 7]. They rely on representing long chains by graphs, assigning $g(R)$ and $u(R)$ functions to each graph point [6, 7].

While obtaining gasoline from petroleum in rectification columns is the best known and spectacular landmark of the petroleum industry, there are other stages of the petroleum production operations. Extensive work on improvement of most these stages, often by using polymers, is being conducted by Elizabete Lucas and her colleagues at the Federal University of Rio de Janeiro [8–14]. Following Lucas and coworkers, we can make a list of these stages:

- Drilling of oil wells begins only after analysis of information generated by exploratory research in the prospecting stage, based on geology and geophysics;
- Cementing: when the drilling reaches a certain depth, a steel casing is introduced into the borehole, starting the cementing step. Important here is successfully displacing drilling fluid from the annulus and properly conditioning the annular surfaces to accept and bond with the cement;
- Completion: after the well has been cemented, the communication between the well and the hydrocarbon-bearing formation must be established by perforating the casing and cement. There are also other operations at this stage, including controlling production of sand or water and cleaning the well and perforation blockages. Nature is not so nice to those who want to get materials out of the Earth… Petroleum can come out containing sand or water—which then have to be separated. Sand can plug the well or else cause erosion of the well walls and of inside surfaces of equipment. Void formation behind the casing is possible, which in the worst case scenario can lead to collapse of the formation and loss of the well. These scenarios can be mitigated by putting a gravel pack at the bottom of the well and using certain polymers such as xanthan gum;
- Production: includes operations facilitating the movement of petroleum within the porous rock medium, its access to the well, and upward movement. One uses here techniques of acidizing and hydraulic fracturing—the fracturing of rock by a pressurized liquid;
- Treatment of petroleum such as dehydration. Emulsions are another important feature associated with crude oil recovery and petroleum processing. In fact, in oil fields,

emulsions are present in reservoirs, producing wells, intermediate production equipment, pipelines, primary processing installations, and flowlines. Emulsions are particularly an issue when water is often co-produced with crude oil, as a result of recovery techniques. Asphaltenes often stabilize emulsions owing to their amphiphilic nature. To counter this and other factors, it becomes necessary to break up emulsions in order to remove water content from petroleum. Polymers are typically used in this regard as demulsifiers, destabilizing emulsions and allowing the phases to separate.

Among the variety of organic compounds found in liquid petroleum are polycyclic aromatic hydrocarbons (PAHs). Polycyclic aromatic hydrocarbons are compounds with two or more fused aromatic rings, such as some of the compounds shown in Figure 8.3. They occur naturally in coal and petroleum; additionally they can be released from combustion of fossil fuels or from degradation of manufactured materials such as lubricating oils, dyes, detergents, and plastics. Polycyclic aromatic hydrocarbons have been found *throughout the universe*, specifically in carbonaceous chondrite meteorites, Martian meteorites, and interplanetary dust particles. These compounds have a relatively low solubility in water but are highly lipophilic (soluble in fats, oils, lipids, and non-polar solvents such as hexane or toluene). Most PAHs are toxic, while many of those with four or more rings are specifically carcinogenic. Thus PAHs are monitored worldwide in drinking water,

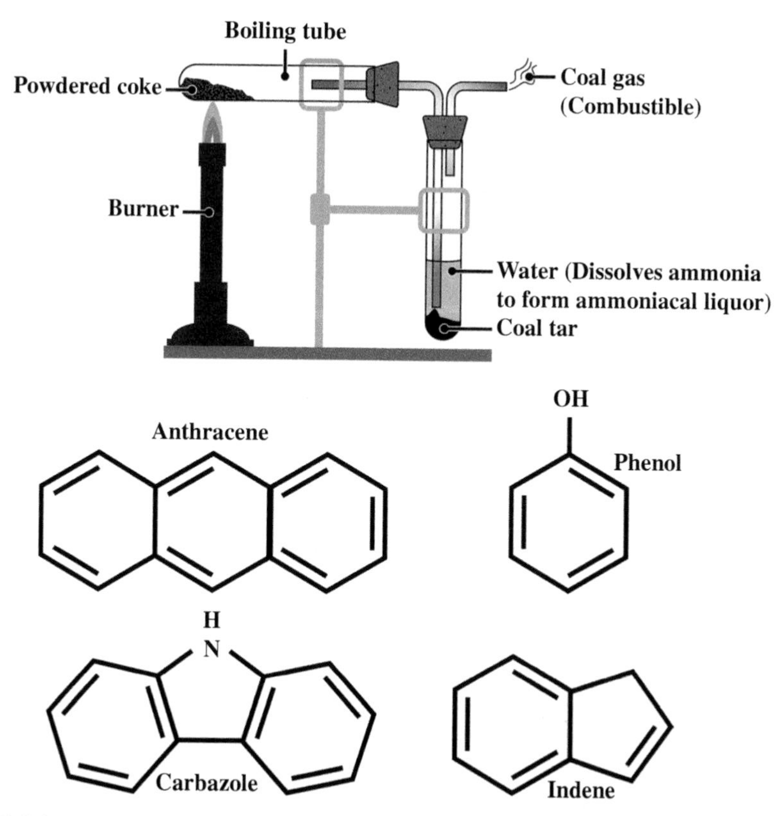

FIGURE 8.3 Schematic of the basic process for coal tar production (top) and structures of some common chemical products derived from coal tar (bottom).

waste water, furnace emissions, soil, hazardous waste extracts, and air over major cities. The typical method for detection of PAHs in spent lubricating oils has been liquid-liquid extraction (LLE), a time-consuming method involving large amounts of organic solvents. However, a method developed by Areshidze and coworkers [15] does not suffer from these drawbacks and thus offers a better way to isolate, identify, and measure the levels of PAHs in our environment.

We need to name one more technique of dealing with petroleum, namely, **hydrocracking** [16]. As we know, petroleum contains a variety of hydrocarbons. Here is an example of a hydrocracking reaction:

$$C_{16}H_{34} + 2H_2 \rightarrow 2C_8H_{18}$$

Hexadecane, $C_{16}H_{34}$, is the highest paraffinic hydrocarbon in the liquid state; $C_{17}H_{36}$ is a solid. We do not need to point out that octane, C_8H_{18}, is a component of gasoline while hexadecane is not.

Let us comment briefly on so much activity in the petroleum area in contemporary Ukraine. There are historic reasons for this. A pharmacist named Ignacy Łukasiewicz in Lviv, then the capital of the Kingdom of Galicia, in 1853 invented the kerosene lamp. The same year he put up the first modern street lamps in Europe in his city, while a year later he constructed the first modern oil well [17]. In 1856 he opened the first industrial petroleum refinery. The original refinery he started in Drohobych in Ukraine is still in operation, expanding and modernizing now in the 21st century. Possibly because of a complex history of that region since 1854, the fact that Łukasiewicz is the father of the modern petroleum industry is not widely known, even among those working in that industry.

Interestingly, while polymers (and synthetic plastics) are end products of petroleum processing, they are also used throughout nearly all stages of the recovery and processing of organic raw materials [12]. Polymers are used in solution and as solid-state materials. For instance, plastics, fibers, and elastomers are used on offshore platforms and in construction of pipelines and floating structures. In solution, polymers are used to modify the properties of fluids and formulations in operations named earlier, including: drilling, cementing, completion, oil production itself, and oil and water treatment.

Polymers—in the form of porous network structures—are also applied in the unfortunate situation of oil spills, as these materials can selectively absorb oil from an aqueous environment (e.g., the ocean) and subsequently release the trapped oil [18–20]. Effective porous polymers of this type allow repetitive operations of absorbing and desorbing oil.

8.4 COAL AND COAL TAR

Coal is another important organic raw material. It is a black or brownish-black sedimentary rock present in layers or veins called coal beds or coal seams. Coal contains mostly carbon but also hydrogen, oxygen, nitrogen, and sulfur. When significant amounts of sulfur are present, *desulfurization* is needed. So-called **acid rain** in the region of Great Lakes in North America occurs because burning coal containing sulfur leads to the formation of sulfur oxides. We shall return briefly to this topic in Section 20.3.

As with natural gas and petroleum, the composition of coal depends on location. Coal formation is a multistage process; dead plant matter is converted into peat, then lignite, sub-bituminous coal, bituminous coal, and finally anthracite. The content of carbon increases during the process. Thus, lignite contains 60–75 wt.% C. If a given deposit has reached the stage of anthracite, the carbon content exceeds 91.5%.

So-called **clean coal technologies** are being developed to mitigate pollution, such as acid rain, dust formation, and release of carbon dioxide in the atmosphere, created by coal burning. Coal gasification was already mentioned in Section 8.2. Capture of CO_2 formed in coal combustion using solvents, sorbents, or membranes is another activity along these lines.

Coal tar is a byproduct of coal carbonization. The main purpose of heating coal above 700 K is to produce coke; however coal gas and coal tar—which includes some sulfur- and nitrogen-containing compounds—are simultaneously obtained in the process. Constituents of coal gas include: hydrogen, carbon monoxide, methane, ethane, ethylene, and smaller quantities of their higher homologs.

Coal tar differs from petroleum chiefly in chemical composition, noted in particular by the carbon/hydrogen ratio. Both are multicomponent hydrocarbon mixtures; the C/H ratio for petroleum falls between 0.41 and 0.45 while for coal tar it averages 1.45. Consequently the profile of industrial products derived from coal tar is distinct from petroleum products. This is evident in the following profile of coal tar products (a few representative structures are given in Figure 8.3):

1. Polycyclic aromatic compounds such as naphthalene, anthracene, pyrene, and acenaphthene.
2. Nitrogen- and oxygen-containing heterocyclic chemicals such as pyridine, quinoline, carbazole, indole, or diphenylene oxide.
3. Phenolic compounds such as phenol itself, cresols, and diethyl phenols.
4. Carbochemical hydrocarbon resins based on indene.
5. Technical aromatic oils such as impregnating oil used for treating timber.
6. Pitches used in the manufacture of electrodes, in particular for the aluminum industry and also the electric steelmaking industry.
7. Technical carbon, including carbon black.

The types and uses of coal tar products are further explained by Figure 8.4. Some of the products in this list are obtained exclusively from coal tar; others can be obtained from both coal tar and petroleum. The C/H ratio is decisive in determining the source of primary chemicals. Note that certain petroleum-based industrial organic chemical processes, while leading to important products and semi-products, also lead to some compounds present naturally in raw coal tar. An example from industrial production illustrates the concept; see Figure 8.5.

Naphthalene—an important substrate for dye, tanning, insecticide, and textile industry—is obtained almost exclusively from coal tar. Likewise anthracene is obtained primarily from coal tar, but for the reason that—due to no lack of effort—it is not economical to produce it from petroleum.

As it is with petroleum, for coal tar the primary industrial method for separating components is column rectification. This procedure is often combined with extraction and crystallization processes owing to the different chemical profile of coal tar. Chemically neutral, acidic, and basic compounds are present in coal tar; hence the consecutive use of H_2SO_4 (sulfuric acid) and NaOH (sodium hydroxide) allows their separation from one

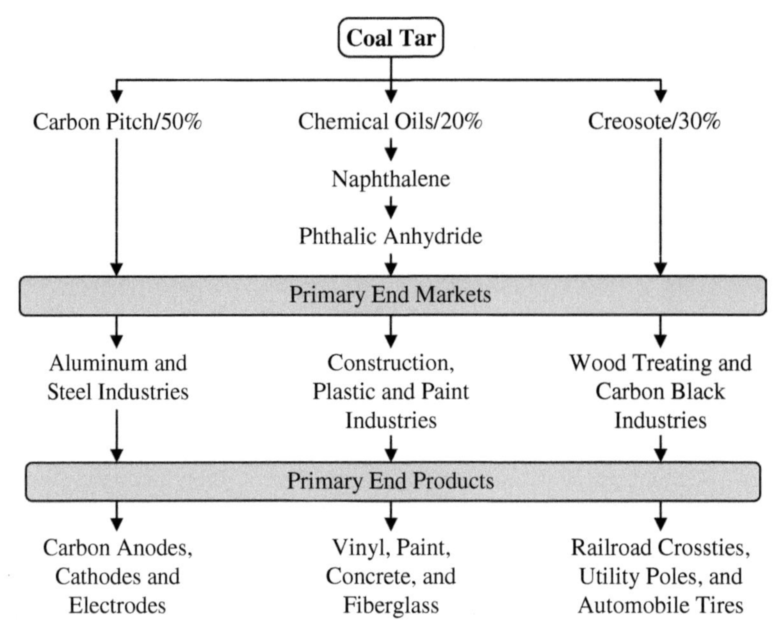

FIGURE 8.4 Delineation of coal tar markets and products.

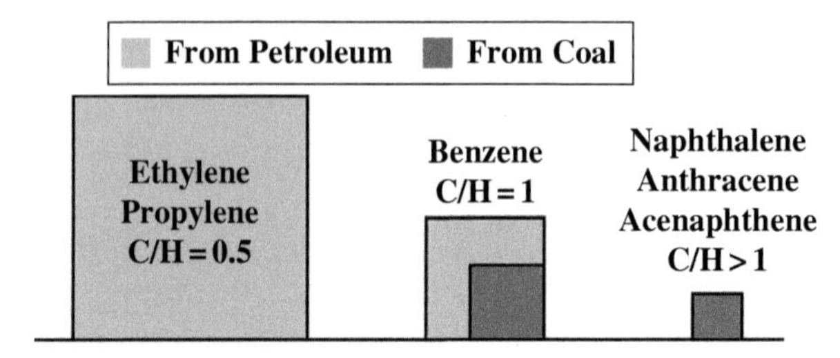

FIGURE 8.5 Comparison of chemicals and associated C/H ratios typically produced from petroleum and coal tar. *Source*: Adapted from [21].

another. If such separation is neglected or only partially completed, the distillation process is complicated by polyazeotropy. The phenomenon was discussed in the second track section above with regard to petroleum but assumes more importance with coal tar. The structures of coal tar components are related to those of petroleum, however specific interactions such as those between weak acids and bases in coal tar are a complication.

8.5 GENERAL REMARKS

As always, we cover the most essential factors. We have already noted that acid rain contains higher than "normal" amounts of nitric and sulfuric acids. Acid rain formation results from both natural sources, such as volcanoes and decaying vegetation, and man-made sources, primarily emissions of sulfur dioxide (SO_2) and nitrogen oxides (NO_x) from fossil fuel combustion. Therefore, methods of desulfurization, for instance of diesel petroleum fractions [22] and of coal [23], are being developed. The effects of pollutants on the Earth's atmosphere will be discussed more in Chapter 20.

The economies of most countries around the world are strongly related to the prices of two materials: petroleum and natural gas. These organic raw materials are therefore important globally, and it matters significantly how we use them. Not only do petroleum and natural gas provide energy, they also provide the starting materials for much chemical synthesis and for plastics production. One develops methods of obtaining synthetic replacements of natural gas, so called **syngas**. Those technologies change quickly, with no clear winner seen yet. Widely used organic raw materials are non-renewable, hence the usual connection between production costs and market price does not apply here; the production costs are only a small fraction of the price.

Finally—and sadly—we need to note fuel explosions. In Tenerife there was a collision at the airport of two passenger planes in 1977. One plane was trying to take off while a second one was standing on the runway. The speed of the moving plane was not that high, but 583 people died, mostly from the deadly mist of the burning aviation fuel. Scientists and engineers tried to at least mitigate the problem by adding high molecular mass polymers (see the next Chapter) to stop fuel droplets from breaking up into a fine mist. Larger burning droplets release energy more slowly. However, it turned out that the polymeric additive was short lived.

An effective solution to the problem came much later, from the group of Julia Kornfield at the California Institute of Technology [24]. In a nice order of steps taken, they first used statistical mechanics to define what kind of materials would do the job. Then they have developed what they call **megasupramolecules** consisting of relatively short building blocks that do not undergo chain scission easily. Moreover, the structures are kept together by hydrogen bonds, so they could later reform. Kornfield and her collaborators fired a projectile with the velocity of 140 miles per hour at a fuel container, with three propane torches in the path of the mist formed after the collision. The fuel with a quite low concentration of the megasupramolecule additive (0.3 wt.%) controlled misting as required. Other uses of the new fuel additive are under investigation, including automobile fuel and so called improvised explosive devices (IEDs).

8.6 SELF-ASSESSMENT QUESTIONS

1. Are enthalpies of combustion a sufficient criterion for choosing a fuel for a given application?
2. Combustion is an important metric for organic raw materials. Consider other naturally occurring materials: would rocks burn (i.e., combust) or melt, given sufficient heat?
3. Discuss the stages of petroleum production operations.
4. We have petroleum together with sand in an oil well. What can happen and what shall we do?
5. Explain the meaning of the terms mixture and solution, and explain what are the components of the latter.
6. Differentiate between natural gas, petroleum, and coal tar.
7. How are organic raw materials related to polymers and the plastics industry?

REFERENCES

1. W.T. Lippincott, *J. Chem. Ed.* **1976**, *53*, 203.
2. J.H. Ay & M. Sichel, *Combust. Flame* **1976**, *26*, 1.
3. B. Weinhold, *Environ. Health Perspectives* **2012**, *120*, A274.
4. A.L. Waddams, *Chemicals from Petroleum*, 4th edn., Murray: London **1978**.
5. W. Swietoslawski, *Azeotropy and Polyazeotropy*, PWN-Macmillan: Warsaw/New York **1963**.
6. W. Brostow, D.M. McEachern, & S. Pérez-Gutiérrez, *J. Chem. Phys.* **1979**, *71*, 2716.
7. W. Brostow, *Phys. & Chem. Liquids* **1981**, *10*, 27.
8. E.F. Lucas & L.S. Spinelli, *J. Mater. Ed.* **2005**, *27*, 43.
9. A. Middea, M.B. de Mello Monte, & E.F. Lucas, *Chem. & Chem. Tech.* **2008**, *2*, 91.
10. M. de Melo & E.F. Lucas, *Chem. & Chem. Tech.* **2008**, *2*, 295.
11. J.B. Ramalho, N. Ramos, & E.F. Lucas, *Chem. & Chem. Tech.* **2009**, *3*, 53.
12. E.F. Lucas, C.R.E. Mansur, L. Spinelli, & Y.G.C. Queirós, *Pure & Appl. Chem.* **2009**, *81*, 473.
13. V.F. Pacheco, L. Spinelli, E.F. Lucas, & C.R.E. Mansur, *Energy & Fuels* **2011**, *25*, 1659.
14. C. da Silva, C. Barros, Y. Queiros, L. Marques, A.M. Louvisse, & E.F. Lucas, *Chem. & Chem. Tech.* **2012**, *6*, 415.
15. G. Areshidze, K. Barbakadze, W. Brostow, T. Datashvili, O. Gencel, E. Lekveihsvili, & N. Lekishvili, *Mater. Sci. Medziagotyra* **2010**, *16*, 170.
16. O. Machinskiy & P. Topilnitskiy, *Hidrokreking*, Lvivska Politechnika Publishers: Lviv **2011** (in Ukrainian).
17. Wikipedia, *Ignacy Łukasiewicz*, http://en.wikipedia.org/wiki/Ignacy_%C5%81ukasiewicz, accessed February 10, **2016**.
18. A.M. Atta, W. Brostow, T. Datashvili, R.A. El-Ghazawy, H.E. Hagg Lobland, A.-R.M. Hasan, & J.M. Perez, *Polymer Internat* **2013**, *62*, 116.
19. A.M. Atta, W. Brostow, H.E. Hagg Lobland, A.-R.M. Hasan, & J.M. Perez, *Polymer Internat* **2013**, *62*, 1225.
20. A.M. Atta, W. Brostow, H.E. Hagg Lobland, A.-R.M. Hasan, & J.M. Perez, *RSC Advances* **2013**, *3*, 25849.
21. R. Oberkobusch, *Erdöl & Köhle* **1975**, *28*, 558.
22. O. Lazorko, M. Bratychak, & S. Pyshev, *Chem. & Chem. Tech.* **2008**, *4*, 309.
23. S. Pyshev, V. Gunka, O. Astakhova, Y. Prysiazhnyi, & M. Bratychak, *Chem. & Chem. Tech.* **2012**, *6*, 443.
24. M.-H. Wei, B. Li, R.L.A. David, S.C. Jones, V. Sarohia, J.A. Schmitigal, & J.A. Kornfield, *Science* **2015**, *350*, 72.

9

POLYMERS

Oligopoem
for Sylvie Coyaud

Their Problem
Propylene,
    Propylene...
How to relieve
    this boredom
of coupling
    with identical
partners?

Self Cure
Do forward things:
    twist here, twist
there, the reactive
    end bites ... itself,
it happens, in novelas,
    and the ring, well
it says kaput
    to fruitful
propagation.

Humans Are So Unimaginative
A problem? Try
    a second partner!

Materials: Introduction and Applications, First Edition. Witold Brostow and Haley E. Hagg Lobland.

Each time the live end
loses its head,
relentlessly
opting for the other,
stuck in the eternal
fickleness
of copolymerization.

<u>*Ours*</u>
On the dizzy chain
from Sade
to Ziegler, Natta,
we're into
control; we want
them strong (or is
it weak?), we want
teflon, and epoxy,
all in a day. Lately,
in a morbid mood,
we've wanted
the spent ones
to just fall apart.

<u>*Reptation*</u>
Polythiophenes,
anguilles à la Bilbao—
entangled, constricted,
how else to move
in their crowded Eden?

<u>*Mono, oligo, poly*</u>
If they could sing
(I mean beyond
the quantum strum,
past C-O stretch
and hindered rotation),
if they could sing
it would be Leadbelly's
tune; of cousins,
of the hard labor
of a protein, the
memory of DNA—
a gang-chained folk,
the utilitarian refrain.

—The author of the above poem [1] is Roald Hoffmann, Frank H.T. Rhodes Professor of Humane Letters at Cornell University in Ithaca, New York. Born in Złoczów, Poland, now Zolochiv, Ukraine, as a small boy he was hiding from the Hitlerites together with his mother thanks to an Ukrainian schoolteacher Mykola Diuk and his family in Univ. He then went to school in Cracow, later via Czechoslovakia, Austria, and Germany, moved to the United States, and obtained his PhD in Chemical Physics at Harvard. He is the recipient of the 1981 Nobel Prize in Chemistry (shared with Kenichi Fukui) for his work on what he calls applied theoretical chemistry. Roald Hoffmann writes not only poems but also theater plays such as: *Oxygen* (joint with Carl Djerassi); *Should've*; and *Something That Belongs to You*.

9.1 POLYMERS AMONG OTHER CLASSES OF MATERIALS

Just look around and you will see how many objects are made from polymers or from polymer-based materials (PBMs), such as fiber-containing plastic composites. Already in 1976 the annual production of metals became equal to that of polymeric materials. Since then the annual production of the latter increases at a much faster rate than that of the former. Some historians declare that we are now in the Plastics Age. The reason for the acronym PBM is important: as composite materials become increasingly prominent, it is useful to distinguish those based on polymer components. In the all-composite airplanes, polymers serve more and more as constituents in composite structures—together with metals, ceramics, and other classes of materials.

You do not have to believe the declaration of historians; simply imagine that suddenly everything made from PBMs has disappeared. Little girls would be crying over the disappearance of their plastic dolls; computers would lose their housings; and all-composite airplanes (e.g., Boeing Dreamliner, with more of such planes expected) would disappear too. This is just to name a few; you can make a list nearly endless of items made from PBMs that we would sorely miss if permanently gone. Hence, even though Polymer Science and Engineering may be young compared to the study of metals and ceramics, we see that polymer-based materials have become essential features in most 21st century cultures.

Polymers can be:

- soft or hard
- rigid or elastic, brittle or deformable
- insulators or conductors
- swelling or water-repellent
- continuous solids or porous foams
- thermoplastics or thermosets: on temperature increase they either melt (as all other materials do) *or harden* and then never melt but burn instead. Those that melt are called thermoplastics, and those that never melt are thermosets
- any color: transparent or opaque, polymers come in all colors. Imagine a plastic doll manufacturer who makes plastic only in the color of navy blue; not many buyers expected...
- used in a very broad range of applications from medicine (e.g., artificial joints and tissue, pharmaceutical preparations) to automotive and airplane manufacturing (increasingly so) to construction industry (for instance, polymer containing or polymer based concretes).

One thing which polymers cannot be: fully crystalline. Polymers are either *semicrystalline* or *amorphous*.

Let us now make a quick survey of the classes of materials and the characteristic ways in which they are used.

1. Metals are used primarily for three reasons: their mechanical strength, high electrical conductivity, and high thermal conductivity. Metallurgists also talk about ease of fabrication, but there are some other materials fabricated easily (and even more so) as well.
2. Ceramics are also used primarily for three reasons: their high melting points (i.e., refractory materials), low electrical conductivity (insulators), and rigidity.
3. Semiconductors are used for one reason alone: their electrical conductivity within a certain range (nobody would use them as construction materials...).
4. Organic raw materials are used for two reasons: high enthalpy of combustion (as fuels) and as starting materials for polymer synthesis.
5. Composites, including polymer-based composites, are used largely because of their mechanical properties and often low weight.
6. Smart materials are used because of their response to stimuli and frequently are comprised of polymers.
7. Polymers by themselves are used for all the reasons named above and more. For instance, polymers in fact require much less energy for processing than do metals. Moreover, processing of polymers is easy. Their versatility is evident. Note that categories 1–3 listed above are typically thought of as alternatives to polymers while 5 and 6 can contain polymers or be comprised exclusively of polymers.

There are disciplines of Polymer Science and Engineering, Polymer Chemistry, Polymer Physics, and more. On one hand, this makes sense; methods of polymer processing alone can fill a fat volume. However, Roald Hoffmann [2], John Baglin [3], and Rustum Roy [4], among others, have called for integration of science and engineering, as opposed to undesired *fragmentation*. Such integration can take place in a natural way. A good textbook of Polymer Physics by Ulf Gedde [5] discusses among other issues measurement of rheological properties of molten polymers—knowledge needed also for Polymer Processing.

9.2 INORGANIC AND ORGANIC POLYMERS

The name polymer comes from ancient Greek: *polus* = many; *meros* = part, or segment. Let us now go back once more to Figure 1.1: the macroscopic properties and behavior of any physical system (i.e., material) are determined by structures and interactions at the microscopic and molecular (atomic) level. Full knowledge of any two vertices of the triangle (*hardly possible* to achieve) provides us with the third vertex. So let us look now at some molecular structures of polymers.

In Section 7.1 we talked about inorganic and organic materials (and also about the fact that this classification is not perfect...). How do we create an inorganic polymer? If we pour molten sulfur into cold water, the result is polymeric sulfur:

$$\cdots S-S-S-S-S-S\cdots$$

It is soft, translucent, and highly elastic. Polymers need not consist of only one element or contain only single bonds, for example poly(dichlorophosphazene), precursor to some elastomeric (rubber-like, discussed later in this chapter) phosphazene polymers:

$$\begin{array}{cccc} \text{Cl} & \text{Cl} & \text{Cl} & \text{Cl} \\ | & | & | & | \\ -\text{P}{=}\text{N}- & \text{P}{=}\text{N}- & \text{P}{=}\text{N}- & \text{P}{=}\text{N}- \\ | & | & | & | \\ \text{Cl} & \text{Cl} & \text{Cl} & \text{Cl} \end{array}$$

How about creating an organic polymer? By one route we begin with natural gas which contains ethane C_2H_6, or else we obtain ethane by cracking of petroleum. Then we perform dehydrogenation of ethane and obtain **ethylene** $CH_2{=}CH_2$. Ethylene is susceptible to polymerization; the result is **polyethylene**:

$$\cdots-CH_2-CH_2-CH_2-CH_2-\cdots$$

Note that the "mer" or single unit in polyethylene is $CH_2{-}CH_2$ and not CH_2. This is because the "mer"—more accurately termed **monomer**—corresponds to the non-polymerized unit from which it came, namely, ethylene with its two carbons and four hydrogens. At both ends of the polymeric chain we have methyl groups CH_3 rather than methylene groups CH_2. Polymers are also commonly denoted with the repeating monomer unit in brackets, for instance, we write the formula for polyethylene (PE) as $(CH_2{-}CH_2)_n$ where n = the **degree of polymerization**, which is essentially how many monomer units are contained in the polymer chain.

The polymers shown above are **homopolymers**; this means that there is only one kind or type of building block (mer) repeated in a macromolecule. A homopolymer chain can be straight or branching (with short chain branches off the main chain). However, we also have **copolymers** in which there is more than one kind of building block. As a consequence, several types of structures are possible; these are shown schematically in Figure 9.1. Different copolymers can be obtained from the same monomers by changing their sequence or statistical distribution. Thus, employment of the various options seen in Figure 9.1 allows us to tune specific material properties (such as flexibility or hydrophobicity) to a predefined application.

In Figure 9.1 we still have a tacit assumption: two kinds of monomers. However, **terpolymers** exist also, for instance, acrylonitrile-butadiene-styrene (ABS), a synthetic elastomer, shown here as an alternating one:

$$\cdots-ABS-ABS-ABS-\cdots$$

Incorporating more than three types of monomers into a polymer is not impossible in principle but quite difficult in practice.

9.3 THERMOPLASTICS AND THERMOSETS

We have said above that thermoplastics behave as "decent" materials do, that is the solids melt upon heating. This is not a complete story. When we have a molten polymer, upon further heating it will *not* go into the gas phase. This is unless we declare that butane

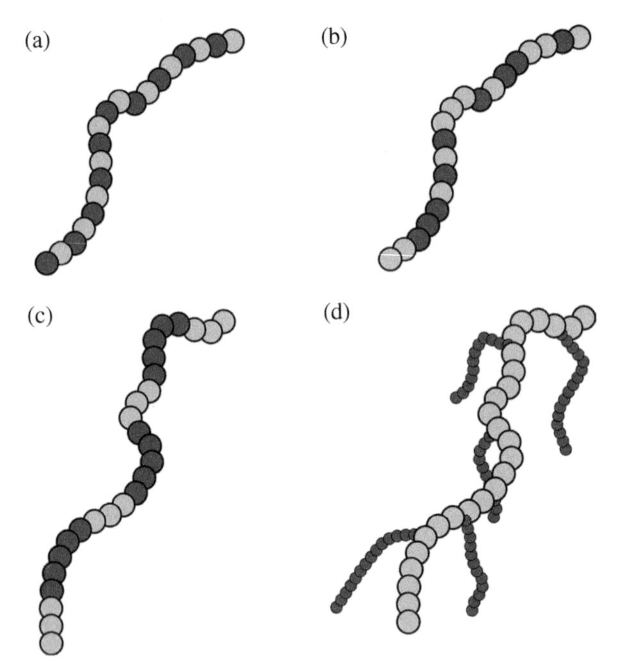

FIGURE 9.1 Types of copolymers based on structural assembly of the monomer units; (a) is alternating, (b) is random, (c) is block, and (d) is grafted. Dark and light spheres represent different monomer building blocks.

CH_3—CH_2—CH_2—CH_3 is a polymer. It is to avoid such a problem that Paul Flory [6] talks about high polymers:

> Such characteristic properties as high viscosity, long-range elasticity, and high strength are direct consequences of the size and constitution of the covalent structures of high polymers. Intermolecular forces profoundly influence the properties of high polymers, just as they do those of monomeric compounds, but they are not primarily responsible for the characteristics which distinguish polymers from their molecularly simple analogs. [6]

Others—like Roald Hoffmann in his poem—distinguish monomers, **oligomers** (short chains, solid at room temperature), and polymers (equivalent to Paul Flory's high polymers).

How about thermosets? First, a short story. When Great Britain had under its power the present country of Malaysia, large amounts of natural rubber were being brought to Britain by ships. In 1824 an enterprising Scottish gentleman named Charles Macintosh (no connection to computers or to fruits) dipped a dozen overcoats in a vat of natural rubber, thus creating the first impermeable raincoats (still called macs in the United Kingdom of today). However, a disaster ensued on a first-rainy-but-later-sunny day: the impregnation melted in the sun, and a proud owner of the new expensive overcoat had rubber on his suit, on his shoes, and also on the street pavement. This was before 1839 when the American Charles Goodyear invented the **vulcanization** process.

What exactly happens during vulcanization? Goodyear's invention consisted in adding sulfur to natural rubber and thus causing **crosslinking** of the polymeric chains; see

```
 H   H   H   H   H   H   H   H
 |   |   |   |   |   |   |   |
-C - C = C - C - C - C = C - C -
 |           |   |           |
 H           H   H           H
   Sulfur→         ↘
      H           H   H           H
      |           |   |           |
     -C - C = C - C - C - C = C - C -
      |   |   |   |   |   |   |   |
      H   H ↑ H   H   H   H   H   H
            └── Break unsaturated bond

 H   H   H   H   H   H   H   H
 |   |   |   |   |   |   |   |
-C - C = C - C - C - C = C - C -
 |           |   |           |
 H           H   S           H
                 |
                 S
                 |
                 S
                 |
             H   S   H   H   H   H   H   H
             |   |   |   |   |   |   |   |
            -C - C - C - C - C - C = C - C -
             |   |   |   |   |   |   |   |
             H   H   H   H   H   H   H   H
```

FIGURE 9.2 Vulcanization of natural rubber.

FIGURE 9.3 Rubber bicycle tire. *Source*: Parhamr, https://commons.wikimedia.org/wiki/File:Bicycle_tire_-_Cheng_Shin.jpg. Public Domain.

Figure 9.2. No matter how many polymeric chains we had to start with, after successful vulcanization we have only one. Now let us look at a car tire (called *tyre* in the United Kingdom) in Figure 9.3. The color comes from 30 wt.% of carbon black filler, not from rubber. Laymen, for instance, changing tires mostly do not realize that they have in their hands just one polymeric molecule, or one **macromolecule**.

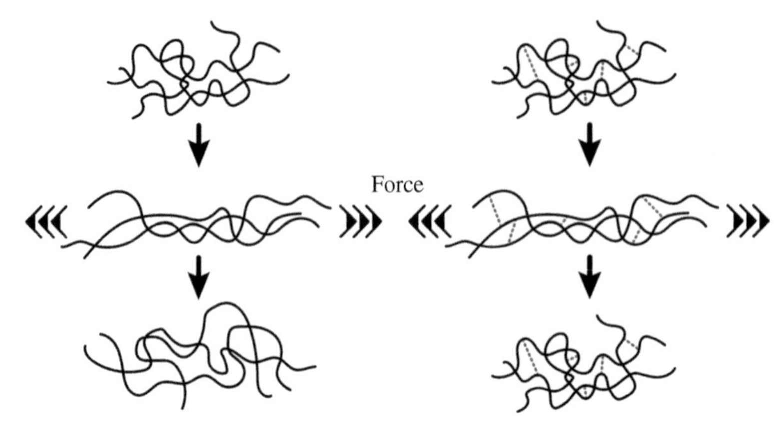

FIGURE 9.4 Thermoplastic (left) and thermoset (right) subjected to a horizontal tensile force.

Is vulcanization of natural rubber an exceptional process? Actually, it is just one example of **polymer crosslinking**, also called **polymer curing**. Always a crosslinking agent—such as sulfur for natural rubber—is needed. When one wants to buy a strong glue and goes to a hardware store, one usually will come out with an epoxy—in *two* bottles. One of them contains a curing agent.

How does crosslinking affect properties? Consider Figure 9.4 where we are showing on the left a thermoplastic and on the right a thermoset. We see in that Figure a difference in behavior. After application of a tensile force, the thermoset returns to the original size and shape (almost or even fully), while a thermoplastic does not.

Now we can come back to the issue raised in the beginning of this section: what happens when we heat a thermoset, trying to melt it? When a non-polymeric material is heated, at some temperature the thermal energy of individual molecules becomes higher than the energy of attractive interactions (the basic MSE triangle at work again). The molecules "break loose" and begin moving around individually. It is at this point that the **molten state** appears; *each molecule can "swim" in a different direction*. The attractive forces still exist and prevent the molecules from running at high speeds into the air—which is why we have a liquid and not a gas. Melting is not possible in a thermoset; there is only one molecule. If we recklessly keep on heating a thermoset trying to melt it at any price, we shall eventually *burn it* instead.

Is crosslinking (curing, vulcanization) with a chemical agent (sulfur, epoxy hardener) the only way to obtain thermosets? There are at least three more routes:

1. irradiation, such as gamma rays or ultraviolet (UV) light. The irradiation intensity as well as exposure time have to be optimized. Too strong or too long exposure can result in chain scission *in addition* to crosslinking;
2. microwaves;
3. laser light.

Finally, we need to consider **elastomers**, or polymers having high "elasticity" and very high failure strain compared to other materials. Historically these were always thermosets

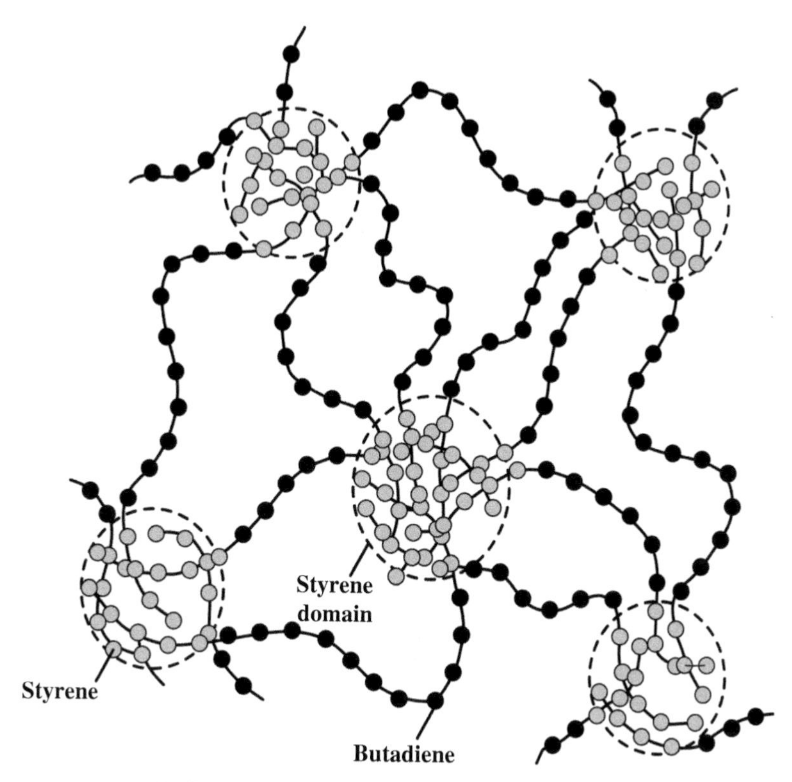

FIGURE 9.5 Styrene/butadiene copolymer.

like rubber, but there are increasingly more **thermoplastic elastomers**. It is understandable from Figure 9.4 how the crosslinks in thermosets enable a material, after stretching, to return to its original shape or configuration. To understand how the same phenomenon can occur in thermoplastic polymers, which like a plate of spaghetti consist of many unconnected strands or chains, we must refer to Figure 9.5. Since the copolymer in Figure 9.5 is a thermoplastic, there are no crosslinks. There are, however, regions of multiple entanglements (shaded circles in the Figure). When we apply a force to the material, for instance, a tensile one, *the entanglements behave as if they were crosslinks* preventing too large an extension and facilitating a recovery. There is in this scenario an added advantage from the point of view of manufacturing: scrap from a bad run can be made molten and a new thermoplastic elastomer made. Again, thermoplastic elastomers behave in operation similarly as thermosets, illustrated in the right hand side of Figure 9.4. There is another advantage; conventional rubber regimes require two processing steps: shaping (molding) and vulcanization. Thermoplastic elastomers require only one.

Note that while all thermosets are crosslinked and behave as illustrated in Figure 9.4, not all thermosets are elastomers. That classification is reserved for polymers which can undergo very large extension (or strain) without simultaneous permanent deformation. Thus we are familiar with various types of rubber as elastomers, while epoxies are typically rather brittle and not elastomeric.

9.4 POLYMERIZATION PROCESSES

Polymerization—the process of synthesizing a polymer from starting materials—can be performed in a variety of media:

- in the gas phase;
- on fluidized beds: there is a gas present as in the preceding option but it is flowing. Also usually a catalyst sits on the bed;
- in the bulk, only monomer present;
- in aqueous solutions, allowing easy temperature control. See Figure 9.6;
- in non-aqueous solutions. The mixing of miscible solvents is key for control of polymerization kinetics. Care in the use of solvents in non-aqueous solutions is not trivial owing to environmental concerns, for instance, to avoid the breathing of solvent fumes by plant workers. For example, this comes into play with Kevlar, which is made in solution of a very corrosive solvent;
- in suspension, typically aqueous, with the suspension (for instance, oil-type) particles "swimming" in water;
- in emulsion, a suspension consisting of an immiscible liquid dispersed and held in another liquid by an emulsifier. Low overall viscosity is an advantage. See a more advanced description in the shaded box.

By these various polymerization methods we obtain a variety of polymer types. The major categories are listed in Figure 9.7, which shows also the chemical bases for the different polymer types and names some familiar applications. Note that Figure 9.7 provides a comprehensive though not exhaustive list of known polymer types.

We shall now briefly discuss Nylons and then Kevlar. **Nylons** are a family of **aliphatic polyamides**, with two representatives shown in Figure 9.8. A Nylon was first produced on February 28, 1935, by Wallace Carothers at the E.I. du Pont de Nemours & Co. Experimental Station, Wilmington, Delaware. Soon thereafter a young researcher named Paul J. Flory became a collaborator of Carothers. Later Flory went to Cornell, finally settling at Stanford, where he received a Nobel Prize in Chemistry for providing much understanding of polymer behavior; in turn, one of us (W.B.) was a student of Flory at Stanford. During World War II, Nylon was first used for parachutes and flak vests since silk was scarce.

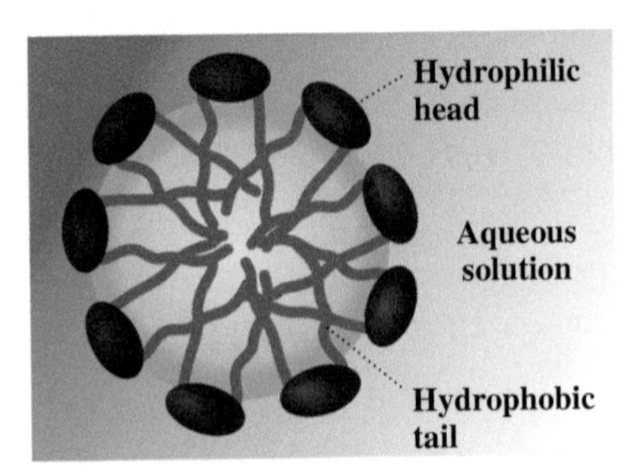

FIGURE 9.6 Scheme of a micelle formed by surfactant molecules in an aqueous solution. *Source*: After SuperManu, https://commons.wikimedia.org/wiki/File:Micelle_scheme-nl.svg. Used under CC-BY-SA-3.0, 2.5, 2.0, 1.0, GFDL.

Polymer	Functional Units	Typical Applications
Phenolics	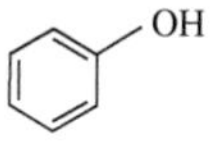 Phenol (C_6H_5OH)	Adhesives, coating, laminates
Amines	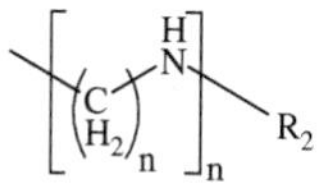 $-[(CH_2)_n-NH]_n-R_2$	Adhesive, cookware, electrical moldings
Polyesters	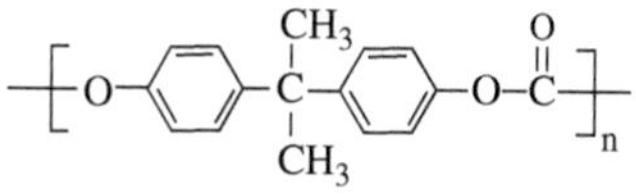 $-[O-C_6H_4-C(CH_3)_2-C_6H_4-O-C(=O)]_n-$	Electrical molding, decorative laminates, polymer matrix in fiberglass
Epoxies	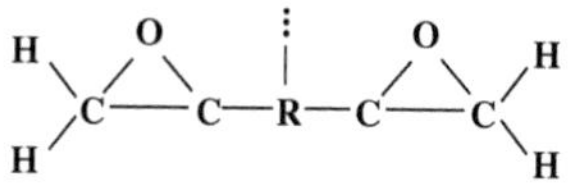 $H_2C(O)C-C-R-C-C(O)H_2$ (epoxide rings on both ends)	Adhesives, electrical moldings, matrix for composites
Urethanes	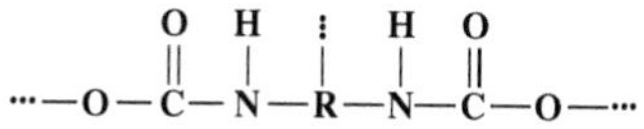 $\cdots-O-C(=O)-NH-R-NH-C(=O)-O-\cdots$	Fibers, coatings, foams, insulation
Silicone	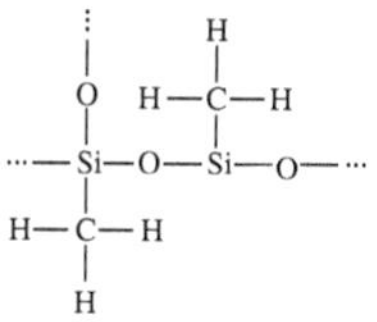 $\cdots-Si(O\cdots)(CH_3)-O-Si(CH_3)-O-\cdots$	Adhesives gaskets, sealants
Polyolefins	alkenes, C_nH_{2n}	Electrical insulation sleeves, rash guards, flexible foams, seat cushing, roofing
Polyamides	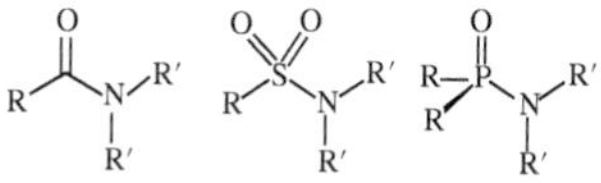 $R-C(=O)-NR'R'$; $R-S(=O)_2-NR'R'$; $R_2P(=O)-NR'R'$	(e.g., silk, wool, nylon) Textiles, carpets, ropes, parachutes, instrument strings
Teflon®≡ poly(tetrafluoro-ethylene) (PTFE)	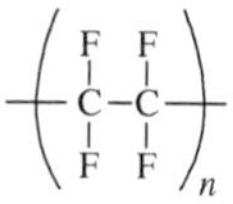 $-(CF_2-CF_2)_n-$	Low friction surfaces and coatings
Poly(ethylene-terephthalate) (PET)	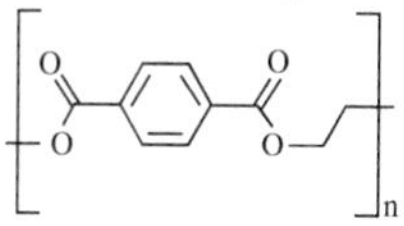 $-[O-C(=O)-C_6H_4-C(=O)-O-CH_2CH_2]_n-$	Fibers, beverage containers
Polypropylene (PP)	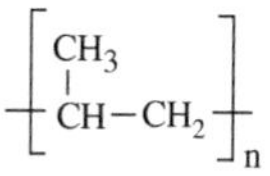 $-[CH(CH_3)-CH_2]_n-$	(a polyolefin) Pipes, furniture, cables, hinges, storage containers
Poly(vinyl chloride) (PVC)	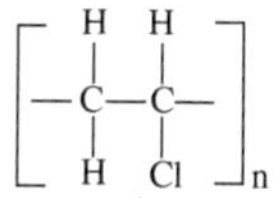 $[-CH_2-CHCl-]_n$	Pipes, construction materials, clothing
Polystyrene (PS)	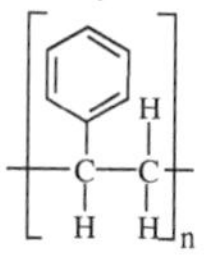 $-[CH(C_6H_5)-CH_2]_n-$	Insulating foams, disposable cutlery and dinnerware
Poly(vinyl alcohol) (PVA)	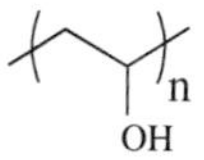 $-[CH_2-CH(OH)]_n-$	Paper adhesive, carbon dioxide barrier, mold release

FIGURE 9.7 Polymers or polymer classes, representative functional units, and typical applications.

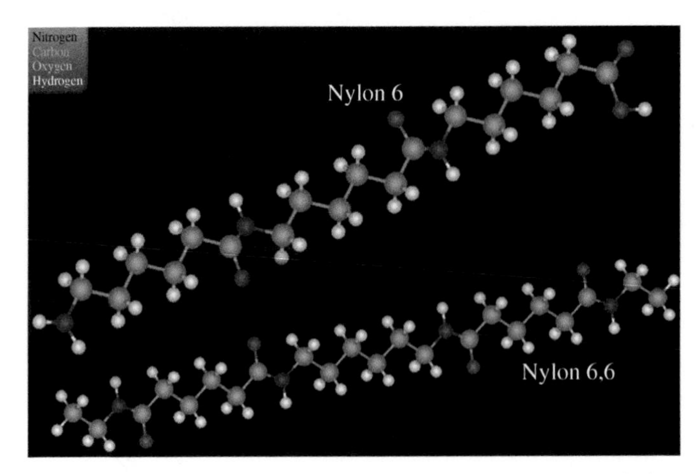

FIGURE 9.8 Structures of two Nylons. *Source*: Michael Ströck (mstroek) https://commons.wikimedia.org/wiki/File:Nylon6_and_Nylon_66.png. Used under CC-BY-SA-3.0.

After the War, Nylon was used for women's stockings. Now various Nylons are used for mechanical parts such as machine screws, gears, and other low- to medium-stress components previously cast in metal. Nylon fibers are used in clothing fabrics, packaging paper, carpets, musical strings, pipes, and ropes.

As for **Kevlar**, Stephanie Kwolek created it in 1965 also at the E.I. du Pont de Nemours Experimental Station in Wilmington, DE [7]. In her childhood Stephanie's father Jan Chwałek, an immigrant from Poland, instilled in her interest in natural sciences. Kwolek's own reminiscence of her discovery: "The solution was unusually low viscosity, turbid, stir-opalescent and buttermilk in appearance. Conventional polymer solutions are usually clear or translucent and have the viscosity of molasses, more or less. The solution that I prepared looked like a dispersion but was totally filterable through a fine pore filter. This was a liquid crystalline solution, but I did not know it at the time." We shall discuss liquid crystals in Chapter 12. This sort of cloudy solution that Stephanie Kwolek obtained was in those times usually thrown away. However, she persuaded technician Charles Smullen, who ran a spinneret, to test her solution. She was amazed to find that the new fiber would not break when Nylon typically would. The reaction leading to Kevlar is shown in Figure 9.9. Kevlar is five times stronger than steel.

We can guess why right after its creation Kevlar was compared to Nylon. It seems that since the discovery by Wallace Carothers everything made at du Pont was compared to Nylon. To the credit of the du Pont Company, they practically immediately assigned to Stephanie Kwolek a team of researchers to work under her direction on her new material. Very few women in those times participated in research in MSE—and practically never in managerial positions.

Kevlar is used in bulletproof vests, underwater cables, brake linings, space vehicles, boats, parachutes, skis, and building materials. The first application is the best known. "When you think about what she has done, it's incredible. There's literally thousands and thousands of people alive because of her," said Ron McBride, former manager of the Kevlar Survivors' Club, a not-for-profit partnership between DuPont and the International Association of Chiefs of Police [8]. The group has documented 3,200 lives saved through use of Kevlar in body armor.

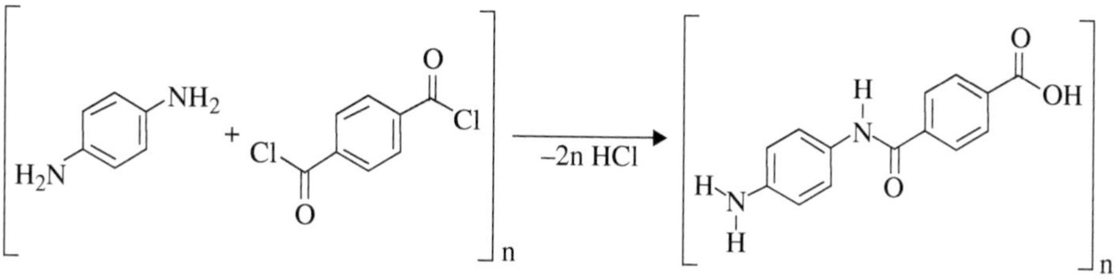

FIGURE 9.9 Synthesis of Kevlar from the monomers 1,4-phenylene-diamine (*para-phenylenediamine*) and terephthaloyl chloride. Hydrochloric acid is produced as a byproduct of the step growth (condensation) polymerization reaction. *Source*: cacycle, https://commons.wikimedia.org/wiki/File:Kevlar_chemical_synthesis.png. Used under CC-BY-SA-3.0.

We need to say also that polymers are often classified by the *users*, e.g., as "plastics" (typically thermoplastics), rubbers, glues, paints, etc.

There are two basic types of polymerization reactions:

- Step growth. All the chains grow at approximately the same rate. Therefore, there are no long chains in the beginning of the polymerization process. Nor are there short chains when the reaction is close to completion. Synthesis of polyesters, which occurs in concert with water formation, is an example. Hence the old name: *condensation polymerization.* A step-growth mechanism involves functional monomers which react to form dimers, trimers, etc. For example, carboxyl termination and amine termination link two monomers forming amide functionality.
- Chain growth. In this case there are several stages: initiation, propagation and termination. Polymer chains grow with the subsequent addition of monomers to the chain end. The chain end carries either a charge (positive or negative) or a radical. Each chain is initiated individually and grows very rapidly to a high molecular mass until its growth is terminated; see again the poem by Roald Hoffmann above. **Radical polymerization** is an example. A radical starts the chain formation and then runs along the chain—until it is destroyed and then the chain growth ends. However, there are several kinds of chain growth polymerization reactions: radical, cationic, anionic, or coordination. In all cases we might have some long chains after only one second (or less) as well as some short ones even at the reaction's endpoint (i.e., the radical got killed before a chain grew long). Synthesis of vinyl polymers (more on such polymers in Sections 9.6 and 9.7) such as poly(vinyl chloride) (PVC) is an example.

EMULSION POLYMERIZATION

Recall our discussion of interfaces in Section 2.5. Energetically speaking, interfaces are important because the enthalpy and entropy associated with molecular rearrangements and reactions at an interface are different from those of the bulk. A review of effects of interfaces on properties of multiphase polymer systems has been provided by Kopczynska and Ehrenstein [9].

Before we apply this to the case of emulsion polymerization, consider the nature of a surfactant, for instance, a detergent. The hydrophobic tail of a molecule of detergent attracts the dirt and greasy grime from our hands. The polar head, an ionic salt, dissolves in water and takes with it the nonpolar tail and attached grime. As a further thought exercise, ask yourself what will happen to a detergent in ocean water?

Now, when surfactant molecules congregate with their tails inwards and polar head groups outwards, a structure called a micelle forms. See Figure 9.6 again for a schematic. A micelle can therefore be suspended in an aqueous solution, in spite of the hydrophobicity of the molecular tails. This has useful consequences, since a hydrophobic monomer or other nonpolar compound can be held inside the micelle.

Now we can explain better the technique of emulsion polymerization, which requires dissolving a surfactant in water until reaching the critical micelle concentration. Emulsion polymerization occurs only by free radical initiation. The key ingredients needed are: water, a surfactant, a water-soluble initiator, and a water-insoluble monomer capable of polymerization by free radicals (a step-growth process).

Once everything is thrown in the pot, monomer can be found in a few different places, mainly in large droplets floating around aimlessly or localized in micelles. Initiation of the reaction occurs when the initiator diffuses into the micelle and reacts with a monomer. At this point, the micelle is referred to as a particle. The polymer particles can grow, while monomer from the large droplets migrates into the micellar particles to feed the reaction. Each micelle can be considered as a miniature bulk polymerization, while the water serves as a heat sink during the reactions. Unlike bulk polymerization, decreasing the initiator concentration increases the molecular weight and rate of polymerization. A synthetic latex is defined as a colloidal dispersion of *polymer* particles in an aqueous medium. Of course the term first referred to the white sticky sap from the rubber tree. Latex paints are a familiar example of the synthetic variety; they contain between 40% and 60% polymer solids. Poly(vinyl acetate), polychloroprene, polymethacrylates, poly(vinyl chloride), polyacrylamide, and copolymers of polystyrene, polybutadiene, and polyacrylonitrile are made commercially by emulsion polymerization.

The termination step in chain growth polymers is not insignificant. Various factors affect the energy associated with this stage of the reaction. For example, diffusion plays a role in the terminating reaction of radicals located at chain ends [10]. The primary termination processes are combination, disproportionation, and chain transfer. Reaction with impurities can also (rarely) terminate chain growth. Combination and disproportionation occur in radical polymerization processes; the former is when radical chain ends couple together, and the latter is when the pair of radical chain ends forms two inactive molecules via hydrogen transfer. Chain transfer termination occurs when an active end is transferred to another species (chain, monomer, initiator, solvent).

Directing a polymer synthesis reaction in the desired direction can be tricky; let us provide a now "classical" example. Penczek, Slazak, and Duda [11] performed anionic copolymerization of propylene sulfide with elementary sulfur to obtain polysulfides with the structure $-CH_2-CH(CH_3)-S_x-$; they reported $x=12$. Aliev, Krentsel, and coworkers [12] tried something similar, but they always got $x \leq 2$. Therefore, the latter claimed that the elementary analysis performed by Penczek and coworkers gave false results because of the presence of elementary sulfur. Penczek and coworkers replied [13] that they have obtained transparent films of their polymer, while the presence of sulfur would render the films opaque. Moreover, the Raman spectra demonstrated the absence of elementary sulfur in their polysulfides.

Polymerization reactions can also be initiated and controlled by light; such procedures are referred to as photoinitiated and photocontrolled reactions. In the former, light initiates polymer growth. In the latter, light mediates each monomer addition. As with other types of reactions, photoinitiated polymerizations can involve radicals, cations, or anions. Photocontrolled reactions typically utilize one of three modes of activation: chain-end, monomer, and catalyst. The possibilities for controlling the rate of polymerization and the resultant molar mass and mass distribution of the synthesized products are among the key attractions of such techniques.

9.5 MOLECULAR MASS DISTRIBUTION

Before dealing with macromolecules, consider a set of circular discs: 9 discs with the diameter $d = 1.0$ cm and 1 disc with the diameter $d = 91.0$ cm. Thus, the average diameter $d_{av} = 10.0$ cm. From that average alone, one would expect disks with diameters such as 9, 11, or 13 cm. However, none of the discs has its diameter anywhere near the average of 10 cm. The respective square value is $d_{av}^2 = 100\ \text{cm}^2$. However, the average of the diameters square $d_{av}^2 = (9 \cdot 1^2 + 1 \cdot 91^2)/10 = 829\ \text{cm}^2$, very far from $100\ \text{cm}^2$. It is only from the *combination* of the two values d_{av} and d_{av}^2 that we get an idea that actual values of the diameters are not close to 10 cm.

What applies to disc diameters applies also to lengths of macromolecular chains. We are now better prepared to deal with the following definitions:

- The **number-average molecular mass** is

$$M_n = \frac{\sum_i N_i M_i}{\sum_i N_i} \tag{9.1}$$

  Here N_i is the number of molecules with the molar mass M_i; N_i is the numerical fraction of molecules with the molar mass M_i. Similar to the scenario of discs with different diameters, the number-average is sensitive to the concentration of species having low molecular weights.

- The **weight-average (mass average) molecular mass** is

$$M_w = \frac{\sum_i N_i M_i^2}{\sum_i N_i M_i} \tag{9.2}$$

  Thus Eq. (9.2) gives us the average of the squares.

- There are also higher averages such as the so-called **Z-average molecular mass**

$$M_z = \frac{\sum_i N_i M_i^3}{\sum_i N_i M_i^2} \tag{9.3}$$

By a similar pattern, using higher powers of the molecular mass, one can define the $(z+1)$, $(z+2)$ etc. averages. The more averages we have, the better we can characterize a given polymer in terms of the distribution of its molecular masses (or of chain lengths, or of degrees of polymerization). If we have the first two, we define the **dispersity** as

$$D = \frac{M_w}{M_n} \tag{9.4}$$

The value of D is greater than or equal to 1, where for all chains of equal length we get $D=1$. The old name of dispersity: polydispersity index. Let us now go back to the two types of polymerization reactions, step growth and chain growth. You can figure out which results in higher dispersity values.

9.6 MOLECULAR STRUCTURES OF IMPORTANT POLYMERS

The variety of polymers and classes of polymers are extensive. Roald Hoffmann talks in his poem about polypropylene, Teflon, epoxy, polythiophenes, DNA… Let us now have a look at important examples of synthetic polymers, listed in Figure 9.7. The last four polymers are known as **vinyls**; they have a general formula $(CH_2{-}CHR)_n$.

Among polymers listed is poly(ethylene terephthalate) (PET), used in soft drink bottles since it provides a barrier material. What is a carbon dioxide barrier? The soft drink contained in a 2L bottle made of PET would become "flat," that is CO_2 would escape through the walls, if not for the presence of a **barrier material**.

9.7 SPATIAL STRUCTURES OF MACROMOLECULES AND ASSOCIATED PROPERTIES

Consider a chain consisting of carbon atoms (with hydrogen or some other atoms attached too, since carbon has valence=4; that is, it forms chemical bonds with four other atoms). Carbon atoms can also form double bonds in compounds called alkenes or triple bonds in compounds called alkynes. As we expect from Section 2.2, C—C bonds are the longest, C═C bonds are shorter, and C≡C bonds are the shortest.

For simplicity, let us look now at single bonds only, as in the Nylon chains seen in Figure 9.8. Look at those segments where we have only carbon atoms in the main chain—with covalent bonds between the atoms. We see that such chains are <u>not linear</u>. Textbooks of Organic Chemistry tell us that, when fully bonded to other atoms, the four bonds of the carbon atom are directed to the corners of a tetrahedron and make angles of about 109.5° with each other. Is this true? Well, most of the time it is. If a mechanical force causes a deformation, that angle can change. Steric hindrance, also a topic of Organic Chemistry, can modify that angle slightly as well.

The fact that the angle between consecutive carbon atoms in a chain is at or not far from 109.5° has consequences. Consider the increasing length of chain shown in Figure 9.10. Let us look at part (a) of the Figure 9.10. No mechanical force is applied. If the left and middle atoms do not move, the right atom can be found on the circle marked by a dotted line. In a polymeric chain we have many segments; the situation shown in Figure 9.10b is possible, but situations such as Figure 9.10c when there is no approximate straight line running through the centers of the atoms is more probable.

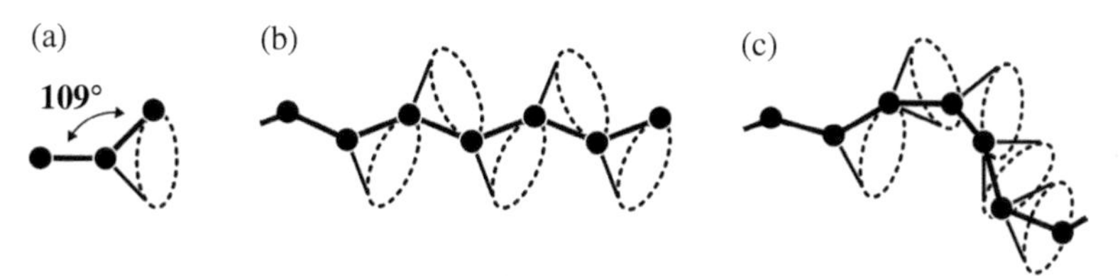

FIGURE 9.10 Carbonic chains with single bonds only. Schematics (b) and (c) illustrate possible structural formations given the natural bond angle shown in (a) for carbon in single bonds.

The longer the chain, the more conformations it can have. *Reversals* of chain direction are possible too, such that the chain essentially loops back on itself. We now need to consider the **random walk problem**, also called **the drunk walk problem** [5, 6, 14]. The second name comes from the fact that a drunkard just out of a restaurant door constitutes a *memoryless system*. He or she does not know anymore where home is. Therefore, the probability of making the first step to the right equals the probability of making the first step to the left which is equal to ½. Sometime later the drunk was found hanging to a street lamp, apparently unable to walk any further. The distance from the restaurant door to that street lamp is much shorter than the sum of the steps taken by the drunk. If we pursue the path of the drunk—and also remember that a polymer chain behaves similarly—we get an important result. The average **end-to-end distance** r_e for a given polymer is not proportional to the average degree of polymerization n, but to *the square root of* n (where n is the number of monomeric units in a polymer). We already know the reason for this: reversals of direction—which occur to the drunk as well as to polymer chains.

The above discussion along with Figure 9.10 deals with conformations of polymeric chains. **Conformational** changes involve rotations around single bonds. A conformation of segments (usually chemical groups) characterizes the geometric state of the molecule. Changes in conformation arise from variations in factors such as stress, temperature, or pressure. We have to distinguish conformations from **configurational** changes which involve "permanent" differences and cannot be altered without breaking chemical bonds. Often double (or sometimes) triple bonds are the reason that a given structure cannot change conformation without bond breaking.

As an example, let us now consider **gutta-percha**, a term used both for a family of trees and for sap of these trees. As a chemical compound, the sap is *trans*-1,4-polyisoprene:

CH_3 CH_3 CH_3 CH_3

It is evident that the **trans** conformation of 1,4-polyisoprene is a regular structure that tends towards linearity. The gutta-percha chains are practically at maximum extension; they cannot be made any longer. As a consequence of these structural features, gutta-percha is predominantly crystalline and is a fairly rigid material. Heavy use of gutta-percha during the second half of the 19th century, particularly as insulation for underwater telegraph cables, led to unsustainable harvesting and a collapse of the supply.

We spoke earlier of rubber in the context of its elasticity and of its use for Macintosh raincoats. Natural rubber, derived primarily from the Pará rubber tree, is *cis*-1,4-polyisoprene:

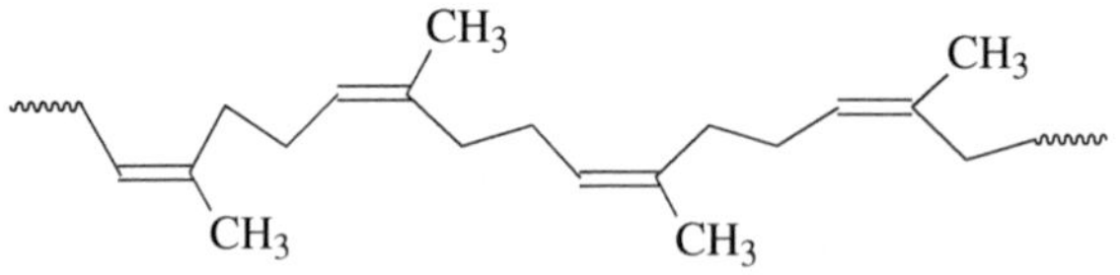

We see the irregularity of the structure and that the *cis* form can be extended—as anybody who has dealt with natural rubber knows. Thus the same chemical formula but in a different spatial arrangement results in different properties. We know that gutta-percha had its glory period; by contrast, the uses of natural rubber are expanding. Chin Han Chan and her colleagues [15] are working on solid polymer electrolytes based on modified natural rubber.

In connection with the polymers listed in Figure 9.7, we have talked above about vinyl polymers $(CH_2{-}CHR)_n$. Imagine a projection from three dimensions of such a polymer onto a two-dimensional plane. If all R groups are at the top of such a figure (or all at the bottom), we have an **isotactic** vinyl polymer. If the R substituents alternate top-bottom-top-bottom etc., we have a **syndiotactic** polymer. If the spatial distribution of the substituents is random, we have an **atactic** vinyl polymer; polystyrene is almost always atactic. The tacticity of polymers is shown schematically by 2D projections in Figure 9.11. You can figure out which of these three forms can crystallize.

There is a category of polymeric structures called **dendrimers** (dendron means tree in ancient Greek) whose general formation is illustrated by the schematic in Figure 9.12. Clearly the molecular structure of dendrimers is closer to **hyperbranched polymers** (see Figure 9.13) than to polymeric snakes (see short discussion in Section 9.9). Both dendrimers and hyperbranched polymers (a sub-class of dendrimers) are densely branched polymers with a large number of end-groups. While dendrimers have completely branched star-like topologies, hyperbranched polymers are characterized by imperfectly branched irregular structures.

Given the radial-type symmetry of dendrimers, with every "shell" (technically referred to as a generation), there are more and more functional groups on the outside, and those

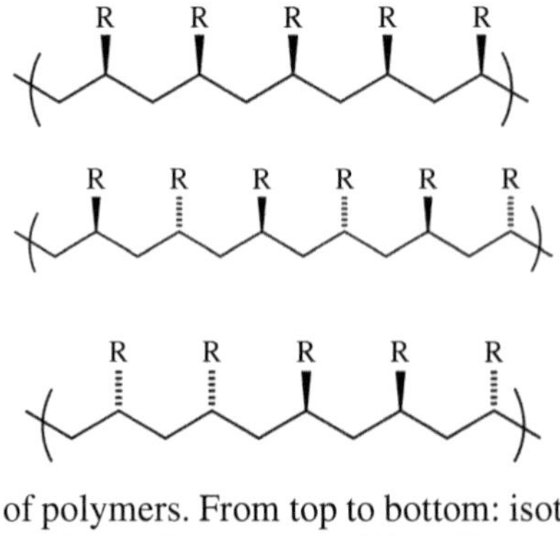

FIGURE 9.11 Tacticity of polymers. From top to bottom: isotactic, syndiotactic, atactic. *Source*: Benjah-bmmm27, https://commons.wikimedia.org/wiki/File:Isotactic-A-2D-skeletal.png, https://commons.wikimedia.org/wiki/File:Syndiotactic-2D-skeletal.png, and https://commons.wikimedia.org/wiki/File:Atactic-2D-skeletal.png. Public Domain.

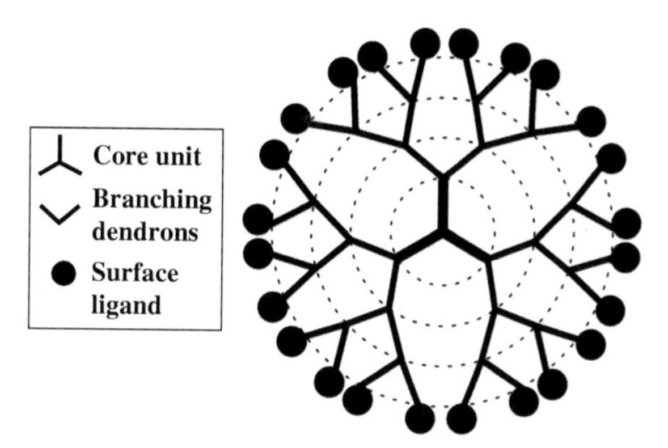

FIGURE 9.12 A dendrimer polymer; note symmetry with respect to the center.

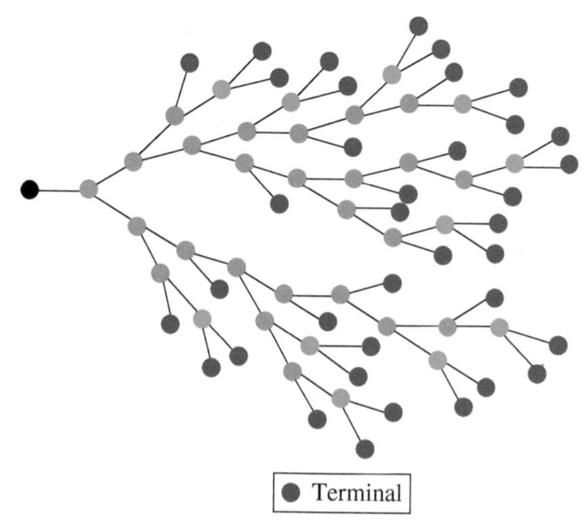

FIGURE 9.13 A hyperbranched polymer chain. *Source*: Reprinted by permission from Macmillan Publishers Ltd.: M.-A. Kakimoto, S.J. Grunzinger, & T. Hayakawa, "Hyperbranched poly(ether sulfone)s: preparation and application to ion-exchange membranes", *Polymer J.* **2010**, *42*, 697–705.

groups can react with a variety of agents. Thus, dendrimers are sought after for use as targeting materials, detecting materials (for instance, in dyes), imaging agents, and active pharmaceutical compounds. Synthesis of the first such polymer is credited to Buchleier, Wehner, and Vögtle [16], with further significant developments by Tomalla and coworkers [17] and also by Feast and coworkers [18].

Dendrimers were not synthesized before 1978 owing to the difficulty associated with the synthesis. Unusual tricks have to be applied to achieve the symmetry seen in Figure 9.12. Feast, Rannard, and Stoddard describe synthesis of aliphatic polyurethane dendrimers [18]. Above all, an appropriate core molecule has to be used. Because of the high probability of unwanted reactions during synthesis, molecules of diethylenetriamine have to be protected at the outset from those reactions. Later in the process the amine functionality has to be "liberated" [18] for the desired reaction; or in other words the protection has to be removed.

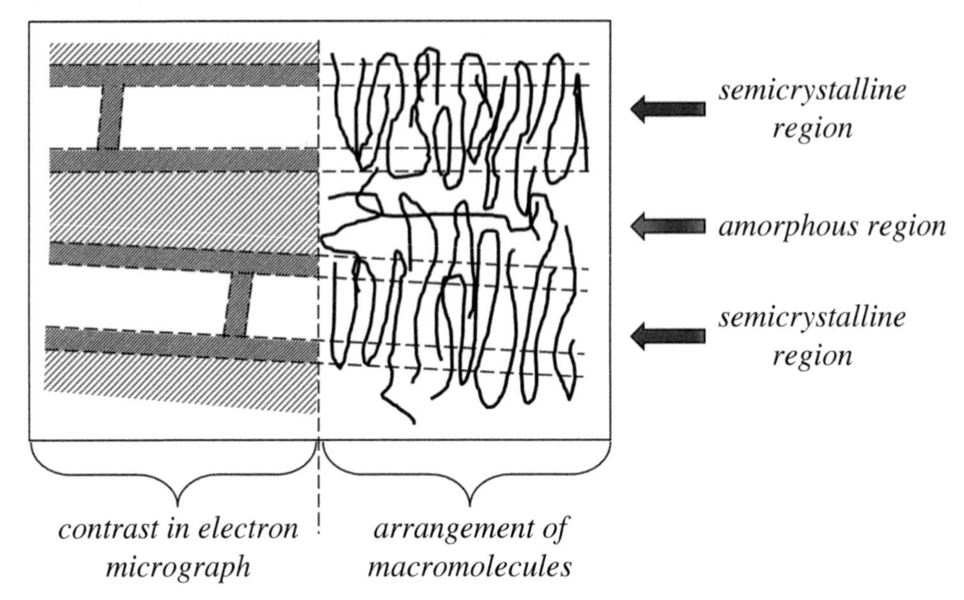

FIGURE 9.14 Structure of a semicrystalline polymer. Left: contrast in electron micrograph. Right: two crystalline regions with an amorphous region in the middle.

We have noted above that polymeric materials are never fully crystalline. Considering the bottom scenario of Figure 9.11, one can imagine why. It would be quite difficult to "persuade" a set of polymeric chains to create identical crystal cells. The results are that some polymers are fully amorphous and some semi-crystalline. There is a whole book by Michler and Baltá-Calleja [19] devoted to nano- and micro- structures and mechanics of polymers. In semi-crystalline polymers we have crystalline lamellae embedded in an amorphous matrix; see Figure 9.14. Actually, crystalline regions can form not only flat lamellae. A biodegradable polyester called polycaprolactone normally forms *six-sided crystals* [20], but in a solution of a supercritical carbon dioxide and chloroform, they can form rectangular crystals [21]. Crystalline regions can also form circular shapes called **spherulites** [21]. Analogous to different crystalline phases of metals (α, β, γ, and so on), there are different phases of crystalline polymers, for instance, α and β forms of isotactic polypropylene [22].

Still more complicated structures exist. Above we have carefully contrasted thermoplastics and thermosets. Vulcanized rubber—and other crosslinked polymers—can be considered as a polymer network. However, we have also **interpenetrating polymer networks** (IPNs), materials comprising two or more networks which are at least partially interlaced but not covalently bonded to each other. An IPN cannot be separated unless chemical bonds are broken. **Semi-interpenetrating polymer networks** (SIPNs) contain linear or branched polymers plus at least one network. Thus, they can be considered as polymer blends such that, in principle, the constituent polymer network(s) can be separated without breaking chemical bonds. The advantage of SIPNs is that they exhibit dual thermoplastic and thermoset nature [23]. **Gels** are an important subset of network polymers. These jelly-like materials consist of a solid continuous-phase network containing a dispersed liquid phase throughout. The term **hydrogel** refers to gels in which water is the swelling agent.

In Chapter 7 we described the sol-gel technique as a method to fabricate metal oxides. Polymers are often used in the sol gel process to facilitate gelation of the liquid sol. The sol is a colloidal suspension—of nanoparticles suspended in a liquid—that acts as a precursor for an integrated network (gel) of the discrete particles. Often metal alkoxides, formed by the reaction of monomers with the solid metal particles, serve as the basic unit for network formation. The transition from sol (liquid) to gel (immobile, non-flowing) is called gelation. The point in time when the particle network extends across the entire volume of the liquid causing it to immobilize is called the **gel point**. The time required for a gel to form after mixing all the components together is called the **gel time**. When the solid network is isolated from the liquid medium, we have an **aerogel**: a solid with air pockets dispersed throughout.

We have mentioned **swelling** in Section 9.1; it means typically absorption of a liquid or more rarely of a vapor causing an increase of a material volume. **Die swelling** can occur in processing when an extruded material leaves a die, the main cause being stresses imposed, while the material was inside of the die. As in so many other issues concerning polymers, important understanding of the phenomenon has been provided by Flory [6, 24]. He assumed that the thermodynamic effects represented by the change on mixing of the Gibbs function G^{mix} or of the Helmholtz function A^{mix} are independent of the mechanical effects (in which elasticity plays a role). This assumption provides good results when confronted with experiments [24, 25]. In particular, in binary natural rubber + organic solvents systems, the number of network chains per unit volume is independent of swelling liquid [25]. We have mentioned barrier properties above; diffusivity of n-hexane in natural rubber and in polyethylene has been related to swelling [26].

9.8 COMPUTER SIMULATION OF POLYMERS

In the chapters on metals and on ceramics we have noted the possibilities of explaining materials behavior by computer simulations. Polymers are more difficult to simulate since one has to create complicated chain structures [27–31] before subjecting them, for instance, to a tensile force. Apart from molecular dynamics, the **Monte Carlo** (MC) method [21, 31–33] is also used. The latter [32] was originally devised to make one move at a time, but it can simulate essentially any process consisting of random or biased/random events, such as creating a truly amorphous polymer system (a start for yet another MC or an MD simulation) or a random walk of a diffusing molecule. The MC method has an interesting history. In the late 1930s a group of mathematicians with Stanislaw Ulam in the city now known as Lviv tried to devise a method of scientifically "breaking the bank" in the casino in Monaco. They clearly have not succeeded, since the casino is still in business. Escaping from the Hitlerites, Ulam eventually landed in the Los Alamos National Laboratory in New Mexico [32], then a center of scientific effort to aid the allies in winning World War II. There the Monte Carlo method was deemed to have military importance; hence Ulam was not able to discuss it outside of the Laboratory walls. In 1941 the first programmable computer was built in Berlin by Konrad Zuse, a civil engineer, called Turing-complete Z3. Probably without knowledge of Zuse's work, in 1943 the first US made computer called ENIAC was built in Philadelphia; it contained 17,468 vacuum tubes, 7,200 crystal diodes, 1,500 relays, 70,000 resistors, and 10,000 capacitors. Even before Z3 or ENIAC became operational, Ulam perceived ahead of others capabilities that computers can provide, in general and in particular the capacity to utilize the Monte Carlo method. The ban on publication was eventually lifted, and the first description of the Monte Carlo simulation method was published in 1949 by Metropolis and Ulam [34].

Computer simulations of polymeric materials can serve a variety of objectives. With molecular dynamics one can address any process with a "random" or biased/random event, everything from creating a truly amorphous polymer system (a start for yet another MC simulation or an MD simulation) to a random walk of a diffusing molecule. We have mentioned above the importance of barrier properties. Gedde, Hedenqvist, and their colleagues studied those properties using both Monte Carlo and molecular dynamics methods [21, 35–39] and additionally confronting the results with experiments. In these studies, penetrants were molecules such as hexane. Among other results, penetrant-induced loosening of the interfacial structure between the crystalline and amorphous regions was found.

Let us now go back to Figure 9.14. As discussed by Nilsson *et al.* [21], the chains in the amorphous regions can be in:

- tight folds ("about face" turns at the surface of the crystalline region),
- statistical loops,
- loose chain ends called **cilia**,
- chains staying completely inside the amorphous region,
- **tie chains** connecting two crystalline regions, or
- trapped entanglements, chains connected via permanently entangled loops to the adjacent crystal.

It turns out that the concentration of trapped entanglements is higher than the concentration of tie chains; in earlier work trapped entanglements were neglected. Moreover, Monte Carlo simulations have shown that the thickness of the amorphous layer has a larger effect on the tie chain concentration than does the crystal thickness [21].

9.9 POLYMER SOLUTIONS

In Section 9.4 we looked at the variety of media in which polymers are synthesized. Solution methods constitute one route to polymer processing; other modes of processing will be discussed in the next Section. Kevlar production is expensive because of the difficulties that arise from using concentrated sulfuric acid, needed to keep the water-insoluble polymer in solution during its synthesis and spinning.

Beyond the aspects of processing, important characteristics can be determined from polymers in solutions. We discussed the connection between the average chain end-to-end distance r_e and the degree of polymerization n in Section 9.7. If experiments in solution, for instance, light scattering experiments, show that the end-to-end distance is larger than what is predicted by that relationship, then the meaning is that we are dealing with *semi-rigid chains*—in contrast to those presented in Figure 9.10. That is, some of these conformations, particularly those shown in Figure 9.10c, cannot be realized. Double bonds are here the usual suspect.

As demonstrated by Sabine Enders and her collaborators [40], the shape of macromolecular chains affects their behavior in liquid phases. Consider in particular the leaf-shaped hyperbranched polymers shown in Figure 9.13. Various theoretical models, usually expressed in terms of the Helmholtz function, are evaluated from the point of view of the ranges of their validity. On this basis, Enders and coworkers predict limited miscibility (existence of critical solution temperatures, see Section 3.12) not only in linear polymer+hyperbranched polymers systems but also in mixtures of two hyperbranched polymers. We have discussed hyperbranched polymers and dendrimers in Section 9.7. Both these classes are distinguished from long polymer chains, with or without branches; in terms of graph theory, chains without branches are called **snakes**.

At gasoline stations one can buy motor oil such as 10W-40. It is praised by the seller as having its viscosity η independent of temperature T. If true, how is this effect achieved? Viscosity is a function of temperature; as a result, oils undergo changes in viscosity with varying weather conditions. The changing viscosity is problematic, but can be overcome by the addition of "viscosity index improvers." These are typically polymers that, when added to oils, reduce the dependence of viscosity on T. At higher temperatures these viscous polymer compounds dissolve in the oil and thus increase the viscosity—counteracting the inherent decrease that occurs at elevated temperatures. The higher T, the more they dissolve, keeping $\eta \approx$ constant. At low temperatures the polymer precipitates out. Therefore the main oil constituents are responsible for the overall viscosity; and since the temperature is low, η is fairly high—and thus comparable to η at higher T with the polymer dissolved. Foundations of knowledge on polymer solutions are the subject of a book edited by William Forsman [41].

9.10 POLYMER PROCESSING AND THE ROLE OF ADDITIVES

There are many books and publications about polymer processing. Processing via solution for the thermoset Kevlar was already discussed in the previous Section. A summary of important methods for processing thermoplastics is provided in Figure 9.15. The process of blow molding, used especially for fabrication of beverage bottles (see the link in [42], for instance, for a short video of the process), is illustrated in more detail in Figure 9.16. The molds used for many polymer fabrication techniques are machined from metals. Familiar products ranging from cell phone housings to children's toys are made by injection molding, which in industry is a high throughput automated fabrication method. In general, the energy costs of heating polymers to the molten state are a primary factor in production expenditures. There are also a number of specialized techniques such as spin coating and dip coating used, for example, to apply coatings in microelectronics or other small device fabrication.

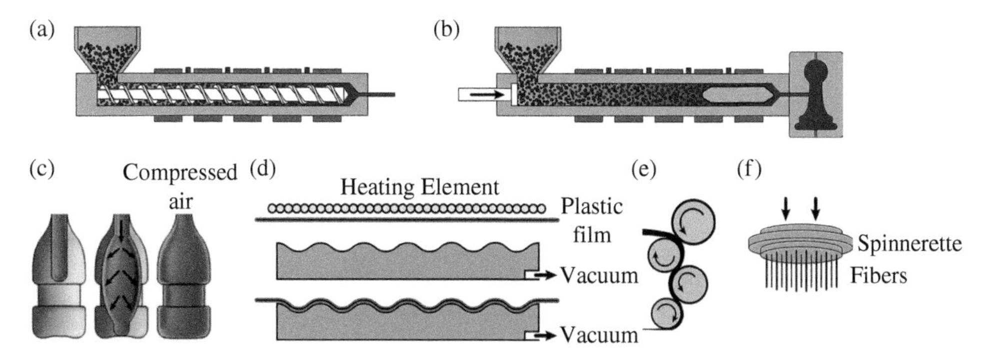

FIGURE 9.15 Key processing methods for thermoplastics (not to scale): (a) extrusion; (b) injection molding; (c) blow molding; (d) thermoforming; (e) calendaring; and (f) spinning.

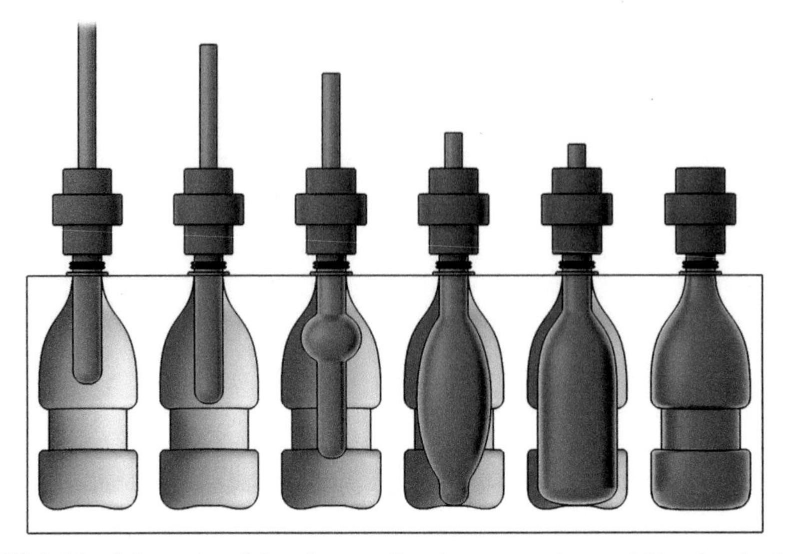

FIGURE 9.16 Schematic of the phases of parison extrusion and blowing in the blow molding process.

An important, but not yet mentioned, role in polymer processing is played by **additives** [43, 44]. They are used not only as processing aids. Here is a list of important classes of additives:

- Fillers (considered to be additives if their concentration is <2 wt.%)
- Pigments
- Stabilizers
- Antistatic Agents
- Flame Retardants
- Plasticizers
- Antiplasticizers
- Reinforcements
- Catalysts
- Slip and antiblock agents
- Nucleating agents
- Clarifiers
- Additives for hygiene
- Additives for aesthetic reasons

Most of the additives listed above are used for obvious reasons, but plasticizers and antiplasticizers require an explanation. If, during processing, a polymer has too high viscosity, one can either increase the temperature—with obvious energy costs—or else add a **plasticizer** to reduce the viscosity and facilitate flow of the material through the processing equipment. Some polymers in the molten state "flow like water," in which case **antiplasticizers** help to shape the products and maintain a manageable viscosity. As for additives used for hygiene, blood bags are made of PVC, and here an additive is used to *slow down the oozing out of a plasticizer from the bag walls into blood.*

Plasticizers are also frequently used in PVC. Rigid PVC pipes are very familiar owing to their frequent use in plumbing, among other applications. However, it may come as a surprise to know that flexible lawn and garden water hoses are also made of PVC. How can PVC be both rigid and flexible? The answer is plasticizers, which when added to PVC lower the glass transition temperature so that at ambient temperatures a PVC water hose is above the T_g and therefore in a rubbery state.

Among additives, there is a distinction based on geometry. Additives are typically categorized as particulate or fibrous. There is of course also some variety of particulate materials; they may be spheres, platelets, oblong, and so on, while fibers are distinguished by having fairly large length to width ratios. Additives are not to be confused with composites (the topic of Chapter 10). Additives are introduced into a material to modify one or more properties without significantly changing the overall characteristics of the material. On the other hand, the components of composites are combined specifically to get a new kind of material with properties better than either component alone. In composites, each component plays a major role in defining the properties and behavior of the whole, as explained with more detail in Chapter 10.

9.11 APPLICATIONS OF SPECIALTY POLYMERS

While Figure 9.7 provides information about application of massively used polymers, those polymers which are manufactured in smaller quantities are important as well. Some of them are manufactured in the powder form [45]. The majority of polymers are electrical insulators, but polymers (often multicomponent materials) that are *good electricity conductors* have their role. Pioneering work in this area was done by Gerhard Wegner [46, 47]; three other people were awarded the Nobel Prize for conducting polymers [48]. In some cases, quite low concentrations of a conducting additive suffice to impart conductivity to a polymer. Binary and ternary blends containing polyaniline dispersed in a matrix of one or two polymers have high electric conductivities [49] at concentrations of only 0.5 wt.% polyaniline. Carbon black can also provide conductivity to polymers [50].

In Section 9.7 we discussed polymer networks, including interpenetrating ones. **Polymeric foams** are also made on the basis of such networks [51–54], including inorganic-organic networks. As discussed by Jackovich and her colleagues [55], polymeric foams can be used as thermal and vibration insulators and also as energy absorption materials. They are also used in various biomedical applications.

Polymers are used extensively as coatings—for a variety of applications. One familiar kind of **polymeric coating** is Teflon (polytetrafluoroethylene) on frying pans. Protective coatings are used on appliances, as automotive primers, on pipes, on wires carrying electricity [56], as responsive film for the release of particles or molecules, on steel to lower friction and enhance wear resistance [57], as replacement of welding and riveting in aircraft and automobiles, as bonding materials for dental uses, in construction industry as flooring, paving, and airport runway repair, and also as force sensors [58].

As already mentioned, the performance of polymeric coatings seems determined by the **interphase**, that is the first few layers between the substrate and the bulk polymer, with the thickness of a few nm. Balzer, Micciulla, and their colleagues have determined the spatially resolved adhesion properties of the interphase in polyelectrolyte multilayers (PEMs) by the technique of desorbing a single polymer covalently bound to an atomic force microscope cantilever tip from PEMs with varying thickness [58]. The PEMs are so-called because each layer is a double layer consisting of two different polyelectrolytes. They have found that the adhesion of the first few layers (up to three double layers) is dominated by the surface potential of the substrate. Thicker PEMs are controlled by cohesion in between the PEM polymers, while in turn cohesion is determined by the local film conformation. These findings have been generalized by utilizing oligoelectrolyte multilayers—hydrophilic as well as hydrophobic—as coatings [58].

A subcategory of network polymers known as **superabsorbent polymers** (SAPs) is becoming increasingly prevalent. These polymers can absorb many times their own weight in water. Applications range from agricultural (e.g., for irrigation) to hygienic (e.g., disposable diapers) with many others in between. At the other end of the spectrum are superhydrophobic polymers—biologically inspired, for example, by the lotus leaf—used for extreme water-repellency.

Polymers are used as membranes for separation and filtration of liquids. There are also polymers that can be used as sensors for detection of organic liquid solvents and of their vapors [53]. Knowledge of liquid transport properties is required for this application.

We have to be careful in reading publications and reports on polymers since more than one name for the same term exists; in a few cases we have provided above "old names". Michael Hess has spearheaded an effort by the International Union of Pure & Applied Chemistry (IUPAC) to provide uniform terminology [59–61].

Given the variety of properties and versatility of polymers discussed in Section 9.1, the number of applications of polymers and PBMs *can only increase in the future*. At the same time, they are distinct from other classes of materials owing to their viscoelasticity. While metals and ceramics are primarily thought of as elastic materials, polymers exhibit simultaneously both viscous and elastic behavior. These distinctions are further explained in Chapters 13 and 14.

9.12 SELF-ASSESSMENT QUESTIONS

1. Discuss media in which polymerization can take place.
2. Discuss basic types of polymerization reactions. What are the differences in molecular mass distributions?
3. Why does epoxy glue always come in two bottles?
4. Why does Boeing Corp. make such a commotion about the plane they call Dreamliner?
5. Why can a car tire not be melted to make new tires?
6. What does "vinyl" mean?
7. Explain how multi-viscosity motor oil approximately preserves its viscosity in spite of temperature changes.

REFERENCES

1. R. Hoffmann, Oligopoem, in *Soliton*, Truman State University Press: Kirksville, MO **2002**.
2. R. Hoffmann, *Some Reasons to be Interested in Carbides, Invited Lecture at the 13th Annual POLYCHAR World Forum on Advanced Materials*, Singapore, July 3–8, **2005**.
3. R. Roy, Interdisciplinary materials research: The reluctant reformer of Western science, *Internat. Union Mater. Res. Societies Facets* **2005**, *4* (2), 18.
4. J.E.E. Baglin, Forum on Materials Education: Future directions, *Internat. Union Mater. Res. Societies Facets* **2005**, *4* (2), 23.
5. U.W. Gedde, *Polymer Physics*, 2nd edn., Springer/Kluwer: Dordrecht **2001**.

6. P.J. Flory, *Principles of Polymer Chemistry*, Cornell University Press: Ithaca, NY **1953**.
7. M. Bellis, *Kevlar—Stephanie Kwolek*, http://inventors.about.com/library/inventors/blkevlar.htm, accessed September 12, **2014**.
8. A. Nathans, Kevlar inventor Stephanie Kwolek, 90, dies, *News Journal*, http://www.usatoday.com/story/money/business/2014/06/20/kevlar-inventor-stephanie-kwolek-dies/11133717/, accessed September 17, **2014**.
9. A. Kopczynska & G.W. Ehrenstein, *J. Mater. Ed.* **2007**, *29*, 325.
10. M. Strukelj, J.M.G. Martinho, M.A. Winnik, & R.P. Quirk, *Macromolecules* **1991**, *24*, 2488.
11. S. Penczek, R. Slazak, & A. Duda, *Nature* **1978**, *273*, 738.
12. A.D. Aliev, Zh. Zhumbaev, & B.A. Krentsel, *Nature* **1979**, *280*, 846.
13. S. Penczek, R. Slazak, & A. Duda, *Nature* **1979**, *280*, 846.
14. P.J. Flory, *Statistical Mechanics of Chain Molecules*, Wiley-Interscience: New York/London/Sydney/Toronto **1969**.
15. C.H. Chan, H.W. Kammer, S.N.H. Mohd Yusoff, L.H. Sim, & T. Winie, *Solubility of Li Salt in Solid Polymer Electrolytes Based on Modified Natural Rubber, presented at the 21st POLYCHAR World Forum on Advanced Materials*, Gwang-Ju, March 11–15, **2013**.
16. E. Buchleier, W. Wehner, & F. Vögtle, *Synthesis* **1978**, *2*, 155.
17. D.A. Tomalla, H. Baker, J. Dewald, M. Hall, G. Kallos, S. Martin, J. Roeck, J. Ryder, & P. Smith, *Polymer J.* **1985**, *17*, 117.
18. W.J. Feast, S.P. Rannard, & A. Stoddard, *Macromolecules* **2003**, *36*, 9704.
19. G.H. Michler & F.J. Baltá-Calleja, *Nano- and Micromechanics of Polymers: Structure Modification and Improvement of Properties*, Hanser: Munich/Cincinnati, OH **2012**.
20. N. Sanandaji, L. Ovaskainen, M. Klein Gunnewiek, G.J. Vancso, M.S. Hedenqvist, S. Yu, L. Eriksson, S.V. Roth, & U.W. Gedde, *Polymer* **2013**, *54*, 1497.
21. F. Nilsson, U.W. Gedde, & M.S. Hedenqvist, *Eur. Polymer J.* **2009**, *45*, 3409.
22. A.J. Lovinger, J.O. Chua, & C.C. Gryte, *J. Polymer Sci. Phys.* **1977**, *15*, 641.
23. N. Nemirowski, M.S. Silverstein, & M. Narkis, *Polymers Adv. Tech.* **1999**, *7*, 247.
24. B.E. Eichinger & P.J. Flory, *Trans. Faraday Soc.* **1968**, *64*, 2035.
25. W. Brostow, *Macromolecules* **1971**, *4*, 742.
26. M.S. Hedenqvist & U.W. Gedde, *Polymer* **1999**, *40*, 2381.
27. S. Blonski, W. Brostow, & J. Kubát, *Phys. Rev. B* **1994**, *49*, 6494.
28. S. Fossey, Chapter 4: Computer simulation of mechanical properties, in *Performance of Plastics*, ed. W. Brostow, Hanser: Munich/Cincinnati, OH **2000**.
29. W. Brostow, M. Donahue III, C.E. Karashin, & R. Simões, *Mater. Res. Innovat.* **2001**, *4*, 75.
30. W. Brostow, J.A. Hinze, & R. Simões, *J. Mater. Res.* **2004**, *19*, 851.
31. W. Brostow & R. Simões, *J. Mater. Ed.* **2005**, *27*, 19.
32. S.M. Ulam, *Adventures of a Mathematician*, University of California Press: Berkeley, CA **1991**.
33. F. Nilsson, X. Lan, T. Gkourmpis, M.S. Hedenqvist, & U.W. Gedde, *Polymer* **2012**, *53*, 3594.
34. N. Metropolis & S. Ulam, *J. Am. Statist. Assoc.* **1949**, *44*, 335.
35. A. Mattozzi, P. Serralunga, M.S. Hedenqvist, & U.W. Gedde, *Polymer* **2006**, *47*, 5588.

36. J. Ritums, B. Neway, F. Doghieri, G. Bergman, U.W. Gedde, & M.S. Hedenqvist, *J. Polymer Sci. Phys.* **2007**, *45*, 723.
37. A. Mattozzi, M. Minelli, M.S. Hedenqvist, & U.W. Gedde, *Polymer* **2007**, *48*, 2453.
38. A. Mattozzi, M.S. Hedenqvist, & U.W. Gedde, *Polymer* **2007**, *48*, 5174.
39. F. Nilsson, M.S. Hedenqvist, & U.W. Gedde, *Macromol. Symp.* **2010**, *298*, 108.
40. S. Enders, K. Langenbach, Ph. Schrader, & T. Zeiner, *Polymers* **2012**, *4*, 72.
41. W. Forsman, ed., *Polymers in Solution*, Plenum Press: New York **1986**.
42. YouTube, *Blow Molding, PolytekMX*, published April 11, **2013**, https://www.youtube.com/watch?v=NE4c1gwzPb4, accessed February 27, 2016.
43. R. Gächter, H. Miller, & P.P. Klemchuk, eds., *Plastics Additives Handbook*, 4th edn., Hanser: Munich/Cincinnati, OH **1996**.
44. H. Zweifel, R.D. Maier, & M. Schiller, eds., *Plastics Additives Handbook*, Hanser: Munich/Cincinnati, OH **2009**.
45. M. Narkis & N. Rosenzweig, *Polymer Powder Technology*, John Wiley & Sons: Chichester/New York **1995**.
46. G. Wegner, *Angew. Chem. Internat. Ed.* **1981**, *20*, 361.
47. M. Monkenbusch & G. Wegner, *Makromol. Chem. Rapid Comm.* **1984**, *5*, 157.
48. B. Nordén & E. Krutmeijer, *The Nobel Prize in Chemistry, 2000: Conductive Polymers*, Kungl. Vetenskapsakademien, The Royal Swedish Academy of Sciences, http://www.nobelprize.org/nobel_prizes/chemistry/laureates/2000/advanced-chemistryprize2000.pdf, accessed February 27, **2016**.
49. M. Narkis, Y. Haba, E. Segal, M. Zilberman, G.I. Titelman, & A. Siegman, *Polymers Adv. Tech.* **2000**, *11*, 665.
50. I. Mironi-Harpaz & M. Narkis, *J. Appl. Polymer Sci.* **2001**, *81*, 104.
51. H. Tai, A. Sergenko, & M.S. Silverstein, *Polymer* **2001**, *42*, 4473.
52. A. Sergienko & M.S. Silverstein, *Polymer Eng. & Sci.* **2001**, *41*, 1540.
53. M.S. Silverstein, H. Tai, A. Sergienko, Y. Lumelsky, & S. Pavlovsky, *Polymer* **2005**, *46*, 6682.
54. M. Narkis, S. Srivastava, R. Tchoudakov, & O. Breuer, *Synthetic Metals* **2000**, *113*, 29.
55. D. Jackovich, B. O'Toole, M. Cameron Hawkins, & L. Sapochak, *J. Cellular Plastics* **2005**, *41*, 153.
56. C. Antoinette, W.T. Bigbee, W. Brostow, G. Granowski, H.E. Hagg Lobland, N. Hnatchuk, R. Pahler, A. Richards, A. Spagnolo, J. Wahrmund, & W. Xie, *Mater. Res. Innovat.* **2013**, *17*, 537.
57. W. Brostow, M. Dutta, & P. Rusek, *Eur. Polymer J.* **2010**, *46*, 2181.
58. B.N. Balzer, S. Micciulla, S. Dodoo, M. Zerball, M. Gallei, M. Rehahn, R. von Klitzing, & T. Hugel, *ACS Appl. Mater. & Interfaces* **2013**, *5*, 6300.
59. M. Baron, K.-H. Hellwich, M. Hess, K. Horie, A.D. Jenkins, R.G. Jones, J. Kahovec, P. Kratochvil, W.V. Metanomski, W. Mormann, R.F.T. Stepto, J. Vohlidal, & E.S. Wilks, *Pure & Appl. Chem.* **2009**, *81*, 1131.
60. R.G. Jones, J. Kahovec, R. Stepto, E.S. Wilks, M. Hess, T. Kitayama, & W.V. Metanomski, *Compendium on Polymer Terminology and Nomenclature*, RSC Publishing: Cambridge **2009**.
61. M. Vert, Y. Doi, K.-H. Hellwich, M. Hess, P. Hodge, P. Kubisa, M. Rinaudo, & F. Schué, *Pure & Appl. Chem.* **2012**, *84*, 377.

10

COMPOSITES

At a restaurant in New York City, a waiter eager to please, asked Groucho Marx: "How did you find your steak, sir?" "Quite by accident," Groucho replied. "I moved that little piece of tomato, and there it was underneath."

10.1 INTRODUCTION

Looking at a composite surface does not tell us what is underneath—like Groucho's steak under a tomato slice. Another simple example from everyday life: a box of chocolates; for each treat in the box there is chocolate coating on the outside, but what is inside each of the small chocolates? See Figure 10.1 showing such a box from Vilniaus Pergalē; the text in Lithuanian says that the chocolates contain some coffee... and now that we have whet our appetites with such delectable analogies, let us dig in deeper on the topic of composite materials.

As a class of materials, composites can be defined as a combination of two or more dissimilar materials. Composite materials are manufactured on a wide-scale in the 21st century to meet industrial demand for materials with very specific property profiles that cannot be met by single phase materials. The design possibilities are much broader with composites than with metals alone, or ceramics alone, or polymers alone. It is largely for this reason (among others) that composites now garner so much attention from materials scientists and engineers.

Because composites are comprised of dissimilar materials, they possess a profile of properties different from the individual components used to make them. The characteristic properties of each constituent interact, resulting in the collective behavior of the composite. For example, due to the combination of superhard ceramics and ductile metals, both

Materials: Introduction and Applications, First Edition. Witold Brostow and Haley E. Hagg Lobland.

FIGURE 10.1 A box of chocolates (made in Lithuania by Vilniaus Pergalė). (http://www.pergale.lt/en/box-of-sweets-pupa-3/.)

extremely hard and tough composites can be fabricated, or polymers can be significantly strengthened by long ceramic fibers. Some properties—such as density, specific heat, thermal and electrical conductivities—can typically be predicted based on the **rule of mixtures**, wherein the composite properties are weighted sums of the individual constituent values. In many composites, however, the components interact in such a way that the end result is a materials system with properties far exceeding those of the constituents alone. Clearly this category of composites is of great importance to scientists and engineers.

Nature has likewise produced its own composites. Wood and bone, for instance, are examples of nature's composites (and we shall discuss them in Section 10.10). Nevertheless, when we speak of composites, we typically are referring to artificially made multiphase materials. (By design and by the presence of two or more distinct components, composites are inherently multiphase materials.)

The terms **matrix** and **dispersed phase** are common descriptors for composites. The case when we do *not* have matrices are laminar composites, where the distribution of each phase is 2-dimensional and varies layer by layer. Thus, **laminates** achieve their desired properties because of the differences in composition and structure between multiple layers. A laminate is usually assembled in a permanent structure by application of heat, pressure, or a welding operation, or else by adhesives.

Otherwise the matrix is continuous and surrounds the other phase, called the dispersed phase. The dispersed phase may be in the form of particles, fibers, whiskers, flakes, etc. The matrix is normally, but not always, the majority component. Furthermore, the components in a composite usually perform different functions. In particular, in fiber-reinforced composites the fiber carries the load while the matrix distributes the load.

The prediction of properties of composites is more difficult than for "pure" materials. The properties depend on: properties of each phase; geometry of the dispersed phase; volume fraction of each phase. The variety of possible composites is endless. That said, we have a finite space to discuss them. We shall discuss important classes of composites including fiber-reinforced composites, cermets and other metal matrix composites (MMCs), ceramic matrix composites (CMCs), carbon–carbon composites, polymer-matrix composites (PMCs), hybrid composites, laminar and sandwich composites, concretes, and asphalts. Finally, we shall spend a little time in description of natural composites. As in previous chapters, the structure of the materials and how it relates to their properties is an important feature.

10.2 FIBER REINFORCED COMPOSITES

The goal of fiber reinforcement is usually to obtain higher strength and stiffness in the matrix material. Materials in this category of composites are typically named either by the type of matrix (e.g., polymer matrix composite) or by the type of reinforcing fiber (e.g., glass-reinforced plastic). In any case, various microscale and macroscale parameters of the fibers strongly influence the resultant properties of the composite. Notwithstanding this, the matrix is important as it protects the fibers from surface damage and other environmental conditions. We shall introduce here several types of fiber reinforced composites (FRCs) and discuss aspects of fiber type, length, orientation, and concentration.

Fibers are prepared from a variety of material types. They include:

1. **Glass**—glass is inexpensive and the most common fiber. Usually used for reinforcement of polymer matrices, it has high tensile strength and moderately low density ($\approx 2.5\,g\,mL^{-1}$).
2. **Carbon**—carbon fibers are a high-performance material used in advanced polymer matrix composites. Carbon fibers consist of crystalline graphitic regions as well as amorphous regions. Since carbon is less dense than glass, its stiffness is higher; moreover its strength can exceed that of construction steel. Owing to the graphitic structure, the properties of carbon fibers are anisotropic.
3. **Polymer**—high molecular weight polymers such as Kevlar and ultra-high-molecular-weight polyethylene (UHMWPE) can increase toughness in brittle matrices. They perform better in tension than compression.
4. **Ceramic** other than glass—owing to poor properties in tension and shear, ceramic fibers are small and typically fall into the category of whiskers (which have large length-to-diameter ratios). Fibers of materials such as silicon nitride, alumina, and silicon carbide are useful for composites exposed to high temperatures or harsh environment.
5. **Metal**—metallic fibers typically have larger diameters and are considered as wires. Strength is high, but high density mitigates use in weight-critical applications. Steel, Mo, and W are characteristic materials in this respect. Metallic wires are used in radial steel reinforcement of automobile tires and in wire-wound high-pressure hoses.

The extent of strengthening provided by fibers depends both on the properties of the fiber and on how well the applied load is transferred from the matrix to the fibers. In our classification of fiber types above, we have mentioned the characteristic properties of different fibers. The load transmittance, however, is governed primarily by the interface between fiber and matrix. A first requirement for the interface is that the matrix must "**wet**" or coat the fiber. Sometimes (for incompatible materials) a coupling agent is required to achieve suitable wettability (a **coupling agent** is a multi-functional compound having moieties that adhere to two different material types). To achieve desirable properties in a composite, the load has to be effectively transferred from matrix to fiber via the interface. Depending on the application, **debonding**—or failure at the interface—may or may not be desirable. The size of the interface as well as strength of the bond between fiber and matrix will impact whether debonding occurs under stress and fracture. The comparative strength of bonding within the matrix (either weak van der Waals forces or strong covalent bonds) also plays a role here. Interfacial strength is not merely a concept but is measurable by standard tests. The fibers (especially steel fibers) can have different tensile strengths; the same applies to the matrix. The desired load carrying capacity can only be achieved by understanding the interaction between matrix and fibers. If the fiber has a high tensile strength while the matrix has a low strength, the fibers are pulled out from the matrix under load. In the opposite case, when the matrix is strong and the fibers are weak, the fibers are broken without transferring enough energy under loading. We have already mentioned possible debonding. Steel fibers and polymeric fibers have a mechanical bond with the matrix—while glass fibers have both mechanical and chemical bonding with the matrix.

From the list of fiber types above, it is evident that fiber length and aspect ratio are important parameters for composite behavior. There exists a critical fiber length to achieve worthwhile strengthening and stiffening in a composite. The value of the critical length l_{crit} is dependent on fiber diameter d_F, fiber strength at failure σ_F, and interfacial shear strength τ_c. A relationship for calculating the critical fiber length is

$$l_{crit} = \frac{\sigma_F d_F}{2\tau_c} \tag{10.1}$$

According to Ehrenstein [1], however, this equation is insufficient for determining l_{crit} for the reason that it is the actual shear stress distribution along the fiber surface that determines the magnitude of the transmitted forces, not the shear strength τ_c. A simple test can be used for such determination, while Eq. (10.1) provides a good guideline for estimating the value.

There is likewise a critical value of the fiber content that must be exceeded to achieve strengthening of the matrix. Moreover, orientation of the fibers will alter that value. On one hand, fibers may all be aligned in a single direction, parallel with respect to their longitudinal axis. On the other hand, fibers may be oriented randomly throughout. In between these two extremes is the possibility of partial alignment. Regardless of alignment, a uniform distribution of fibers is desirable. Various possible types of reinforcement are displayed in Figure 10.2, and various possible orientations for layers of fiber-reinforced materials are seen in Figure 10.3.

The outcome of fiber alignment is **anisotropic** properties, meaning the properties are dependent on the direction in which they are measured. A consequence of parallel fiber alignment is the achievement of significant reinforcement in the longitudinal direction. Although reinforcement in the transverse direction is non-existent, the high degree of strengthening along the fiber orientation axis is desirable for many applications.

This category may be the most widely recognized of synthetic composites, primarily because of so-called fiberglass, which is really **glass-fiber reinforced plastic (GRP)**, familiar even to most non-engineers. Both thermoplastic and thermosetting resins are used in GRPs, yielding materials with good overall properties at low cost (since they do not require high curing temperatures or pressures). Common uses of glass-fiber reinforced plastics are for car bodies and boat hulls, though they may even be found in small products

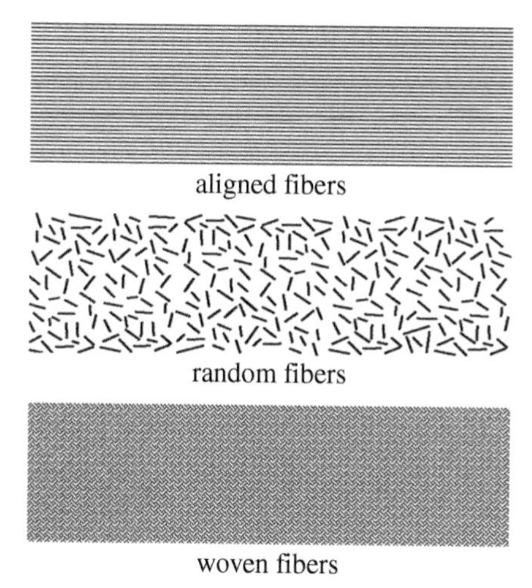

FIGURE 10.2 Typologies of fiber-reinforced composites.

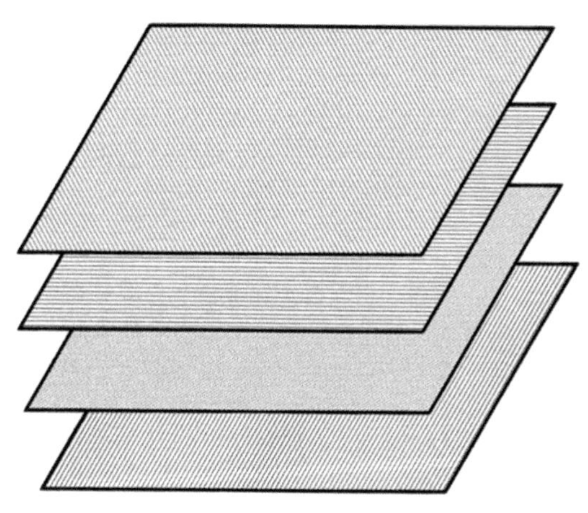

FIGURE 10.3 Four layers of a material reinforced with continuous fibers; the angle of orientation is rotated for each subsequent layer. Laminar composites have the direction of high-strength orientation varied between layers.

such as circuit boards. Glass-fiber reinforced plastics comprise the majority of fiber reinforced **polymer-matrix composites (PMCs)**.

Another common type of PMC is **carbon-fiber reinforced plastic**. Carbon fibers are a high strength, high performance material. In addition to excellent mechanical properties, carbon fibers possess good resistance to moisture and other chemicals. These fibers consist of amorphous and crystalline regions. They are manufactured from rayon, polyacrylonitrile (PAN), or pitch (a coal tar product, cf. Section 8.3) by various techniques. Carbon-fiber reinforced plastics are used in sports and recreational equipment (bicycle frames, golf clubs), demanding aircraft structural components, and also wind turbine blades. Later in this section we discuss methods of processing such materials.

A third type of fiber commonly used to reinforce plastic is **aramid fibers** such as Kevlar and Nomex; we have talked about Kevlar in the previous chapter. These polymer fibers generally have a good strength-to-weight ratio, making them more desirable than metals for many applications. They are also known for toughness, impact resistance, and resistance to creep and fatigue failure. Aramid-fiber reinforced plastics are used for structural applications as well as in large ropes and some ballistic products.

Together, glass, carbon, and aramids are used in the majority of fiber reinforced plastics (FRPs). Other types of fibers (ceramic or metallic) are used much less frequently. As a group, FRPs offer several advantages including weight savings (owing to lower density materials) and easier processing. Beyond these, each fiber type offers additional improvements to the chosen polymer matrix. Most applications utilize products with continuously aligned fibers; filament winding, pultrusion, stitched preforms, and the so-called prepreg process are common manufacturing processes for production of continuous fiber polymer matrix composites.

In **pultrusion** fibers are pulled through a resin, possibly followed by a separate preforming system, and into a heated die, where the resin undergoes polymerization. A constant cross-section of the composite is obtained and the method is applied to both thermoplastics and thermosets.

The term **prepreg** tells us that the fibers in the composite are already pre-impregnated. This can be done on a flat surface, which is convenient. The fibers can form a weave or else are unidirectional. In a **weave** two distinct sets of fibers, yarns, or threads are interlaced at right angles. In the prepreg process the resin is first only partially cured to allow easy handling, we have a so-called **B-Stage material** which requires cold storage to prevent complete curing. After cold storage, composite structures built of pre-pregs are put into an oven or an **autoclave** (a pressurized and heated vessel) to complete the polymerization.

Stitched preforms—stitched with thermoplastic yarns—are used to make carbon FRCs for the aircraft industry with epoxy matrices [2]. Low-melting temperature yarns based on polyamide and phenoxy are used, providing prestabilization of the dry preforms by thermobonding. Liquid composite molding mitigates laminate disturbances typical for standard polyester yarns. Thus, composite strength properties are better than in other procedures [2].

By contrast, the properties of composites with randomly oriented fibers are **isotropic** (not dependent on the direction of measurement). However, the level of reinforcement that can be achieved is also somewhat less (for a given fiber type and concentration) than if the fibers were aligned. Nevertheless, the range of possibilities for orientation of fibers allows one to tailor the design of FRCs to meet the needs of different applications. Furthermore, once the critical concentration has been defined, a concentration level may be selected that minimizes cost and maximizes performance.

Altogether the technique of fiber reinforcement yields an important class of materials. By imparting designed and organized structures into chosen matrices, strength can be improved along specific directions. Moreover, many properties can be maximized through carefully-selected combinations of fiber and matrix.

10.3 CERMETS AND OTHER METAL MATRIX COMPOSITES (MMCs)

In MMCs we have one phase mechanically strengthened by dislocation impediments that arise from the second phase. Processing methods involve two stages: first is introducing the reinforcement into the matrix; second is transforming the composite into a finished component. Metals and alloys as matrices resist attack by organic fluids, thus protecting the reinforcement.

There are a variety of MMCs [3]. The most widely used in this class are cermets, the name is derived from their composition: a combination of ceramic and metal. **Cermets** may also be referred to as cemented carbides, hard metals, or sintered carbides.

First developed in the 1920s as materials for cutting tools, cermets were superior to plain carbon steels as tool materials for demanding applications. At high cutting speeds exceeding 30 m min^{-1}, temperatures may well surpass 725°C, which is beyond the range sustainable by alloyed tool steels containing Mn or W. The primary disadvantage of such steel tools is that their hardness decreases rapidly with temperature above 225°C. By contrast, cermet cutting tools can withstand cutting speeds as high as 300 m min^{-1}; and they have better wear and creep resistance as well as toughness.

The constituents of several common cermets are given in Table 10.1. Cermets consisting of tungsten carbide (WC) plus 6–20% Co have long been used. The manufacturing process for such a material is based on sintering (cf. Section 7.3), where fine powders of WC and Co are heated together until a liquid phase appears. The liquid wets the solid WC particles and fills in the gaps between them. Upon cooling the whole, a continuous matrix is thus formed.

Recent interest in cermets is focused less on cutting tools and more on other applications such as mining drills and ballistic protection materials. The most prominent application of cermets is in aerospace industry where high-temperature performance is essential. Cermets can be used for thermal protection shrouds and rocket nozzles as well as for protection against micrometeorite bombardment.

While cermets have a long tradition, new such materials are being created with ingenuity, such as TiB_2 + Cu cermets [4–6]. At high TiB_2 concentrations, such cermets have properties competitive with WC + Co cermets listed at the top of our list in Table 10.1. TiB_2 is apparently harder than commonly used WC. Thus, we are dealing with a class of very high performing metal matrix composites. We note that the final properties of a cermet are determined not only by the dispersed phase but also by the matrix. The dispersed particles should maintain the cohesion-based stability of the matrix. Otherwise hard but rather brittle intermetallic matrix may be formed; in this situation the matrix does not perform its role of a ductile metallic binder [5]. It was demonstrated that cobalt, a commonly used matrix for WC, is unstable in the presence of TiB_2 since the matrix is now a ternary compound [4]. That means that universal recipes do not exist, every system has to be considered separately.

Aluminum is frequently used as a matrix in MMCs because of its low melting point, which simplifies processing and reduces costs. Among the materials used as reinforcements in aluminum-matrix composites are boron, silicon carbide, graphite, and alumina. Other matrix materials include magnesium, titanium, copper, and superalloys. Each type has its own defining features and target areas of application.

TABLE 10.1 Characteristic Compositions (Ceramic + Metal) of Cermets

Cermet Components
WC + Co
TaC + Co
TiC + Co
MoC + Co
ZrC + Fe
Cr_3C + Ni
MgO + Ni
Mo_2B + Ni
Al_2O_3 + Cr

One cannot assume that introduction of a reinforcement will necessarily bring about mechanical strengthening of a metal. Christman, Needleman, and Suresh [7] studied tensile behavior of aluminum alloys as matrices with SiC whiskers as the reinforcement. Numerical calculations using the finite-elements methods were confronted with experimental results. Void formation turned out to be a problem, caused by triaxial stresses. It was also concluded that strong adhesion of the reinforcement to the matrix might not be sufficient, since matrix debonding in the immediate vicinity of the whiskers is possible.

Considering a reinforcement for a metallic matrix, one has to choose the shape. A finite elements numerical study by Needleman and coworkers [8] has led to the following order of shapes with increasing load carrying capacity: double-cone $\mapsto$ sphere $\mapsto$ truncated cylinder $\mapsto$ unit cylinder $\mapsto$ whisker.

10.4 CERAMIC MATRIX COMPOSITES (CMCs)

Like metal matrix composites, ceramic matrix composites are desired for their ability to retain strength and other properties at high temperatures. One example of a CMC is silicon-carbide fiber reinforced alumina. One realizes that it belongs both to fiber-reinforced composites and to CMCs. Reinforcement with continuous fibers can help overcome the natural brittleness of ceramics. Low-cost, low-density ceramic fibers have been developed, such as Siboramic ($SiBN_3C$ from the Max Planck Institute for Solid State Research, Stuttgart, see Figure 10.4), for use in CMCs (and other matrices) with the hope of eventually replacing

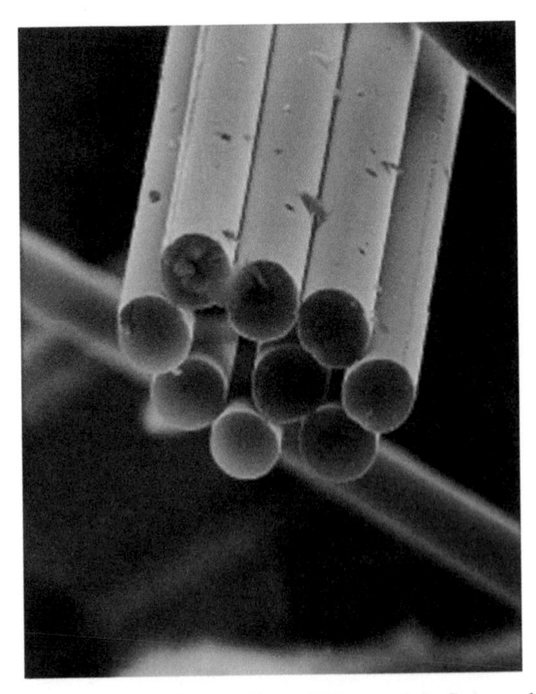

FIGURE 10.4 SEM picture of $SiBN_3C$ fibers (Siboramic). *Source*: http://www.fkf.mpg.de/124596/Applications.

of traditional ceramics in aeronautic and power-generation components. Ceramic matrix composites containing silicon nitride and silicon carbide fiber reinforcement may one day be used instead of nickel-based superalloys for high-temperature applications in aircraft engines. The improved toughness and reduced creep rate imparted by addition of Si_3N_4 to molybdenum disilicide ($MoSi_2$) is impressive. The addition of SiC fibers further improves toughness and impact resistance.

A ceramic matrix composite usable at very high temperatures consists of TiB_2 in a boron carbide matrix [9]. The melting temperatures of individual components are 3200 and 2450°C, respectively, for TiB_2 and B_4C. The eutectic temperature in this system is above 2200°C.

How do we know that there is no damage accumulation in a CMC? Smith, Morscher, and Xia [10] have developed a method based on determination of electrical resistivity. In a typical CMC, the fibers are conductive, while the matrix is an electric insulator. However, even if the matrix and the fibers are made of the same material such as SiC, changes in electrical resistivity are seen after loading [10]. Loading causes crack formation in the matrix so that *in-situ* damage detection is possible by monitoring changes in electrical resistivity.

10.5 CARBON–CARBON COMPOSITES

In this unique category of composite materials, carbon fibers reinforce a carbon matrix. Carbon–carbon composites are noted for their very light weight and high strength—which make them good candidates for aircraft and automobile applications. They are produced from a carbon fabric consisting of carbon fibers—arranged in specified patterns and orientations—in an organic (polymeric) resin. The fabric is placed in a mold and then pyrolyzed: the high temperature of pyrolysis drives off non-carbon elements leaving behind a carbon matrix. Further heating converts the amorphous carbon to a graphite structure and causes densification (and thereby strengthening) of the matrix.

New processing techniques and applications for carbon–carbon composites are being developed. For instance, they are being used to replace graphite electrodes, which in some applications are problematic. With these advancements, carbon–carbon composites are now used in a range of applications including racing car brakes and clutches, hot glass transfer elements, protective shielding, vacuum/inert gas furnace insulation, hot pressing molds, metal sintering trays, and nuclear applications.

10.6 POLYMER MATRIX COMPOSITES (PMCs)

A category of polymer-based composites was already discussed in Section 10.2: glass reinforced plastics. Here we talk about PMCs with fillers such as powders which are not elongated. They are created for various purposes, often to reinforce polymers so as to go to some extent towards metals which are much stronger.

Putting a ceramic filler into a polymer matrix is easier said than done when the constituents "do not like each other". In such cases **compatibilizing agents** can be used. An example is polypropylene + ethylene-propylene-diene rubber filled + a thermal-shock-resistant ceramic filler which contains α-Al_2O_3, mullite ($3Al_2O_3{\cdot}2SiO_2$ or $2Al_2O_3SiO_2$), β-spodumene ($LiAl(SiO_3)_2$), and aluminum titanate [11, 12]. Three distinct compatibilizing agents have been used. Such composites are made not only to provide improved mechanical properties to polymers but also to improve their tribological properties, for instance, to enhance resistance to wear or to lower friction.

So-called barrier properties are important. In some cases we wish as little as possible diffusion of gases through the polymer, such as in soft drink bottles in which we want to retain carbon dioxide—rather than let it escape through the plastic bottle walls. In other cases we might wish to increase diffusion across polymeric walls. Böhning, Schönhals, and their colleagues [13] have dispersed plasma synthesized silicon carbide nanoparticles in dichloromethane/poly(bisphenol-A-carbonate) solutions by ultrasonication. Samples were then prepared by film casting. They have used dielectric relaxation spectroscopy (see Section 17.6) and then determined diffusion of carbon dioxide through their films. Such polycarbonate based nanocomposites show much lower plasticization than neat polycarbonate. The authors explain their results in terms of enhanced molecular mobility in the vicinity of the dispersed SiC nanoparticles.

Once the components are compatible, one has to provide appropriate processing and manufacturing procedures. Development of *models* can be useful here. As an example, partly crosslinked polyethylene can be processed as a thermoplastic, with an evident capability of making composites on that basis. Bonten and Schmachtenberg [14] have developed models explaining mechanics inside a welded joint of semicrystalline polyethylene.

Polymer-matrix composites do not necessarily have the polymeric matrix as the majority component. In 1994 Moshe Narkis with coworkers created an artificial marble called Caesarstone [15] based on a polyester (no less than 6 wt.%) and 92% quartz. Thus, the filler is a nearly overwhelming majority component. Earlier artificial marbles included $CaCO_3$ rather than quartz. Attempts to combine polyester with silica led to materials with poor quality. Narkis used polyester and quartz but included also less than 1% vinylsilane—a key component that improved the properties dramatically. Caesarstone is used around the world. In 2014 an improvement consisting of the inclusion of an antimicrobial substance (0.002% dry weight) was patented in the United States [16].

10.7 HYBRID COMPOSITES

A class known as **hybrid composites** is also recognized. These contain fibers of two or more different types, usually in a polymeric resin. The fibers may be combined in a variety of ways; the structure may also be laminar, involving the third dimension (see the next section for more on laminar composites). Because there are many possible combinations for hybrid composites, the property profiles are also varied and thus there is a range of applications for such materials.

10.8 LAMINAR AND SANDWICH COMPOSITES

As their names imply, laminar and sandwich composites are defined by having layered structures. **Laminar composites** are assembled from materials having a preferred high strength direction, such as wood or aligned fiber-reinforced composites. The layers are assembled such that the orientation varies (e.g., by angles of 0°, 45°, 90°) between adjacent layers (see again Figure 10.3). A common example is plywood, wherein the grain of the wood varies by right angles between adjacent plies. A few examples of laminar composites are:

1. Clad metals, such as copper-clad stainless steel, already mentioned in Chapter 6.
2. Paper- and plastic-based laminates. Examples are paper laminated with plastic film or metal foil; nylon fabric laminated with metal layers; roofing paper.
3. Safety glass, which has a layer of poly(vinyl butyral) bonded between two layers of glass.
4. Clothing, especially for high-performance situations. For example, raingear that allows sweat vapor to escape but keeps rainwater out utilizes layering of fabrics, each with a different set of properties.

Sandwich composites can be considered as a subset of laminar composites. Used primarily for structural components, their defining characteristic is a high strength-to-weight ratio. This is achieved by having outer surfaces, or **facings**, made of material higher in density and stiffness than the **core**, which supports the facings (Figure 10.5). While the facings are typically selected from among strong and stiff materials, it is possible to make the facings very thin because they are supported by the core material. Facings are often made of aluminum, wood, fiber-reinforced plastic, or steel. Core materials are generally cellular materials, having configurations such as honeycomb, waffle, corrugated, tube, and cone (Figure 10.6). Such configurations provide a rigid structure without the weight of a

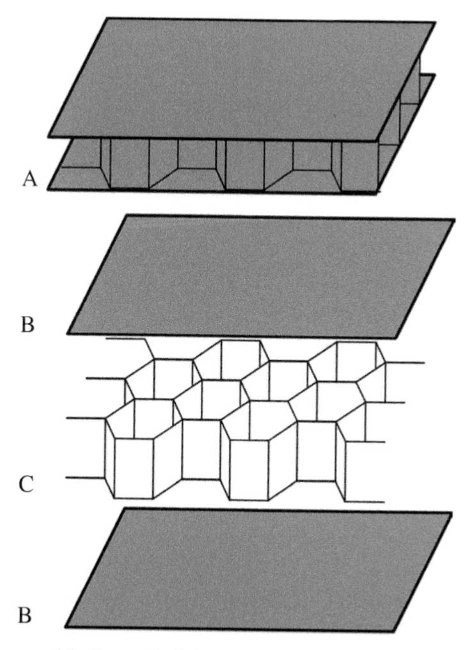

FIGURE 10.5 An assembled sandwich composite panel A. Beneath it are the facing layers B and honeycomb core layer C, shown unassembled. *Source*: George William Herbert, https://commons.wikimedia.org/wiki/File:CompositeSandwich.png. Used under CC-BY-SA-2.5.

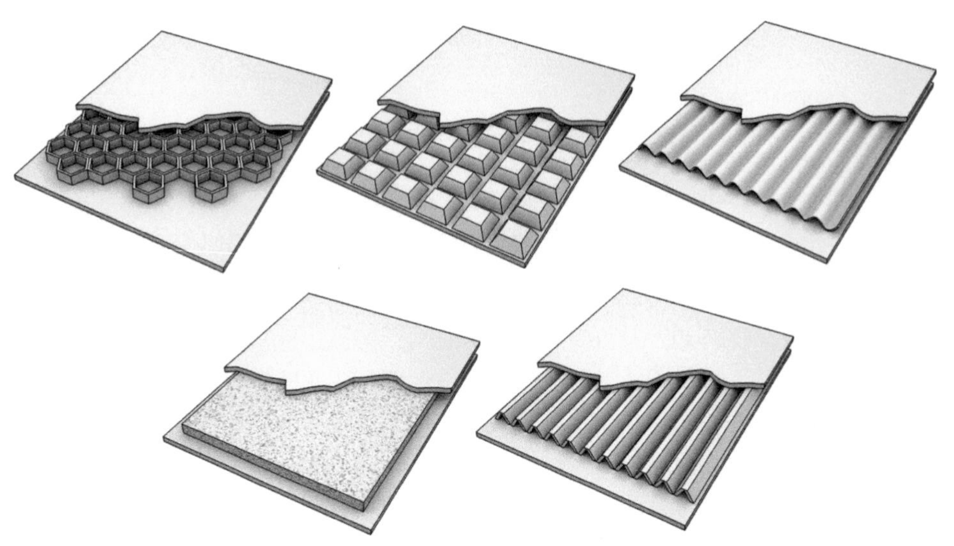

FIGURE 10.6 Sandwich core structures: honeycomb, waffle, corrugated, foam, prism.

solid mass spanning the same volume. Solid and foamed materials—including plywood, plastics, and ceramics—are also used as core materials.

The **honeycomb cell structure** is unique: mathematically it is the strongest possible shape for a collection of adjacent cells. DuPont's Nomex aramid is used as a honeycomb core in sandwich composites for a host of demanding applications. Kevlar and aluminum honeycombs are also high performing core materials in sandwich composites.

While laminates typically include multiple layers, we need to distinguish **coatings** where there is one layer on the surface different from the interior. Coated metals noted in Section 6.6 belong here, for example silver deposited on brass and copper shapes [17]. Cermets based on WC or else on Ni + Cr + Si + B have been deposited as coatings on various substrates to enhance the wear resistance [18].

10.9 CONCRETES AND ASPHALTS

While bricks made from mud and straw may be the oldest known composite, concrete also has a long venerable history. Concrete is mainly made of three components: rock and/or gravel (the course aggregate) + sand (the fine aggregate) + hydrated Portland cement. The coarse aggregate makes up the bulk of the material; the sand fills in some of the voids that always arise in the material; and the cement binds the whole together. Often various admixtures (divided into "mineral" or "chemical") and/or fibers are added in function of specific applications. Particle packing has a significant impact on the resultant concrete properties [19]. Portland cement concrete (PCC) is known worldwide for its high strength and good insulating capacity. On the other hand, low tensile strength, poor environmental durability (e.g., resistance to freeze-thaw), and susceptibility to sulfate and acid attack impose limits on its performance. Like ceramics, PCC is considered brittle, and compared to polymers it has low flexural strength and low toughness. Addition of fibers into the concrete compensates for these shortcomings up to a certain level. Steel fibers increase the flexural strength of the concrete by some 200%. An air-entrance agent increases the freeze-thaw resistance, while plasticizers increase the compressive strength.

The final properties of concrete depend on composition, handling methods, and environmental conditions. Specifically, resultant properties depend on:

- Ratio of constituents. A common mixture is 4 parts course aggregate + 2 parts sand + 1 part cement powder. In self-compacting concretes one puts more fine aggregate and more filler into the matrix.
- Water to cement ratio. The cement must be hydrated (see Section 7.8); however, too much water weakens the material, resulting in the lower level of calcium silica hydrate (CSH) gels in the concrete. The excess water either escapes, leaving voids, or gets trapped in capillaries inside the bulk concrete.
- Geometry of the aggregate: apparently sharp-edged particles result in a better product than mostly round-edged particles. Texture and porosity significantly affect the concrete performance.
- Mixing and laying techniques. Over- or under-mixing yields a poor product. When laying concrete, *vibrating* the concrete into place rather just *pouring* it into place gives better results. Except for self-compacting concrete, concretes do not flow; large air gaps are likely to remain unless the material is vibrated.
- Curing or setting time (cf. Section 7.6). While the curing process may continue for years, the majority of the process is completed within hours. Sufficient humidity is required if the concrete is set in air not under water.

Although PCC—and details of its nature, composition, and use—is well known, there exist other mineral concretes. There are **alkali-activated cements** with compositions falling in the Me_2O—MeO—Me_2O_3—SiO_2—H_2O system [20]. The named reference reviews their history of development and discusses their status of use. Such cements are environmentally friendly, making use of substantial amounts of by-product and waste materials, thereby consuming less energy and generating less waste. Comparisons to Portland cements show that the properties of alkali-activated cements are superior [20]. We also have **geopolymers**, essentially polysialates, which are used not only in concretes but also in high-temperature ceramics, new binders for fire-resistant fiber composites, and for toxic and radioactive waste encapsulation. Geopolymers are typically inorganic chemical compounds or mixtures of compounds consisting of repeating units, for example silico-oxide (—Si—O—Si—O—), silico-aluminate (—Si—O—Al—O—), ferro-silico-aluminate (—Fe—O—Si—O—Al—O—), or alumino-phosphate (—Al—O—P—O—). Raw materials used in the synthesis of silicon-based polymers are mainly rock-forming minerals of geological origin, hence the name *geopolymer* proposed by Joseph Davidovits in 1978 and described later by him in a review article [21].

There are still further factors which enter into choosing a concrete composition. **Fly ash** is a by-product of coal-fired electric generating plants. It is the finely divided residue resulting from the combustion of ground or powdered coal, which is transported from the firebox through the boiler by flue gases. In Texas alone 13 million tons of coal ash are produced each year. The respective number is 18 million tons in Turkey and 80 million tons in India while globally 600 million tons are produced annually. Instead of being an environmental nuisance, fly ash can be incorporated in concrete [22] to improve workability, reduce water demand, reduce segregation and bleeding, and lower the enthalpy of hydration. This has been done for **self-compacting concrete** (SCC) [23, 24] that is able to flow under its own weight, completely filling formwork and achieving full compaction without vibration. Furthermore, fly ash reduces permeability, reduces corrosion of reinforcing steel, and increases sulfate resistance. However, fly ash reaches its maximum strength more slowly than concrete made with Portland cement only.

Given notorious nuclear power plant disasters (e.g., Fukushima), the composition of concrete used in construction of such plants is important. **Hematite** has an advantage here since it absorbs radiation well. Therefore, mineral concretes containing hematite are used [25]. There are also additives to concrete such as cationic surfactants [26] aimed at creating **porous concretes**. Such concretes have low weight per unit volume and are good thermal insulators.

Polymer concretes (PCts) boast impressive properties including higher strength and shorter curing time than Portland cement [27]. Polymer concrete is a composite of mineral aggregates (e.g., sand or gravel) with a monomer, usually of a thermoset polymer resin. The final concrete product, in which we have a polymeric matrix as the continuous phase and dispersed particles as a discrete phase, is formed by curing the monomer to form a polymer network. The curing can be done within one or two hours. The service life of PCt is typically longer than PCC and largely free of maintenance. Furthermore, PCt is marked by improved chemical resistance in harsh conditions, better freeze-thaw resistance, and increased flexural, compressive, and tensile strengths compared to PCC. Polymer concretes are used in highway pavements, underground wastewater pipes, for manufacturing thin overlays, as precast components for bridge panels, parking garage decks, industrial floors, and more. Polymer concretes also tend to have very good abrasion resistance. Polymer concretes are used to repair cracks in mineral concretes since they are easily inserted and provide better protection against the crack revival and propagation than do mineral inserts.

There is ongoing work to improve the methods for processing and reinforcing polymer concretes. Flexible matrices allow inclusion of a variety of fillers [28–32] including vermiculite [28, 31], waste marble [29], pumice [30], silica fume, furnace slag, fly ash, natural sand, and boron [32]. Other modifications exist such as inclusion of polypropylene fibers [31] or gamma irradiation—either of PP fibers or of the whole concrete after inclusion of fillers. Steel fibers have also been used. For that matter, similar fillers are used also in mineral concretes [33, 34]. We have already noted fly ash above. The use of fillers such as waste marble, furnace slag, or fly ash has three advantages: the desired reinforcement is achieved; these fillers would otherwise become waste contaminating the environment; and since they are byproducts of industrial processes, they are cheap. Thus, it is possible to achieve lowering of the price along with property improvement—while in so many cases property improvement requires *more expensive* additives.

Portland cement concrete is widely used in building construction. It is occasionally used in road construction; however, in that scenario high rigidity of concrete can be a disadvantage. Softer asphalt pavings—another composite material—are often a more suitable material for roads. Asphalt comprises the matrix; rock aggregate and sand comprise the dispersed phase. **Asphalt** itself consists of solid high molecular weight hydrocarbons called asphaltenes and of certain oily residues not well described by scientific studies. The service behavior of asphalt pavings depend not only on the component materials but also on the underlying soil base.

10.10 NATURAL COMPOSITES

Long before humans were designing composites in the laboratory, and even before humans gathered raw natural materials and assembled them into composites, nature boasted many of its own composites. While fiber-reinforced composites or other types of composites may incorporate natural materials (e.g., plant fibers), no synthetic composite has matched the properties of naturally occurring composites. Noteworthy examples are wood, bones, and teeth. Among natural composites, the majority are fiber composites.

By the simplest description, **wood** consists of **cellulose** fibers in lignin [35, 36]. This simplistic definition, however, does not fully relate the complexity of wood composition and structure. Examine a cross-section of raw wood, as shown in Figure 10.7; it is an artistic version by Raymond H. Pahler, but highly accurate. In Muir Woods National Monument north of San Francisco, one can see very old sequoia and other trees. One of them has been cut and it looks like our Figure 10.7. One can see marked the ring for the year 1215 when King John of England signed *Magna Carta Libertatum* limiting somewhat his previously excessive powers. In Figure 10.7 we can distinguish layers: outer bark, inner bark, cambium, ray cells, sapwood, heartwood, and pith. The rings are formed at the rate of one per year. The sapwood and heartwood make up the bulk of a tree's trunk and consequently are the main components of wood as we know it. The microstructure consists of long tubular cells whose walls are bundles of cellulose chains. The hollow centers, called lumens, of the tubular cells contain water or sap; the cellulose chains of the cell wall, with their local crystalline structure, are held together in a network of non-crystalline **lignin**. Dry wood contains about 50% cellulose, 25% hemicelluloses, and 25% lignin. The density of wood varies according to its exact makeup, and properties of wood are largely dependent on the density. Pure cellulose has a specific gravity of 1.5. The specific gravity of lignum vitae, the hardest wood, is 1.3, while the value for balsa wood is 0.15. Most other woods

FIGURE 10.7 *Twig Cross Section* by Raymond H. Pahler, mixed media. *Source*: Used with permission, Copyright (1984) R.H. Pahler.

have a specific density less than 1; and note that these values are for the dried wood (a tree may contain a weight of water equal to its dry weight).

The chemical building block of cellulose is the saccharide glucose, which may be considered as the monomer unit. The glucose molecules are connected by what is known as a beta acetal linkage. (The type of linkage is important: starch is also comprised of glucose monomers but with a different linkage, resulting in different chain structure and other properties.) Cellulose chains are rigid and arrange themselves into crystalline structures. Owing to polymorphisms, there are at least five distinct crystal forms of cellulose [35].

We also mentioned earlier that besides cellulose, hemicelluloses and lignin were the other main constituents of wood. **Hemicelluloses** refer to a variety of carbohydrate compounds with different degrees of branching and including some carbohydrate acetates. They are found also in plant bulbs and some seeds. One example is mannan, a crystalline carbohydrate found in the ivory nut. In wood, the hemicelluloses are located between the crystallites, yet still oriented parallel to the crystal chains.

Lignin refers to a large group of non-crystalline plant polymers. A simple formula for lignin cannot be easily written. Moreover, while cellulose occurs in nature without lignins, lignins never occur in the absence of cellulose and other carbohydrates. They are therefore defined by characteristic properties. They are comprised primarily (if not completely) of phenylpropane units. The majority of the methoxyl content of wood is associated with lignin. In reference to chemical activity, lignins are essentially non-hydrolyzable by acids, readily oxidizable, soluble in hot alkali, and will react easily with phenols and thio compounds. An additional characteristic of wood lignin is that when oxidized with nitrobenzene in alkali, the soft-wood lignins produce vanillin while the hard-wood lignins yield both vanillin and syringaldehyde. Large quantities of lignin are produced as a waste byproduct from paper production. In spite of continued efforts towards innovative uses for lignin, lignin is still used only in limited applications and therefore does not garner much profit.

POLYSACCHARIDES IN NATURAL COMPOSITES

Polymeric molecules can be roughly divided into *oligomers*, consisting of small numbers of monomer units, and *high polymers*. We shall utilize this distinction in our discussion of polysaccharide crystals. The unit cell in crystals of low-molecular-weight compounds contains entire molecules. Thus, in oligosaccharides the end-groups of the molecules constitute an important part of the unit cell as they must be located in order to characterize the crystal structure. In Section 9.8, we discussed the localizing of end groups in solid amorphous polymers. Now consider that as polysaccharides increase in chain length, the concentration of end groups is reduced, yielding a smaller influence of these groups upon the structure. The outcome is formation of a *macromolecular lattice* (or *sublattice*)—contrasted with the usual molecular lattice—in which the end groups are not integral nor do they even show in diffractometric diagrams. Therefore, polysaccharides, including cellulose, tend to form crystalline arrays of parallel chain molecules. Consequently, the crystal structure depends not on the chain length but on the nature of the repeating unit and interunit linkage and on the lateral forces acting between chains.

It is interesting to observe polysaccharide structures as they occur in nature. The crystal formation of parallel chains is restricted in the direction perpendicular to the chain length; thus the girth of a crystalline bunch of chains, called a crystallite, is on the order of 10 nm. The spaces between crystallites are non-crystalline but also longitudinal in shape. Bundles of crystallites form fibers with a longitudinal fiber axis in a similar way as for synthetic fibers mentioned in Section 10.4.

We shall conclude this section with a discussion of bone composition and structure. Such a discussion also provides a natural segue to Chapter 11 on Biomaterials. As with other composites (and in particular FRCs), the mechanical properties depend on the arrangement of the components and on the bond between the fibers and matrix. There exist different types of bone, each with its own arrangement of fibers and therefore with distinct properties. One can read extensively about bone in a multitude of resources. Thus, we will limit ourselves to what is applicable under the topic of this chapter.

CONFORMATIONAL ANALYSIS

Analysis of structure of lignin continues since such structures are difficult to evaluate. Two mutually complementary approaches are used: conformational analysis and diffractometry. We know what a conformation is, while diffractometry has been noted before. In **conformational analysis** one minimizes iteratively the system energy as a function of relative positions of atoms and atom groups. Foundations of this approach have been developed by Flory [37]. Three types of contributions to the configurational (=interactional) energy of the system can be distinguished: intrinsic torsional potentials attributed to the bonds; van der Waals repulsions between nonbonded atoms and groups of atoms; and dispersion type attractions between nonbonded atoms. When dealing with carbohydrates, we need to remember also electrostatic interactions.

Bone is a composite of organic and inorganic materials. At the nanometer-scale it is comprised of type I collagen fibers reinforced with calcium phosphate crystals. The inorganic phase is basically carbonated **hydroxyapatite** $Ca_{10}(PO_4)_6(OH)_2$ that appears as small crystals. Hydroxyapatite accounts for most of the weight of dry bone. While the mineral phase gives stiffness to bone, the collagen accounts for most of its toughness or capacity to absorb energy. **Collagen** itself is a macromolecule and is the most abundant protein in mammals. It is formed from three polypeptide chains that form a triple helix molecule. The collagen molecules in bone arrange themselves into fibrils that come together in a parallel array forming larger fibers. Apatite crystals are located in gaps between collagen molecules; they are deposited in the form of flat plates aligned with the fiber axis. Thus, the properties of bone are anisotropic. The interface between the mineral and protein phases is as yet not fully understood.

Bone has a hierarchical structure. The collagen fibers with their crystalline mineral deposits are assembled into larger structures ultimately forming bone with its commonly recognized macrostructure. As a composite, bone has a complex structure and a remarkable set of properties; and these occur by design in nature. The task of creating a synthetic composite to mimic bone is a challenge—one discussed further in the next chapter.

10.11 A COMPARISON OF COMPOSITES

We have outlined representatives of the main categories of composites, including particle-reinforced (cermets and concretes), fiber-reinforced, and structural (sandwich and laminar). Considering load distribution among reinforced composites, we have observed the following distinctions that arise from the structure:

1. Fiber-Reinforced—fiber is the primary load-bearing component.
2. Particle-Reinforced—load is shared by the matrix and the particles.

Some make a further distinction of dispersion-strengthened composites wherein the reinforcing particles are smaller in size than in typical particle-reinforced composites. Such a distinction applies really only to metals containing hard particles such as an oxide. In this case the matrix is the major load-bearing component while the particles help the matrix to resist deformation, making the material harder and stronger.

Regarding isotropy and anisotropy in composites:

1. Fiber-reinforced composites are usually anisotropic: some properties vary depending upon the geometric axis or plane in which they are measured.
2. To obtain isotropy in a specific property, all reinforcing elements—whether fibers or particles—must be randomly oriented. This is difficult to achieve even for short fibers since most processing methods tend to impart a certain orientation to the fibers.
3. Continuous fibers are typically used deliberately to render a composite anisotropic in the direction known to be the principally loaded axis or plane.

As mentioned in multiple sections above, the interfacial bond between the matrix and the dispersed phase is critical in determining the behavior of composites. Important aspects regarding interfaces are summarized as follows:

1. The interface is a boundary surface or region where a discontinuity (physical, mechanical, chemical, etc.) occurs.
2. For good adhesion, the matrix material must "wet" the particle or fiber. Coupling agents are sometimes used to improve wettability.
3. In FRCs, the applied load is transferred from the matrix to the fibers via the interface. The effectiveness of load transfer is dependent on size of the interface (determined in part by wettability) and bond strength between fiber and matrix. Failure at the interface (called debonding) may or may not be desirable.
4. Bonding at the interface can be through weak van der Waals forces, strong covalent bonds, or something in between.
5. Interfacial strength can be measured by inducing adhesive failure between the fibers and the matrix. The three-point bending test is commonly used for this purpose.

While the largest amount of work on composites is aimed at improvement of mechanical properties, there is still more to it. Thus, complex nanoparticles with the formula Pt-Fe_xO_y

are created and studied by Easterday and her colleagues [38, 39] for use as catalysts. One finds that alloys are not formed, the resulting materials consist of domains with different compositions. A new kind of morphology called "cluster-in-cluster" is seen [40]. While iron oxidation was expected from the start, formation of *amorphous* iron oxide is another unusual finding. Another combination of such nanoparticles includes Ru, RuO_2, and iron oxides—as catalysts for hydrogenation of nitrobenzene [40].

Finally, we have seen in natural composites an affirmation of many of the concepts set down above and simultaneously a complexity that still exceeds our full understanding. Our knowledge regarding composite materials is indeed expansive; and it continues to profit us in the development of new materials. At the same time nature continues to provide inspiration for new developments in composite materials.

10.12 SELF-ASSESSMENT QUESTIONS

1. Compare cermets with carbon steels for cutting tools applications.
2. Compare clad composites, sandwich composites, and laminates.
3. Why are geopolymers so named, and what are some applications?
4. What are the similarities and differences between mineral and polymer concretes?
5. Name a natural composite and describe its structure.

REFERENCES

1. G.W. Ehrenstein, *Polymeric Materials: Structure, Properties, Applications*, Carl Hanser Publishers: Munich/Cincinnati, OH **2001**, p. 129.
2. U. Beier, F. Wolff-Fabris, F. Fischer, J.K.W. Sandler, V. Altstädt, G. Hülder, E. Schmachtenberg, H. Spanner, C. Weimer, T. Roser, & W. Buchs, *Composites A* **2008**, *39*, 1572.
3. K.U. Kainer (ed.), *Metal Matrix Composites: Custom-made Materials for Automotive and Aerospace Engineering*, Wiley-VCH: Weinheim **2006**.
4. M. Ziemnicka-Sylwester, K. Matsuura, & M. Ohno, *ISIJ Internat.* **2012**, *52*, 1698.
5. M. Ziemnicka-Sylwester, *Adv. Sci. & Tech.* **2013**, *77*, 146.
6. M. Ziemnicka-Sylwester, *Materials & Design* **2014**, *53*, 758.
7. T. Christman, A. Needleman, & S. Suresh, *Acta Metall.* **1989**, *11*, 3029.
8. Y.-L. Shen, M. Finot, A. Needleman, & S. Suresh, *Acta Metall. & Mater.* **1985**, *43*, 1701.
9. M. Ziemnicka-Sylwester, *Materials* **2013**, *6*, 1903.
10. C.E. Smith, G.N. Morscher, & Z.H. Xia, *Scripta Mater.* **2008**, *59*, 463.
11. W. Brostow, T. Datashvili, J. Geodakyan, & J. Lou, *J. Mater. Sci.* **2011**, *46*, 2445.
12. W. Brostow, T. Datashvili, & J. Geodakyan, *Polymer Internat.* **2012**, *61*, 1362.
13. M. Böhning, H. Goering, N. Hao, R. Mach, & A. Schönhals, *Polymers Adv. Tech.* **2005**, *16*, 262.
14. C. Bonten & E. Schmachtenberg, *Polymer Eng. & Sci.* **2001**, *41*, 475.
15. V. Kominar, M. Narkis, A. Siegmann, & O. Breuer, *Sci. & Eng. Compos. Mater.* **1994**, *3*, 61.
16. M. Narkis & R. Simhony, Artificial marble and methods. US Patent 8729151 B2, Caesarstone Sdot-Yam Ltd., May 20, **2014**.

17. H. Lille, J. Koo, & A. Ryabchikov, *Mater. Sci. Medziagotyra* **2008**, *14*, 226.
18. (a) P. Kulu, I. Hussainova, & R. Veinthal, *Wear* **2005**, *258*, 488; (b) A. Zikin, M. Antonov, I. Hussainova, L. Katona, & A. Gavrilovic, *Tribology Internat.* **2013**, *68*, 45.
19. D.M. Roy, B.E. Scheetz, & M.R. Silsbee, *J. Mater. Ed.* **1993**, *15*, 1.
20. D.M. Roy, *Cement & Concrete Res.* **1999**, *29*, 249.
21. J. Davidovits, *J. Mater. Ed.* **1994**, *16*, 91.
22. O. Gencel, W. Brostow, T. Datashvili, & M. Thedford, *Composite Interfaces* **2011**, *18*, 169.
23. O. Gencel, W. Brostow, J.J. del Col Diaz, G. Martinez-Barrera, & A. Beycioglu, *Mater. Res. Innovat.* **2013**, *17*, 382.
24. O. Gencel, F. Köksal, G. Martinez-Barrera, W. Brostow, & H. Polat, *Mater. Sci. Medziagotyra* **2013**, *19*, 203.
25. O. Gencel, F. Köksal, & W. Brostow, *Mater. Res. Innovat.* **2013**, *17*, 92.
26. M. Kligys, A. Laukaitis, M. Sinica, & G. Sezemanas, *Mater. Sci. Medziagotyra* **2007**, *13*, 310.
27. G. Martinez-Barrera, E. Vigueras-Santiago, O. Gencel, & H.E. Hagg Lobland, *J. Mater. Ed.* **2011**, *33*, 37.
28. F. Koksal, O. Gencel, W. Brostow, & H.E. Hagg Lobland, *Mater. Res. Innovat.* **2012**, *16*, 7.
29. O. Gencel, C. Ozel, F. Koksal, E. Erdogmus, G. Martinez-Barrera, & W. Brostow, *J. Cleaner Production* **2012**, *21*, 62.
30. T. Uygunoglu, W. Brostow, O. Gencel, & I.B. Topcu, *Polymer Composites* **2013**, *34*, 2125.
31. O. Gencel, J.J. del Coz Dias, M. Sutcu, F. Koksal, F.P. Alvarez Rabanal, G. Martinez-Barrera, & W. Brostow, *Energy & Buildings* **2014**, *70*, 135.
32. W. Brostow, N. Chetuya, N. Hnatchuk, & T. Uygunoglu, *J. Cleaner Prod.* **2016**, *112*, 2243.
33. A. Beycioglu, O. Gencel, H.Y. Aruntas, W. Brostow, & H.E. Hagg Lobland, *Mater. Sci. Medziagotyra.* **2016**, *22*, http://dx.doi.org/10.5755/j01.ms.22.4.13354.
34. I. Tekin, M.Y. Durgun, O. Gencel, T. Bilir, W. Brostow, & H.E. Hagg Lobland, submitted to *Constr. Build. Mater.*
35. R.H. Marchessault & A. Sarko, *Adv. Carbohydr. Chem.* **1967**, *22*, 421.
36. W. Brostow, T. Datashvili, & H. Miller, *J. Mater. Ed.* **2010**, *32*, 125.
37. P.J. Flory, *Statistical Mechanics of Chain Molecules*, Wiley-Interscience: New York/London/Sydney/Toronto **1969**.
38. R. Easterday, O. Sanchez-Felix, B.M. Stein, D.G. Morgan, M. Pink, Y. Losovyj, & L.M. Bronstein, *J. Phys. Chem. C* **2014**, *118*, 24769.
39. R. Easterday, C. Leonard, O. Sanchez-Felix, Y. Losovyj, M. Pink, B.D. Stein, D.G. Morgan, N.A. Lyubimova, L.Zh. Nikoshvili, E.M. Sulman, W.E. Mahmoud, A.A. Al-Ghamdi, & L.M. Bronstein, *ACS Appl. Mater. & Interfaces* **2014**, *6*, 21652.
40. R. Yesterday, O. Sanchez-Felix, Y. Losovyj, M. Pink, B.D. Stein, D.G. Morgan, M. Rakitin, V.Yu. Doluda, M.G. Sulman, W.E. Mahmoud, A.A. Al-Ghamdi, & L.M. Bronstein, *Catal. Sci. & Tech.* **2015**, *5*, 1902.

11

BIOMATERIALS

Over the last several decades, an increase in longevity and life expectancy has raised the average age of the world's population. Among the countries currently classified by the United Nations as more developed (with a total population of 1.2 billion in 2005), the overall median age rose from 29.0 in 1950 to 37.3 in 2000 and is forecast to rise to 45.5 by 2050. This worldwide increase in the average age of the population has, in turn, led to a rapidly increasing number of surgical procedures involving prosthesis implantation, because as the human body ages, the load-bearing joints become more prone to ailments. This has resulted in an urgent need for improved biomaterials and processing technologies for implants.

—Soumya Nag and Rajarshi Banerjee in 2012 [1].

11.1 DEFINITIONS

A book unit describing the major categories of materials would be incomplete without a chapter on biomaterials. Biomaterials are comprised of a wide range of material types from the categories already discussed (see Figure 11.1). They serve distinct purposes, and they constitute a large sector of technology.

What is a biomaterial? By the consensus of experts in the field, a **biomaterial** is "A material intended to interface with biological systems to evaluate, treat, augment, or replace any tissue, organ, or function in the body" [2]. Hollinger notes that just as medical professionals must vow to "do no harm" to the patient, likewise biomaterials must "do no harm." Based on this definition, biomaterials are *not* equivalent to *natural materials* (which we have mentioned in previous chapters), though some biomaterials may be or incorporate natural materials. We shall discuss both types of materials in this chapter.

Materials: Introduction and Applications, First Edition. Witold Brostow and Haley E. Hagg Lobland.

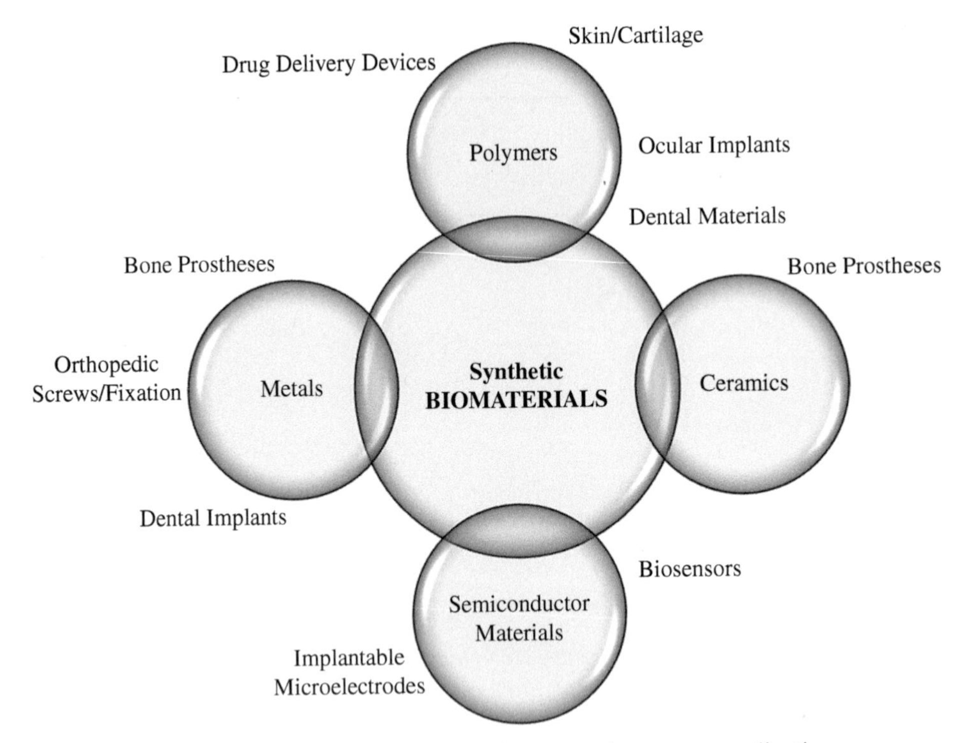

FIGURE 11.1 Biomaterials constituents and common applications.

Biocompatibility is "the ability of a material to perform with an appropriate host response in a specific application" [2]. The host response can be considered as "the reaction of a living system to the presence of a material" [2]. Hollinger further adds several descriptors that seem appropriate, yielding the following working definition of biocompatibility: "the ability of a material to function *in situ* with an appropriate and predictable host response in a specific application" [2]. As seen in other locations in this book, predictability of materials properties is desirable though not always easily attained; thus the addition of the descriptor "predictable" to the definition of biocompatibility is a sage one. Again, the corollary to "do no harm" applies here as to the definition of biomaterials. Moreover, it must do no harm as it is implanted, over time, or as it degrades (if it is biodegradable in the body). Alternatively, for good biocompatibility a material ought to have low **immunogenicity**, that is, the capacity to induce an immune response in the body. As an example, certain α-amino acid based bioanalogous polymers in the head-to-tail orientation work well as therapeutic materials; however the same polymers in other conformations—head-to-head and tail-to-tail—have a certain amount of immunogenicity [3]. Such polymers are similar in form to surfactants, pictured in Figure 9.6 and featured in the Chapter 9 discussion of emulsion polymerization. With surfactants we saw also molecules with structures described by a head and tail having different chemical functionalities.

You have probably observed that the prefix **bio-** is frequently used to indicate that a material contains a component of biological origin, irrespective of whether or not the material is intended for a biomedical application. In the broadest sense, these materials too are biomaterials. Biopolymers and other bio-based materials are discussed in various chapters throughout this book. This particular chapter contains three main divisions.

The first focuses on biomaterials as defined above, namely, those used in biomedical applications (subdivided and discussed in Sections 11.3 through 11.7). The second division defines and describes biological or natural materials (Section 11.8), and the third is about what we call bio-based materials (Section 11.9).

Before going on, we shall quickly note a class of material structures that shows up often in biomaterials, namely, gels. A **gel** (from the Latin *gelu*—freezing, cold, ice or *gelatus*—frozen, immobile) is characterized as a jelly-like material, however gels range in type from soft and weak to hard and tough. Gels are typically comprised of networked polymers, which we described in Section 9.7, although the network may also be colloidal in nature. The distinguishing feature of a gel is that its structure is expanded through the whole volume by a fluid. The term **hydrogel** is used when the fluid is water. Such fluid-containing three-dimensional crosslinked networks exhibit no flow in the steady-state. It is not surprising therefore that though gels are mostly liquid by weight, they behave like solids, deriving mechanical strength from the crosslinking. Hence, one can consider gels as a dispersion of liquid molecules within a solid such that the solid is the continuous phase and the liquid is the discontinuous phase.

11.2 OVERVIEW OF BIOMATERIALS AND APPLICATIONS

In 2000, Harvard University conducted a survey among professors of medicine in the United States. There was only one question: Which development in the 20th century was the most important for the progress of Medicine? More than 700 replies, with a large variety of answers, were received. Nevertheless, the most frequent answer was: *new materials*. This answer makes good sense. The discipline of Medicine has been in existence for some thousands of years, including the aspect of surgery. Because of this, it is known how to remove many kinds of damaged parts from a human body. Owing to advancements in materials technology, we now have increasing options for using materials in the body not just to replace damaged components for other purposes, as well. Refer to Table 11.1 for a partial list of common applications of biomaterials in human medicine. Among these, we can distinguish materials which are used externally, such as artificial limbs, from those used internally, where biocompatibility is critical.

There are interesting analogies between structure and function of water pipelines and blood vessels [4]. One can read exclusively about ceramic biomaterials [5], while Buddy

TABLE 11.1 Common Biomedical Applications of Materials

Joint replacements
Dental materials
Ophthalmological applications
Drug delivery devices
Biosensors
Cardiovascular devices
Cardiac assist devices
Artificial red blood cell substitutes
Burn dressings and skin substitutes
Sutures
Cochlear prostheses

Early biomaterials	Gold: Malleable, inert metal (does not oxidize); used in dentistry by Chinese, Aztecs, and Romans—dates 2000 years Iron, brass: High strength metals; rejoin fractured femur (1775) Glass: Hard ceramic; used to replace eye (purely cosmetic) Wood: Natural composite; high strength to weight; used for limb prostheses and artificial teeth Bone: Natural composite; uses: needles, decorative piercings Sausage casing: cellulose membrane used for early dialysis Other: Ant pincers. Central American Indians used to suture wounds
1860s	Lister develops aseptic surgical technique
early 1900s	Bone plates used to fix fractures
1930s	Introduction of stainless steel, cobalt chromium alloys
1938	First total hip prosthesis
1940s	Polymers in medicine: PMMA bone repair; cellulose for dialysis; nylon sutures
1952	Mechanical heart valve
1953	Dacron (polymer fiber) vascular grafts
1958	Cemented (PMMA) joint replacement
1959	Implantable pacemaker
1960	First commercial heart valves
1960s	Liposomes described
1970s	PEO (poly(ethylene oxide)) protein resistant thin film coating
1976	FDA amendment governing testing & production of biomaterials/devices
1976	Artificial heart (W. Kolff, Prof. Emeritus, University of Utah)

FIGURE 11.2 Abbreviated timeline of advances in biomaterials through the 1970s. Recent advances are too numerous to list and span many specialties in biomedicine.

Ratner has compiled a detailed history of biomaterials in his textbook dedicated to the topic [6]. A brief timeline of advances in biomaterials is given in Figure 11.2.

Although many readers of this book may find the long history of biomaterials quite interesting, in light of the goal of this book, we shall now focus primarily on the *structure, interactions, and properties of biomaterials*.

11.3 JOINT REPLACEMENTS

Among other reasons, the increasing life expectancy of the population, the hard use of joints in athletics, and even the underuse of muscles (through sedentary lifestyles) has brought an ever-increasing need for materials specifically suited for bio-implantation in the joints. Already in the year 2000, some $12 billion was spent on biomaterials for joint replacements; this number is increasing each year. Several hundred thousand hip replacement procedures are conducted every year in the United States. In spite of the clinical success of these replacements, the implanting of an artificial joint remains undeniably a highly invasive procedure.

Case Study: Hip Joint Replacement

(a) *What Is the Need?*

The human hip joint undergoes high levels of mechanical stress, and in the course of a lifetime suffers considerable abuse. Hip (and knee) replacements may be among the most familiar orthopedic operations of the present day. Not only has the number of replacement surgeries increased, but also the number of revision surgeries of these implants has risen because of higher life expectancy. Therefore, development of materials for long life implantation in human body has become a priority.

This is not an easy task by any means. For instance, testing the wear, strength, etc. of implant materials must be based on reliable data and ISO standards. As testing and modeling methods have improved, decades old data informing ISO standards needs replacement. Bergmann and his coworkers at Charité in Berlin [7] have developed instrumented knee implants that enable measurement of tibio-femoral contact forces *in vivo*. This is important because the development of better implant materials and surgical procedures depends on the acquisition of real-time data and force acquisition during normal human activities. By the methods utilized in Bergmann's study [7], more and better data on loads during various kinds of movement—walking, ascending stairs, standing up, jogging, etc.—were collected, assessed, and correlated to percent body weight, thereby providing realistic guidelines on what can be considered an average, high, or extreme load.

(b) *What Materials Can We Use?*

Hip-joint prostheses have been fabricated from titanium, stainless steel, high-strength metal alloys, ceramics, composites, and polymers. The dominant materials used for orthopedic applications are: titanium alloy, Co–Cr–Mo alloy, stainless steel, poly(methyl methacrylate) (PMMA), ultrahigh-molecular-weight polyethylene (UHMWPE), alumina (Al_2O_3), and zirconia (ZrO_2). Each of these materials has been declared biocompatible, but their properties vary rather widely; none of them matches the properties of bone *exactly*. We have already seen (in Chapter 10) that *bone is a complex composite material*, thus each material type offers particular advantages and disadvantages that must be weighed for the benefit of a given patient. Hip prosthetics consist of multiple parts—cup, head, and stem—with the cup and head components simulating the ball and socket joint. Different combinations of materials may be used for each of these components, as illustrated in Figure 11.3. We

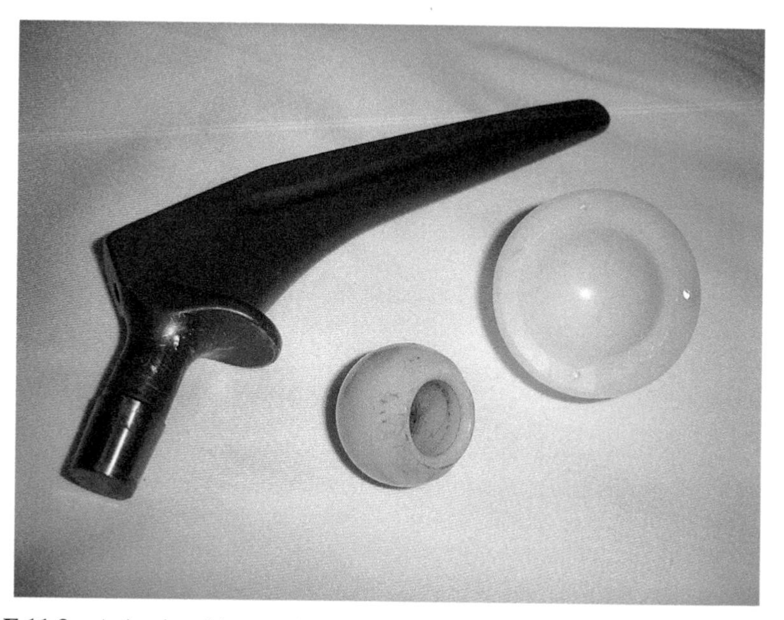

FIGURE 11.3 A titanium hip prosthesis (stem), with a ceramic head and polyethylene acetabular cup. *Source*: After Nuno Nogueira, https://en.wikipedia.org/wiki/Hip_replacement#/media/File:Hip_prosthesis.jpg. Public Domain.

have discussed in earlier chapters the importance of interfaces; in this case also the behavior at the bone-biomaterial interface is important for healing [8].

(c) *What Properties Do We Have to Consider?*

Biomedical implants must meet the multiple and varied demands of the human body. One must consider both biomechanical properties (stiffness, strength, fracture toughness, wear resistance, fatigue strength) and biomedical properties (toxicity, surface state, osseointegration, corrosion resistance). Biomechanical properties should match those of human tissues without adverse effects (see definition of biocompatibility). Good processability (e.g., by casting, plastic forming, powder metallurgy, machining, welding, etc.) is also highly desirable to reduce the cost.

Stiffness. In designing joint replacements, which substitute for bone, matching of Young's modulus (also called elastic modulus and closely linked to stiffness, explained fully in Chapter 14) between the implant and the bone is one of the most important requirements. When an alloy with a stiffness mismatch to bone is inserted into a load-bearing bone, the implant takes over a considerable part of the load, which shields the bone from the necessary stressing required to maintain its strength, density, and healthy structure. This effect, called stress shielding, eventually causes bone resorption by the body, which may in turn lead to complications such as bone fracture, implant loosening, and premature failure. Those alloys with high stiffness are not generally recommended for implants used for long term implantation (more than 10 years) in load-bearing bones. Therefore, to decrease the number of revision surgeries, researchers have aimed to decrease the Young's modulus of implant materials. The elastic modulus of human cortical bone and cancellous bone are approximately 4–30 GPa and 0.2–2 GPa, respectively. Titanium alloys are among the best at fulfilling the desired combination of high strength and low modulus (~55 GPa) to match the mechanical properties of real human bones.

Tribological properties. Besides the mechanical stresses applied at the joint, there is considerable friction involved in joint movement, which can cause wear on bone surfaces. The fatigue due to normal activity may in some persons cause the materials of the natural joint to wear out. Likewise, the movement of an artificial hip joint produces billions of microscopic particles that are rubbed off during motions. These particles are trapped inside the tissues of the joint capsule and may lead to unwanted foreign body reactions. *Wear* and *abrasion resistance* (discussed in Chapter 19) mainly determine the service period of an implant because low wear resistance causes implant loosening due to osteolysis and inflammatory reactions from wear debris.

Corrosion resistance. Additionally, corrosion represents an important aspect of wear in orthopedic implant materials. Materials implanted *in vivo* initially come in contact with extracellular body fluids, such as blood and interstitial fluids which contain a high concentration of chloride ions, amino acids, and proteins. This corrosive environment can cause the release of metal ions into the body fluid for a prolonged period of time; ions may combine with biomolecules, such as proteins and enzymes, leading to allergic and toxic reactions. Some titanium alloys are considered to be among the most biocompatible and corrosion-resistant of implant materials. When in contact with body fluids having close to neutral pH, titanium alloys exhibit corrosion rates that are extremely low. This is primarily due to the chemical stability and structure of the native titanium dioxide film (TiO_2) that grows spontaneously on the material's surface upon exposure to air and that protects the metal from further oxidation. Although titanium alloy implants tend to show no visible signs of corrosion, they often contain potentially harmful metals, such as Al and V (Ti-6Al-4V), as alloying elements. Shape-memory alloys (see Section 12.6), which are used in surgical and other biomedical procedures, typically contain nickel (in TiNi), another element known to cause allergic reactions.

(d) *How Can We Make Some Improvements?*
To achieve an even lower Young's modulus that would enable homogeneous stress transfer between implant and bone, several methods such as thermomechanical treatments (TMT), control of microstructure by composition, and design of porous alloys are being investigated.

Various approaches have also been suggested to mitigate the possible release of potentially toxic metals into the body. One research thrust in this area is surface modification of titanium alloys [9]. Surface modifications can be designed not only to improve corrosion resistance but also to form a barrier that prevents leakage of metal ions from the bulk. Developments in surface modification include, for example, a novel Power Immersed Reaction Assisted Coating (PIRAC) nitriding method that can be applied even to complex shapes characteristic of implants [9–11]. For example, the PIRAC nitriding method has been shown to improve the corrosion-resistance for nickel-titanium (NiTi, Nitinol) shape memory alloy [12–15]. Moreover, such nitride coatings can improve other aspects of wear, such as from fretting that occurs in joint replacements [13].

Nevertheless, there are still unanswered questions and unresolved problems relating even to commonly used orthopedic biomaterials. For instance, the *wear* of UHMWPE and of metals has to be mitigated, either by improved materials, improved design, or otherwise. Wear of materials is an important topic discussed in Section 19.6. Also, as there are increasing numbers of patients with metallic implants, more information needs to be acquired on the clinical significance of elevated metal content in body fluids and remote organs [16]; this is in addition to the need for assessing the effects of internal metal objects on organ function via disruption of acupuncture meridians.

(e) *Can We Think Out of the Box and Find Alternative Solutions?*
Strategies involving **tissue engineering**—defined as using the regenerative capacity of the body to develop a tissue or an organ within the body—are an established means to improve bone repair [17]. Important here are **scaffolds**, that is frameworks or structural elements that hold cells or tissues together maintaining their shape. Artificial scaffolds are often porous, allowing in-growth of the natural cells.

One of the options in hip (and other joint) replacement involves *injectable tissue engineering*. This entails various approaches that use injectable, *in situ* gel-forming systems. A flowable, polymeric gel-forming matrix may incorporate therapeutic agents for improved efficacy; additionally the gel can fill any shape of defect, does not necessarily contain residual solvents (as some preformed scaffolds may), and importantly does not require open surgical procedure for placement. Another new avenue in tissue engineering—not limited in application to joint replacements only—is the use of liposomes tagged with magnetic nanoparticles [18]. Castro and Mano [18] describe how magnetic force can be utilized in this technology to produce functional tissues.

11.4 DENTAL MATERIALS

Dental materials are utilized as **obturations** for filling cavities and as **implants**, substituting for entire teeth (once a crown is attached to the implant). Polymeric resins and composites are primarily used for the former, metals for the latter. In both applications, formation of a tight seal to prevent bacterial invasion into the gum (gingiva) is a critical requirement. The design of suitable dental materials is in fact quite challenging, and it remains therefore an active area of research as there is still room for improvements upon the current commercially available materials. Mercury amalgam, which was long used in tooth fillings, is now known to be toxic and is therefore not as widely used. Polymers and ceramics each have certain advantages as dental obturations but are still limited by various drawbacks. Hybrid (composite) materials are already in use and will likely be the predominant materials for dental obturations and implants in the future.

Since the natural tooth consists of both inorganic and organic constituents, synthetic obturation materials typically mimic this combination. Hard polymer resins are combined with ceramic powders—even hydroxyapatite, which occurs naturally in enamel—to make solid or porous materials for tooth repair. Polymers and ceramics—especially hydroxyapatite (HAp) and tricalcium phosphate—are also used individually as dental implant materials. In spite of the fact that HAp is a naturally-occurring component of teeth, its high level of performance cannot be perfectly mimicked in synthetic HAp implants or obturations. Nevertheless, hydroxyapatite—and also titanium—are desirable because their capacity for integration with bone is predictable. Titanium has reasonable stiffness and strength, which renders it suitable for implants, while HAp can be applied as a coating to improve biocompatibility and biointegration. Man-made HAp is not that good for load-bearing applications, but calcium-deficient HAp can be reacted with hexamethylene diisocyanate, treated with polylactide, and then pressure applied. The resulting composites have compressive strength (see Section 14.2) suitable for load bearing orthopedic applications [19].

ENVIRONMENT AND SPECIAL PROPERTIES OF DENTAL MATERIALS

The mouth is a rather inhospitable environment. Apart from exposure to food, drink, saliva, and bacteria, the process of mastication (chewing) imposes repetitive mechanical stress on teeth. Tooth enamel *is the hardest substance in the human body*. Thus dental materials must have sufficient strength and stiffness to withstand the mechanical stress of chewing as well as resistance to degradation or staining by food and drink.

Surface properties of obturation materials are important because both mastication and tooth brushing expose the tooth surface to repetitive scratching and abrasion [20]. Synthetic obturation materials must therefore exhibit very high wear resistance [21–23]. Surface properties are important also because there appears to be a correlation between surface roughness and bacterial adhesion [24]. Plaque formation is also affected by surface characteristics. Furthermore, dental biomaterials must interact with the dentin and underlying tissue. If dental materials are not incorporated well into the existing gum and tooth structures, the materials will loosen over time, compromising their mechanical performance and perhaps inviting bacterial infection. One avenue to minimize this problem is to design materials that are better able to bond to the biological tissues. Another avenue pursued by some is creation of porous materials that are not only capable of bonding to underlying tissue but also allow **vascularization** (the formation of blood vessels) through the porous network [23]. The porosity in these and in some bone implant materials may also be designed to encourage **biomineralization** (*in vivo* mineralization) of the implanted material, thereby advancing its integration with the pre-existing tooth or bone structure.

Zirconia (ZrO_2) ceramics (see Section 7.9 for more on zirconia and advanced ceramics) are emerging as another viable material for dental implants, particularly as they offer improved aesthetics. Zirconia can be deposited onto Ti, or entire implants may be composed of the material [24]. Additionally, adhesives and sealants are important biomaterials utilized in dental applications [25]. These include tooth and bone cements and filling materials.

The inhospitable mouth environment becomes still much worse if **periodontitis** (non-inflammatory degradation of the tooth bone tissue leading to loosening and eventually loss of teeth) takes place. An interesting attempt to deal with periodontitis has been developed by Bottino, Janowski, and their colleagues [26, 27]. They create *membranes with graded structures*, stable for a length of time that is sufficient for periodontal regeneration. A core layer comprised of neat poly(DL-lactide-*co*-ε-caprolactone) is surrounded by two composite layers containing a protein, hydroxyapatite, and an anti-pathogen component. Individual layers are deposited by electrospinning under carefully optimized parameters. The grading of structures by such a method provides the means to tailor the time stability and improve periodontal outcome.

11.5 VASCULARIZATION IN CARDIAC AND OTHER APPLICATIONS

Needless to say, vascularization—the formation of blood vessels—is important not only to obtain high performing dental materials but also for biomedical components used in conjuncture with the vasculature of other parts of the body. A number of disciplines come together here: Chemistry (preparation of materials), Mechanical Engineering (mechanical performance), Biology (biological stability—or lack of it), and Rheology (an aspect of MSE dealing with fluid flow).

One class of materials used in this area are **polyurethanes** (PUs), which are polymers containing **urethane** –NH–(C=O)–O– groups as links between the molecular units. Polyurethanes can be soft or hard, thermoplastics or thermosets (more often); hence a wide range of properties and applications is possible. However, PUs are subject to biodegradation—by hydrolysis, oxidation, or both—in the body. Burriesci, Seifalian, and coworkers [28] have solved this problem by using a POSS (polyhedral oligomeric silsesquioxane) with the chemical formula $RSiO_{3/2}$, where R is a functional group such as alkyl or alkene and the molecule has the spatial structure of a cage. They have attached polyhedral oligomeric nanoparticles (the POSS unit) as pendant chain functional groups to the backbone of a poly(carbonate urea) urethane (PCU). The resulting nanocomposite is called POSS-PCU or else UCL-Nano™ since it was created at the University College London. Polyurethanes containing **siloxane** SiOSi groups have higher biostability than other PUs.

POSS-PCU is being used to manufacture small-caliber vascular bypass grafts. Moreover, the material is also implanted in humans as lower limb bypass grafts, lacrimal duct conduit, and recently as a tracheal replacement. The **lacrimal duct** carries tears from the lacrimal sac into the nasal cavity. The **trachea**, popularly called **windpipe**, is a tube that allows the passage of air to the lungs, hence it is present in all air-breathing animals with lungs.

POSS-PCU lacrimal ducts [29] are a replacement for rigid tubes that have been in use since the 1960s. The rigid tubes suffer from complications: displacement and blockage of the tube (requiring regular checkups) as well as irritation of the surrounding tissue including the nose and the eye. Biocompatible POSS-PCU tubes do not have these problems. Creation of the POSS-PCU structures involved a variety of processing techniques: ultrasonic atomization spraying, electrohydrodynamic atomization spraying/spinning, extrusion-coagulation, and high-pressure coagulation by autoclave and casting. Eventually, coagulation and casting techniques were chosen as the best [29].

Another topic that falls under the heading of vascularization is **heart valves** (HVs); HVs should provide unidirectional smooth blood flow through the heart chambers and through the main vessels by regular opening and closing throughout each cardiac cycle—this while maximizing flow rate and minimizing flow resistance. This vital function can be affected by valvular heart diseases, leading to either obstruction in the blood flow (stenosis) or backward leakage of the blood (regurgitation), or both in the worst-case scenario. Actually, slight backward leakage is always present and provides a good way to evaluate the quality of artificial valves.

At least two options to remedy the situation exist: the first is **tissue engineering**, mentioned above in Section 11.3, which utilizes the regenerative capacity of the body to develop a tissue or an organ within the body. Consider the vessels through which blood flows. The **endothelium** is the thin layer of cells that lines the interior surface of blood vessels (or of lymphatic vessels for that matter); these cells are at the interface between circulating blood and the rest of the vessel wall. In the cardiovascular system, **endothelial progenitor cells** (EPCs) are endothelial cells with known regenerative capability. They potentially can be used in developing partially or totally viable tissue engineering devices such as heart valves. Synthetic biodegradable polymers are used as scaffold materials [30]. Here also POSS-PCU is particularly useful due to its capability to retain EPCs on its surface and endorse their proliferation and differentiation to mature endothelial cells [31, 32]. See now Figure 11.4; biofunctionalization of the polymer surface provides cell-specific adhesion motifs that aid attachment of EPCs. This promotes in situ **endothelialization** (formation of endothelial tissue) and significantly impacts the processes of neovascularization and **angiogenesis** (formation of new blood vessels from existing ones).

It turns out that in synthetic material interactions with cells the topography of the interfaces is important. Lensen and her coworkers work on cellular responses of synthetic biomaterials. They have studied hydrophilic star-shaped poly(ethylene glycol) (PEG) and hydrophobic linear perfluorinated polyether (PFPE) [33]. Smooth unstructured elastomers showed no cell adhesion; apparently PEG was too hydrophilic and PFPE too hydrophobic. However, a UV-based imprinting process can provide various micrometer size topological patterns on the PEG and PFPE surfaces. The imparting of new topography resulted in strong adhesion of mouse fibroblast cells with line patterns, having cells elongated along the line rims. Pillar patterns caused "floating" of cells on tops of such surfaces.

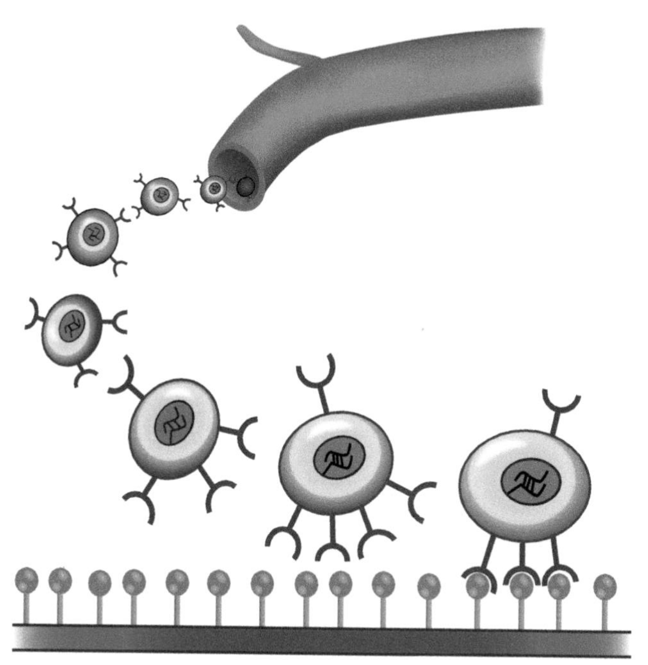

FIGURE 11.4 Schematic illustration of surface modification of synthetic materials by biofunctional peptides and nanotopographic features. Biofunctionalization of the polymer surface (shown as a flat blue plane) can facilitate the attachment of endothelial progenitor cells (EPCs, shown as blue cells) in the circulatory blood by providing cell-specific adhesion motifs (shown brownish gold) on the surface of the synthetic biomaterial that promote in situ endothelialization. Endothelial progenitor cells can proliferate and differentiate to mature endothelial cells and have an important role in neovascularization and angiogenesis. Potentially, they can be used in the endothelialization process of biomaterials. *Source*: Sarkar *et al.* [28]. Reprinted with permission from Elsevier.

We said above that dealing with defective heart valves we have at least two options. A variation on the first option is the use of swine heart valves for installation in humans. As a second option, instead of tissue engineering, we can create fully **synthetic heart valves**, including so-called mechanical heart valves. While this option was tried for a long time, it was for a long time unsuccessful, with high mortality rates. Two developments have now made synthetic heart valves into a viable alternative. First are new materials—including nanomaterials such as UCL-Nano discussed above—and second is improved *design*. A novel design for polymeric heart valves, namely, a semi-stented wire-reinforced valve characterized by suture loops for attachment at the sino-tubular region of the aortic root, has been developed by Burriesci, Seifalian, and collaborators [32]. Using finite element analysis followed by experimental validation, the proposed valve design was optimized to reduce the energy absorbed during the operating cycle, a feature that results in high hydrodynamic performances and reduced stress levels. The leaflet design of this semi-stented aortic valve (SSAV) facilitates easy surgical insertion with minimal damage to the native tissue and improved hemodynamic performance. A schematic of the SSAV device is shown in Figure 11.5. The polymer leaflets of the valve are reinforced with a supportive medical-grade titanium wire frame (wire diameter 0.55 mm). The SSAV is less obstructive and more flexible than other

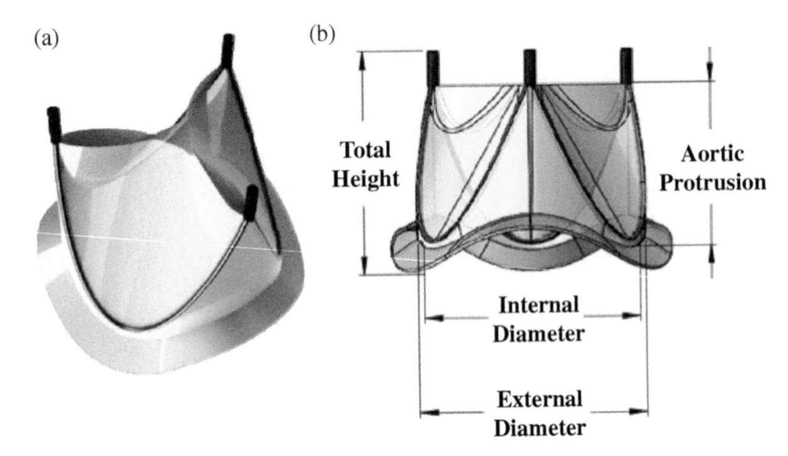

FIGURE 11.5 Representation of the SSAV (semi-stented aortic valve): (a) isometric view and (b) technical specifications used: total height=20mm, aortic protrusion=14, internal diameter=21mm, and external diameter=22mm. The SSAV is based on a design strategy aimed at reducing the energy absorbed during the operating cycle, which results in high hydrodynamic performances and reduced stress levels. The valve design also aims to facilitate easy surgical procedure with minimal damage to the native tissue and improved hemodynamic performance. *Source*: Ghanbari *et al.* [30]. Reprinted with permission from Elsevier.

valves, features that result in less stress concentration on the components. Thus, it is clear how the SSAV results from a combination of material advancements and novel design features. Important in such design is avoidance of stagnation areas in the valve vicinity.

Sutures used in the aorta region are of course a special case. In general, a **suture** is a row of stitches (or a single stitch) holding together the edges of a wound, a trauma, or a surgical incision. Installation of sutures, as we know, involves using a needle with a thread attached to it, while experience has clearly shown that sutures accelerate healing. We can classify sutures into three categories: those that are dissolved by the body; those that are removed after a period of time, such as 3–5 days for facial wounds; and those that are intended to remain in the body, called permanent sutures.

Ingenuity of materials scientists and engineers is important for sutures also. Alejandro Müller, Robert Prud'homme, and their colleagues have studied the hydrolytic degradation of sutures made from high molecular weight poly(*p*-dioxanone) (PPDX) [34]. As we know from Chapter 9, polymers are either amorphous or semicrystalline. Poly(*p*-dioxanone) is semicrystalline. In distilled water or in a phosphate buffer at 37°C, the hydrolytic degradation occurs in two stages; the amorphous regions are attacked first, the crystalline regions later. The crystalline phase consists of spherulites. One can affect their morphology and the thickness of the crystalline regions by varying the supercooling rate during crystallization [34].

11.6 INTRAOCULAR LENSES AND CONTACT LENSES

Intraocular lenses (IOLs) are used to replace the natural lens of the eye after cataract formation. Intraocular lenses are typically made of poly(methyl methacrylate) (PMMA), soft acrylic polymers, silicone, or hydrogels (see Section 9.7 and discussion of gels in

Section 11.1). By most parameters, the latter three (newer materials) are better than PMMA lenses, which have long been the standard.

Hydrogel lenses contain polyhydroxyethyl-methacrylate (polyHEMA) as a neat polymer or co-polymerized with another acrylic monomer. Non-HEMA soft acrylic lenses are varied in their monomer composition. Intraocular silicones are deformable solid silicone elastomers. These silicones are of high molecular weight and are extensively cross-linked. While PMMA is only flexible when it is thinner than 0.12 mm, soft acrylics, silicones, and hydrogels are foldable lenses that can be inserted through a small incision in the eye. The ideal IOL material would not incite any immune response, but in fact they all do incite some response, though biocompatibility may be somewhat better with hydrogel or silicone.

Soft contact lens materials—contrasted with IOLs—are typically made of flexible hydrogels (containing from around 24% to 70% water content). However, since the 1930s, there has been a rather continuous evolution in the development of soft contact lens materials. Contact lenses are in intimate contact with the cornea, but unlike IOLs, they are not permanent. The first commercial contact lenses were made of PMMA, a polymer resin with optical clarity better than glass. In response to improved understanding of the physiological needs of the cornea, hydrogel lenses were developed and have since altered forever the contact lens industry. Siloxane-rich hydrogel lenses now provide good oxygen permeability to the eye while correcting vision with good biocompatibility.

SURFACE PROPERTIES AND PERMEABILITY OF CONTACT LENSES

The cornea uses oxygen to maintain its clarity, structural integrity, and overall function. The needed oxygen is taken from the air, thus permeability of oxygen through the lens is a key requirement for suitable contact lens materials—an area in which early contact lenses, including PMMA, performed poorly. Compatibility with the eye also requires that the material "maintain a stable, continuous tear film for clear vision", be "resistant to deposition of tear film components", sustain "normal hydration", be "permeable to ions", and "be non-irritating and comfortable" [31]. Clearly then surface characteristics of contact lens materials are critical. The materials should be neither hydrophobic nor lipophilic while yet possessing a suitable polymer composition and appropriate morphology. Further reviews of soft contact lens materials are available elsewhere [35, 36].

The crown achievement is improved oxygen permeability: modern contact lenses do not quickly deprive the eye of its health by oxygen starvation but rather provide improved comfort and allow for longer wear with high oxygen availability. Contrasted with the first hydrogel lenses that were homogeneous, modern contact lenses are heterogeneous hydrogels: they consist of water-rich hydrophilic phase with a water-poor hydrophobic phase of siloxane moieties dispersed throughout. In the early lenses, water was the primary means of oxygen permeability. In newer lenses, oxygen permeability occurs mainly through the siloxane-rich phase. To render the lens surface hydrophilic, the lenses must first be treated with radiofrequency (RF)-plasma (or by another suitable procedure) since a hydrophobic surface is not tolerated by the eye.

11.7 DRUG DELIVERY SYSTEMS

Drug delivery systems are used for the controlled release of pharmaceutical compounds to specific target tissues. Specialized materials are used to serve as such systems. Polymers are the primary material type used as drug carriers. Diffusion, water penetration, and polymer degradation are the three major mechanisms that control the rate of drug release [37].

Monolithic drug delivery systems that are *diffusion-controlled* consist of a polymer matrix in which a pharmaceutical compound is dissolved or dispersed. The rate of drug release is dependent on how it diffuses out of the polymer. Thus, important factors are degradation rate of the polymer matrix, solubility of drug in the matrix, porosity, and geometry of the polymer matrix. There are three possible scenarios of monolithic diffusion-controlled systems: (1) a dissolved drug diffusing through the polymer; (2) a dispersed drug diffusing through the polymer; and (3) a dispersed drug diffusing through channels. In the first two cases, diffusion through the polymer matrix is the rate-limiting step. The impetus for drug release is the drug concentration gradient between the matrix and its environment [32]. For this to hold true in cases (1) and (2), the polymer should be non-degradable or else degrade slowly so that its integrity is maintained for the duration of drug diffusion. In scenario (1), the drug is loaded—often via simply immersing the polymer in a drug-containing solution—at a concentration level below saturation. By contrast, in scenario (2) the drug is loaded to a concentration exceeding saturation, and the delivery device may be assembled by compression molding or solvent casting. The mathematical model for the mass of drug released over time assumes that the drug is uniformly suspended as minute particles, which dissolve and diffuse away, moving the interface between dissolved and dispersed particles continuously towards the interior of the system. A good way to disperse the drug is via extrusion of the polymer + drug material [38]. Whether this approach is applicable, depends on the miscibility of the two components. In turn, the miscibility can be evaluated from determination of dependence of the glass transition temperature T_g on the drug concentration [38].

Case (3) is similar to (2) except that channels are created in the wake of drug particles that have dissolved or diffused away. The presence of plasma in the channels, the number of channels, and the tortuosity of the channels all affect the diffusivity of the system. The rate of release in these monolithic systems tends to decrease with time. Therefore rate-controlling membranes can be used to alter the drug-release kinetics.

Apart from diffusion-controlled systems, there are *water penetration-controlled systems* and *polymer degradation-controlled systems*. In the former category are osmotic pumps, which expel drug from a front compartment as water enters and expands a compartment on the back side of a moveable semi-permeable membrane. Also in this category are swelling-controlled systems. We already know from our study of polymers that swelling is possible. One system consists of the drug in a dehydrated hydrophilic crystalline polymer matrix. Water penetration into the system induces a transition from the crystalline to an amorphous phase thereby accelerating drug release from the system. Another type of swelling-controlled system allows for a delay period followed by a rapid drug release [37]. In such a device, a drug and polymer are encapsulated within a semipermeable membrane allowing water penetration. Eventually the swelling causes the membrane to rupture, releasing the drug all at once.

Polymer degradation-controlled systems are based on the biodegradability of the constituent polymers. Under physiological conditions, the polymers undergo hydrolysis into nontoxic molecules. Like other systems already mentioned, these may consist of a drug dispersed in a polymer matrix. Water penetration, however, causes degradation of the bonds between monomers, thereby allowing the drug to escape. Alternatively, a drug compound may be bonded as a pendent on the polymer backbone. In this case, hydrolysis of the bond between the polymer chain and drug agent causes release. The capacity for high drug loading is an advantage of this type of system.

Heller and Hoffman discuss additional drug delivery systems including a variety of responsive systems and particulate systems [39]. For instance, bone defect reconstruction can be benefited from nanoparticulate systems based on hydroxyapatite (HAp). Scientists from Belgrade, San Francisco, and Nis have prepared HAp-based multifunctional nanoparticulate systems for "rapid and sustained local delivery of cholecalciferol," an important regulator of ion concentration, and for the secondary healing effect of the HAp carrier itself [40].

Most drug delivery systems utilize macromolecular materials, and owing to the variety of system designs and treatment options they form an interesting area of polymer research. For example, an injectable, thermally responsive adhesive gel used for sustained protein delivery spans the boundary of biomaterial and smart material [41]; refer to Chapter 12 for more on smart materials. The modes of targeting drug delivery systems to specific tissues are also varied. For example, the liposome pictured in Figure 11.6 has peptides on its surface that are recognized by specific cell types. The drug itself is contained within the liposome and is released when the liposome gets incorporated into the target cell.

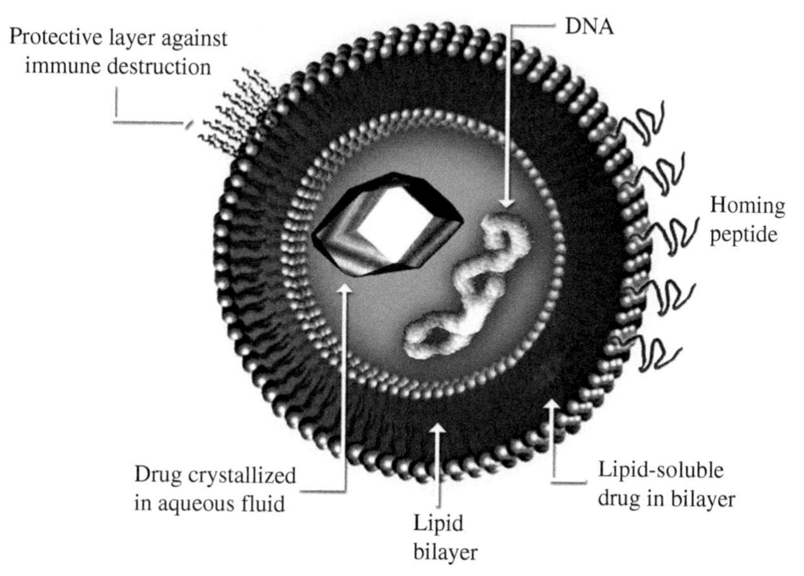

FIGURE 11.6 Schematic of a liposome used as a drug delivery system. *Source*: After https://en.wikipedia.org/wiki/File:Liposome.jpg. Public Domain.

POLYMER CARRIER SYSTEMS FOR CANCER TREATMENT

A practical application of polymer carriers is the treatment of cancers. Chemotherapy is well-known not only as an effective way to destroy tumors by intravenous injections of highly toxic drugs but also, unfortunately, for its side effects (nausea, hair loss, increased susceptibility to illness, cardiotoxicity, and so on). Owing to the negative side effects of chemotherapy, medical practitioners often restrict the dose and frequency of injections to a level that is not sufficient to inhibit a tumor's growth. It has been proposed that water-soluble polymers could provide a means to increase drug circulation time, improve drug solubility, prolong drug residence time in a tumor, and reduce toxicity [42, 43]. In this kind of system, the drugs are covalently attached to the polymers via reversible linkages. Cytotoxic drugs attached this way more effectively target tumor tissue than the drugs alone. It is suggested that improved therapeutic effectiveness can be achieved through inhibiting the unwanted passage of drugs through certain pores by controlling and optimizing the following properties: hydrodynamic volume (or molecular weight), molecular conformation, chain flexibility (which affects passage through pores of the kidneys, much smaller than the ones of tumors), branching, and location of the attached drug.

GENE THERAPY

Human hereditary diseases are written in the DNA. Once the chromosomal defects are identified, a cure might consist in correcting the genetic defects by "transferring cloned and functionally activated genes into afflicted cells" [41]. Although viruses are very efficient vectors, the potential risks in using them have not been fully explored and dispelled.

Cationic polymers have the ability to interact electrostatically with the phosphates on DNA to form a compact particle. Therefore, synthetic DNA-delivery systems, using cationic polymers such as polylysine, intact or fractured polyamidoamine (PAMAM) dendrimers (synthesized from methyl acrylate and ethylenediamine), and polyethylenimine, are being extensively studied [42]. Their cytotoxicity, complex formation, and transfection efficiency are the main parameters under investigation. Transfection efficiency can sometimes be improved by incorporation of other chemical agents.

These cationic polymers interact with DNA by the formation of a complex that has a toroid morphology. An interesting observation is that the polymer structure (linear, radial branched, random branched…) does not drive the condensation and compaction of DNA into the toroidal form. The structure of the resulting complexes is only determined by the electrostatics of interaction.

11.8 BIOLOGICAL AND NATURAL MATERIALS

In the field of MSE one considers an object of study from two perspectives: both science and engineering. The area of biomimetics is particularly a playground for the enthusiastic engineer. We are basically talking about turning to nature for inspiration in creating novel materials and components. To be precise, the technology of **biomimetics** seeks to duplicate the processes that living organisms use to construct bones, tissues, shells, webs, and other materials. This entails the engineer asking questions about how Mother Nature does things. Why does a blade of grass not break when stepped on? What makes the adhesive produced by barnacles so amazingly strong? The capacity to duplicate such materials in the laboratory could result in significant advancements in our synthetic materials.

As discussed by Ebenstein and Wahl [44], **spider silk** is a natural material that is both strong and extensible. NASA used knowledge acquired from their study of the spider webs of a space-borne spider to design tennis racquets with greater power, feel, and control. Some spiders have up to seven silk glands, which can produce silks that perform different functions in the web. The **dragline silk fibers** have the highest combination of strength and toughness. These are the fibers that support the spider's weight as it hangs in midair and that form the framework for the web. They have typically a 2 μm diameter. To understand these silk structures, Ebenstein and Wahl carried out indentation tests. It turns out that the fibers consist of microfibrils with diameters ranging from 80 to 230 nm, aligned with the drawing axis of the fiber [45]. This is but one example of the great potential in biomimetics.

We find in nature a variety of strategies to provide permanent or temporary adhesion to surfaces, both wet and dry. Geckos, spiders, and flies have created hierarchical nanostructures for the purpose of adhering to some surface; the mechanisms of such are based largely on van der Waals interactions [46]. **Gastropods** produce gels that modify film thickness and viscosity; this is how snails can climb up vertical surfaces. **Barnacles** secrete protein rich cement that cures underwater. Charles Darwin had already studied these creatures in 1854, but understanding of the mechanism of barnacle adhesion came only in 2012, provided by Kathryn Wahl and coworkers [46]. It turns out that acorn barnacle *Balanus amphitrite* adheres to ship and other surfaces by a two-stage cement secretion. The cement secreted in the second stage (autofluorescent as it happens) provides a 2-fold increase in the adhesive strength as compared to the adhesion of the first stage cement. The two secretions show differences in morphology, protein conformation, and chemical functionality. Before this was understood, the same group had already found that barnacles resist removal by crack trapping [47] and that nanofibrils are the major component in the barnacle adhesive plaque [48]. Needless to say, mitigation of **biofouling**—the accumulation of microorganisms, plants, algae, or animals on a wetted surface—has to be based on such understanding.

NATURE'S STAR CERAMIC

The **nacre** that lines abalone shells would be a star among ceramics for its combined strength and toughness (or lack of brittleness). Scientists have long been trying to design materials based on natural materials such as nacre, also known as mother-of-pearl, but the fineness of structure in natural materials is difficult to mimic. A search on the topic of synthetic nacre returns numerous studies. Although no one has fully reproduced the precise complexity of the naturally occurring substance, most synthetic versions can boast of a variety of impressive properties.

In spite of the difficulty in reproducing nature's materials, there is much we can learn from investigating them; and sometimes using that knowledge even in part can lead to improvements in materials design and technology. Investigation of the sheep crab shell reveals that it is an extremely strong and tough material although composed of rather weak constituents [49]. Indeed, that is often the case for biological materials. Thus (as we expect based on the MSE triangle in Figure 1.1), it is the structural assembly of those components and the interactions between them that is responsible for the mechanical properties of the sheep crab exoskeleton and of numerous other materials found in nature.

Bone—a hierarchical composite—and the complexity of bone have already been mentioned (above and in Chapter 10). In humans (and in other vertebrates), **bones** constitute a dense connective tissue that supports and protects the various parts of the body. This is by no means the only function; bones also store minerals and produce red and white blood cells. In Section 10.10 we described bone composition and identified the main components as hydroxyapatite, with the general formula $Ca_{10}(PO_4)_6(OH)_2$ (a mineral also found in coral reefs), and collagen—an elastic protein. Strictly speaking, the main component of bone is a deficient, carbonated hydroxyapatite. As far as composites go, bone is a relatively hard and lightweight one. Perhaps surprising is that bones, including teeth, are viscoelastic; hence they undergo recovery in scratch resistance testing [16].

Bone is an adaptive material, and there are property variations depending on function and on species. Elk antler bone, for instance, has drawn the attention of Joanna McKittrick and her collaborators [50] because of its toughness. They have carefully investigated the fracture behavior to determine the causative mechanisms [50]. The same group has studied seahorses, whose extraordinary flexibility has been fascinating people for ages. It turns out that the seahorse skeleton consists of a number of small bony plates overlapping along the length of the fish [51, 52]; see Figure 11.7. Between the bony plates are thick collagen layers. Seahorses

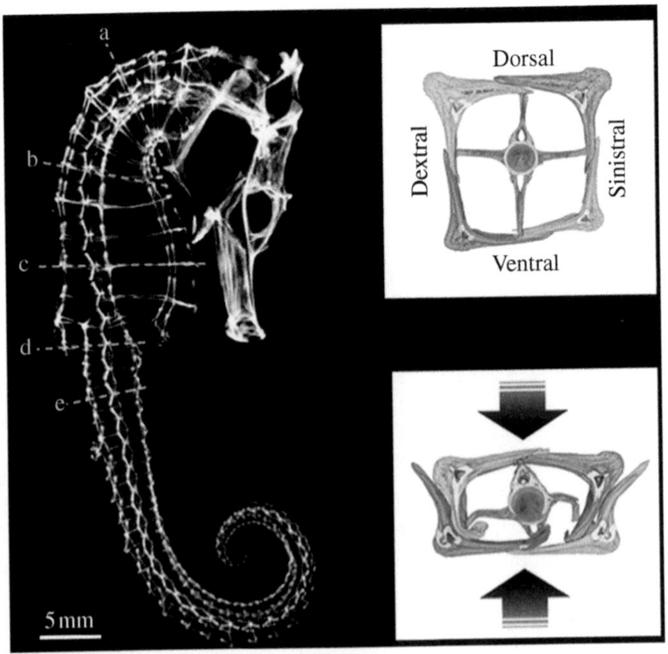

FIGURE 11.7 Skeletal structure of a seahorse. *Source*: Porter *et al.* [51]. Reprinted with permission from Elsevier.

thus have joint flexibility protecting them against impact and crushing. Typically animal tails have circular or oval cross-sections, but the seahorse tail plate cross-section is approximately square. The plates can slide, in contrast to circular plates which can both slide and rotate. The consequence: the square plates absorb more energy before failure than the circular ones. Moreover, the square plates are also better for prehension (grasping ability) [52].

We looked earlier at drug-delivery systems (Section 11.7); now consider the drug compounds themselves. Most pharmaceutical drugs are based on naturally occurring compounds. Therefore, it is not surprising that historically all medicinal compounds were derived from natural sources, primarily plants. These medicinal compounds—from animal, vegetal, and mineral sources—constitute a sub-class of natural materials. (Modern pharmaceutical drugs are largely synthetic variants of natural compounds, for which reason they have added side-effects.)

Many textiles are composed of natural materials. These include cotton, linen, wool, and silk. Cotton and linen are cellulosic materials; of course wood (discussed in Chapter 10) and other plant fibers are also in this class, being composed largely of the compound cellulose. Wool and silk are protein fibers. Under a microscope, each type of fiber has its own characteristic morphology; Figure 11.8 provides a schematic for comparison of fiber types. Natural fibers are used in textiles largely in their native form, although low-temperature plasma modification of textile materials is gaining popularity. Such modification is praised as clean, solvent-free, time-saving, and environmentally friendly [53]. In the next section, however, we shall see how natural fibers and other biological materials are incorporated into other materials or else used to synthesize new materials.

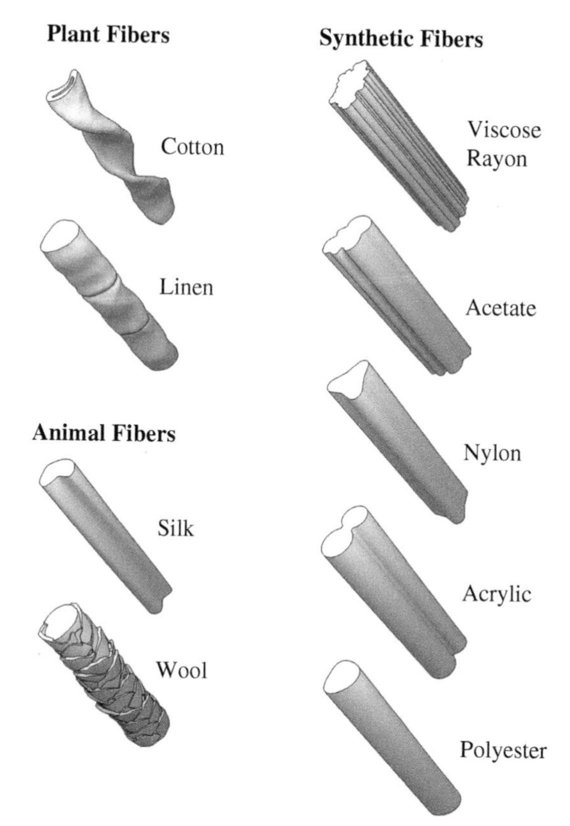

FIGURE 11.8 Schematic illustration of natural and synthetic fiber structures.

SCAFFOLDS FOR BONE TISSUE ENGINEERING

At the intersection of synthetic biomaterial and natural material is a synthetic polycaprolactone scaffold impregnated with natural compounds extracted from Patagonian *Usnea* lichen [54]. The extraction process from the lichen entails use of supercritical carbon dioxide; recall our discussion of the supercritical state in Chapter 3. The polymer, PCL (polycaprolactone), is a semicrystalline polymer that is biodegradable. Fanovich, Ivanovich, and their colleagues define conditions for the polymer scaffold [54]:

> An ideal scaffold for bone tissue engineering should be biocompatible, biodegradable, osteoconductive and it should provide structural support to the new formed bone. These materials, obtained from natural or synthetic compounds, must have a proper degradation rate and some required characteristics related with the morphological feature for good performance. A scaffold should have an interconnected porous structure, sufficient mechanical strength and good cell–scaffold interaction. Pore size, pore morphology and degree of porosity are very important parameters in tissue engineering. An interpenetrating network of pores in the range of 100–500 μm is required to allow vascularization and tissue in-growth.

Moreover, the PCL scaffold created by the group (working from Mar del Plata, Belgrade, and Hamburg) has also antibacterial activity [54]. From the MSE perspective, it is clear that such a development has required the integration of knowledge and techniques from multiple areas of science and engineering.

A different approach to scaffolds consists in creation of extracellular matrices (ECMs) [55]. In such a matrix the cells can migrate, interact with each other, and proliferate. This approach provides the capability to pre-define structures of the matrices. Marga Lensen and her coworkers created ECMs consisting of parallel channels by uniaxial freezing of bulk hydrogels [55]. A result is unusually high mobility of the cells in the channels; cells can travel several mm along a channel. Moreover, the longevity of cells in such structures is much higher than in un-oriented structures without porosity.

NATURALLY ANTIMICROBIAL MATERIALS

There are, of course, naturally occurring materials with antibacterial activity. In society, the imparting of antibacterial or antimicrobial properties to a wide range of commercially available products of various material types has become quite common. Products ranging from clothing to shoes to kitchen tools to toys are now imbued with antibacterial agents. Some suggest that in public transportation antibacterial agents ought to be incorporated into the materials most contaminated by passengers. Scientists at Auburn University [56] have tested lysosome, a powerful natural antibacterial protein, for such a purpose as it is applicable even in fabrication processes where heat treatment is involved. For products that undergo extensive use (e.g., handrails), mechanically robust materials are needed. The Auburn group has demonstrated the possibility of using single walled carbon nanotubes (SWCNTs, see Section 7.5) to create coatings by bottom-up layer-by-layer assembly. The coatings were shown to have verifiable long-term antimicrobial activity while thickness was controlled within 1.6 nm, not a small technological feat [56]. Since the overuse of synthetic antibacterial agents is a serious concern, the development of a strategy incorporating a natural antimicrobial agent into materials is promising for better human health and safety.

From the mention of textiles we are reminded that the use of biological and natural materials is obviously not new. Since ancient times man has used animal skins and bladders for containers, bone for cutting tools and sewing needles, plant matter for ropes, and both animal and vegetal matter for clothing. The Biblical record, spanning many centuries, names such uses—as do some other ancient texts. Somewhat more recently, Georgius Everhardus Rumphius compiled in the 17th Century the *Ambonese Herbal*: six volumes describing in remarkable detail more than 2000 plants, their habitats, and their economic and medicinal uses in Indonesia, especially on Ambon Island and its archipelago. As a consequence of his presentation of the plants and their multiple uses, we now have a cultural and scientific treasury of unparalleled value for today's botanists, anthropologists, ethnobotanists, science historians, medicinal chemists, and other scholars. This too plays into the history of natural and biological materials. Papyrus, another ancient material, is an outstanding example of a natural material utilized for human purposes over many centuries. The use of silk—already discussed in this chapter—also has a long history of use by humans, while many new uses for it are described in a presentation recorded in 2011 [57] (available for viewing on the internet). From here we proceed to the final section on biomaterials of the type that span and combine many of the varieties named above.

11.9 BIO-BASED MATERIALS

Buzz-words such as "green", "biodegradable", and "eco-friendly" may in part correspond to bio-based materials. In fact, this label is broad, but it generally refers to one of two scenarios: (1) materials that incorporate a biological material or (2) materials in which one or more components are derived from natural materials.

A primary example of the first category is natural fiber-reinforced composites. For instance, there is concrete reinforced with natural fibers including: sisal from *Agave Sisalana* in Mexico [58]; coir from coconut husks [59]; bamboo [60, 61]; jute, grown in India, Bangladesh, China, and Thailand [60]; akwara, predominantly from Nigeria [58]; elephant grass [58], and kenaf (*Hibiscus cannabinus* L.) growing in tropical climates [60]. Kenaf is a promising crop for supplying emerging fiber markets; the kenaf fibers are bundles of cells that undergo extensive cell-wall thickening during maturation [60]. There exist also a large variety of polymer matrix composites reinforced with natural fibers. The matrices themselves may be either petrochemical or natural in origin. The study of natural fiber reinforced composites is a large and active one [61–64], particularly owing to the need to use renewable resources—whenever feasible—for sustainability (see Chapter 20) of material fabrication. For instance, Alexander Bismarck and his colleagues describe the biosynthesis and bioprocessing of bacterial cellulose for applications in advanced fiber composites [65]. Jiasong He and his colleagues have demonstrated that cellulose can be dissolved in ionic liquids and then esterified [66]. The esterification provides versatility in creation of a variety of cellulose-based materials, including fibers and films.

Another example of the first type of bio-based materials is medium density fiberboard (MDF), a widely used construction material composed of wood fibers bound in a polymer resin. Other scrap can also be used in place of wood as the filler. For instance, rice husk, a waste byproduct of rice production, can be powdered and used to prepare MDF [67]. The composition of resin can also be improved over the typically used toxic formaldehyde resins; for instance, one can bind the filler with less harmful synthetic polymeric resins or with naturally occurring polymers.

In the second category of bio-based materials are those that have one or more components derived from a natural material (other than petroleum). These are not natural materials, strictly speaking, but derivatives; and while this may improve sustainability of production, it does not dictate whether the product will be biodegradable or have low environmental impact (see Chapter 20 for more on these topics). One example is a biodegradable plastic fabricated from sodium montmorillonite aerogels filled with a polymer that has been synthesized from milk casein [68]. Casein is also used in other plastics, glue, some paints, and protective coatings. Others have used keratin from chicken feathers to create bio-fibers [69] or, in combination with polyurethane, to create hybrid membranes for removal of chromium from water [70]. Flocculants—used for water purification—that are synthesized by chemically modifying plant-derived starch (a natural polymer) provide another example of a bio-based material.

Polylactic acid (PLA) is a polymer that may be produced by industrial fermentation or by traditional synthetic routes from plant derived starting materials. Polylactic acid is a biodegradable polyester that breaks down into innocuous lactic acid. Polylactic acid is used in both biomedical and industrial applications.

Soybeans—and in fact the entire soy plant—provide the basis for a variety of materials; compounds derived from the soy plant are found in plastics, inks, and biofuel. Now consider another plant called switchgrass, a grass growing in North America and providing nesting and cover for pheasants, quail, and rabbits. A bacterium called *Caldicellulosiruptor bescii* can solubilize in five days $\approx 25\%$ of biomass of switchgrass, including hard-to-degrade lignin, and thus provide biofuel [71, 72]. This contradicts an opinion popular for a long time that "you can make anything out of lignin except for a profit".

Evaluation of the structures of all types of biological and natural materials is of course needed for development of even better bio-based materials. Experimental techniques are being developed for this particular purpose. For instance, a group at Cornell University [73] has developed a technique of multi-spin electron spin resonance for investigating complex dynamical behavior of proteins in solution and in heterogeneous systems such as lipid bilayers. (Lipids are a group of naturally occurring molecules that include fat-soluble vitamins, fats, waxes, and various glycerides.) Likely there shall continue for some time a need to discovering ways to modify existing techniques or else to invent new techniques to analyze and better characterize an ever increasing variety of materials under investigation. We have only uncovered the tip of the iceberg with regard to number of bio-based materials under development.

The possibilities for using natural materials are as wide as our imaginations. There is in nature inspiration for the development of countless new materials. This is why one talks about biomimetics—sometimes called **biomimicry**—as the imitation of nature for the purpose of solving problems faced by humans. Yet in the midst of burgeoning technological innovation, there remains a call for responsible use even of those materials that are considered "renewable".

11.10 OTHER ASPECTS OF BIOMATERIALS

Degradation, which we have mentioned several times in this chapter, is an important feature of biomaterials. While the issue of biocompatibility may appear foremost, the stability and longevity of materials in contact with the human body is critical. Some biomaterials are designed to degrade quickly; others are intended to last for years. In either case there are byproducts—from chemical degradation, wear debris, or chemical reactions—released into the body. In the design process of biomaterials, these byproducts have to be identified and their effects evaluated. Note that biomaterials are not the only materials that undergo biodegradation. In Chapter 20 we discuss the biodegradation of materials in the context of reducing environmental contamination from materials waste. This is discussed also in Section 11.9.

Advances in biomaterials abound. As interdisciplinary approaches are applied to the field, many different avenues are being followed for the development of improved materials. Methods of tissue engineering are certainly advancing, while biomimetics (defined just above) is also a rising focus among investigators. Other methods and approaches have gathered speed from our advancing understanding of materials already at hand. However, challenges for the future are not confined simply to creating new materials. The sterilization of biomaterials, whether synthetic or natural in origin, is required for biomedical use. There is a need for appropriate (and economical) sterilization techniques for materials that originate in natural environments: for example proteins from cattle, chitin from crab shells, or alginate from seaweed. Additional challenges in working with biomaterials are dealing with swelling (movement of biological fluids into the material), leaching (dissolution of material components into biological tissue, introducing porosity in the material), corrosion (mainly affecting metallic biomaterials), and particulate release (causing prosthetic loosening, periprosthetic bone resorption, and a variety of host reactions and cellular harm).

BIOSENSORS

The study of sensors typically lies in the realms of physics and electronics; however a convergence of knowledge, technology, and advances in biomaterials has paved the way for increased development of sensors for biomedical applications. There are both physical sensors and chemical sensors. **Physical sensing** involves pressure, volume, flow, electrical potential, and temperature. A few examples of physical sensors are thermocouples, thermistors, and pressure monitors. **Chemical sensing** typically deals with determination of the concentration of a chemical species in a volume of gas, liquid, or tissue. Chemical sensors may be used to measure pH, ion concentrations, glucose, or blood gases (O_2 and CO_2), to name a few. In order to detect more complex biomolecules, viruses, bacteria, and parasites *in vivo*, **biosensors** must operate based on biological molecules, tissues, organisms, or principles [74]. An example is thermally responsive polymers that can be used to detect Gram negative bacterium [75]. In Chapter 9 we mentioned the importance of end groups in affecting polymer behavior; likewise end groups are an important determinant in the functionality of these thermo-responsive polymers.

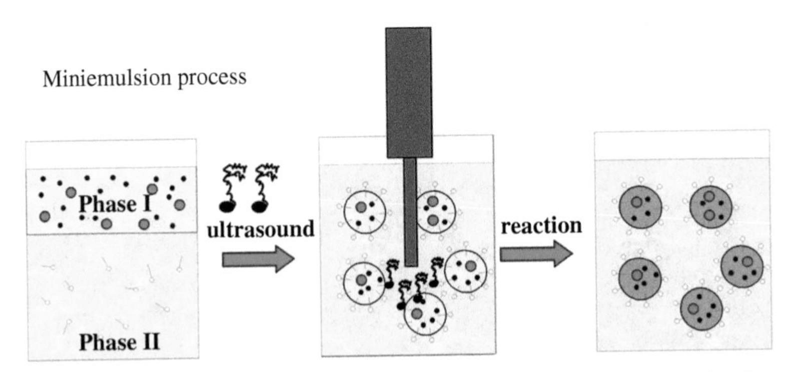

FIGURE 11.9 Schematic of the miniemulsion process. *Source*: Reprinted (adapted) with permission from Mailänder and Landfester [77]. Copyright (2009) American Chemical Society.

It is seen throughout this book that often one class of materials can have applications in vastly different areas. Nanoparticles provide an instructive example. Katharina Landfester and coworkers [76–79] describe the possibilities for synthesizing nanoparticles with numerous and varied and surface modifications based on a **miniemulsion** process, illustrated schematically in Figure 11.9 [77]. Emulsion processes, of course, are designed to join aqueous and oily components; recall discussions of immiscibility of such phases in Section 3.2 and emulsion polymerization in Section 9.4. The miniemulsion process illustrated in Figure 11.9 shows that oil droplets are formed by stirring and ultrasonification. The resultant nanodroplets, shown suspended in the center of the figure, are stabilized by a surfactant (drawn in black with a non-polar head region and squiggly polar tail region). The resulting droplets have the diameters between 30 and 500 nm. The constituents of the oily phase determine what happens in the final reaction. For instance, monomers in the oily phase can be polymerized, while the size and geometry of the droplets is preserved throughout the reaction.

Therefore, based on the miniemulsion process, it is possible to obtain a wide variety of particles; these can incorporate different polymers and enclosed substances (e.g., drugs, marker compounds, etc.), be formed into different geometries (e.g., nanoparticles, nanocapsules, etc.), and contain different surface functionalities. Figure 11.10 shows how such modifications can be designed and utilized for various applications.

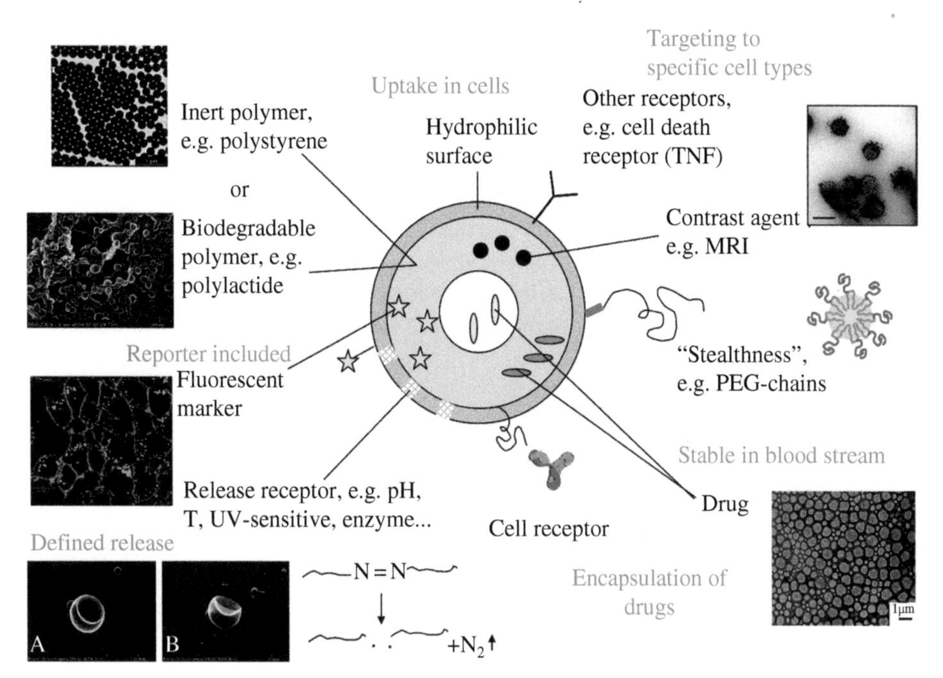

FIGURE 11.10 Scheme of possible applications of nanoparticles and nanocapsules prepared by the miniemulsion process. *Source*: Reprinted with permission from Mailänder and Landfester [77]. Copyright (2009) American Chemical Society.

Possibly the most fascinating use of this technology is creation of materials that can cross the **blood-brain barrier** (BBB). The barrier consists of a dense layer of endothelial cells facing the blood flow, connected by tight junctions, with an interposed membrane. Only hydrophilic compounds with molar mass $M < 150\,\text{g·mol}^{-1}$ and hydrophobic ones with $M < 400\,\text{g·mol}^{-1}$ can pass through the blood-brain barrier. This is an important defense mechanism of the human body: *do not let into the brain substances that are not recognized as needed there*. The same team based in Mainz has created nanoparticles that have been shown in rats to go across the blood-brain barrier [79]. This suggests a possible new route for drugs that can be "smuggled" into the brain.

If we now consider the whole of the contents of this chapter, we find that the biomaterials discussed could each be classified under the headings of previous chapters—yet each type of biomaterial has also certain defining features that distinguish it from others. Smart materials, which we discuss in the next chapter, may also be fabricated as biomaterials. Biocompatibility is by necessity a predominant feature to be evaluated for biomaterials. Nevertheless, biomaterials are subject to the same rigorous investigations of thermodynamic, mechanical, and other properties as are other materials. For the materials scientist and engineer, an understanding of the main material types will improve his study of a select material type. This is especially true as more and more materials consist of hybrids or composites of differing material types. Our discussion of biomaterials in the present Chapter has called out salient features of materials—such as stiffness, wear, and rheology—that shall be called out again in the third unit of this book.

11.11 SELF-ASSESSMENT QUESTIONS

1. Distinguish between biomedical materials (i.e., biomaterials according to a widely accepted definition provided in Section 11.1), biological/natural materials, and bio-based materials.
2. Which type of biomaterials do you personally consider the most important? Why?
3. Explain biodegradability and its relevance to different types and applications of biomaterials. Materials and applications of biomedical, natural, and bio-based materials should be included in the discussion.
4. What is biomimetics and how important do you think it is in the development of new materials?
5. What are natural fibers and what they can be used for?
6. Are bio-based materials necessarily better for the environment? Explain.

REFERENCES

1. S. Nag & R. Banerjee, Fundamentals of medical implant materials, in *Materials for Medical Devices ASM Handbook*, Vol. *23*, (R. Narayan, Ed.), ASM International, Materials Park, OH **2012**, pp. 6–17.
2. J.O. Hollinger, Ed., *An Introduction to Biomaterials*, Second Edition, CRC Press, Boca Raton, FL **2011**, pp. 2–3.
3. R. Katsarava & D. Tugushi, Chapter 7, in *Unique Properties of Polymers and Composites*, Vol. *1*, (Yu.N. Bubnov, V.A. Vasnev, A.A. Askadskii & G.E. Zaikov, Eds.), NOVA Science Publishers, New York **2011**, pp. 113–131.
4. M. Schorr, B. Valdez, E. Valdez, A. Eliezer, N. Lotan & M. Carillo, *J. Mater. Ed.* **2012**, *34*, 59.
5. A. Saenz, E. Rivera-Muñoz, W. Brostow & V.M. Castaño, *J. Mater. Ed.* **1999**, *21*, 297.
6. B.D. Ratner, A history of biomaterials, in *Biomaterials Science: An Introduction to Materials in Medicine*, Second Edition, (B.D. Ratner, A.S. Hoffman, F.J. Schoen & J.E. Lemons, Eds.), Elsevier Academic Press, New York **2004**, pp. 10–19.
7. G. Bergmann, A. Bender, F. Graichen, J. Dymke, A. Rohlmann, A. Trepczynski, M.O. Heller & I. Kutzner, *PLoS One* **2014**, *9*, e86035.
8. D.A. Puleo & A. Nanci, *Biomater.* **1999**, *20*, 2311.
9. A. Shenhar, I. Gotman, E.Y. Gutmanas & P. Ducheyne, *Mater. Sci. & Eng. A* **1999**, *268*, 40.
10. D. Starosvetsky, A. Shenhar & I. Gotman, *J. Mater. Sci. Med.* **2001**, *12*, 145.
11. T. Bonello, J.C. Avelar-Batista Wilson, J. Housden, E.Y. Gutmanas, I. Gottman, A. Matthews, A. Leyland & G. Cassar, *Mater. Sci. & Eng. A* **2015**, *619*, 300.
12. D. Starosvetsky & I. Gotman, *Biomater.* **2001**, *22*, 1853.
13. A. Shenhar, I. Gotman, S. Radin, P. Ducheyne & E.Y. Gutmanas, *Surf. Coat. Tech.* **2000**, *126*, 210.
14. I. Gotman & E.Y. Gutmanas, *Adv. Biomater. & Devices in Med.* **2014**, *1*, 25.
15. J.C. Avelar-Batista Wilson, S. Wu, I. Gotman, J. Housden & E.Y. Gutmanas, *Mater. Lett.* **2015**, *157*, 45.

16. N.J. Hallab, J.J. Jacobs & J.L. Katz, Orthopedic applications, in *Biomaterials Science: An Introduction to Materials in Medicine*, Second Edition, (B.D. Ratner, A.S. Hoffman, F.J. Schoen & J.E. Lemons, Eds.), Elsevier Academic Press, New York **2004**, pp. 527–555.
17. S. Wang, L. Lu, B.L. Currier & M.J. Yaszemski, Orthopedic prostheses and join implants, in *An Introduction to Biomaterials*, (S.A. Guelcher & J.O. Hollinger, Eds.), CRC Press, Boca Raton, FL **2006**, pp. 369–393.
18. E. Castro & J.F. Mano, *J. Biomed. Tech.* **2013**, *9*, 1129.
19. A. Rakovsky, E.Y. Gutmanas & I. Gotman, *J. Mater. Sci.* **2010**, *45*, 6339.
20. A. de la Isla, W. Brostow, B. Bujard, M. Estevez, J.R. Rodriguez, S. Vargas & V.M. Castaño, *Mater. Res. Innovat.* **2003**, *7*, 110.
21. M. Estévez, S. Vargas, V.M. Castaño, J.R. Rodríguez, H.E. Hagg Lobland & W. Brostow, *Mater. Letters* **2007**, *61*, 3025.
22. M. Estevez, J.R. Rodriguez, S. Vargas, J.A. Guerra, H.E. Hagg Lobland & W. Brostow, *J. Nanosci. & Nanotech.* **2013**, *13*, 4446.
23. S. Vargas, M. Estevez, A. Hernandez, J.C. Laiz, W. Brostow, H.E. Hagg Lobland & J.R. Rodriguez, *Mater. Res. Innovat.* **2013**, *17*, 154.
24. B.A. Doll, A. Kukreja, A. Seyedain & T.W. Braun, Dental implants, in *An Introduction to Biomaterials*, (S.A. Guelcher & J.O. Hollinger, Eds.), CRC Press, Boca Raton, FL **2006**, pp. 395–416.
25. D.C. Smith, Adhesives and sealants, in *Biomaterials Science: An Introduction to Materials in Medicine*, Second Edition, (B.D. Ratner, A.S. Hoffman, F.J. Schoen & J.E. Lemons, Eds.), Elsevier Academic Press, New York **2004**, pp. 573–583.
26. M.C. Bottino, V. Thomas & G.M. Janowski, *Acta Biomater.* **2011**, *7*, 216.
27. M.C. Bottino, V. Thomas, G. Schmidt, Y.K. Vohra, T.-M.G. Chu, M.J. Kowolik & G.M. Janowski, *Dental Mater.* **2012**, *28*, 703.
28. S. Sarkar, G. Burriesci, A. Wojcik, N. Aresti, G. Hamilton & A.M. Seifalian, *J. Biomech.* **2009**, *42*, 722.
29. K. Chaloupka, M. Motwani & A.M. Seifalian, *Biotechnol. Appl. Biochem.* **2011**, *58*, 363.
30. H. Ghanbari, H. Viatge, A.G. Kidane, G. Burriesci, M. Tavakoli & A.M. Seifalian, *Trends Biotechnol.* **2009**, *27*, 359.
31. H. Ghanbari, A.G. Kidane, G. Burriesci, B. Ramesh, A. Darbyshire & A.M. Seifalian, *Acta Biomater.* **2010**, *6*, 4249.
32. B. Rahmani, S. Tzamtzis, H. Ghanbari, G. Burriesci & A.M. Seifalian, *J. Biomech.* **2012**, *45*, 1205.
33. M.C. Lensen, V.A. Schulte, J. Salber, M. Diez, F. Menges & M. Möller, *Pure & Appl. Chem.* **2008**, *80*, 2479.
34. M.A. Sabino, J. Albuerne, A.J. Müller, J. Brisson & R.E. Prud'homme, *Biomacromol.* **2004**, *5*, 358.
35. P.C. Nicolson & J. Vogt, *Biomater.* **2001**, *22*, 3273.
36. M.F. Refojo, Ophthalmological applications, in *Biomaterials Science: An Introduction to Materials in Medicine*, Second Edition, (B.D. Ratner, A.S. Hoffman, F.J. Schoen & J.E. Lemons, Eds.), Elsevier Academic Press, New York **2004**, pp. 583–591.

37. Y. Loo & K.W. Leong, Biomaterials for drug and gene delivery, in *An Introduction to Biomaterials*, (S.A. Guelcher & J.O. Hollinger, Eds.), CRC Press, Boca Raton, FL **2006**, pp.341–367.
38. R.J. Babu, W. Brostow, O. Fasina, I.M. Kalogeras, S. Sathigar & A. Vassilikou-Dova, *Polymer Eng. & Sci.* **2011**, *51*, 1456.
39. J. Heller & A.S. Hoffman Drug delivery systems, in *Biomaterials Science: An Introduction to Materials in Medicine*, Second Edition, (B.D. Ratner, A.S. Hoffman, F.J. Schoen & J.E. Lemons, Eds.), Elsevier Academic Press, New York **2004**, pp. 628–648.
40. N. Ignjatović, V. Uskoković, Z. Ajduković & D. Uskoković, *Mater. Sci. & Eng. C* **2013**, *33*, 943.
41. L. Mayol, M. Biondi, F. Quaglia, S. Fusco, A. Borzacchiello, L. Ambrosio & M.I. La Rotonda, *Biomacromol.* **2011**, *12*, 28.
42. C.C. Lee, J.A. MacKay, J.M.J. Fréchet & F.C. Szoka, *Nature Biotech.* **2005**, *23*, 1517.
43. M.E. Fox, F.C. Szoka & J.M.J. Fréchet, *Acc. Chem. Res.* **2009**, *42*, 1141.
44. C. Garman, N. Bindert, A. Sunkara, L. Paliulis & D.M. Ebenstein, *MRS Proceedings* **2009**, *1185*. 10.1557/PROC-1185-II04-02.
45. D.M. Ebenstein & K.J. Wahl, *J. Mater. Res.* **2006**, *21*, 2035.
46. D.K. Burden, D.E. Barlow, C.M. Spillmann, B. Orihuela, D. Rittschof, R.K. Everett & K.J. Wahl, *Langmuir* **2012**, *28*, 13364.
47. C.-Y. Hui, R. Long, K.J. Wahl & R.K. Everett, *Royal Soc. Interface* **2011**. 10.1098/rsif.2010.0567.
48. D.E. Barlow, G.H. Dickinson, B. Orihuela, J.L. Kulp III, D. Rittschof & K.J. Wahl, *Langmuir* **2010**, *26*, 6549.
49. P.-Y. Chen, A.Y.-M. Lin, J. McKittrick & M.A. Meyers, *Acta Biomater.* **2008**, *4*, 587.
50. M.E. Launey, P.-Y. Chen, J. McKittrick & R.O. Ritchie, *Acta Biomater.* **2010**, *6*, 1505.
51. M.M. Porter, E. Novitskaya, A.B. Castro-Ceseña, M.A. Meyers & J. McKittrick, *Acta Biomater.* **2013**, *9*, 6763.
52. M.M. Porter, D. Adriaens, R.L. Hatton, M.A. Meyers & J. McKittrick, *Science* **2015**, *349*, aaa6683.
53. R. Abd Jelil, *J. Mater. Sci.* **2015**, *50*, 5913.
54. M.A. Fanovich, J. Ivanovic, D. Misic, M.V. Alvarez, P. Jaeger, I. Zizovic & R. Eggers, *J. Supercrit. Fluids*, **2013**, *78*, 42.
55. V.A. Schulte, D.V. Alves, P.P. Dalton, M. Moeller, M.C. Lensen & P. Mela, *Macromol. Biosci.* **2013**, *13*, 562.
56. D. Nepal, S. Balasubramanian, A.L. Simonian & V.A. Davis, *Nano Letters* **2008**, *8*, 1896.
57. TED2011, *Fiorenzo Omenetto: Silk, The Ancient Material of the Future*, TED Talks, Filmed March **2011**. http://www.ted.com/talks/fiorenzo_omenetto_silk_the_ancient_material_of_the_future.html, accessed March 29, 2016.
58. J. Castro & N.E. Naaman, *ACI Mater. J.* **1981**, *78*, 69.
59. P. Balaguru, *Internat. J. Development Tech.* **1985**, *3*, 87.
60. B.G. Ayre, K. Stevens, K.D. Chapman, C.L. Webber III, K.L. Dagnon & N.A. D'Souza, *Textile Res. J.* **2009**, *79*, 973.

61. K. Ghavami, *Cement & Concrete Composites* **2005**, *27*, 637.
62. C.S. Rodrigues, K. Ghavami & P. Stroeven, *J. Mater. Sci.* **2006**, *41*, 6925.
63. O. Faruk, A.K. Bledzki, H.-P. Fink & M. Sain, *Macromol. Mater. & Eng.* **2014**, *299*, 9.
64. W. Brostow, T. Datashvili & H. Miller, *J. Mater. Ed.* **2010**, *32*, 125.
65. K.Y. Lee, G. Buldum, A. Mantalaris & A. Bismarck, *Macromol. Biosci.* **2014**, *14*, 10.
66. J. Zhang, W. Chen, Y. Feng, J. Wu, J. Yu, J. He & J. Zhang, *Polymer Internat.* **2015**, *64*, 963.
67. S. Vargas, J.R. Rodriguez, H.E. Hagg Lobland, K. Piechowicz & W. Brostow, *Macromol. Mater. & Eng.* **2014**, *299*, 807.
68. T. Pojanavaraphan, R. Magaraphan, B.-S. Chiou & D.A. Schiraldi, *Biomacromol.* **2010**, *11*, 2640.
69. A.L. Martínez-Hernández, C. Velasco-Santos, M. de Icaza & V.M. Castaño, *Composites B* **2007**, *38*, 405.
70. V. Saucedo-Rivalcoba, A.L. Martínez-Hernández, G. Martínez-Barrera, C. Velasco-Santos, J.L. Rivera-Armenta & V.M. Castaño, *Water, Air & Soil Pollution* **2012**, *218*, 557.
71. I. Kataeva, M.B. Foston, S.-J. Yang, S. Pattathil, A.K. Biswal, F.L. Poole II, M. Basen, A.M. Rhaesa, T.P. Thomas, P. Azadi, V. Olman, T.D. Saffold, K.E. Mohler, D.L. Lewis, C. Doeppke, Y. Zeng, T.J. Tschaplinski, W.S. York, M. Davis, D. Mohnen, Y. Xu, A.J. Ragauskas, S.Y. Ding, R.M. Kelly, M.G. Hahn & M.W.W. Adams, *Energy & Environ. Sci.* **2013**, *6*, 2186.
72. J.T. Andres, *Chemistry World* **2013**, *10* (8), 32.
73. K.A. Earle, B. Dzikovski, W. Hofbauer, J.K. Moscicki & J.H. Freed, *Magn. Reson. Chem.* **2005**, *43*, S256.
74. P. Yager, Biomedical sensors and biosensors, in *Biomaterials Science: An Introduction to Materials in Medicine*, Second Edition, (B.D. Ratner, A.S. Hoffman, F.J. Schoen & J.E. Lemons, Eds.), Elsevier Academic Press, New York **2004**, pp. 669–684.
75. P. Sarker, J. Shepherd, K. Swindells, I. Douglas, S. MacNeil, L. Swanson & S. Rimmer, *Biomacromol.* **2011**, *12*, 1.
76. K. Landfester, *Ann. Rev. Mater. Res.* **2006**, *36*, 231.
77. V. Mailänder & K. Landfester, *Biomacromol.* **2009**, *10*, 2379.
78. O. Lunov, T. Syrovets, C. Loos, J. Beil, M. Delacher, K. Tron, G.U. Nienhaus, A. Musyanovich, V. Mailänder, K. Landfester & T. Simmet, *ACS Nano* **2011**, *5*, 1657.
79. C.K. Weiss, M.V. Kohnle, K. Landfester, T. Hauk, D. Fischer, J. Schmitz-Wienke & V. Mailänder, *Chem. Med. Chem.* **2008**, *3*, 1395.

12

LIQUID CRYSTALS AND SMART MATERIALS

La belleza es la otra forma de la verdad.

—Alejandro Casona, Spanish playwright [1]; it means: *beauty is the other form of truth.*

12.1 INTRODUCTION

Up to now, we have seen what may be referred to as *passive* materials. Polymer fabrics, wooden beams, concrete blocks, or metal sheets may be cut, shaped, or formed into desired products or components. Such materials are required primarily to exist—and to do that without changing much via swelling, bending, corroding, etc. These are contrasted with a newer generation of **smart materials** that are *active*: rather than simply *be*, they *do* things. Smart materials are designed to undergo purposeful changes. These active materials respond in a significant and predictable manner to external stimuli such as temperature change, pressure change, magnetic field imposition (or change), electric field imposition (or change), and so on. In essence, smart materials are designed to mimic biological organisms and to create a material structure that encompasses a control and feedback system. Such materials will be the topic of this final chapter in Part 2; and we shall see here examples from each major class of materials already discussed: metal, ceramic, polymer, composite, and biomaterial. Lately, the term **functional materials** has also been used to describe this class of materials possessing a specific functionality (e.g., piezoelectricity, magnetoresistance, etc.).

We need only look to Nature, however, for the ultimate smart materials. Animals and plants clearly have the ability to adapt to the environment in real time. Plants can adapt their shape, allowing leaf surfaces to track sunlight. Leaves can be considered as a type of

Materials: Introduction and Applications, First Edition. Witold Brostow and Haley E. Hagg Lobland.

solar cell capable of self-repair. Not only that, leaves are structural materials, and they automatically regulate gas intake and release. Wood possesses some type of strain-sensing mechanism that enables it to reinforce itself in the needed location as it grows. Limping is a real-time response that causes a change in the load path or distribution through the structure (a body) to avoid overload of the damaged region. More examples of smart materials in nature are: electrodetection systems in fish; sonar used by dolphins; actuators in the form of spring-loaded seed pods; manufacturing systems, as in materials that carry their own factories, for instance, for protein synthesis.

In the remainder of this chapter we shall explain the nature and workings of synthetic smart materials. We begin with liquid crystals, a class of materials exhibiting very unique properties. **Liquid crystals (LCs)** are smart materials, but they are used both as passive and as active materials. For this reason we shall talk about them first. Liquid crystals are also colorful and many of them may be called beautiful; in fact they were discovered because of their colors; see again the motto of this chapter. Subsequent sections will treat field responsive composites, functional materials including electrochromic and piezoelectric materials, and finally shape memory materials. As already noted, the properties of smart materials are sharply responsive to the environment. Their functionality often hinges on phase transitions associated with a critical temperature or pressure. Thus smart materials are especially useful in sensors, actuators, and control and information storage devices. Moreover, as we shall see in this and the chapters of Part 3, the behavior of smart materials is correlated with spin, charge, orbital, and lattice degrees of freedom.

12.2 LIQUID CRYSTALS

To begin, liquid crystals can be classified in two categories: (1) **monomer liquid crystals** (MLCs), irrespective of whether they are able to polymerize, and (2) **polymer liquid crystals** (PLCs). This simple classification scheme is due to Edward T. Samulski [2]. (Those who are not familiar with his terminology use unnecessarily long names, such as low molecular mass liquid crystals or LMMLCs for MLCs.) Liquid crystals were discovered by the Austrian botanist Friedrich Reinitzer, who published his discovery in *Monatshefte für Chemie* in 1888 [3]. Not everybody believed him—that a flowing liquid has a structure as if it were a crystal—so he sent some of his samples to the German microscopist Otto Lehmann in Freiburg im Breisgau. The results of Reinitzer were confirmed by Lehmann, who published a paper in *Zeitschrift für physikalische Chemie* in 1889 [4]. (One hundred years later, there were big celebrations: in Austria in 1988 and in Germany in 1989.) Thus, the defining feature of LCs is that they flow (like liquid) yet do not have random structures.

The different structures of liquid crystals dictate their uses. To understand the uses, we need to discuss structures first. We can consider the structure of LCs at three levels: molecular, microscopic, and macroscopic. The microscopic *textures* of LCs are often stunning; see Figure 12.1. At the molecular scale, LC structures range in type from something like isotropic liquids to solid crystals. For both MLCs and PLCs, three distinct phase structures are defined: nematic, cholesteric, and smectic. A schematic diagram of each is shown in Figure 12.2; the term **mesogen** refers to an LC molecule. In the **nematic** phase, which is most like an isotropic liquid, mesogens have a preferred orientation along an axis in space called the *director*. Different textures are possible for the nematic phase, for example the schlieren texture—shown in Figure 12.1—and also a marble texture. The **cholesteric** phase

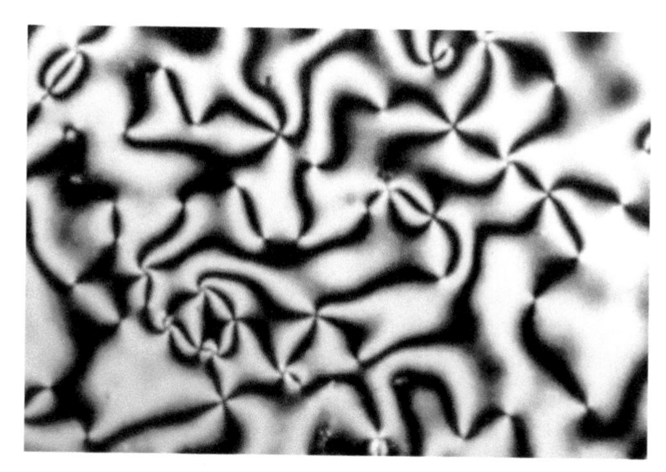

FIGURE 12.1 Schlieren-texture of a nematic phase liquid crystal: Schlieren-Textur der nematischen Phase eines kalamitischen Flüssigkristalls—1,5-hexandiol-bis{4-[4-(4-*n*-octyloxy-benzoyloxy)benzylidenamino]benzoat} bei 250°C. *Source*: Minutemen, https://commons.wikimedia.org/wiki/File:Nematische_Phase_Schlierentextur.jpg. Used under CC-BY-SA-3.0, 2.5.

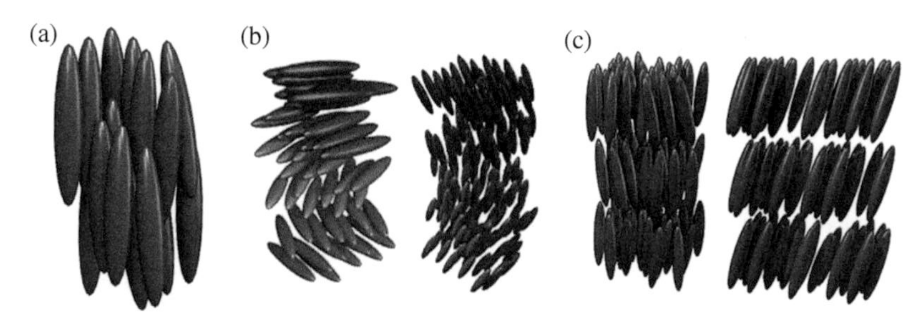

FIGURE 12.2 (a) Schematic of mesogen alignment in a liquid crystal nematic phase. (b) Schematic of mesogen ordering in chiral liquid crystal phases. The chiral nematic phase (left), also called the cholesteric phase, and the smectic C* phase (right). The asterisk denotes a chiral phase. (c) Schematic of mesogen ordering in the smectic liquid crystal phases: smectic-A (layered) and smectic-C (layered and tilted). *Source*: After Kebes, https://commons.wikimedia.org/wiki/File:LiquidCrystal-MesogenOrder-Nematic.jpg, https://commons.wikimedia.org/wiki/File:LiquidCrystal-MesogenOrder-ChiralPhases.jpg, https://commons.wikimedia.org/wiki/File:LiquidCrystal-MesogenOrder-SmecticPhases.jpg. Used under CC BY-SA-3.0, GFDL.

consists of layers of nematic phases with the director changing from layer to layer; thus a cholesteric phase is chiral. The secondary structure of the cholesteric is characterized by the distance along the twist axis over which the director rotates through a full circle. This distance is called the pitch. In Figure 12.3 we see layers spanning a half of the pitch. The pitch can be quite sensitive to an applied magnetic or electric field. Finally, there are a variety of **smectic** phases (denoted A, B, C, and so on); these have a director, layers, as well as additional features of symmetry or long-range order. In the smectic-A phase, the director is perpendicular to the smectic plane (see Figure 12.2); within each plane there is no further

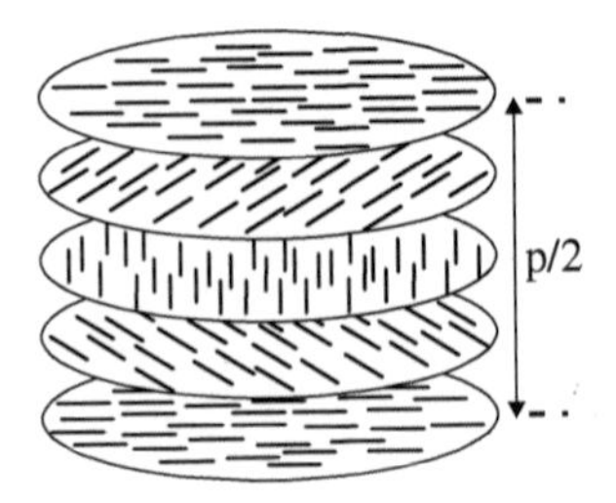

FIGURE 12.3 Changes of the director between consecutive layers in a cholesteric liquid crystal. Total thickness shown here is half the pitch (denoted p). *Source*: de:Benutzer: Heimoponnath, https://commons.wikimedia.org/wiki/File:Cholesterinisch.png. Used under CC-BY-SA-3.0, GFDL.

positional order. The smectic-B phase is similar, except that within each layer the molecules are positioned in a hexagonal lattice. The smectic-C phase differs from phase A in that the director is tilted at a constant angle from the plane normal. The director field is subject to distortion and can be aligned by magnetic and electric fields, and by properly prepared surfaces. The last specific phase we shall describe is smectic-G; it is characterized by a two-dimensional hexagonal lattice and tilting of the director with respect to the plane normal. Thus it exhibits herringbone symmetry.

Upon heating, each LC phase becomes eventually an isotropic liquid at the **clearing temperature**. A given LC compound may exist in several different phases depending on the temperature; this is the meaning of **thermotropic**. An example is 4,4′-di-*n*-heptyloxyazooxybenzene, which passes through smectic and nematic phases:

$$\text{Solid} \xleftrightarrow{74°\text{C}} \text{smectic} - \text{C} \xleftrightarrow{95°\text{C}} \text{nematic} \xleftrightarrow{124°\text{C}} \text{isotropic} \qquad (12.1)$$

We see that the clearing temperature is here 124°C. Since the different phases have different properties, each phase may be utilized by manipulation of the operating temperature. Not surprisingly, optical properties of LCs may also change with temperature variation or phase change. For instance, when viewed under a polarizing microscope, a color change is observed with the phase change of some liquid crystals. The color of light reflected from a cholesteric crystal depends on the temperature.

As said above, the types of phases formed by MLCs and PLCs are the same. Their applications are different, however. Let us discuss first the applications of MLCs. Given the changes of color just discussed, a thin layer of a cholesteric phase covering a machine or reactor provides an immediate estimation of temperature at any spot on the surface. Liquid crystals such as this are called **thermochromic**. Another application is in medicine. When a part of the human body affected by fever exceeds a certain temperature, a liquid crystal paint on that part changes color, prompting action by medical personnel. Consider also that such monitoring does not require attachment to a machine for detection while additionally it might easily catch the awareness of a patient who might otherwise be ignorant of digital monitoring and machine beeps that are all too common in the hospital setting.

Typically thermochromic LCs are cholesteric. In general, most LCs are optically anisotropic and exhibit double refraction of light. Likewise, dielectric properties of liquid crystals are anisotropic. Two dielectric constants are distinguished for nematic phases: that is, for measurement parallel and perpendicular to the molecular orientation.

Optoelectronic liquid crystal elements are likely the most familiar application of MLCs. Devices such as watches, laptop computer screens, digital clocks, microwaves, and CD players utilize a **liquid crystal display (LCD)** for the display of information—primarily of simple alphanumeric shapes, except in the case of computer screens. The operation of LCDs relies on the following features: light can be polarized, LCs can transmit polarized light, the structure of LCs can be altered by electric current, and there are available electrically conductive transparent substances. Thus, an LCD can be assembled from a twisted nematic liquid crystal layer sandwiched between positive and negative electrodes, polarized glass filters, and front and back covers—typically a glass cover in front and a mirror in back. This assembly creates a **reflective LCD**, in which the mirror or other reflective surface reflects light, now polarized, that entered the front of the device. Liquid crystals, untwisted in response to electrode signals, block the reflected light, creating dark spots—forming alphanumeric or other symbols—on the display. A schematic of a reflective LCD is shown in Figure 12.4. A **backlit LCD**, as in a laptop computer, has its light supplied by fluorescent tubes above, beside, and sometimes behind the LCD. An added panel behind the LCD serves to diffuse the light evenly behind the entire display.

The LCD shown in Figure 12.4 is based on a common-plane electrode, which is useful in devices that show the same information over and over again (e.g., watches). The hexagonal bar shape is the most common form of electrode arrangement in such devices, but almost any shape is possible. Building on this simple electrode layer, one gets more complex displays. Pixels in passive and active matrix LCDs, for instance, are governed by a multitude of integrated circuits or thin-film transistors, respectively. Fine control of the

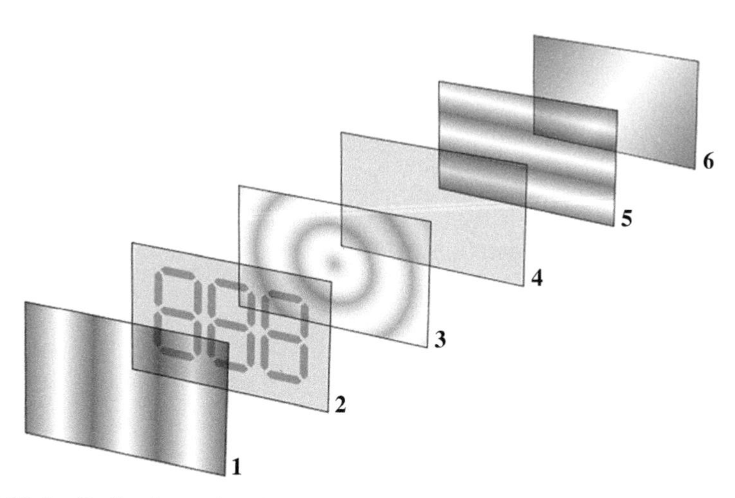

FIGURE 12.4 Reflective twisted nematic liquid crystal display (LCD). 1. Vertical filter film to polarize the light as it enters. 2. Glass substrate with ITO (indium tin oxide) electrodes. The shapes of these electrodes will determine the dark shapes that will appear when the LCD is turned on. Vertical ridges are etched on the surface so the liquid crystals are in line with the polarized light. 3. Twisted nematic liquid crystals. 4. Glass substrate with common electrode film (ITO) with horizontal ridges to line up with the horizontal filter. 5. Horizontal filter film to block/allow through light. 6. Reflective surface to send light back to viewer. *Source*: After ed g2s, https://commons.wikimedia.org/wiki/File:LCD_layers.svg. Used under CC-BY-SA-3.0, 2.5, 2.0, 1.0, GFDL.

voltage supplied to liquid crystals in active matrix LCDs allows partial untwisting of mesogens for precise modification of the amount of transmitted light. A color LCD has three subpixels with red, green, and blue color filters.

We have mentioned that LCs respond to temperature and to electric current, but these are not the only stimuli that elicit a useful response from liquid crystals. A cholesteric PLC + nematic MLC sandwiched between two silicon strips can be used as a deformation sensor. In the un-stretched state it is one color; upon stretching (i.e., deformation), the color changes, as shown in Figure 12.5. The effect is explained by Petr Shibaev and his coworkers [5] in terms of the Helmholtz function of the material.

Helicenes, which are polycyclic aromatic compounds in which the benzene rings are connected in the ortho position leading to screw-shaped molecules, are interesting materials. An LC helicene cooled from the melt can organize itself into LC fibers that are visible under an optical microscope [6]; this occurs with no external force acting. The fibers are comprised of lamellar arrays of the helical columns (hence the name of this class of compounds). The same helicene has an unusually strong ability to rotate the plane of polarization of plane-polarized light [6].

Let us discuss now PLCs and their applications. Polymer liquid crystals have multilevel structures, and a claim has been made that their smaller structures determine larger structures and thus eventually macroscopic properties [7]. A hierarchy of order in liquid crystals has been described wherein so-called polycaps can spontaneously self-organize into polymeric liquid crystals and then into micrometer-scale fiber assemblies [8].

Gedde and co-workers have described the synthesis and properties of pyroelectric PLCs with second-order non-linear optical susceptibility [9–11]. The optical features of such materials are discussed further in Chapter 16, while pyroelectricity is explained in Section 12.5 of this Chapter.

An interesting phase structure not seen in other materials is the **quasi-liquid** (q-l) phase of PLCs [12, 13]. This is the non-crystalline part of the semicrystalline polymer at

FIGURE 12.5 Surface images of a cholesteric PLC + a nematic MLC sandwiched between 2 Si strips. Top: red surface of the unstretched material. Bottom: color changes to green when stretched 20%. Thus the material can be used as a sensor of deformation. *Source*: Shibaev *et al.* [5]. Reprinted with permission of Taylor & Francis Ltd, http://www.tandfonline.com.

temperatures between the glass transition and melting. The mobility of this phase is lower than that of an ordinary isotropic liquid because of the presence of the crystalline component and because of the presence of rigid mesogens. Further, the notion of a liquid is associated with a material that upon heating can vaporize or, if it is a polymer melt, will have no further transition at all. By contrast, the material containing a q-l phase has to undergo at least two more phase transitions: melting and clearing points. The quasi-liquid phase can serve as a medium for a process known as cold crystallization—although this process can take place also in non-LC polymers between their glass and the melting transitions. We discuss cold crystallization in Section 15.3. Furthermore, the quasi-liquid phase is similar to the "leathery" state in elastomers to be discussed in Chapter 14. Both q-l and leathery phases appear directly above glass transitions. Both exhibit *retarded* responses to application of external forces.

Apart from the value of active responses elicited by external stimuli, PLCs are attractive materials based on other thermal and mechanical properties. Polymer liquid crystals can be used alone or as reinforcements in another polymer matrix. PLCs are often stronger than engineering polymers. PLCs have lower isobaric expansivity than engineering polymers; related to this, they also have higher flexibility than solids but lower compressibility than isotropic liquids, a feature that makes them advantageous for a variety of applications. Polymer liquid crystals have a considerably large temperature service range, extending to higher temperatures than for the average engineering polymer. Polymer liquid crystals can also be used to improve processing by lowering the viscosity of the polymer to which they are added. The addition of as little as 5 wt.% PLC to polypropylene results in a significant difference in viscosity [13]; see Figure 12.6 and note

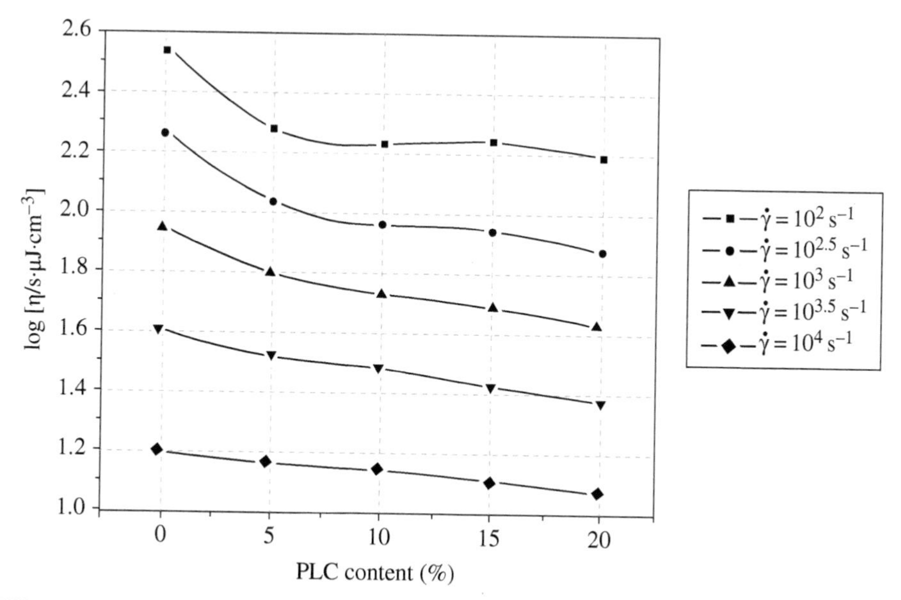

FIGURE 12.6 Changes in melt viscosity of polypropylene as a function of concentration of a PLC (PET/0.6PHB where PET = poly(ethylene terephthalate) and 0.6 is the mole fraction of *p*-hydroxybenzoic acid in the PLC copolymer). *Source*: [13]. Reprinted with permission from Elsevier.

that the scale of viscosity for various shear rates is logarithmic. This fact is important also because PLCs are more expensive than typical engineering polymers; lower processing temperature makes quite a difference to energy costs. The cost of 5% PLC is much less than the cost of achieving a higher temperature needed for effective processing of the neat polymer without the PLC.

There are now thousands of known liquid crystal compounds. The number of compounds is likely proportionate to the number of applications, a fact that belies their comparative unfamiliarity to the general populace. Nevertheless, liquid crystals provide a distinct set of properties—interesting in their own right—that are useful for a host of applications.

12.3 FIELD-RESPONSIVE COMPOSITES

Traditional composites (discussed in Chapter 10), especially polymer composites, can be tailored and designed to meet most of the needs for specialty materials in aerospace, automotive, biomedical, and construction industries. Those materials are passive. There are composites, however, that are *field-responsive*: these materials have rheological properties that can be continuously, rapidly, and reversibly varied by applying electric or magnetic fields [14]. Such materials consist of *polarizable particles* in a *non-polarizable medium*; they may be in the liquid or solid phase. The rheological behavior of field-responsive composites yields materials with stiffness and dissipative characteristics that can be varied in real time. New applications for this particular class of smart materials are expanding.

Before we examine the particulars, let us take a quick look at rheology (more detail in Chapter 13). Simply, rheology is the study of flow and deformation of materials. More precisely to the scientist or engineer, rheology is "the study of the flow of materials that behave in an interesting or unusual manner" [15]. The answer to why toothpaste does not drip out of the tube unless you squeeze it, or why cake batter creeps up the electric beaters as it is mixed, lies within the field of rheology. Thus, the defining feature of the rheological fluids and elastomers that will be discussed in the next three subsections is that they undergo a change in *flow* in response to electric or magnetic fields.

In fluids, the application of an external field induces polarization of the suspended particles. The subsequent induced dipoles lead the particles to form a columnar structure aligned with the applied field. As a consequence, the motion of the fluid is restricted, thereby increasing viscosity of the suspension. Altering the strength of the applied field then alters the yield stress of the fluid. Put another way, the liquid transitions from a free-flowing, linearly viscous liquid with Newtonian behavior to a semi-solid with Bingham behavior in which flow does not begin until a certain critical stress is imposed.

Multiple aspects of composition, structure, and environment affect the properties of controllable fluids. These include: concentration and density of particles, particle size and shape distribution, viscosity and density of the carrier, temperature, and applied field [14]. The interplay between these factors is complex, and elaboration on those features is beyond the scope of this textbook. Both **magnetorheological (MR)** fluids and **electrorheological (ER)** fluids develop a yield stress when a field is applied; but the magnitude of the stress is higher for magnetorheological fluids. The operating temperature range of MR fluids is broader than for ER fluids, for which the operating temperature depends on the type of polarization mechanism. Unlike MR fluids, ER fluids are susceptible to electrical breakdown over time and to negative effects of cycling.

12.3.1 Magnetorheological Fluids

Magnetorheological fluids respond to an applied magnetic field with a dramatic change in rheological behavior [16, 17]. As already indicated, they have the ability to reversibly change from a free-flowing, linear viscous liquid to a semi-solid with controllable yield strength. Moreover, this change can occur in *milliseconds* when the MR fluid is exposed to a magnetic field. In the "off-state", particles are randomly dispersed and exhibit Newtonian behavior. In the "on-state", particles are aligned into fibrous structures and exhibit Bingham plastic behavior, making the fluid act more viscous. Particle polarization increases with the field until a point of saturation. The behavior of a magnetorheological fluid is illustrated by the schematic shown in Figure 12.7.

Magnetorheological fluids are not the same as *ferrofluids*, differing principally in particle size: 1–10 nm for ferrofluids, and typically greater than 1 μm for MR fluids [14]. Yet much like ferrofluids, they are composed of ferromagnetic particles in a non-magnetic medium. Particle volume fractions are typically between 0.1 and 0.5. Iron has the highest saturation magnetization among the elements, although certain alloys may have slightly higher saturation. The magnetic particles may be suspended in a variety of carrier oils or other non-magnetic media, commonly including: petroleum based oils, silicone, mineral oils, polyesters, polyethers, water, and synthetic hydrocarbon oils. Polar media are also sometimes used. Carrier oils are typically chosen on the basis of their temperature stability, and their rheological, lubricating, and oxidative properties. There are additives used to inhibit sedimentation and agglomeration [14].

To illustrate the nature and usefulness of MR fluids, let us consider two practical applications: controlling vibration of a household washing machine with MR fluid damping and use of a suspension damper in automobiles, including brake and shock absorber systems. Magnetic dampers inside a washing machine can decrease noise and vibration. Similarly MR fluids can be used to absorb the shock in a vehicle driving down the road. The stiffness

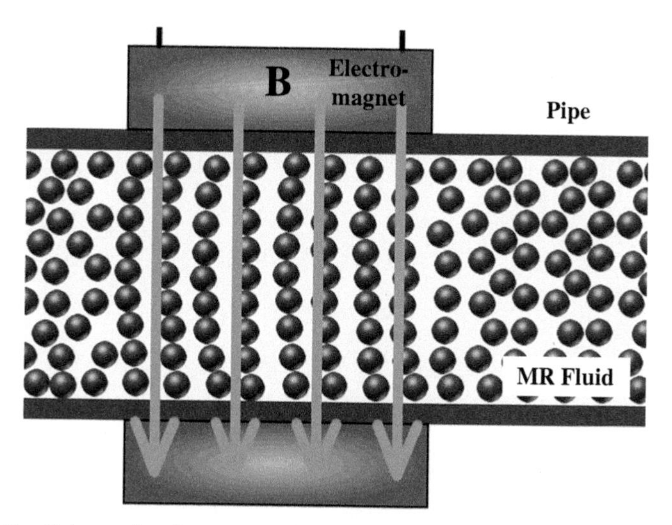

FIGURE 12.7 Schematic of a magnetorheological fluid—magnetic particles suspended in a carrier oil—solidifying in response to an external magnetic field and thereby blocking a pipe. *Source*: Modified from INVENTUS Engineering GmbH, https://commons.wikimedia.org/wiki/File:MRF-Effekt-static-crop.png. Used under CC-BY-SA-3.0.

of magnetic shock absorbers can be electronically adjusted thousands of times per second, providing a remarkably smooth ride. Magnetorheological fluids can generate large forces quickly and flexibly. Magnetorheological fluids are used also in prosthetics, for mitigation of seismic and wind vibrations in buildings and bridges, and also for polishing (high polishing rates and smooth surface finishes) [17].

Magnetorheological fluids operate in different modes in different applications. The three main modes of operation, illustrated in Figure 12.8, are flow (or valve) mode, direct shear mode, and squeeze mode. In each of these cases, the magnetic field is perpendicular to the planes of the plates, thereby restricting fluid flow in the direction parallel to the plates.

In valve mode the fluid is located between a pair of stationary plates (magnetic poles). Fluid flows as a result of a pressure gradient between the plates; and the resistance to fluid flow is controlled by modifying the magnetic field between the poles, in a direction perpendicular to the flow (Figure 12.8). Devices using this mode of operation include servo-valves, dampers, shock absorbers, and actuators. For instance, the movement of an automobile forces MR fluid through channels across which a magnetic field is applied.

In shear mode the MR fluid is located between a pair of moving plates (translational or rotational motion). Relative displacement is parallel to the plates; see Figure 12.8. The apparent viscosity, and thus the "drag force" applied by the fluid to the moving surfaces, is controlled by the magnetic field between the plates. Devices operating in this mode include clutches, brakes, locking devices, and dampers. Figure 12.9 illustrates the most popular brake geometries, so one can better visualize how an MR-fluid is utilized in these devices.

Squeeze mode is the most recent of the three MR fluid operational modes to be defined and utilized. As illustrated in Figure 12.8, the MR fluid is located between a pair of moving poles, and relative displacement is perpendicular to the direction of the fluid flow.

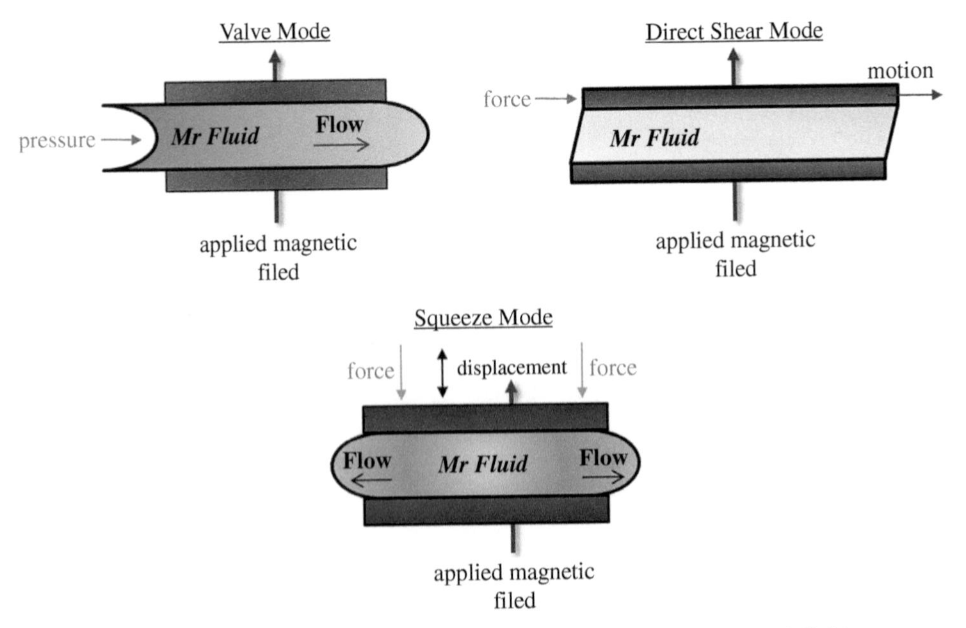

FIGURE 12.8 Three modes of operation of magnetorheological fluids.

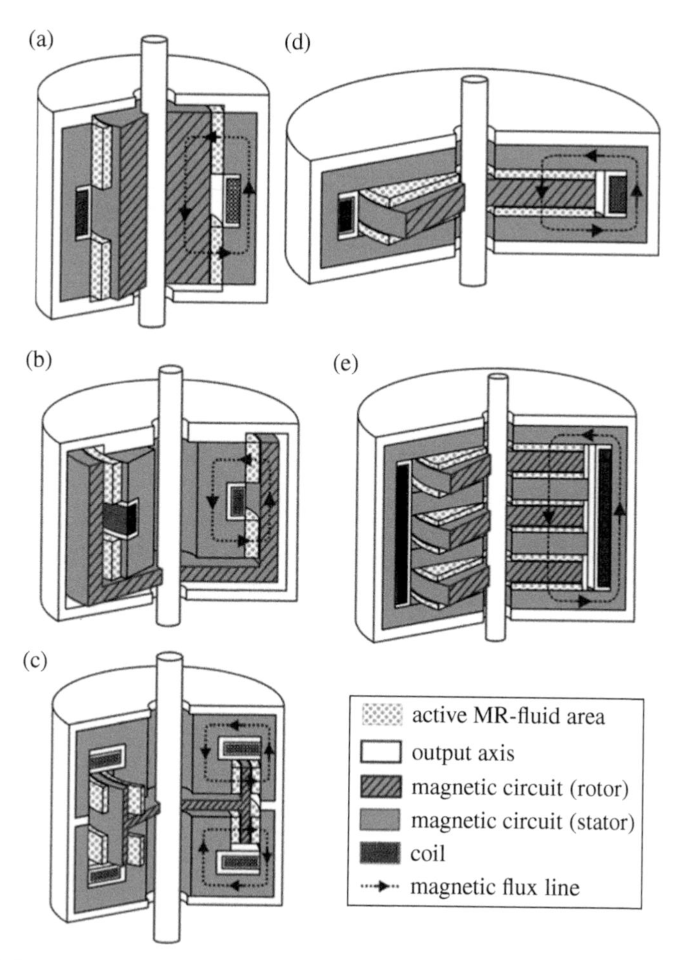

FIGURE 12.9 Various MR-brake designs: (a) drum, (b) inverted drum, (c) T-shaped rotor, (d) disk, (e) multiple disks. *Source*: André Preumont, *Vibration Control of Active Structures, An Introduction*, 3rd Ed., Springer Science+Business Media B.V.: Dordrecht© 2011, p. 455. Republished with permission of Springer; permission conveyed through Copyright Clearance Center, Inc.

The compression force applied to the fluid varies periodically. Compared to other modes, displacements are small, on the order of millimeters, although the resistive forces are high. Note that for the two other modes, the magnitude of resistive forces can be controlled by modifying the magnetic field between the poles. In spite of being less understood than the other modes, the squeeze mode is being explored for use in small amplitude vibration and impact dampers.

Although the potential applications for smart fluids are many, commercial feasibility is limited for the following reasons:

- High density, due to presence of iron, makes them heavy. However, operating volumes are typically small, so while this is a problem it is not insurmountable.
- High-quality fluids are expensive.

- Sedimentation of soft-magnetic particles such as carbonyl iron in the MR fluid. This can be prevented by coating the particle surfaces with polymers, or else by additives such as carbon nanotubes [17, 18].
- Settling of ferroparticles can be a problem for some applications.

Therefore, commercial applications will remain relatively few until these hurdles—especially cost—are overcome. Already existing applications include *physical fitness machines*.

12.3.2 Electrorheological (ER) Fluids

Electrorheological fluids consist of electrically polarizable particles suspended in an insulating medium. Whereas polarization of MR fluids arises from a magnetic field, particle polarization in ER fluids results from an electric field. The process involves ionic or electronic conduction and atomic mechanisms. As with MR fluids, the outcome is a dramatic change in rheological behavior [19, 20].

In electrorheological fluids we observe similar behavior of the suspended particles as with MR fluids. Refer to Figure 12.10, which illustrates effects of electric field (E) imposition on structure and on shear stress [21]. In the off-field state, there is low viscosity. Under field imposition, the ER fluid does not flow unless the yield stress threshold is exceeded. One challenge with ER fluid suspensions is that in time the particles may settle out. One solution may be matching the densities of the solid and liquid components; another is to use nanoparticles.

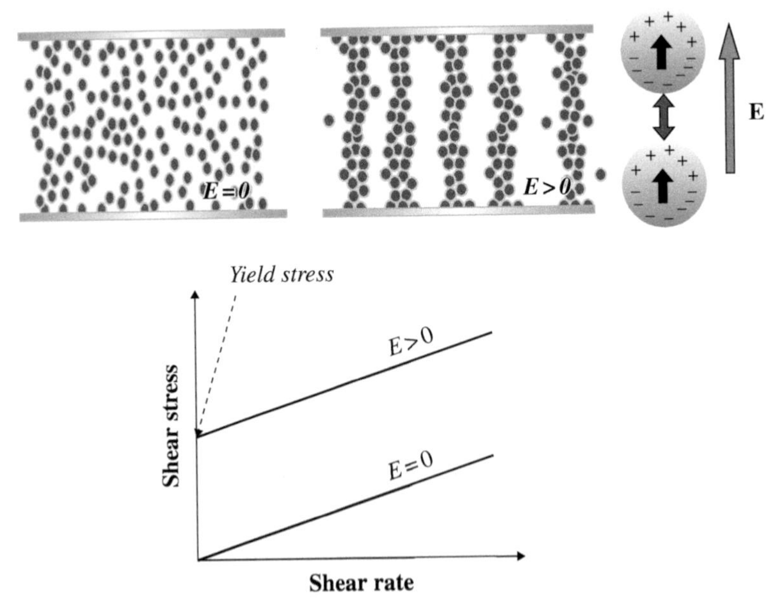

FIGURE 12.10 Behavior of an electrorheological fluid in response to an applied electric field (E). Columnar formation (top) of suspended particles increases viscosity or stiffness of the fluid and the corresponding effect on shear stress (bottom). *Source*: Courtesy of Anna Krzton-Maziopa, Warsaw University of Technology.

Krzton-Maziopa and her coworkers developed a different approach [19]: composite microspheres consisting of polymer electrolyte shell with pre-defined shell thickness and inorganic hollow cores. A relatively weak microstructure is formed which suppresses the sedimentation of the solid phase in the off-electric field state. Additional advantages are lower power consumption than in ordinary polymer based ER fluids as well as structures more stable with respect to temperature changes.

One application of ER fluids is building construction. It used to be that construction materials were designed simply to *withstand* environmental attack (from earthquakes, tornadoes, etc.). With smart materials, a new goal is to design materials that *adapt* to the stresses and strains of such events. Electrorheological fluids can be used to this end and have been employed, for example, in the Kajima Shizuoka Building in Shizuoka, Japan, to protect against earthquakes; see Figure 12.11. Electrorheological fluid in the frame of the building provides hydraulic damping against seismic vibrations. This technology has been developed after thousands of years of doing just the opposite: building thick walls, assuming that the thicker and less mobile they are, the better will they resist the seismic vibrations.

The action of ER fluids is described by two main theories: the interfacial tension or "water bridge" theory [22] and the electrostatic theory. The water bridge theory assumes a three phase system in which the particles contain the third phase, which is another liquid (e.g., water) that is immiscible with the main phase liquid (e.g., oil). In the absence of an applied electric field, the third phase is attracted to and held between the particles. Upon application of an electric field, the third phase is driven to one side of the particles, and thereby adjacent particles bind together, becoming immovable. Thus the ER fluid becomes solid-like.

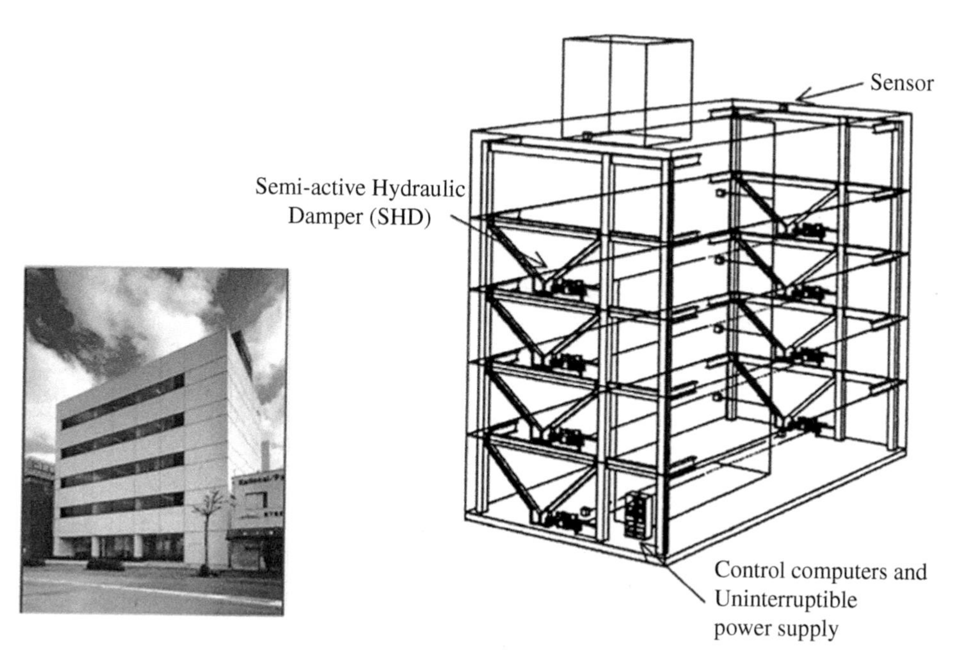

FIGURE 12.11 The ER fluid frame (right) provides hydraulic dampers against seismic vibrations in the Kajima Shizuoka Building in Shizuoka, Japan (left). *Source*: http://article.sapub.org/10.5923.c.jce.201401.04.html#Sec1.

ELECTROMAGNETIC ACTUATORS

An actuator is used to control the conversion of energy into motion. There are hydraulic, pneumatic, electric, and mechanical actuators, defined by the energy source that drives them. Electromagnetic actuators are also called solenoids. A specially designed electromagnet consists of a coil and a movable iron core (known as armature). Current flow through the wire creates a magnetic field. As a consequence, the core moves to diminish (if not to eliminate) the air gap inside the electromagnet. The movable core is typically spring-loaded to make possible retraction of the core to the original position at the time the current is switched off. The force generated in the on position is approximately proportional to the square of the current and inversely proportional to the square of the length of the air gap. Magnetorheological fluids—as well as shape memory alloys, discussed later in this Chapter—can be used in actuators, for example for better modulation of the applied force. Therefore, the fundamental activity of an actuator is to convert energy into linear motion as in:

- Automotive electric-start motors
- Electric door locks
- Hydraulic valves
- Speaker/voice coils
- Power relays
- Pinball machines

The electrostatic theory assumes just two phases, with dielectric particles forming aligned chains analogous to magnetic particles in MR fluids under an applied field. Demonstrating this mechanism is an ER fluid incorporating solid phase conducting particles with an insulator coating [23]. Such an ER fluid clearly cannot work by the water bridge model; and though it demonstrates that some ER fluids can work by the electrostatic effect, it does not prove that all ER fluids do so.

The particles in ER fluids are electrically active. They can be ferroelectric, conducting materials coated with an insulator, or electro-osmotically active particles.

Geometry of the electrodes has been shown to influence the performance of ER fluids [24]. In 2003, the giant electrorheological (GER) effect was described [25]. The static yield stress displayed "near-linear dependence on the electric field, in contrast to the quadratic variation usually observed" [25]. Electric resistivity is important in device design since it is directly related to power consumption.

12.3.3 Electrorheological and Magnetorheological Elastomers

As their name implies, ER and MR elastomers, jointly called field responsive elastomers, exhibit the same type of phenomena as the corresponding fluids; however they are solid state materials. Electrorheological elastomers are composites consisting of particles in an elastomer matrix. The matrix material should be insulating in order to prevent excessive current flow and establish a polarizing field. It is also important for the elastomer to have high dielectric breakdown strength (see Chapter 17) to prevent early degradation of the material. The particulate phase is frequently composed of ceramic materials. The formation of columnar particle structures (as in Figure 12.7) within elastomers varies with the applied electric field and results in a field-dependent shear modulus.

Magnetorheological elastomers—solid state analogs of MR fluids—have also been devised. They are used to some extent in dielectric applications. In general, ER and MR elastomers are not as widely used as the corresponding fluids. However, this will almost certainly not be the case for long as our understanding of them improves with continued research in the field.

12.4 ELECTROCHROMIC MATERIALS

Electrochromic smart materials are not only useful but also appealing to the eyes. These materials change color in response to an external voltage. The work of Chunye Xu and her collaborators [26–29] has advanced the development of electrochromic materials and provides several examples of applications that we will present here. Figure 12.12 shows that goggle lenses with a polymeric electrochromic coating can change color with an applied voltage. Similar functionality is possible on windows through the use of electrochromic conducting polymers and carefully designed electrodes. An aircraft window changing color, as illustrated in Figure 12.13, could provide obvious benefits to passengers or the pilot as light conditions outside vary. It is easy to imagine the usefulness of the color change in eyewear or windows. The fact that the change occurs rapidly is an advantage compared to the type of color-changing lenses frequently used in prescription eyeglasses. There is also an interesting option of self-powered electrochromic windows [29], in which electrical energy harvested from some other device function is sufficient to drive the color change. Actually, in the case of windows, energy consumption for operation can be quite low as it is a relatively simple matter to derive some energy from a solar panel strip mounted outside the window. There is a large transmittance difference between the transparent and deep blue states shown in Figure 12.13. Imagine now a house having mostly glass walls and windows, and these with an electrochromic coating (in the self-powered case two face-to-face layers are used [29]). Such a house can be protected from too much sunlight in the

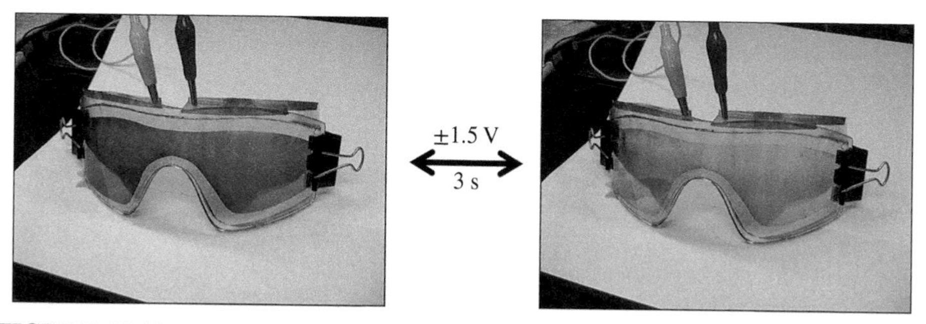

FIGURE 12.12 Electrochromic goggles: tint switches between blue (left) and transparent (right) in 3 seconds at ±1.5 V (+1.5 for oxidation and −1.5 V for reduction). *Source*: Reprinted with permission from Chunye Xu, University of Science and Technology of China, Heifei.

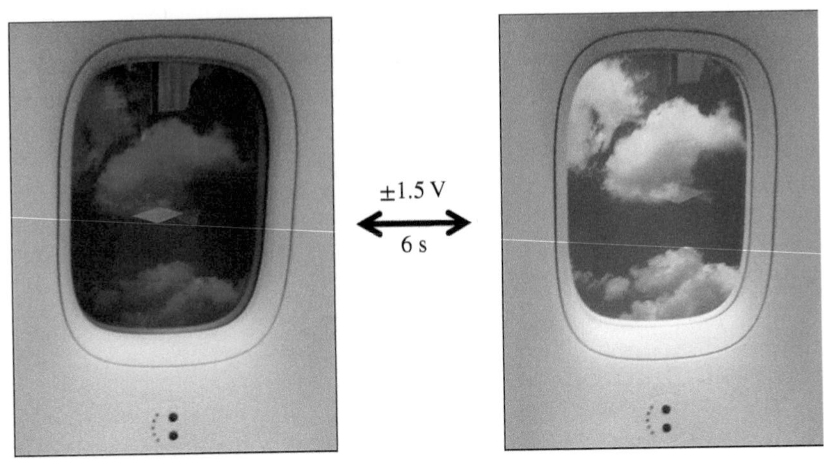

FIGURE 12.13 Electrochromic window in airplane: color switches between blue (left) and transparent (right) at ±1.5V in 6 seconds. *Source*: Reprinted with permission from Chunye Xu, University of Science and Technology of China, Heifei.

summer, while in winter the same windows and walls in the transparent mode will let inside as much light and thermal energy from the sun as they can get.

In Chapter 16 we discuss color, its atomic origins, and related properties in more detail.

12.5 PIEZOELECTRIC AND PYROELECTRIC MATERIALS

Like other smart materials, **piezoelectrics** produce a significant response to some stimulus. The behavior of piezoelectrics involves multiple properties explained further in Chapter 17. Piezoelectrics operate based on a conversion between mechanical energy and electrical energy. The activity is bidirectional; that is a mechanical stress can be converted to an electrical signal or an electrical impulse can yield a mechanical action. For this reason, piezoelectrics are sometimes used as sensors, as are other smart materials mentioned earlier. The accumulation of an electric charge in response to applied mechanical stress occurs not only in various crystals (e.g., quartz, Rochelle salt, topaz, tourmaline) and in certain ceramics but also in biological matter such as bone, DNA, and various proteins.

The piezoelectric effect was first demonstrated by Jacques and Pierre Curie in 1880. Prior investigations of the **pyroelectric effect**—whereby a material generates an electric potential in response to a temperature change—served as a precursor to the discovery of piezoelectricity. The reverse piezoelectric effect—conversion of electricity into a mechanical deformation—was demonstrated in 1881, not long after the initial demonstration by the Curie's. One early use of piezoelectric materials was in crystal radios, wherein quartz crystal was used to select frequencies and thereby tune the radio. During World War I, piezoelectrics were used to develop an ultrasonic submarine detector. Piezoelectric materials continue to be used for the development of sonar technology and applications. The piezoelectric materials now in use are mostly synthetic (though quartz is still used) and are found in devices ranging from speakers, microphones, alarms, pagers, and ignitors to watches, ink-jet printers, vibration dampers, and automobile airbags. Apart from applications already named, piezoelectricity is used as the basis for a number of scientific instrumental techniques having atomic resolution, for example scanning probe microscopies such as scanning tunneling microscopy (STM), atomic force microscopy (AFM), microthermal analysis (MTA), and near-field scanning optical microscopy (NSOM).

Piezoelectrics are solids bonded together by ionic bonds. Thus we can consider the presence of charge centers within the solid structure. A mechanical deformation, such as pressing on the material, results in an offsetting of the centers of electric charge. In effect, this creates dipoles in the material, which in synergy produce a measurable electric potential. The reverse also occurs: application of an electric potential results in a measurable deformation. Thus piezoelectrics have their name (*piezo*) derived from the Greek word meaning "to sit on, or press". Both processes are reversible and all basic scenarios are illustrated in Figure 12.14. An alternating current (AC) would cause a piezoelectric crystal to vibrate. The AC potential can be adjusted to match the natural mechanical frequency of the material, thereby producing the maximum amplitude of vibration (resonance). This behavioral response is used to advantage in developments of sonar technology.

Ferroelectric materials such as barium titanate and lead zirconate titanates, discovered during World War II, were found to be good piezoelectric materials. Rochelle salt, so named after a city in France, has been mentioned already. Materials that display piezoelectricity are non-centrosymmetric crystals, typically polar or chiral crystalline materials, many of them with perovskite structures. Several natural and synthetic piezoelectric materials have already been named. Other synthetic ceramics are lithium niobate ($LiNbO_3$), lithium tantalate ($LiTaO_3$), zinc oxide (ZnO), sodium potassium niobate ($(K,Na)NbO_3$), bismuth ferrite ($BiFeO_3$), and sodium niobate ($NaNbO_3$). The polymer polyvinylidene fluoride (PVDF) exhibits piezoelectricity, indeed better than quartz does. In this case the mechanism involves attraction and repulsion of intertwined macromolecular chains. Combinations of inorganic and organic components have also been used to create piezoelectric materials. A group in Lausanne [30] has used a copolymer of PVDF with

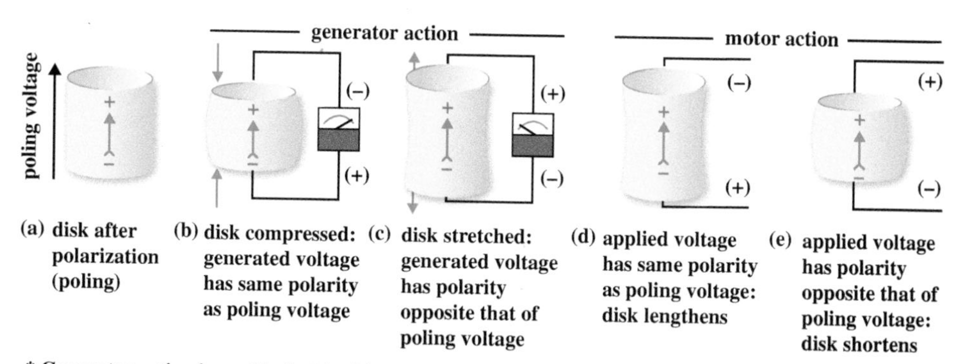

FIGURE 12.14 Generator and motor actions of a piezoelectric element. Possible scenarios of piezoelectric and inverse piezoelectric effects. *Source*: https://www.americanpiezo.com/knowledge-center/piezo-theory/piezoelectricity.html. Reprinted with permission of APC International, Ltd.

trifluoroethylene P(VDF-TrFE) as a matrix for up to 60 vol.% of $BaTiO_3$. The crystalline structure of the matrix turned out to be strongly dependent on the processing route (solvent casting vs. compression molding) because of porosity and inhomogeneity of solvent cast composites—with the resulting changes in composite properties. By contrast, the structure of $BaTiO_3$ was not affected by processing.

Additionally, there are methods of creating piezoelectric composites that utilize fibers to produce I–III composites or that use powders greater than nanoscale in size. Such composites are used in the manufacturing of echography sensors in medicine, for example. So-called electroactive polymers that exhibit a change in size or shape in response to an electric field are in essence piezoelectric materials.

The pyroelectric effect was mentioned earlier in this Section. Pyroelectricity, as stated, refers to the temperature dependence of spontaneous electrical polarization in some anisotropic materials [31]. Pyroelectric materials include the mineral tourmaline, triglycine sulfate single crystals, the ceramic lead zirconate titanate, the polymer polyvinylidene fluoride, and even the biological material collagen.

In pyroelectric crystals we observe thermodynamically reversible interactions among thermal, mechanical, and electrical properties. These are illustrated by the triangular diagram shown in Figure 12.15. (For simplicity, magnetic properties are omitted.) Specific relationships are denoted by each type of line in the schematic. For instance, the three short bold lines connect variables that describe the physical properties of heat capacity, elasticity, and electrical permittivity. These lines designate that a small change in one of the variables produces a corresponding change in the other. Lines joining pairs of circles at different vertices of the triangle indicate coupled effects. The lines denoted primary and secondary correspond to contributions that make up the pyroelectric effect. Accompanying the electric displacement that occurs from a change in temperature (the *primary* pyroelectric effect) is

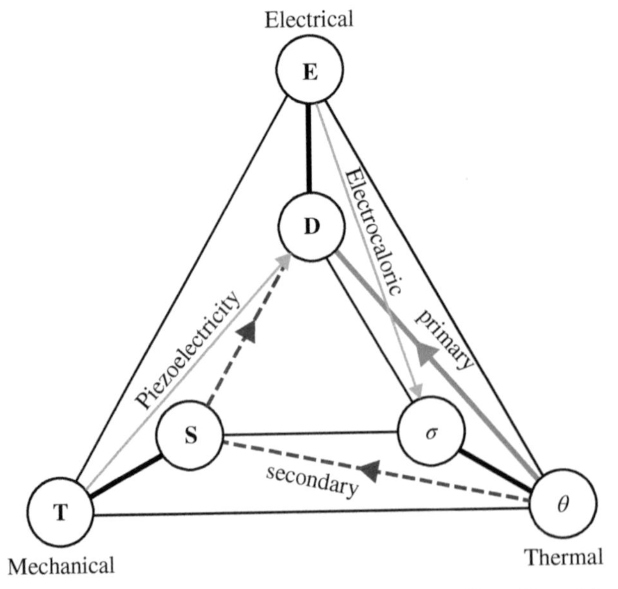

FIGURE 12.15 Crystal properties and the pyroelectric effect. Variables correspond to electrical, mechanical, and thermal properties. *Source*: Reproduced with permission from Georgiades and Oyadiji [24]. Copyright (2005) American Institute of Physics.

an electric displacement due to crystal deformation by thermal expansion (the *secondary* pyroelectric effect, by a piezoelectric process).

Pyroelectric devices can be used to measure small changes in temperature, for instance, in the detection of long-wavelength IR radiation. Topics of research in this area include the development of multilayer materials with increased detection sensitivity and fabrication of thin-film pyroelectrics. In Figure 12.16 we see a schematic for a pyroelectric device used to detect and monitor radiation. The pyroelectric elements are configured to minimize signal noise and to shield one of the elements from radiation exposure. This kind of device consists of single pyroelectric elements, in contrast to others which utilize one- or two-dimensional arrays of elements. Atypical applications of single-element detectors (such as that shown in Figure 12.16) have been in space missions, for instance, in mapping cloud temperatures on Venus (by the *Pioneer Venus Orbiter*). Much more common are IR pyroelectric detectors of the single and multiple element types.

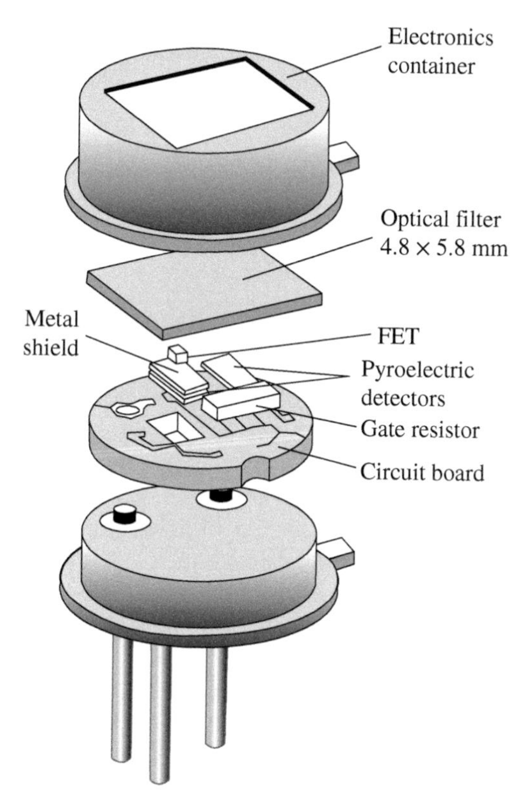

FIGURE 12.16 A single-element pyroelectric detector. In this configuration intended to monitor radiation with a wavelength near 10 μm, two lead titanate pyroelectric elements are used. One detector element is exposed to incoming radiation, and another shielded under a metal strip. Both crystals are electrically connected with opposite polarities to cancel out the effect of any drift in ambient temperature. *Source*: Reproduced with permission from Georgiades and Oyadiji [24]. Copyright (2005) American Institute of Physics.

12.6 SHAPE-MEMORY MATERIALS

The phenomenon of shape memory occurs in metal and polymer systems by different mechanisms. These materials can be deformed and then "remember" and return to their original shape by heating them to a specified temperature. The Russian metallurgists Kurdjumov and Khandros first reported this effect in 1936. In 1951 Chang and Read observed it in In + Ti alloys. Quite by accident the well-known alloy Nitinol (trade name for a nickel titanium alloy, now often used in reference to the general family of nickel titanium alloys) was discovered in the Naval Ordnance Laboratory; hence the name NiTiNOL.

The mechanism of shape memory in metal **shape-memory alloys** (SMAs) relies on a transformation between martensite and austenite phases (refer to Section 6.5). Transition temperatures for SMAs range from 60 to 1450 K. Generally speaking, when the martensite alloy is heated, it transforms to austenite over a temperature range beginning at the *austenite start temperature* (A_s) and ending at the *austenite finish temperature* (A_f). Likewise when austenite material is cooled, it transforms back to martensite over a range of temperatures beginning at the *martensite start temperature* (M_s) and completing the transformation at the *martensite finish temperature* (M_f). These are known as the characteristic transformation temperatures (A_s, A_f, M_s, and M_f).

Alloy composition (such as NiTi or Ni_4Ti_3) and metallurgical treatments, such as cold drawing to produce ultrafine-grained wire or precipitation to improve the superelastic effect [32, 33], significantly impact the *transformation temperatures*. Practically, the SMA can have two different phases: martensite and austenite. In its martensite form, the alloy is soft and ductile and can be easily deformed (somewhat like soft pewter). In the austenitic form, the alloy is first fairly strong and hard and then reaches a point where stress-induced martensitic transformation takes place, leading to large deformations at almost constant strain, called superelastic deformation. Nitinol, a near equiatomic NiTi shape memory alloy, behaves in such a way, with its property expression dependent on the operating temperature [34].

Upon a change in ambient temperature, an SMA is able to convert its shape to a preprogrammed structure. While Nitinol is soft and easily deformable in its low-temperature martensite form, it resumes its original shape and rigidity with heating to the austenite form. This is a **one-way shape-memory effect**, also the most common effect. The capability of SMAs to recover a preset shape when heated above the transformation temperatures and to return to a certain alternative shape upon cooling is known as the **two-way shape memory effect**. Superelasticity (also pseudoelasticity) enables SMAs to undergo elastic deformation (not permanent plastic deformation). Features of the phenomenon are shown schematically in Figure 12.17. This is contrasted with plastic deformation that occurs by slip in ordinary metals, as shown in Figure 12.18. There is another type of SMA, called a **ferromagnetic shape-memory alloy** (FSMA), that changes shape under applied (strong) magnetic fields. Such transitions occur not through a martensitic transformation, but through a movement of the mobile twins to accommodate a change in the lattice orientation to align with the magnetic field [35]. These materials are especially interesting because the magnetic response tends to be faster and more efficient than temperature-induced responses.

Applications in medicine have been devised, for instance, as stents to hold open blood vessels. Consider how this can be accomplished. The stent can be formed into the desired shape at a temperature near that of the body. It can subsequently be deformed to a smaller, thinner structure at room temperature or below. The deformed structure can be more easily inserted into the body, and as the temperature of the stent reaches body temperature it will

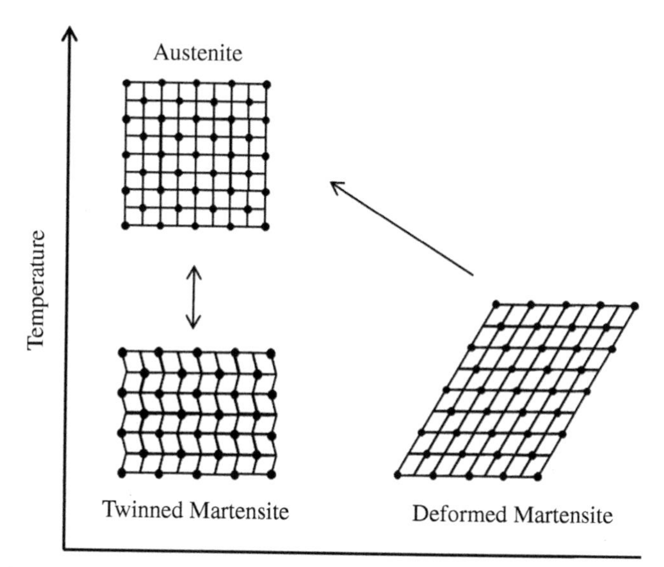

FIGURE 12.17 Transformation mechanism of metal shape-memory alloys. Deformation is elastic.

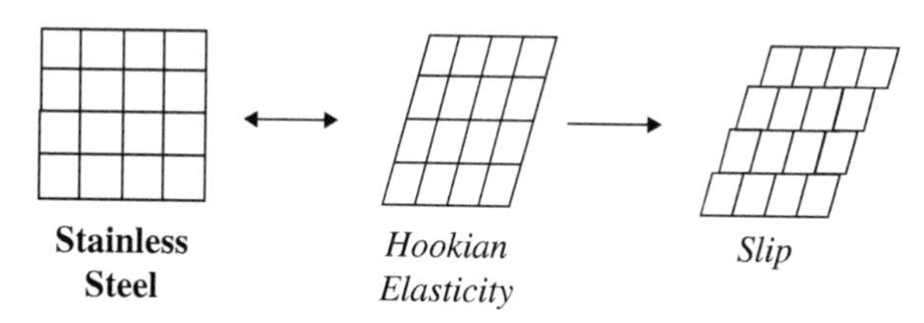

FIGURE 12.18 Slip mechanism of deformation in ordinary metals.

expand into the original desired shape, opening up the vessel and holding itself in place. Other biomedical uses include medical tweezers, glasses frames, orthodontic devices, and surgical retractors [36], while shape-memory alloys are also applied as robotics actuators and micromanipulators [37].

In addition to the shape memory effect and superelasticity, shape memory alloys (SMAs) also exhibit high damping capacity in its martensitic state, which effectively converts mechanical energy into heat. Two primary mechanisms (internal friction and martensitic twin reorientation) are responsible for the high damping capacity in martensitic SMAs. Due to their strong ability of attenuating and dissipating energy, their high yield stress, and their strong resistance to corrosion, SMAs are promising candidates for damping applications when reduction of vibrations and thus reduction in potential structural damage are needed [38].

Nitinol is also extensively used as an actuation device, since the actuation can be activated simply by heating; thus removing the need for a mechanical motor and reducing the potential for mechanical failure [39]. Additionally, though used for some biomedical applications, nitinol is not exactly biocompatible [40–43]—in spite of claims to the contrary.

Let us now contrast SMAs with **shape-memory polymers** (SMPs). Like their metal counterparts, SMPs are able to recover their original shape upon exposure to an external stimulus. As discussed by Behl and Lendlein, the stimuli for activation of SMPs include not only heat and magnetism—as for SMAs—but also light, moisture, and even change in pH [44]. The mechanism is based on a dual-segment system involving the coordination of hard segments acting as net points and flexible segments serving as switches capable of responding to stimuli.

Net points in the polymer network define the permanent shape, and this phase has the highest thermal transition temperature. The flexible components exist in the form of amorphous chain segments. Above the glass transition temperature (refer to Chapter 9), the networks will be elastic. Stretching is accompanied by a loss of entropy (owing to increased chain orientation). Upon release of the external force stretching the material, the original shape is recovered, and the entropy lost before is gained back. Above the transition temperature, chain segments are flexible and the material can be deformed (stretched). This temporary state is fixed by cooling the material to below the transition temperature, thereby initiating strain-induced crystallization of the switching segment.

As said before, a variety of stimuli can be used for activating shape-memory polymers. Thermoresponsive SMPs, activated by a change in temperature, are the most common. Conductive SMPs can be heated by electrical current to induce shape recovery. Light induced SMPs possess photo-sensitive chemical groups that act as molecular switches. With the material in a stretched state, light waves with greater than a certain wavelength initiate crosslinking between the photo-sensitive units. This new shape is retained when the stress is released, but illumination by light of a higher frequency cleaves the crosslinks, allowing a return to the original shape. Through the incorporation of magnetic nanoparticles, there are magnetically induced SMPs. Shape changes can be triggered by inductive heating in alternating magnetic fields. Finally, the actuation of some SMPs can be achieved by immersion in water, through modification of the glass transition temperature. This effect is not necessarily rapid. The thermo-mechanical cycle for shape-memory polymers consists of several steps; the molecular mechanism of the effect is illustrated in Figure 12.19a. The basic procedural steps are shown schematically in Figure 12.19b.

(a)

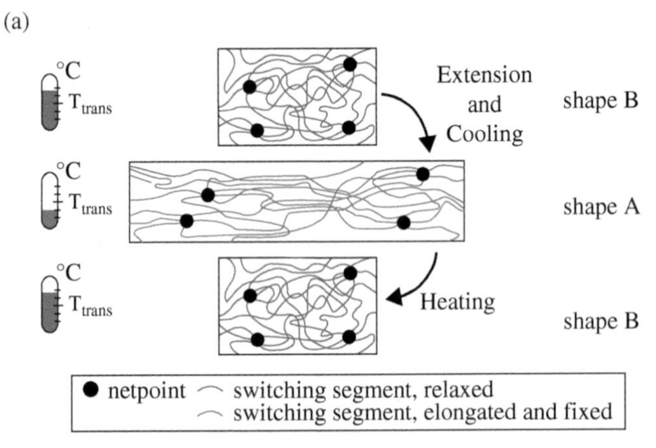

FIGURE 12.19 (a) Molecular mechanism of the thermally induced shape-memory effect. T_{trans} = thermal transition temperature related to the switching phase. *Source*: Reprinted from M. Behl & A. Lendlein, Shape-memory polymers, *Mater. Today* 2006, *10* (4), 20–28, with permission from Elsevier. (b) Schematic illustration of shape changing of a shape-memory element.

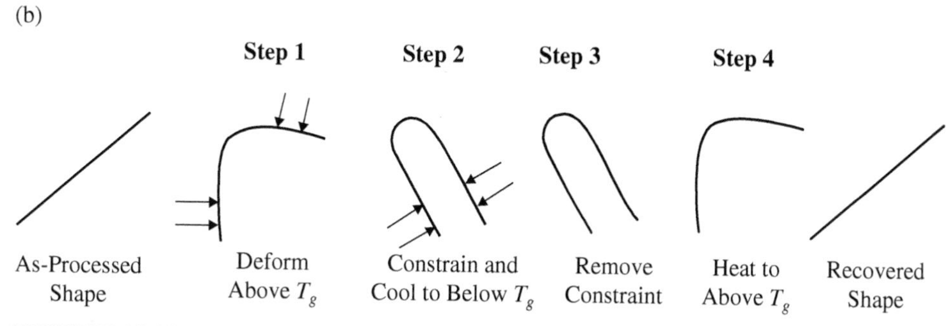

FIGURE 12.19 (Continued)

There are some drawbacks with SMPs. The stiffness is not especially high, which reduces the recovery force after constraint. Fillers and fibers are added to mitigate the problem. Shape-memory polymers are of particular interest to the medical industry. For example, SMPs are sought for use in self-tightening sutures, mechanical removal of blood clots, and aneurysm treatment. Other applications include breathable textiles or as means for active disassembly of parts.

12.7 SELF-ASSESSMENT QUESTIONS

1. Discuss advantages of PLCs in comparison to engineering polymers. Can you explain why the former have lower isobaric expansivities than the latter?
2. Discuss responses of certain living organisms to light and color.
3. What are the advantages of using a magnetorheological fluid in a car suspension damper?
4. Describe a scenario, one not explained in the text, where an electrochromic material would be useful.
5. What is piezoelectricity and how is it useful?
6. Name a practical use for the detection of IR radiation by pyroelectric devices.
7. What applications in the aerospace field can you think of for shape-memory materials?

REFERENCES

1. A. Casona, Proverbia, http://www.proverbia.net/citasautor.asp?autor=175&page=2, accessed August 05, **2016**.
2. E.T. Samulski, *Disc. Faraday Soc.* **1985**, *79*, 7.
3. F. Reinitzer, *Monatshefte Chemie* **1888**, *9*, 421.
4. O. Lehmann, *Z. phys. Chem.* **1889**, *4*, 468.
5. P.V. Shibaev, R. Uhrlass, S. Woodward, C. Schlesier, M.R. Ali & E. Hanelt, *Liq. Crystals* **2010**, *37*, 587.
6. A.J. Lovinger, C. Nuckolls & T.J. Katz, *J. Am. Chem. Soc.* **1998**, *120*, 264.
7. W. Brostow & M. Hess, *Mater. Res. Soc. Symp.* **1992**, *255*, 57.

8. R.K. Castellano, C. Nuckolls, S.H. Eichhorn, M.R. Wood, A.J. Lovinger & J. Rebek Jr., *Angew. Chem. Internat. Ed.* **1999**, *38*, 2603.
9. M. Trollsås, C. Orrenius, F. Sahlén, U.W. Gedde, T. Norin, A. Hult, D. Hermann, P. Rudquist, L. Komitov, S.T. Lagerwall & J. Lindström, *J. Am. Chem. Soc.* **1996**, *118*, 8542.
10. M. Trollsås, F. Sahlén, U.W. Gedde, A. Hult, D. Hermann, P. Rudquist, L. Komitov, S.T. Lagerwall, B. Stebler, J. Lindström & O. Rydlund, *Macromolecules* **1996**, *29*, 2590.
11. M. Lindgren, D.S. Hermann, J. Örtegren, P.-O. Arntzen, U.W. Gedde, A. Hult, L. Komitov, S.T. Lagerwall, P. Rudquist, B. Stebler, F. Sahlén & M. Trollsås, *J. Opt. Soc. Am. B* **1998**, *15*, 914.
12. W. Brostow, M. Hess & B.L. Lopez, *Macromolecules* **1994**, *27*, 2262.
13. W. Brostow, T. Sterzynski & S. Triouleyre, *Polymer* **1996**, *37*, 1561.
14. B.C. Muñoz & M.R. Jolly, Chapter 22: Composites with field responsive rheology, in *Performance of Plastics* (W. Brostow, ed.), Hanser: Munich/Cincinnati, OH **2000**.
15. F.A. Morrison, What is rheology anyway? in *The Industrial Physicist*, American Institute of Physics: College Park, MD **2004**, pp. 29–31.
16. J.D. Carlson, Controlling Vibration with Magnetorheological Fluid Damping, *Sensors*, February 1, **2002**.
17. B.J. Park, F.F. Fang & H.J. Choi, *Soft Matter* **2010**, *6*, 5246.
18. F.F. Fang, H.J. Choi & M.S. Jhon, *Colloids & Surfaces A* **2009**, *351*, 46.
19. A. Krzton-Maziopa, M. Gorkier & J. Plocharski, *Polymers Adv. Tech.* **2012**, *23*, 702.
20. Y.D. Liu & H.J. Choi, *Soft Matter* **2012**, *8*, 11961.
21. A. Krzton-Maziopa, personal communication from the Warsaw University of Technology.
22. J.E. Stangroom, *Physics in Technol.* **1983**, *14*, 290.
23. W.Y. Tam, G.H. Yi, W. Wen, H. Ma & P. Sheng, *Phys. Rev. Letters* **1997**, *78*, 2987.
24. G. Georgiades & S.O. Oyadiji, *J. Intelligent Mater. Systems & Structures* **2003**, *14*, 105.
25. W. Wen, X. Huang, S. Yang, K. Lu & P. Sheng, *Nature Mater.* **2003**, *2*, 727.
26. C. Xu, L. Liu, S.E. Legenski, D. Ning & M. Taya, *J. Mater. Res.* **2004**, *19*, 2072.
27. C. Ma, M. Taya & C. Xu, *Electrochim. Acta* **2008**, *54*, 598.
28. C. Ma, M. Taya & C. Xu, *Polymer Eng. & Sci.* **2008**, *49*, 2224.
29. S. Yang, J. Zheng, M. Li & C. Xu, *Solar Energy Mater. & Solar Cells* **2012**, *97*, 186.
30. S. Dalle Vacche, F. Oliveira, Y. Leterrier, V. Michaud, D. Damjanovic & J.-A. Månson, *J. Mater. Sci.* **2012**, *47*, 4763.
31. S.B. Lang, *Physics Today* **August 2005**, *8*, 31.
32. S. Gollerthan, M.L. Young, A. Baruj, J. Frenzel, W. Schmahl & G. Eggeler, *Acta Mater.* **2009**, *57*, 1015.
33. M.L. Young, M. Frotscher, H. Bei, T. Simon, E.P. George & G. Eggeler, *Internat. J. Mater. Res.* **2012**, *103*, 1434.
34. E. Spārniņš, J. Andersons, V. Michaud & Y. Leterrier, *Smart Mater. & Struct.* **2015**, *24*, 125038.

35. S. Glock, L.P. Canal, C.M. Grize & V. Michaud, *Compos. Sci. & Technol.* **2015**, *114*, 110.
36. G. Kauffman & I. Mayo, Memory Metal, *Chem. Matters*, October **1993**.
37. C.A. Rogers, Intelligent Materials, *Scientific American*, September **1995**, pp. 154–157.
38. K. Yamauchi, I. Ohkata, K. Tsuchiya & S. Myazaki, *Shape Memory and Superelastic Alloys: Applications and Technologies*, Woodhead Publishing: Cambridge **2011**.
39. O. Benafan, J. Brown, F.T. Calkins, P. Kumar, A. Stebner, T. Turner, R. Vaidyanathan, J. Webster & M.L. Young, *Internat. J. Mech. Mater. Design* **2013**, *10999*, 1569.
40. S. Shabalovskaya, J. Anderegg & J. Van Humbeeck, *Acta Biomater.* **2008**, *4*, 447.
41. D. Starosvetsky & I. Gotman, *Biomaterials* **2001**, *22*, 1853.
42. G. Zorn, R. Adadi, R. Brener, V.A. Yakovlev, I. Gotman, E.Y. Gutmanas & C.N. Sukenik, *Chem. Mater.* **2008**, *20*, 5368.
43. H.D. Samaroo, J. Lu & T.J. Webster, *Internat. J. Nanomed.* **2008**, *3*, 75–82.
44. M. Behl & A. Lendlein, *Mater. Today* **2006**, *10* (4), 20.

PART 3

BEHAVIOR AND PROPERTIES

13

RHEOLOGICAL PROPERTIES

Panta rhei

—Heraclitus of Ephesus, assigned also to Simplicius. Ancient Greek, it means: *everything flows*.

13.1 INTRODUCTION

Rheology has been defined as the study of the flow of matter. It is considered most often with regard to matter in the liquid state, however it applies also to the gas and solid states. For instance, there are solids that under certain conditions respond to an applied force with plastic flow rather than deforming elastically. In some sense, tossing a golf ball with dimples, or else softballs or baseballs with fuzz or stitches, involves various aspects of flow. One might even say that a well thrown "curve ball" suggests that a player has acquired a profound knowledge of rheology—though more realistically he has acquired extensive experience through trial-and-error.

The term **soft matter** is applied to solids of real interest to rheology. Rheological analysis is particularly relevant for substances with complex microstructures, such as muds, sludges, suspensions, glass formers in silicates, foams, liquid crystals, emulsions, and gels (see Section 11.1). These categories of materials are mostly self-explanatory or have been discussed before. An **emulsion** consists of at least two immiscible liquid phases in which there is at least one dispersed phase consisting of small droplets. (Refer also to Section 9.4 for additional comments on emulsions.) **Suspensions** are multiphase systems where the main phase is a fluid, and there are also dispersed solid particles large enough to undergo sedimentation (movement to the bottom of the container). One may find foods, biofluids, personal care products, electronic and optical materials, pharmaceuticals, and polymers in

Materials: Introduction and Applications, First Edition. Witold Brostow and Haley E. Hagg Lobland.

most of these categories—for example both pharmaceutical creams and mayonnaise are emulsions; products as varied as mustard, blood, and nail polish are all suspensions. There are entire books on rheology [1] while here as usual we have one chapter to deal with the essentials.

In Chapter 9 we talked about thermoset polymers which do not melt. With this exception, all other classes of materials do melt. There are two ways a material can be found in the liquid state: *melting* and *dissolution*, the latter referring to processes that generate a liquid from solid or gaseous components. What happens to a thermoset that comes in contact with a liquid? Unless it is solvent repellent, it will *swell.*

A layman deals with rheological properties in varied circumstances. Examples include pouring ketchup or honey and squeezing toothpaste from a tube. Other common encounters with phenomena of rheology are in the application of paint to a wall and the blending of cake batter with a mixer.

We have in fact talked about rheological properties in Chapter 12 when discussing electrorheological and magnetorheological fluids. We know these are smart materials, while now we shall be mostly talking about materials which are *not* smart—which comprise the overwhelming majority of materials. It is appropriate to begin Part 3 with a study of rheological properties because rheology determines a material's behavior during many processing operations, while in turn those operations affect a multitude of other properties (mechanical, electrical, optical, and so on). For example, a study of polyurethane polymer foams by Jackovich and her colleagues [2] shows that the elastic modulus strongly depends on the processing temperature—and on the mold size, as well. Apparently the mold size affects the temperature distribution profile inside the mold.

Thus, as said at the start, the central theme of this chapter is fluids, especially fluids in motion, what we call **flow**. There are **classical fluids** that quickly fill a container, taking on its shape. There are also **complex fluids** (such as those named above) of various types that may maintain their shape for some time but eventually flow. In either case, the resistance to flow we call **viscosity**. Classic Newtonian fluids have constant viscosity, while the viscosity of complex fluids depends on the applied strain. To establish a relationship between applied forces and their induced geometrical effects at a point in the fluid is one goal of rheology.

13.2 LAMINAR AND TURBULENT FLOW AND THE MELT FLOW INDEX

Since we are talking about flow, there are two kinds of flow: *laminar* and *turbulent.* To explain the difference, we shall first discuss the **Reynolds Number** (Re). Reynolds number is a dimensionless velocity expressed as the ratio of *inertial* forces (or dynamic pressure) to *viscous* forces (or shearing stress). As such, it indicates the relative importance of these contributions under a given set of flow conditions. A general definition is

$$\mathrm{Re} = \frac{u\rho D}{\eta} \tag{13.1}$$

where u is the mean velocity (in m/s) of the fluid relative to the pipe or surface through which it is flowing; ρ is the density of the fluid (in kg/m^3); D is a characteristic linear dimension (in units of m), such as the hydraulic diameter for a pipe or duct; η is the **dynamic viscosity** of the fluid (in Pa·s or N·s/m^2 or kg/(m·s)). Sometimes one works also with the

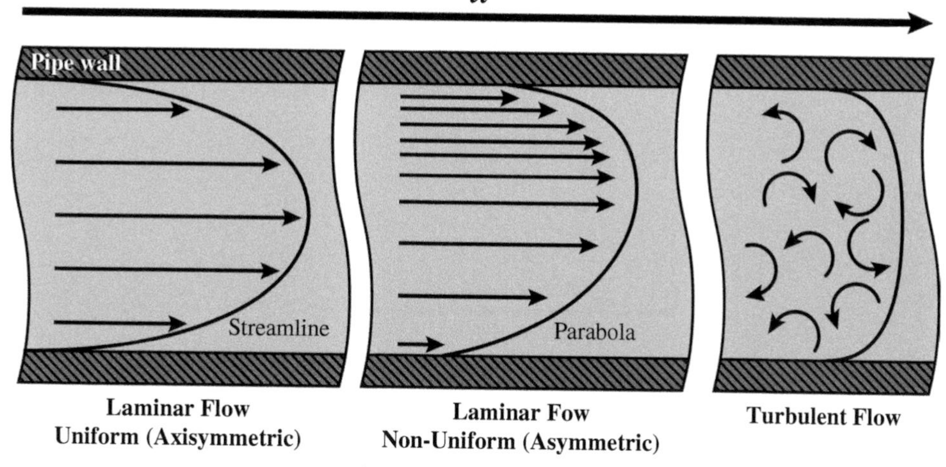

FIGURE 13.1 Types of flow.

kinematic viscosity defined as η/ρ, a quantity that can be substituted into Eq. (13.1) and expressed in m^2/s. The expression of Re for the specific case of flow in a pipe is

$$\mathrm{Re} = \frac{u\rho D_{\mathrm{H}}}{\eta} \tag{13.2}$$

where u is the flow velocity in the pipe and D_{H} is the hydraulic diameter of the pipe.

The smooth, regular flow of a fluid, without perturbations called **eddies**, is known as **laminar flow**. Laminar flow is typical in situations where viscosity is relatively high, the fluid is moving slowly, and the flow channel is fairly small. Examples include oil through a thin tube or blood flow through capillaries. Relative to a stationary plate over which the fluid is moving, the fluid essentially flows in layers (hence the name laminar), with velocity increasing with distance from the stationary plate (or wall). Thus laminar flow in a pipe can be considered as liquid cylinders where the innermost cylinders flow fastest (see Figure 13.1) and those touching the pipe walls do not move at all. Of course, there is variation of the velocity going from the center of the flow to the walls, but the idea of cylinders (see also Figure 13.5) helps somewhat when dealing with such systems. As another illustration, picture in your mind smoke rising from a cigarette and the smoke's movement in a straight path as undergoing laminar flow. Then, after rising a bit, the smoke transitions to *turbulent* flow as it swirls away from its regular path and forms eddies. Laminar flow is typically characterized by $\mathrm{Re}<2100$ [1]. The addition of polymeric chains or suspended solid particles to a liquid tends to increase that critical Reynolds number.

Therefore, at high Reynolds Numbers, the flow is not stable anymore; instead of laminar we now have **turbulent** flow. In pipes, turbulent flow usually occurs when $\mathrm{Re}>4000$. The range $2100<\mathrm{Re}<4000$ is known as the *transition region* or *transient region*, where both laminar and turbulent flow are possible. The eddies and chaotic motion of turbulent flow (see Figure 13.1 on the right) do not contribute (positively) to the flow rate. Rather, because turbulence increases resistance, large increases in pressure are needed to significantly increase the volume flow rate. In surface water systems, the characteristic length scale D is that of the basin, which is typically large—perhaps 1 m to

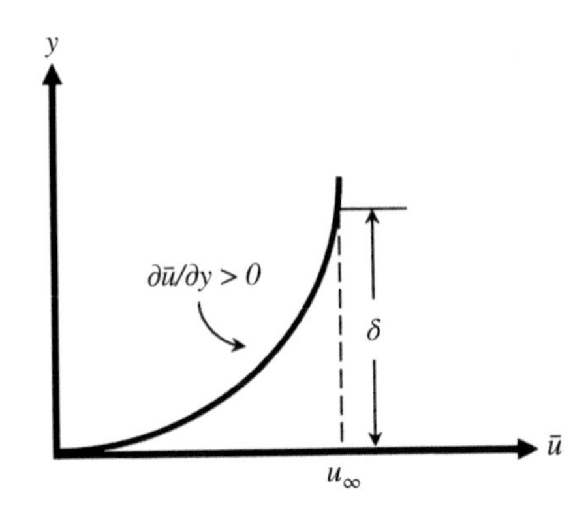

FIGURE 13.2 Velocity profile of a flow as it relates to distance y from the bed. Mean velocity $\bar{u}=0$ at the solid boundary ($y=0$). Velocity reaches a constant value u_∞, the free stream velocity, at some distance above the boundary. δ denotes thickness or depth of the boundary layer.

several hundred km—and results in turbulent flow. By contrast, groundwater systems may have $D<1$ mm, giving flow that is nearly always laminar. Dyes can be used to observe the flow type. After adding a dye to a flow, one may observe a streamlined dye trace in laminar flow or the lateral transport of the dye through turbulent eddies along with some transport along the mean flow path.

The practical outcome of turbulence is to create instability in the velocity of flow. One can make a record of the velocity (longitudinal and latitudinal) at any point in the flow as a function of time; from that, values of the *mean velocity* as well as of *turbulent fluctuations* and of *turbulence strength* can be calculated. There is, however, a distinct behavior of the flow near a solid boundary. In the boundary layer, velocity of the fluid goes to zero. This no-slip condition of the fluid, wherein the fluid velocity has no slip relative to the boundary velocity, is an effect of viscosity—and the force of friction that causes the fluid to "stick" to the boundary. A profile representing the typical variation of velocity as a function of distance (y) from the bed (i.e., solid boundary) is given in Figure 13.2. The spatial variation of velocity over the vertical distance between the bed and the free stream is referred to as **shear**. High shear produces the instability known as turbulence.

In industry and engineering work, calculation of the Reynolds Number is frequent; however it does require means to determine velocity, viscosity, and density if these are not known. What happens if we have a factory without sophisticated equipment to measure Reynolds Numbers, yet we need some information about the behavior of a material that we are processing in the liquid state? If the material is a polymer melt, the **melt flow index (MFI)** can be determined to provide a corresponding parameter; and the test is fairly simple, requiring only inexpensive equipment. Calculated as how many grams of polymer pass through a standardized capillary in ten minutes, the MFI therefore provides a rough comparison of flow characteristics for different materials. Although a capillary rheometer would provide more *accurate* viscosity data, the MFI is suitable for a variety of industrial purposes (e.g., comparing different lots of the same material).

What if we are not dealing with a polymer melt but a different liquid or a gas? One uses so called Pitot tubes, named after French engineer Henri Pitot who first used them in 1732 to measure the water velocity in the Seine river at different depths. The tube is inserted through a hole into the duct and connected to a U-tube gauge providing the difference in pressure ΔP. If there is no flow, we have $\Delta P=0$. When the fluid flows, dynamic pressure $\Delta P \neq 0$ appears, dependent on the flow velocity and also on the density of the fluid. An important application of such devices is in the determination of the air speed of the aircraft on the basis of the dynamic pressure measured. One needs to make sure that the Pitot tube did not get clogged with ice.

We have already talked about viscosity in various contexts. How does one measure viscosity? There are a number of methods and we discuss them in the following Section.

13.3 VISCOSITY AND HOW IT IS MEASURED

Viscosity is commonly thought of as how thick (like honey) or thin (watery) a fluid is. Technically **viscosity** is described as a fluid's resistance to flow. An alternative but related perspective is that viscosity corresponds to a fluid's resistance to the *movement* of an object through that fluid. We intuitively perceive that honey has a higher viscosity than water; it is clear now that our intuition correlates with the technical description. In measuring viscosity, we shall think about *flow* in terms of deformation or strain. Generally speaking, viscous fluids resist shearing deformation. (Shear and other forces are described more fully in Chapter 14.)

In a liquid at rest, there are interactions between molecules, ions, and other particles—as discussed in Chapter 2. These interactions have to be partly overcome when the particles begin to flow—only partly since movement of some particles in unison is also possible. A real fluid flowing in a pipe experiences friction with the pipe walls and friction within the fluid itself. As a consequence, some kinetic energy is converted into thermal energy. We call the frictional forces that inhibit fluid layers from sliding past each other **viscous forces**. As consequence of the internal friction of a fluid in motion and friction effects at the wall (of a pipe or other surface), neighboring fragments of a fluid can move at different velocities. Such behavior was described already for laminar flow. When fluid is forced through a tube, the fluid generally moves faster near the central axis and very slowly near the walls (refer again to Figure 13.1), therefore some stress (such as a pressure difference between the two ends of the tube) is needed to overcome the friction between layers and keep the fluid moving. For the same velocity pattern, the stress required is proportional to the fluid viscosity. Viscosity is therefore the measure of resistance by a fluid to relative motion within the fluid. It can be measured by determining the viscous drag due to a fluid placed between two plates, one stationary and one moving; refer to the schematic in Figure 13.3.

Now let us consider the scenario of Figure 13.3, where one plate is stationary, the other is moving, and the two plates are separated by a fluid layer of thickness d. To obtain usable relationships, we assume that the speed u of the plate is low; this is called **Couette flow**. The velocity profile is a straight line (the diagonal between the two plates). The **shear rate** (in s^{-1}) is defined as

$$\Gamma = \frac{\partial u}{\partial y} \tag{13.3}$$

Since $u=x/t$, we have the equivalent expression for Γ in terms of the ordinate directions x and y:

$$\Gamma = \frac{\partial(\partial x/\partial y)}{\partial t} = \frac{\partial u}{\partial y} \tag{13.4}$$

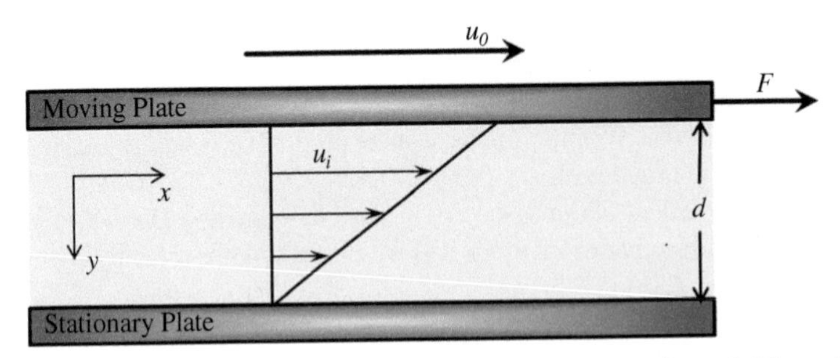

FIGURE 13.3 The shearing force F acts on the top plate as indicated. The velocity u_i of the fluid layer of depth d decreases going down along the vertical vector y since the velocity at the bottom plate is necessarily 0. The velocity profile from the top to bottom is not necessarily a straight line as shown; it can be concave looking from the right (the **capillary effect**, better seen in narrow cylindrical conduits).

The force F needed to maintain constant velocity u_0 of the upper plate can be measured and is proportional to the area A of the plate and to the ratio of velocity u_0 to the distance d between the plates. Moreover, from mechanics (see Chapter 14), we know that F/A is the definition of stress. Therefore, the **shear stress** τ is

$$\tau = \frac{F}{A} = \frac{u_0}{d}\eta = \frac{\partial u}{\partial y}\eta = \Gamma\eta \tag{13.5}$$

The proportionality constant η is called dynamic (or absolute) viscosity. Equation (13.5) was first formulated by Sir Isaac Newton. (Note that often in the scientific literature the dynamic viscosity is represented by the symbol μ.)

Equation (13.5) is obeyed in simple cases that we call **Newtonian flow**. Other cases are also possible; all are presented in Figure 13.4. Analyzing Figure 13.4, consider first the simplest case: Newtonian flow. The proportionality between the shear stress τ and the shear rate Γ is as defined in Eq. (13.5); viscosity is constant and dependent only on temperature. Many familiar fluids including water and mineral oil are Newtonian fluids.

Consider again the following expression for shear stress defined in Eq. (13.5):

$$\tau = \Gamma\eta$$

Now look at the curves for pseudoplastic and dilatant fluids in Figure 13.4. What happens to τ as the shear rate Γ increases? For a **pseudo-plastic**, τ increases faster, thus viscosity decreases with rate of shear. Pseudoplastic fluids are also called **shear-thinning fluids**; examples of shear-thinning fluids are some colloidal substances like paint, shampoo, ketchup, slurries, and some colloidal suspensions. The opposite occurs in **dilatant** fluids: viscosity increases with rate of shear. These fluids are also called **shear-thickening**: examples are quicksand and concentrated mixtures of corn starch and water.

Figure 13.4 shows us also the behavior of Bingham fluids, distinctly different from the other types of fluids shown. What we see in the diagram is that a minimum value of shear stress must be applied before flow commences in a **plastic fluid**. For a **Bingham plastic**, once the minimum shear stress has been achieved, the relationship between τ and Γ is linear.

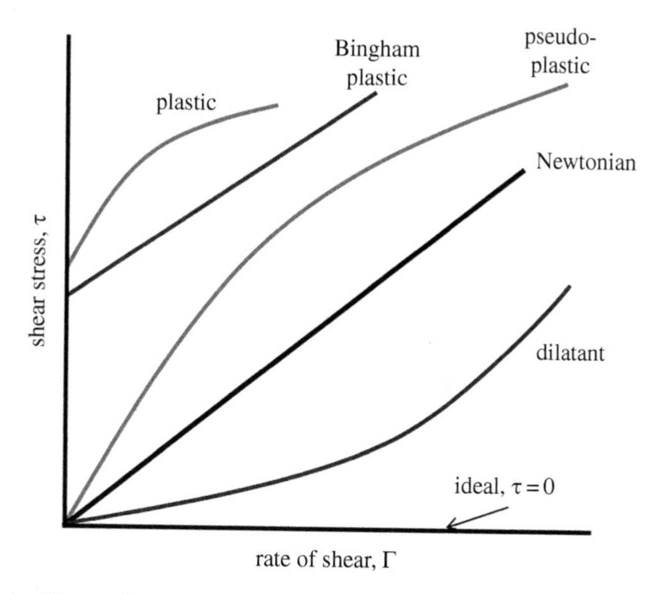

FIGURE 13.4 Newtonian and non-Newtonian fluids. Shear stress is τ; shear rate is $\Gamma = \partial u/\partial y$ on x-axis. Pseudo-plastic fluids are shear thinning; dilatant fluids are shear thickening.

There are various equations, often with a Γ^n term, describing such behavior. Bingham liquids can be described by the formulae

$$\Gamma = 0 \quad \text{for} \quad \tau < \tau_c \tag{13.6a}$$

$$\tau = \tau_c + \Gamma^n \eta \quad \text{for} \quad \tau \geq \tau_c \tag{13.6b}$$

where $n = 1$. For other plastic fluids (not Bingham), n has some value other than 1. Equation (13.6b) reduces to Eq. (13.3) for the case where there is no critical stress τ_c that needs to be exceeded for the flow to start. Examples of Bingham liquids are clay suspensions, drilling mud, toothpaste, mayonnaise, mustard, and avalanches. Imagine how much more dangerous snow-covered mountains would be for humans and animals if avalanches were *not* Bingham liquids.

There are also two categories of non-Newtonian fluids whose behavior is *time dependent* in addition to being dependent on temperature and shear rate: thixotropic and rheopectic fluids. **Thixotropy** refers to time-dependent fluid behavior in which viscosity decreases with shearing time and in which viscosity recovers its starting value upon cessation of shearing; the effect can occur in liquids with a microstructure since it is associated with the time taken to move from one microstructural state to another. Thixotropy was originally defined to describe materials observed to be gel-like at rest but fully liquid once agitated. Such materials, after being stressed, return to their gel-like states over a transition time that may be almost instantaneous or quite long. The process in fact is similar to shear thinning, and in many instances it is difficult to distinguish between the two phenomena. Barnes provides a thorough review, expounding on distinctions of time-scale between normal viscoelastic behavior and thixotropy and the causes of shear-thinning [3].

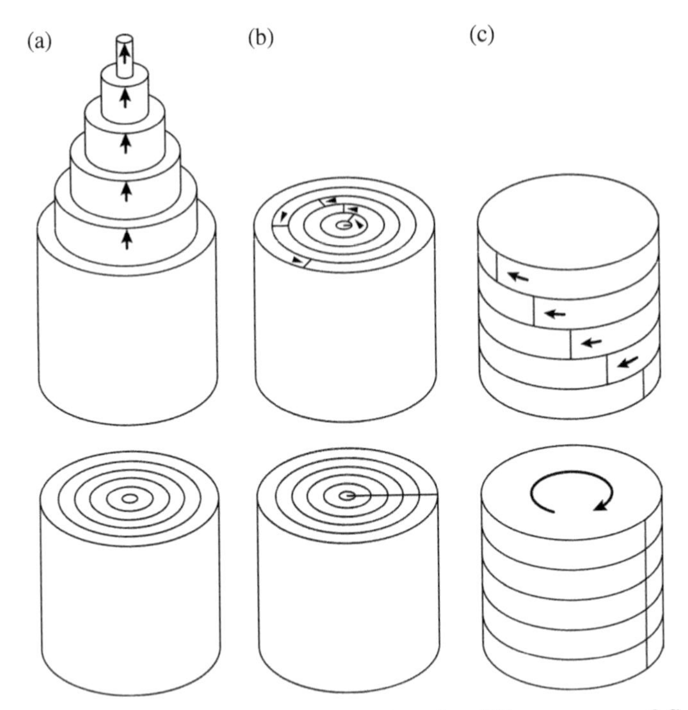

FIGURE 13.5 Tubular cross-sections illustrating the different types of fluid flow: (a) telescoping (laminar); (b) rotational; (c) twisting.

Rheopectic fluids are time-thickening, that is, viscosity increases with time under a shear stress, for instance shaking or stirring. As with dilatancy, rheopecty is characterized by increased viscosity as a result of applied stress. The difference is that rheopectic materials viscosity increases the longer the stress is applied while dilatant materials viscosity increases simply with the increase in stress. Rheopectic fluids are rare. Cream is an example; it becomes stiff only after sustained beating. Other clear examples are some mineral pastes, for instance consisting of components such as kaolin clay, calcium carbonate, and starch in water. The development of new methods to create and utilize rheopectic materials would be desirable for military, sports, and other applications.

Now back to the story of flow and measuring viscosity. We have seen that viscosity can be determined by measuring the force applied to a fluid between a stationary and a mobile plate. We have also characterized the nature of laminar flow and obtained some idea of the shear forces associated with it. If we were to look at the cross section of a tube, we would have a telescopic profile of the velocity for laminar flow, as shown in Figure 13.5a. Other flow options are possible; see the rotational and twisting shear depicted in Figure 13.5b and c. Each of these instances would be characterized by a particular velocity profile.

In rheological testing, equipment can be configured to induce such varied types of flow. Recall from the Introduction that a stated goal of rheology is to draw connections between applied forces and the geometrical effects they induce within a fluid. The cone-and-plate geometry, shown in Figure 13.6, can be used for determination of viscosity. In Figure 13.6, the cone/plate measuring system is compared to two other geometries that are possible in a rotational viscometer. In all three instances, the probe is rotated at a controlled speed to provide a controlled shear rate test. There exist also oscillatory viscometers that apply oscillation, rather than full rotation.

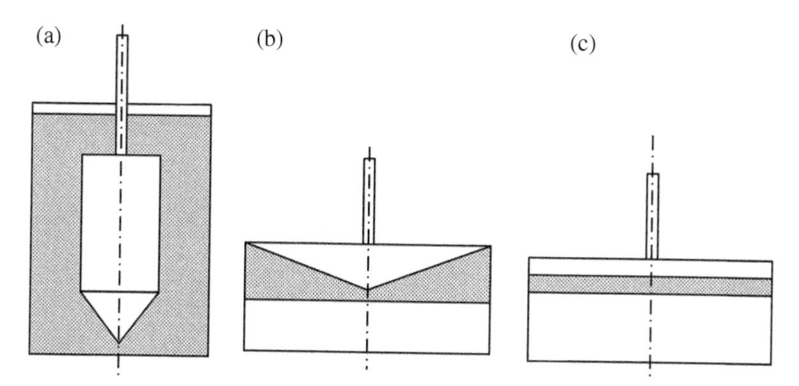

FIGURE 13.6 Testing geometries for a rotational viscometer. (a) cylindrical measuring system; (b) cone/plate measuring system; (c) plate/plate measuring system.

In the geometry shown in Figure 13.6b, the cone is at a shallow angle and in bare contact with a flat plate. In such a system the shear rate beneath the plate is constant to a reasonable degree of precision. Viscosity is obtained from a diagram of shear stress (torque) versus shear rate (angular velocity).

There is also a test method in which a solid sphere of known size and density is allowed to descend through the liquid in a vertical glass tube. Such **glass capillary viscometers** are known as **Ostwald viscometers**, so named after Wilhelm Ostwald who was active in Dorpat—which is now Tartu in Estonia—and later in Leipzig. The time taken for the level of the liquid to pass between two marks in a vertical capillary structure (a precision bore tubing) is proportional to the kinematic viscosity. Most commercial units are provided with a conversion factor, or can be calibrated by a fluid of known properties. Needless to say, if somebody provides us with a value of viscosity of a specific material at a specific temperature, *we need to ask how the viscosity was determined.* Additionally, it is worth emphasizing again that viscosity is a temperature-dependent property. In general, viscosity of a liquid decreases as temperature increases. The reverse is true for gases.

13.4 LINEAR AND NONLINEAR VISCOELASTICITY

We now need to deal with the issue of viscoelasticity. Unlike viscosity, which is a characteristic property of all fluids, viscoelasticity is manifested only in some types of materials. All polymers and polymer based materials are viscoelastic, while only a limited selection of particular materials among the other broad classes of materials also exhibit viscoelasticity. We shall deal with additional aspects of this situation in the next Chapter since viscoelasticity manifests itself also in mechanical properties.

Viscoelasticity is the term used to describe the property of exhibiting both viscous and elastic characteristics during deformation. For viscous fluids (e.g., Newtonian fluids like honey and also non-Newtonian fluids), the *rate* of strain is proportional to the applied stress. Thus, strain is time-dependent. For an elastic solid, *strain* is a function of the applied stress (up to the elastic limit). This strain is independent of the time over which the force is applied, and when the stress is removed, the deformation disappears, with the stretched material quickly returning to its original state. By contrast, a fluid continues to flow (i.e., deform) as long as the force is applied, and it does not recover its original form when the

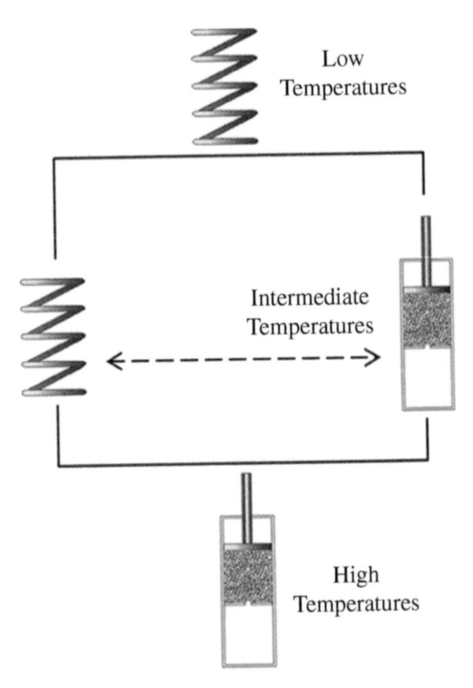

FIGURE 13.7 Model of a viscoelastic material. The spring element represents elastic behavior; the dashpot represents viscous behavior. The elements are connected in series.

stress is removed. **Viscoelastic materials**, therefore, have elements of both of these properties; as such, they exhibit time-dependent strain. Whereas elasticity in an ordered solid is usually the result of bond *stretching* or *conformational* changes, viscosity results from the physical movement of atoms or molecules away from one another within a material.

The popular model shown in Figure 13.7 attempts to demonstrate how a material can have simultaneously properties of an elastic solid and a flowing liquid. In Figure 13.7 the solid elastic behavior is represented by a spring; the viscous flow, or liquid-like behavior, is represented by a dashpot, and the two behaviors are linked with one another. The model illustrates also the effect of temperature. At low temperatures (such as below the glass transition region), we have elastic behavior at small deformations. At high temperatures we have predominantly liquid flow behavior. In-between we have both; the central part of the Figure 13.7 shows equal contribution of both behaviors. Different scenarios in which one or the other type of behavior is predominant are also possible. Loading the dashpot handle with a continuous force caused continuous displacement of the fluid. By contrast the spring reaches its final displacement instantly upon the applied stress. Additionally, if the force to the dashpot handle is applied very quickly, the resistance to motion is greater. Thus deformation in viscoelastic materials is dependent on the rate of loading.

A combination of springs and dashpots can describe any behavior but does not have predictive capabilities. How then do we "sort out" better the rheological behavior of a viscoelastic material into its elastic and viscous components? We apply a *sinusoidal* shearing force. Thus the stress is a function of time t:

$$\tau(t) = \tau_0 \sin(2\pi v t) \tag{13.7}$$

Here ν is the frequency in Hertz, and subscript 0 corresponds to $t=0$ or the start of the test. The outcome of imposing a sinusoidal load is the following behavior of the strain (i.e., deformation), ε:

$$\varepsilon(t)=\varepsilon_0 \sin(2\pi\nu t-\delta) \tag{13.8}$$

In analyzing the two-plate geometry shown in Figure 13.3, strain was in the x direction. If a material is fully elastic, then $\delta=0$. In a viscoelastic material there is always a phase lag between stress and strain so that $\delta \neq 0$. The situation is depicted in Figure 13.8.

We now can define the shear stress in terms of the elastic (denoted with single prime) and viscous (denoted with double prime) components

$$\tau(t)=\tau'(t)+\tau''(t)=\tau_0' \cos(\nu t)+\tau_0'' \sin(\nu t) \tag{13.9}$$

The strain can be defined similarly. Therefore we can define two dynamic modulae:

$$G'=\frac{\tau'}{\varepsilon'} \tag{13.10a}$$

$$G''=\frac{\tau''}{\varepsilon''} \tag{13.10b}$$

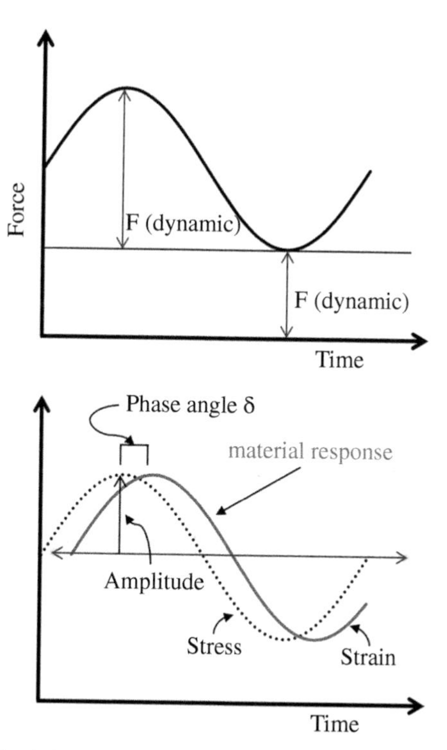

FIGURE 13.8 Viscoelasticity determination for polymer melts. Effects with time are measured while the temperature is constant. Top: force application; bottom: stress and the resulting strain. Stress$=F/A$. Strain$=$deformation (in x or y coordinate depending on test geometry. x for shear).

G' is called the shear **storage modulus**; G'' is the shear **loss modulus**. Hence, G' is the in-phase (elastic, solid-like) modulus while G'' is the out-of-phase (viscous flow, liquid-like) modulus. From Eqs. (13.10), we arrive by geometry at

$$\tan\delta = \frac{G'}{G''} \tag{13.11}$$

The **complex shear modulus** is therefore defined as follows, making use of the imaginary number $i=(-1)^{1/2}$:

$$G^* = G' + iG'' \tag{13.12}$$

Mathematicians would call G'' the imaginary part of the complex modulus G^*. However, G'' is very real; it is a measure of the energy dissipated (mostly as heat) per one cycle of deformation and per unit volume. Namely, the heat H created is

$$H = \pi G'' \varepsilon_0^{\ 2} \tag{13.13}$$

G^* can be calculated as the square root of the sum of the squares of the two dynamic modulae. Alternatively, it can be derived by a simple calculation from the tensile complex modulus. Viscosity can be determined by the relationship

$$\eta = \frac{3G^*}{2\pi v} \tag{13.14}$$

We now see the main objective of the operation which started with imposition of the sinusoidal load according to the Eq. (13.8): we can distinguish the response of a visco-elastic material into its elastic and viscous flow components. Refer again to Figure 13.7: we reiterate that relative contributions of these components depend on the temperature. Knowledge of viscoelasticity of polymer melts allows improved processing. Various scenarios such as flow around a cylinder or through a cross-slot device can be considered [4]. We shall encounter viscoelasticity again when talking about behavior of polymer solids in the next Chapter.

We have discussed the process of crosslinking or curing of polymers in Section 9.3. During the curing process, rheological behavior also changes—as expected—thus, we can pursue the progress of curing by following G^* as a function of time. Maintaining a sinusoidal shearing force (as Eq. 13.7) of a given amplitude, the response of the material changes as curing progresses. This provides the capability of determining the **extent of curing** using an equation that relates G^* to both temperature and time [5].

Effects on polymer chain movement and flow can be significantly altered by confinement and reduction of scale. There is an increasing interest in polymer melts in two-dimensional nanometric confinement. Such situations arise in microelectronics fabrication, in alignment layers in liquid crystal displays (recall Section 12.2), also in thin films of lubricants or adhesives. What happens when the film thickness decreases? There is no space for 3D Gaussian chains, hence formation of 2D disks is expected. In Section 9.7 we talked about the Monte Carlo simulation method and its capabilities to provide understanding of materials behavior that is unavailable experimentally. Termonia [6] performed

Monte Carlo simulations of polymer melts confined between parallel plates—such that the distance between the plates is smaller than the radius of gyration. It turns out that the Gaussian behavior of chains is retained in spite of the confinement. However, there is a dramatic increase in the molecular weight between entanglements [6].

There is also a related effect in platelet nanocomposites: high orientation of the platelets leads to increased confinement and therefore an entropy decrease seen also in Monte Carlo simulations by Termonia [7]. This in turn causes aging, manifested by gradual loss of platelet orientation with time. The aging can be slowed down quite significantly by addition of spherical nanoparticles—an effect seen in calculations as well as in experiments [7].

13.5 DRAG REDUCTION

As noted in Section 13.1, there are two basic ways of getting a material into a liquid state: melting and dissolution. Melting does not necessarily mean raising the temperature. In Section 3.13 we discussed formation of a eutectic between NaCl crystals (melting temperature 801°C) and water (melting temperature at 1 atm pressure known to all) such that we have a binary liquid phase below 0°C. Consider sodium compounds, including NaCl just mentioned; and now for a moment of comic relief, see Figure 13.9.

Frequently, however, we deal with solutions in flow. To a scientist, the word "solution" often brings about an image of a liquid-filled laboratory beaker. To an engineer, it might bring an image of a pipe transporting some fluid in a manufacturing plant. For example, flow occurs in

- oil pipeline conduits,
- oil well operations,
- flood water disposal,
- fire fighting,
- field irrigation,
- transport of suspensions and slurries,
- sewer systems,
- water heating and cooling systems, including district systems (no separate heating or air conditioning in individual buildings),
- airplane tank filling,
- marine systems,
- biomedical systems including blood flow.

We shall discuss suspensions from a different point of view in the next Section. **Drag** is the resistive force acting on a body moving through a fluid. The body does not have to take the form of a solid object. The "no-slip condition" of the boundary layer in laminar flow means that fluid next to the solid surface may become attached to it (having velocity equal to zero,

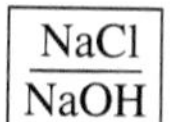

FIGURE 13.9 The *base* is under "*a salt.*"

assuming the solid boundary is a stationary pipe, for example). Since the layers above it are moving, the shear force exerted on the adjacent liquid "layer" acts as a kind of drag force. Therefore, most flows are affected to some degree by drag.

Now consider a fast-growing city; there are many such examples around the world. Each city has a sewer system. With growing population, that system becomes inadequate, leaving city authorities to determine whether they should pull out the sewer pipes from the ground and put in new ones with a larger diameter. This is a very costly operation and a palliative only. With time, the new pipes will become inadequate too.

There is another remedy for this problem: the use of a **drag reducing (DR) agent** to accelerate flow. Such agents are used in the Alaska gas pipeline; indeed they are used in all operations listed above, a fact largely unknown to the general public. At every airport, fuel tank filling involves the use of a drag reducer, otherwise the process would take much more time. The human body produces its own drag reducing agents for blood flow; if it did not, we would require more energy for the pumping of blood. In cases of patients who have atherosclerosis, which narrows the arteries, medical doctors intervene by injecting DR agents, more commonly known as blood thinners.

There are at least two types of drag reducing agents: polymers and surfactants. Consider first polymeric DR agents, which have been discussed thoroughly in an article by Zakin and Hunston [8]. An impressive feature of polymer DR agents is that they are effective at concentrations of only parts per million (ppm). However, there is a complication: in turbulent flow scission of polymeric chains occurs, a process called **mechanical degradation in flow** (MDF).

Taking into account the paper by Zakin and Hunston [8], one can make a list of important observations pertaining to DR by polymers:

(a) Drag reducing is directly proportional to the molecular mass M of the polymer, regardless of liquid (solvent) type;
(b) The concentration of DR agent required for a given level of drag reduction is several times higher in a poor solvent. According to Paul Flory, solvents can be classified as good, theta, and poor; see Figure 13.10;

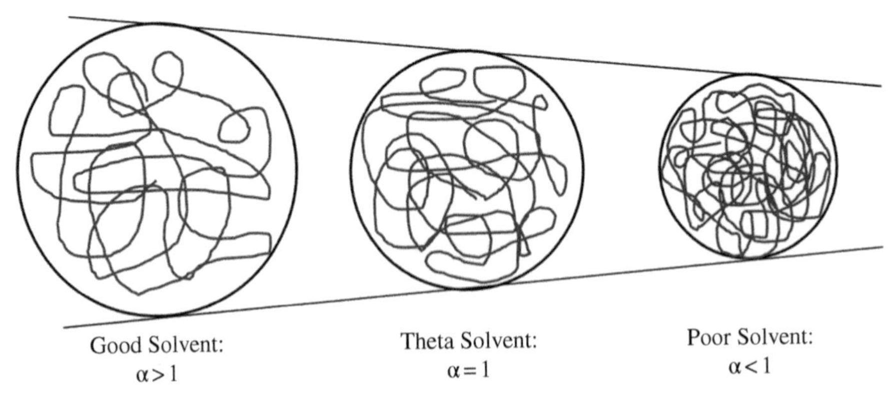

FIGURE 13.10 Effect of solvents on chain dimensions in solution. The chain dimensions in the theta solvent (also called the Flory solvent) are the same in solution as they are (or would be) in the solid (amorphous) state. In a good solvent the chain pervades a higher volume than that; in a poor solvent the chain pervades a lower volume.

(c) Mechanical degradation in flow occurs more frequently in a poor solvent under fixed flow conditions than in a good solvent under the same flow conditions. Mechanical degradation in flow eventually stops, at which point further flow turbulence does not make the chains which already underwent scission any shorter;
(d) Bond scission in flow along chain backbones does not occur exclusively at chain midpoints (as some authors hypothesized) nor is it random;
(e) Mechanical degradation in flow in polydisperse (having non-uniformity in size, shape, and mass distributions) systems occurs mostly by breaking large macromolecules;
(f) Shear degradation at a given shear stress is independent of the viscosity of the solvent;
(g) The degradation rate increases or remains the same as the DR agent concentration is decreased;
(h) In photographs of flowing liquids containing DR agents, Donohue *et al.* [9] observed low speed streaks originating from a conduit wall. They reported that such streaks lift off but after neither oscillating nor bursting return to the wall;
(i) Hunston and Reischman [10] have shown that only polymeric chains above a certain length participate in DR;
(j) Kulicke and coworkers [11] have demonstrated that a few parts per million of single chains inserted into a liquid can also cause DR. This agrees with practical applications of the DR phenomenon which involve sometimes concentrations as low as 10 ppm;
(k) Drag reducing takes place also in laminar flow;
(l) Injecting a DR agent into the center of a pipe results in practically instantaneous drag reduction [12, 13].

Among the experimental findings listed above, (c) was for a considerable time the most puzzling. In a poor solvent each polymer chain pervades the solvent only little, so one believed the chains were then *less* perturbed by turbulence. Actually, (c) became the starting point of a solvation model of DR [14], shown in Figure 13.11, in which the polymer DR agent is considered to be in solution with the flow.

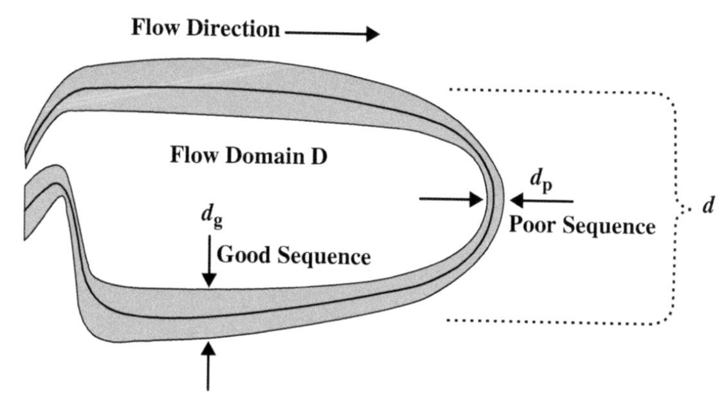

FIGURE 13.11 Solvation model of drag reduction by polymers. Thick line represents a fragment of a polymeric chain. Shaded region is solvated; d_g is the average width of the good sequence, d_p analogously for the poor sequence; d is the average width of the entire domain, here flowing horizontally to the right. *Source*: [15]. Reprinted with permission from Elsevier.

As we see in Figure 13.11, a fluid domain solvated by a polymer chain flows as a unit. There is necessarily some perturbation of its shape during the flow caused by the turbulent eddies along the way. However, the individual solvent molecules solvated by polymer chains are not "attacked" individually by eddies of the turbulence. The solvent molecules defend themselves *collectively*. The solvent molecules inside the flowing domains are protected even more than those which by solvation are attached to macromolecular chains outside. We also see that the model assumes in each polymeric chain two kinds of sequences, good (oriented along the flow direction or close to it and strongly solvated) and poor (oriented more or less perpendicularly to the flow, poorly solvated).

The model presented in Figure 13.11 explains experimental facts listed above. In particular, it explains how very low concentrations of the DR polymer expressed in ppm have such dramatic effects on the turbulent flow. The model was further studied via Brownian dynamics computer simulations of polymer solutions—both static (not moving) and flowing scenarios [16, 17]. For some time the model shown in Figure 13.10 remained a hypothesis—until it was confirmed experimentally. Acoustic measurements of the solvation numbers and of DR efficacy have shown that indeed higher efficacy corresponds to greater solvation [18]. Thus, the lower solvation numbers characteristic with a poor solvent mean that polymer DR molecules are in fact more susceptible to degradation.

DRAG REDUCTION: EFFECTS OF TIME AND DEGRADATION

A problem with reducing drag by polymer solutions has already been noted above: polymeric agents undergo *mechanical degradation in flow*. Every forty miles or so in the Alaska pipeline, more polymer is pumped into the flowing petroleum to compensate for chains shortened by the turbulence. The model described by Figure 13.11, along with experiments, tells us that short polymeric chains do not provide DR.

It would be good to know how DR changes with time t because of MDF and also how DR efficacy λ depends on polymer concentration c. An equation for the dependence of λ on t and c has been developed [19]:

$$\frac{\lambda}{\lambda_0} = \frac{1}{\left\{1+W\left[1-\exp\left(-t/\left(h_0+h_1c+h_2c_2\right)\right)\right]\right\}} \tag{13.15}$$

Here λ_0 is the DR efficacy at $t=0$; W is the average number of vulnerability points per chain, which depends on the good/poor sequences ratio and the turbulent flow velocity ($W=3$ means than on the average each chain will disintegrate into 4 chains); h_i parameters are constants for a given Reynolds number and the polymer + solvent pair. Testing of Eq. (13.15) [15, 20, 21], notably by Choi and coworkers, demonstrates that equation (or its earlier form without the quadratic term) provides good results always, while other $\lambda(t)$ equations do not work. Incidentally, Choi and coworkers have explored the use of the DR phenomenon for cold water piping in an *ocean thermal energy conversion* system [20].

As noted above, there are also **DR surfactants**. Also here, important work is being done by Zakin and his colleagues [22–25]. Cationic surfactants in the presence of counterions form long flexible threadlike micelles which cause DR. Such micelles have actually an advantage over polymeric DR agents in that, when broken up by turbulence, they can re-assemble.

In nature we have observed the use of drag reducing agents by various animals. Dolphins are known for exuding drag reducing agents from their skin, thus swimming faster without using more energy. Similarly, in yacht races one can use a DR agent (such as polyoxymethylene) dispensed from the boat bow—unless the race rules forbid it.

In humans and animals, one's heart beat is connected to properties of turbulent flow as such flow is generated by the heart valve closing. Furthermore, an obstruction in an artery may additionally give rise to acoustic effects called bruit (which is simply noise in French). Since blood is a viscoelastic fluid, what we have said in the present section about DR applies to blood. Recall the blood thinners already mentioned.

HIGHER EFFICIENCY TEMPERATURE CONTROL THROUGH WATER HEATING AND COOLING SYSTEMS

In the list of applications involving a fluid flow, district water heating and cooling systems were named. In such systems, water is temperature-controlled in a central station and then circulated through a district to heat or to cool buildings in a concentrated area, thereby eliminating the need for individual heating or air conditioning systems in each building. As argued by Zakin and his colleagues [23], a central station provides higher energy efficiency with lower cost for maintenance than the sum of many furnaces for heating or air conditioners for cooling. The elimination of these units also frees up space in the buildings, reduces local maintenance, and can minimize peak electrical load periods. Surfactant DR agents are often used in such systems. A problem encountered is the poor heat transfer capabilities of such DR agents, but work on resolving that problem is ongoing [25]. Currently non-surfactant DR agents seem to have more uses than surfactant-based agents.

13.6 SUSPENSIONS, SLURRIES, AND FLOCCULATION

We defined emulsions and suspensions in Section 13.1. There can be also suspensions in gas phases. These definitions may be extended to include cases where the dispersed phase consists of liquid drops (emulsions) or bubbles, provided the deformation of the drops (or bubbles) and processes of break-up and coalescence do not play a major role in the flow behavior.

A **slurry** is any fluid mixture of a pulverized solid with a liquid (usually water); they are often used as a convenient way of handling and moving solids in bulk. Slurries flow under gravity but are also capable of being pumped if not too thick. Comparison of these two definitions suggests that slurries are a subclass of suspensions in which the dispersed particle size is relatively large. Illustrative of slurries is **coal slurry**, which consists of crushed coal and water and is used to move coal inside and outside of mines. Metal ore slurries are similar in this respect. Industrial wastewater is often a slurry.

Milk is a **colloidal suspension** containing micelles (primarily with the protein casein) in the size range 0.04–3 μm. The term **colloid** refers to particles ranging in size from micrometers to nanometers. Particles in a colloidal suspension are larger than those dissolved in a solution but smaller than those in typical suspensions or slurries. Dispersion is on the micro-scale, and particles are not usually visible under an optical microscope. Micelles were described in a second-track feature in Section 9.4. Essentially, a **micelle** is a spherical aggregation of surfactant molecules; thus the inner and outer surfaces will be alternately polar and nonpolar. The larger of these micelles in milk scatter visible light, rendering milk opaque. **Ice cream** is essentially a suspension of microscopic ice crystals and air bubbles dispersed in cream. **Aerosols** or **hydrosols** are suspensions of fine liquid droplets or fine solid particles dispersed in a gas or liquid, respectively. For both aerosols and hydrosols, shelf life stability is important.

Bacterial suspensions can be considered as a special category since we are dealing with "active particles", in contrast to passive particle suspensions. The rheology of algae like *Chloromonas nivalis* or bacteria including *Escherichia coli* and *Bacillus subtilis* is being studied more and more. Actually, the swimming of micro-organisms, and the energy metabolized in this process, implies that such systems are inherently out of equilibrium even in the absence of an imposed flow. Hydrodynamic interactions in groups of microbes may play a crucial role in the formation of biofilms, and therefore in the initiation of infection processes. Modeling the dynamics of micro-organisms found in seawater, such as phytoplankton and algae, which constitute the base of the sea-food chain, evidently has implications for the oceanic ecosystem. Another point of interest is the rheology of algae suspensions that are used for the production of biofuels [26].

Let us now focus on suspensions of solid particles in liquids. In cases of coal or mineral ore slurries, after transport we want to separate the solid particles. The same applies to industrial wastewater. *Water is in fact the most important material used by mankind.* Creating more and more industrial wastewater needs to be at least mitigated by removal of suspended solids to allow purification and possible re-usage of water. The removal can be accomplished by **gravitation** (very slow, but we do not need to do anything), by **coagulation** (dependent on electric charges), and by **flocculation** (not dependent on electric charges) [27, 28].

In the case of **coagulation**, destabilization of colloidal suspensions occurs by the addition of **coagulants** that neutralize the electrostatic forces keeping the suspended particles separated. Aggregates formed in the coagulation process are small and loosely bound; their sedimentation velocities are relatively low, but higher than by gravity separation. Typical coagulants are metallic salts (e.g., aluminum sulfate, ferric sulfate, ferric chloride), though cationic polymers may also be used.

Flocculation, on the other hand, is achieved without significant changes in the particle surface charges, and requires only ppm amounts of a **flocculant** additive. Aggregates, called flocs, are generally larger and more strongly bound than the aggregates from coagulation. Flocculants are used in mineral processing (metals from ores), in oil field operations, in food production (as thickening agents), in cosmetics applications, in paper and textile industries; in viscosifiers and binders. Significant improvement in the settling velocities of solid particles can be obtained through the use of flocculants. Returning to our mining operation example, flocculants can be economically applied to enable re-use of slurried water, thereby reducing the consumption of fresh water and—through the removal of solid particles—minimizing contamination in the wastewater. Similarly, municipal sewage water can be purified before releasing it into rivers or reservoirs.

While **inorganic flocculants** are used in large quantities, they leave large amounts of sludge and are strongly affected by pH changes. **Polymeric flocculants** do not have these drawbacks. How do polymeric flocculants work? The key to answering this question lies in the fact that *polymeric drag reducers are also flocculants*. See again Figure 13.11. Strong solvation of the liquid by polymers in ppm concentrations seen there pushes the solid particles outside the solvated domains [27]. The solid particles aggregate in much smaller regions then available to them. Polymeric flocculants cause formation of large cohesive aggregates (flocs) and are typically inert to pH changes. Both natural and synthetic polymers are used as flocculants. Natural polymers are biodegradable, are effective at large dosages, and are shear stable. (Aspects of water purification and the use flocculants are discussed also in Chapter 20.)

In general, one can distinguish three types of water:

- **potable** or **drinking water**, which becomes more and more scarce;
- **agricultural** water, which is used for field irrigation;
- **industrial wastewater**.

Consider industrial wastewater from which solid particles have been removed by flocculation. Some industrial enterprises recycle such water, that is, they re-use the water for processes in their enterprise. Industrial wastewater, even purified by removal of solids, is not yet pure enough to be used as potable water. It can, however, be used as agricultural water. At least two objectives are thus accomplished: unfiltered industrial wastewater is not contaminating the environment; *freshwater is saved* and not used for irrigation since industrial water is used instead. (A question to ponder: what happens to materials removed from non-potable water?)

13.7 SELF-ASSESSMENT QUESTIONS

1. Generally a higher melt flow index indicates lower viscosity of a material. What difference does it make if (in a factory) the current lot of material being processed by injection molding has a MFI of 10 and a new lot has a MFI of 15?
2. What distinguishes laminar and turbulent flow from one another? What are some practical implications of each?
3. Explain in your own words what viscosity is and why it matters.
4. Explain and differentiate Newtonian and non-Newtonian types of flow.
5. Avalanches are Bingham materials. Explain what this means and what are the consequences.
6. What is the difference between viscosity and viscoelasticity?
7. Explain why short polymeric chains do not provide drag reduction.
8. Explain the connection between drag reducing agents and flocculating agents.

REFERENCES

1. C.W. Macosko, ed., *Rheology: Principles, Measurements and Applications*, Wiley–VCH: Weinheim **1994**.
2. D. Jackovich, B. O'Toole, M. Cameron Hawkins & L. Sapochak, *J. Cellular Plastics* **2005**, *41*, 153.
3. H.A. Barnes, *J. Non-Newtonian Fluid Mech.* **1997**, *70*, 1.
4. W.M.H. Verbeeten, G.W.M. Peters & F.P.T. Baaijens, *J. Non-Newtonian Fluid Mech.* **2002**, *108*, 301.
5. W. Brostow & N.M. Glass, *Mater. Res. Innovat.* **2003**, *7*, 125.
6. Y. Termonia, *Polymer* **2011**, *52*, 5193.
7. Y. Termonia, *Phys. Rev. E* **2013**, *88*, 012603.
8. J.L. Zakin & D.L. Hunston, *J. Appl. Polymer Sci.* **1978**, *22*, 1763.
9. G.L. Donohue, W.G. Tiederman & M.M. Reischman, *J. Fluid Mech.* **1972**, *56*, 559.
10. D.L. Hunston & M.M. Reischman, *Phys. Fluids* **1976**, *18*, 1626.
11. W.-M. Kulicke, M. Kötter & H. Gräger, *Adv. Polymer Sci.* **1989**, *89*, 1.
12. H.W. Bewersdorff, *Rheol. Acta* **1982**, *21*, 587.
13. H.W. Bewersdorff, *Rheol. Acta* **1984**, *23*, 183.
14. W. Brostow, *Polymer* **1983**, *24*, 631.
15. W. Brostow, *J. Ind. & Eng. Chem.* **2008**, *14*, 409.
16. W. Brostow, M. Drewniak & N.N. Medvedev, *Macromol. Rapid Commun.* **1995**, *4*, 745.
17. W. Brostow & M. Drewniak, *J. Chem. Phys.* **1996**, *105*, 7135.
18. W. Brostow, S. Majumdar & R.P. Singh, *Macromol. Rapid Commun.* **1999**, *20*, 144.
19. W. Brostow, H.E. Hagg Lobland, T. Reddy, R.P. Singh & L. White, *J. Mater. Res.* **2007**, *22*, 56.
20. C.A. Kim, J.H. Sung, H.J. Choi, C.B. Kim, W. Chun & M.S. Jhon, *J. Chem. Eng. Japan* **1999**, *32*, 803.
21. S.T. Lim, H.J. Choi, S.Y. Lee, J.S. So & C.K. Chan, *Macromolecules* **2003**, *36*, 5348.
22. J. Myska, Z. Lin, P. Stepanek & J.L. Zakin, *J. Non-Newtonian Fluid Mech.* **2001**, *97*, 251.
23. Y. Zhang, J. Schmidt, Y. Talmon & J.L. Zakin, *J. Colloid & Interface Sci.* **2005**, *286*, 696.
24. W. Ge, E. Kesselman, Y. Talmon, D.J. Hart & J.L. Zakin, *J. Non-Newtonian Fluid Mech.* **2008**, *154*, 1.
25. H. Shi, Y. Wang, B. Fang, Y. Talmon, W. Ge, S.R. Raghavan & J.L. Zakin, *Langmuir* **2011**, *27*, 5806.
26. V.O. Adesanya, D.C. Vadillo & M.R. Mackley, *J. Rheol.* **2012**, *56*, 925.
27. W. Brostow, S. Pal & R.P. Singh, *Mater. Letters* **2007**, *61*, 4381.
28. W. Brostow, H.E. Hagg Lobland, S. Pal & R.P. Singh, *J. Mater. Ed.* **2009**, *31*, 157.

14

MECHANICAL PROPERTIES

Every gun barrel has certain distinctive marks of identification which are placed there by the manufacturer, for instance, the number of grooves, their shapes and dimensions. In addition to that, each rifle barrel has certain peculiarities of its own, minor imperfections in the metal which, in turn, leave tell tale scratches on a bullet which is fired from the barrel with the terrific force of the compressed gases behind it. These marks are as highly individualized as the ridges and whorls on the finger tip of the individual.

—Erle Stanley Gardner in *The Case of the Vagabond Virgin* featuring Perry Mason.

14.1 MECHANICS AT THE FOREFRONT

Mechanics of materials is considered by many as the bottom line of Materials Science and Engineering. If a device, or even one element of a structure, disintegrates into pieces, then *laymen* along with scientists and engineers are affected.

We have a variety of methods of mechanical testing. The predominant testing methods are:

- tension
- compression
- shear
- isostatic deformation
- 3- or 4-point bending
- creep determination

Materials: Introduction and Applications, First Edition. Witold Brostow and Haley E. Hagg Lobland.

- stress relaxation
- dynamic mechanical analysis (DMA)
- impact testing

The meaning of the first four may be easily deduced. In subsequent sections, we shall discuss all of these techniques.

14.2 QUASI-STATIC TESTING

To begin with, we need this important rule: In application of a mechanical force it is important to choose well the **location** for the force application. The most frequent modes of the force application are **tension** and **shear**; see Figure 14.1. **Compression** is a third familiar mode of force application.

Elastic materials were mentioned in the preceding Chapter. By definition, an **elastic material** returns to its original size and shape after the removal of the load. The Robert Hooke equation for tension is

$$\sigma = E\varepsilon \tag{14.1}$$

Here

$$\sigma = \textbf{engineering stress} = \frac{F}{A}$$

where F is the applied force
and A is the cross-sectional area

$$\varepsilon = \textbf{engineering strain} = \frac{(l - l_0)}{l_0} = \frac{\Delta l}{l_0}$$

where l_0 = original length
while l = current length (with $l > l_0$ as a result of tensile strain)

$$E = \textbf{elastic tensile modulus}$$

The elastic tensile modulus E is a measure of the resistance of material to tensile deformation. The elastic tensile modulus is frequently referred to as the Young (or Young's)

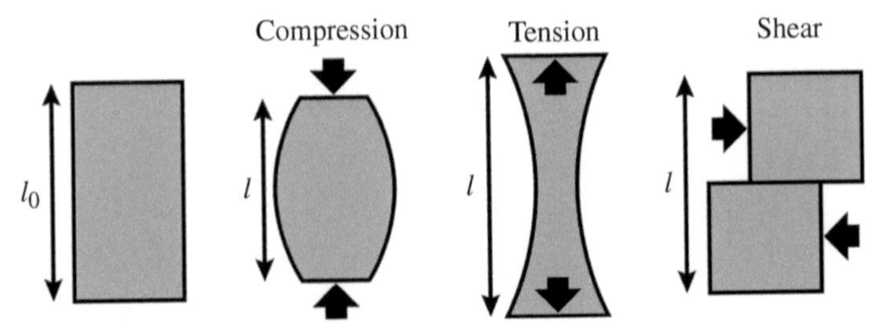

FIGURE 14.1 Common modes of deformation occur from tensile stress, compressive stress, and shear stress. Specimen length denoted by l.

modulus; thus a reference to Young's modulus is likely to pertain to the elastic tensile modulus. An analogous equation applies to compression, only here necessarily $\Delta l < 0$.

For analysis of shear stress and deformation we have another Hooke equation

$$\tau = G_s \varepsilon \tag{14.2}$$

In this case

τ = **shear stress**

G_s = **shear modulus**

We see in Figure 14.1 that the shear stress is applied perpendicularly to the surface, and Δl is perpendicular to l_0.

For all materials for which viscoelasticity can be neglected, we have <u>four</u> modulae independent of time. (Viscoelastic materials will be dealt with in detail later.) We have already seen E and G_s. The third is the **Poisson ratio** ν (so named after Siméon Poisson) which also pertains to tensile testing (or else to compression testing, see the following paragraph). We see in Figure 14.1 that the extension of the length from l_0 to l is accompanied by a reduction in diameter, say, from r_0 to r, so in this case $r - r_0 = \Delta r$ is negative. We define the Poisson ratio as

$$\nu = \frac{-\Delta r}{\Delta l} \tag{14.3}$$

The minus is "built into" the definition because typically $\Delta r < 0$. Thus, in most cases ν is a positive quantity.

Consider now **compression** instead of tension. The definition (14.3) is still valid, but we have the inverse situation: $\Delta r > 0$ with $\Delta l < 0$. Therefore, ν is a positive quantity also in the case of compression. Exceptions exist but are not frequent; some *polymer foams have* $\nu < 0$. If such foams are stretched in one direction, they become thicker in perpendicular directions. Materials with $\nu < 0$ are called **auxetic materials**.

There are some materials which have very small positive ν values. **Corks** are in that category: under compression they show very little lateral expansion, hence very small almost zero ν values. Analyzing the definition (14.3), we find that **incompressive materials**, that is those in which there is no change in volume on deformation, have exactly $\nu = 0.5$. In reality there are no such materials, but often elastomers have ν values close to 0.5.

We said above that there are four basic modulae; the fourth one is the **bulk modulus**. This corresponds to the situation in which force is applied from all sides, the isostatic force, as shown in Figure 14.2. A simple procedure to do so consists in putting the specimen in a

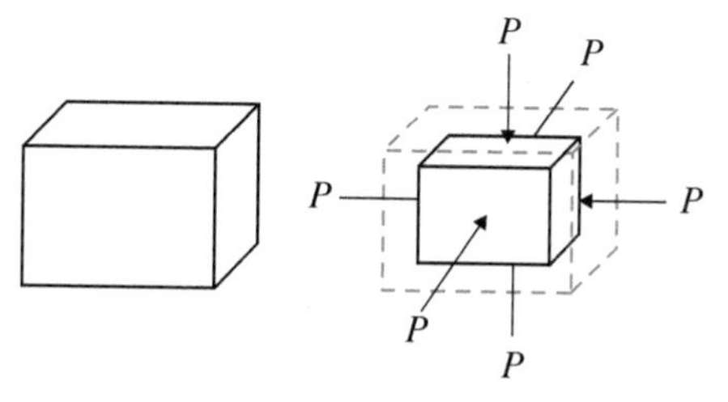

FIGURE 14.2 Isostatic material deformation. *P* represents the applied hydrostatic pressure.

liquid and applying hydrostatic pressure. In contrast to tension, compression, or shear, during this kind of testing *the shape of the specimen is preserved*. We can thus define the bulk modulus as:

$$k_b = V\left(\frac{\partial P}{\partial V}\right)_T \qquad (14.4)$$

Here V is the volume of the specimen, P is the pressure, and T is the temperature.

We have now defined the four modulae, and they are listed together in Table 14.1. There are relations between the modulae, provided in Table 14.2.

TABLE 14.1 Definitions of (Time-Independent) Mechanical Modulae

Name	Symbol	Equation	Variables
Tensile Modulus	E	$E=\sigma/\varepsilon$	engineering stress σ, engineering strain ε
Shear Modulus	G_s	$G_s=\tau/\varepsilon$	shear stress τ, engineering strain ε
Poisson Ratio	ν	$\nu=-\Delta r/\Delta l$	radius r, length l
Bulk Modulus	k_b	$k_b=V(\partial P/\partial V)_T$	volume V, pressure P, temperature T

TABLE 14.2 Relationships between Mechanical Modulae: Tensile Modulus E, Shear Modulus G_s, Poisson Ratio ν, and Bulk Modulus k_b

$k_b=E/[3(1-2\nu)]$
$G_s=E/[2(1+\nu)]$
$\nu=E/(2G_s-1)$

These relationships tell us that only two out of four modulae are independent. Given any two of them, we can calculate the other two. This applies to materials which are not viscoelastic and also are **isotropic**, so that their properties are the same in all directions. If the material is not isotropic, a stress applied to it may be resolved into nine components, three of them tensile (or compressive) and six shearing stresses. An arbitrary solid can have a maximum of 21 modulae and we do not discuss those cases here. Human beings carrying loads might have a non-isotropic distribution of the load, such as the girl with a jug in Figure 14.3.

The all-time favorite of mechanical testing is determination of tensile engineering stress σ as a function of engineering strain ε. Upon application of stress, at some point a behavior known as **necking** begins near the middle of the test specimen. The reason for the name is evident from the illustration in Figure 14.4. Necking is characterized by a weak region having a cross-sectional area *smaller than that of the rest of the specimen.*

Consider now the resultant diagrams, shown in Figure 14.4, for such a test on a metal other than steel, such as copper. Starting from the origin, we first have a region in which stress and strain are proportional to each other; here Eq. (14.1) is valid, and we can calculate the modulus E. At the **yield strength**, *plasticity* manifests itself and the proportionality is lost. Having exceeded the elastic limit, deformation associated with the yield point and higher stresses is permanent. Determining the location of the yield point poses a problem: how do we define a point where a deviation from a straight line appears? In practice what

FIGURE 14.3 Girl with a jug, ca. 1900, by Apoloniusz Kedzierski (1861–1939). *Source*: Dziewczyna z dzbanem, National Museum, Warsaw.

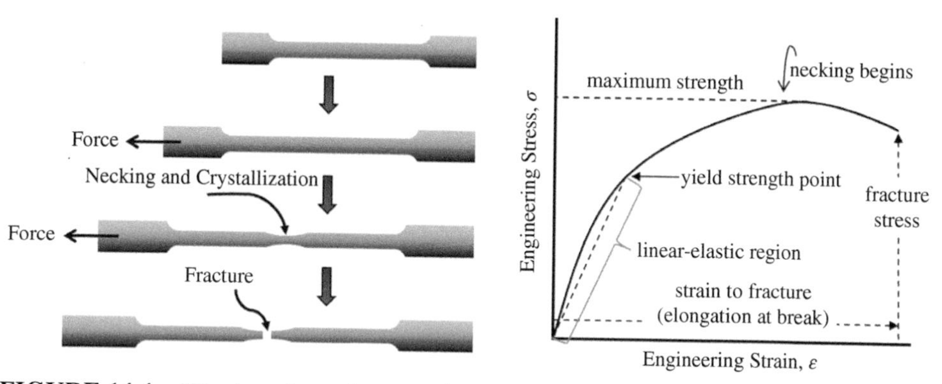

FIGURE 14.4 "Dogbone" specimen evolution during tensile testing (left) and the stress versus strain diagram (right) for a metal other than steel.

one does is define a specific deviation from the straight line, such as 0.2%, thus providing a measurable value not far from the real yield point. Considering both sides of the coin, an elastic material is one that returns to its original length or shape when an applied load is removed while a plastic material deforms easily but does not break. In practice, many materials exhibit some of each type of behavior, with ceramics tending towards 100% elasticity and some polymeric materials displaying mostly plasticity.

When necking occurs, we are at the *maximum* of the $\sigma(\varepsilon)$ diagram. Necking is clearly harmful (this statement applies only to tensile testing, not to relations between humans). The same tensile force acting in the necking region on a smaller area produces a higher stress in that area. It is for this reason that in Figure 14.4 we see that further extension of

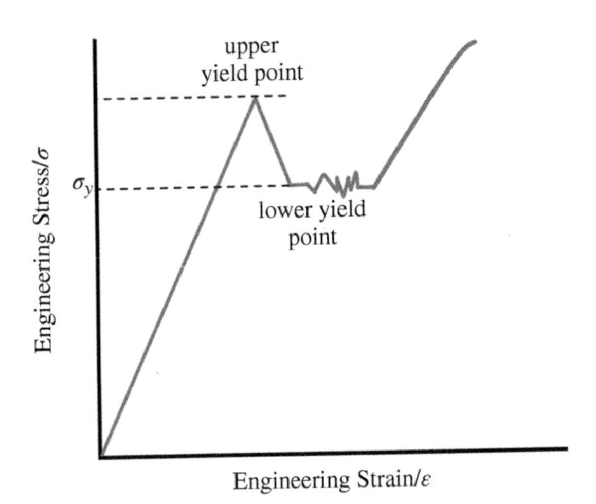

FIGURE 14.5 Stress versus strain curve for mild steel.

the specimen, that is, an increase in strain, requires a lower force than before. We also see in Figure 14.4 that *necking ultimately leads to fracture* (there are no implications here either for relations between humans).

The behavior of steels is different; see Figure 14.5. There might be one "wobble" (between points B and C) early in the diagram, or else multiple wobbles. Whether one or more, the wobbles distinguish the diagrams of steel from those for other materials.

In Section 6.2 we talked about **strain hardening**, also called **work hardening**. We shall now reconsider this issue from the point of view of mechanics. We see in Figure 14.6 that after the first cycle of loading and unloading, the value of the yield point (refer to Figure 14.5) is higher for the second loading. This is the result of strain hardening.

Up to this point, we have been carefully saying "engineering stress" and "engineering strain." There is also **true stress** $\sigma_t = F/A_c$, where A_c is the current or instantaneous cross-sectional area. Given the necking phenomenon, we typically have $\sigma_t \geq \sigma$. Similarly, there is **true strain** ε_t defined as

$$\varepsilon_t = \ln\left(\frac{l_c}{l_0}\right) \tag{14.5}$$

where l_c is the current or instantaneous length. The diagrams of $\sigma(\varepsilon)$ and $\sigma_t(\varepsilon_t)$ for a metal are compared in Figure 14.7. It is much easier to measure $\sigma(\varepsilon)$ than $\sigma_t(\varepsilon_t)$.

So far we have looked at metals only. Consider now the characteristic stress–strain behavior of a ceramic depicted in Figure 14.8. This Figure first confirms what we already know: the slope of the linear part of the $\sigma(\varepsilon)$ diagram provides the modulus of elasticity E according to Eq. (14.1). What else is there? Well, actually nothing. **Ceramics** are brittle materials exhibiting no (or very little) plasticity; typically the end of the linear or elastic region in the diagram is also the fracture point.

We now turn to polymers. The $\sigma(\varepsilon)$ curves for several cases are shown in Figure 14.9. The Figure is nearly self-explanatory, but an important piece of information needs to be added. All polymers are *viscoelastic*, therefore their properties depend significantly on time. For this reason all curves in that Figure have been obtained at a specific time after the

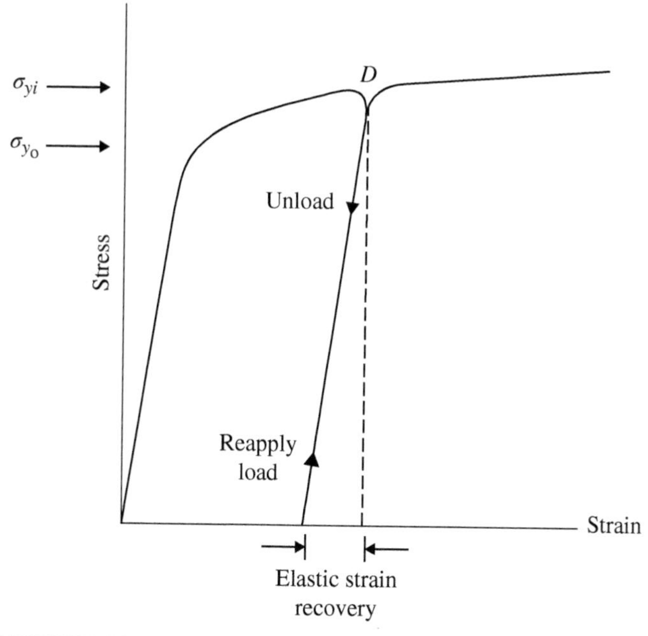

FIGURE 14.6 Strain hardening (work hardening) of a metal.

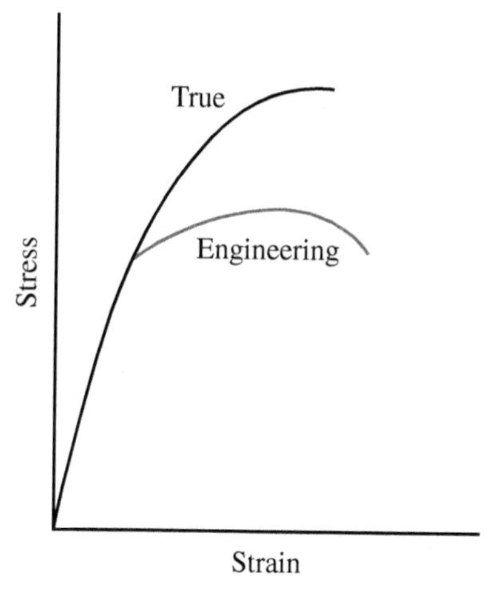

FIGURE 14.7 Diagram of engineering stress versus engineering strain and of true stress versus true strain for a non-ferrous metal.

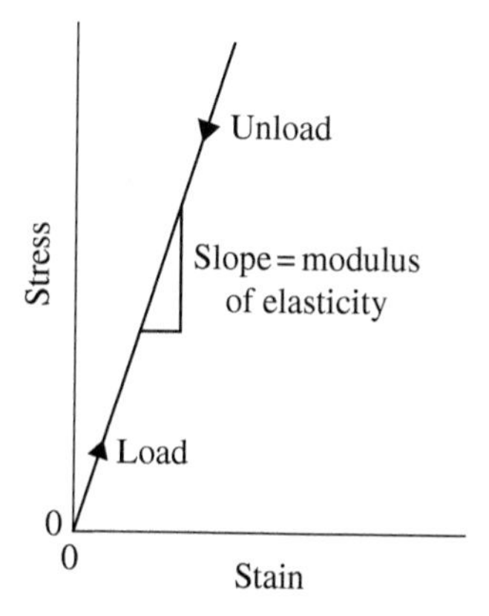

FIGURE 14.8 Stress σ versus strain ε diagram for a ceramic material.

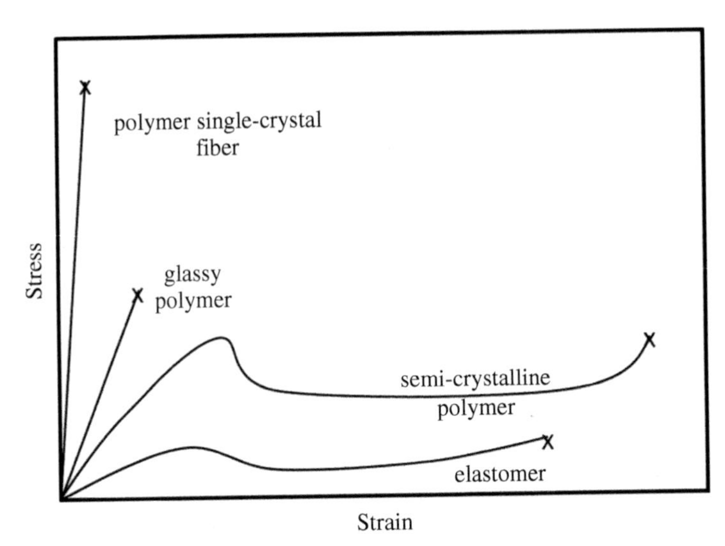

FIGURE 14.9 General stress versus strain diagram for various types of polymers.

imposition of the stress, such as after 5 seconds. The results after a minute or after 5 minutes would be different since chains are undergoing conformational changes which take time.

The rate of force application is here important as well. A good example of the time-dependency of the properties of viscoelastic materials is in the behavior of Silly Putty (a goopy stretchy play substance). If one slowly pulls apart Silly Putty, it will stretch into long thin strands before finally tearing apart. However, if one hits it hard with a hammer, the Silly Putty will fracture. This behavior mirrors the transition of polymers to a glassy, brittle state upon reducing the temperature below the glass transition. One can try a similar experiment on viscoelasticity using a mixture of corn starch and water.

How does the tensile modulus E change with temperature? Refer to Figure 14.10. Examine first the curve for a semi-crystalline material having a measurable melting temperature T_m. Recall that polymers are never fully crystalline. Look next at the diagrammatic curve for a thermoset—a crosslinked material. At low temperatures a thermoset is **glassy**; with increasing temperature it then goes through the glass transition region. Immediately afterwards it is **leathery**—a state of retarded high elasticity. Eventually, it exhibits **high elasticity**, that is, instantaneous elasticity. The third and fourth curves are for amorphous polymers. The B curve corresponds to a material of higher molecular mass than that represented by the A curve. At low temperatures (below the T_g), the behavior of amorphous polymers is similar to that of thermosets. However, beyond the leathery state, amorphous polymers do not exhibit high elasticity.

Referring back to Section 14.1, our list of techniques for determination of mechanical properties includes also 3-point bending and 4-point bending. The geometrical setup and direction of force application for the 3-point bending test are illustrated in Figure 14.11. In some cases one performs bending tests on specimens that are pre-notched; for more on notches, see Section 14.5. Flexural tests by the 3-point or 4-point bending methods are commonly used for assessing the strength of ceramics since they are typically too brittle to allow measurement of mechanical properties by conventional tensile tests. In the microelectronics industry the 4-point bending test is an accepted method for evaluating the adhesion properties of multilayer thin film systems [1].

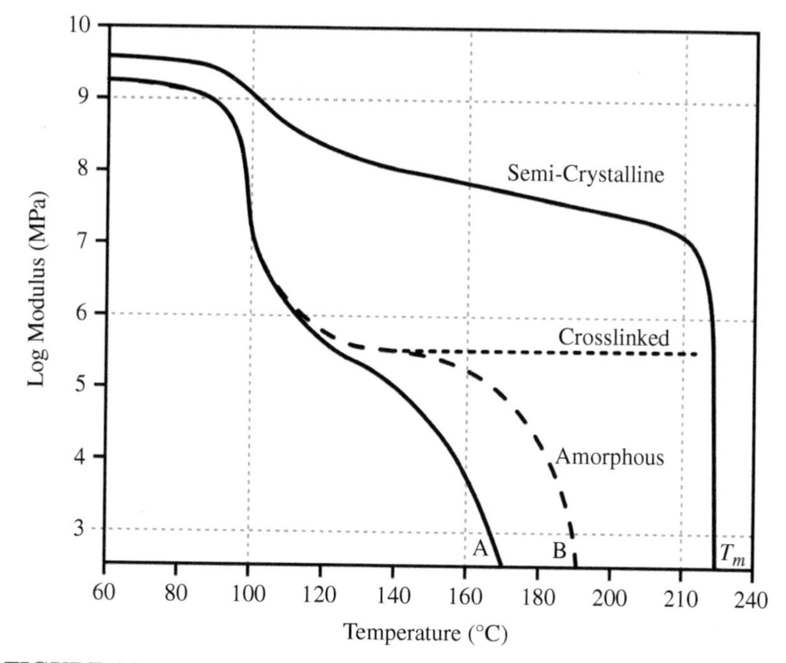

FIGURE 14.10 Dependence of polymer elastic modulus on temperature.

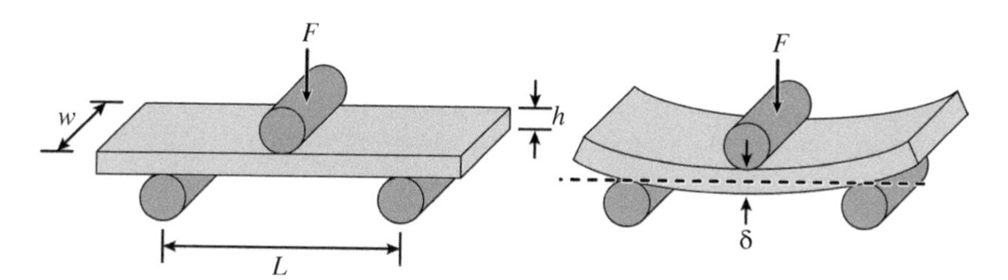

FIGURE 14.11 Three-point bending test: three-point configuration shown at left; typical appearance of deformation from the test shown at right.

Why is "quasi-static testing" given as the title for this Section? Quasi-static testing is the popular name for the first four kinds of tests on our list—which have been discussed above. The name implies that the force application is slow, in contrast, for instance, to dynamic and impact testing, which will be discussed later and in which the force application is rapid. Three-point and four-point bending tests can be conducted in either quasi-static or dynamic modes.

Before moving on, consider what further information can be derived from quasi-static testing. Think about morphology and deformation and also about energy. The energy of an applied force on a material specimen must be dissipated. The varied outcomes of mechanical testing provide some clues as to the relation between energy absorption and resistance to deformation. Visual assessments and analysis of morphology through the aid of microscopes can yield more information about how materials behave in response to various stresses.

14.3 PROPERTIES: STRENGTH, STIFFNESS, AND TOUGHNESS

We are now acquainted with the yield stress, elastic modulus, strain, and so on. How do these quantitative measures correlate to our understanding of a materials strength, stiffness, and toughness? These three descriptors are *not* interchangeable.

A material's **strength** is associated with its yield strength point, the stress value marking the end of the elastic region of the $\sigma(\varepsilon)$ diagram (refer again to Figure 14.4). We are talking here about the maximum load that can be placed on a material before it permanently deforms or breaks. The meaning applies to all types of materials. Put simply, a strong material is not easily broken.

It has been said that "one should not be stressed out over stresses, but instead should be more mindful of displacements and deformations" [2]. **Stiffness** corresponds to the elastic tensile modulus E and tells us about the amount of deflection caused by a load (refer to Table 14.1). A high value of E indicates that it is difficult to change the shape of the material by an applied stress.

Frequently engineers are very concerned with the strength of a material and how high a load it can bear before breaking. However, it is equally important to have the material design be stiff enough to prevent excessive flexing under load. A few examples will make the differences between strength and stiffness more clear. First is Kimball's example of a rubber staircase [2]. We know that steel is stiffer than rubber, and a rubber object will deflect more than a similarly shaped steel object bearing the same load. It is possible to design a rubber staircase without exceeding the rubber's ultimate strength, but walking on such a staircase would be like running through a funhouse. Although the staircase would be acceptable from the perspective of stress and fatigue, it would be unacceptable from the standpoint of stiffness.

Geometry is integral also because the stiffness of a given material under a particular load varies with geometrical shape. For instance, a hollow cylindrical steel pipe undergoes less deflection than a solid round steel pipe of the same dimensions carrying the same load. There is a story—with validity unknown—of a motorcycle company investing large sums of money on the development of a new titanium frame to outperform its winning steel design, but in the end the titanium frame motorcycle could not complete one lap at full power. Engineers had touted its high strength-to-weight ratio and analyzed the stress capacity of every joint. Dimensions of the frame were identical to the previous steel frame, but nobody looked at the fact that titanium has an elastic tensile modulus about half that of steel. Although the titanium frame was lighter and had ultimate strength as high as that of the steel frame, the motorcycle felt spring-like, as though "trying to ride and drive a noodle" [2].

Therefore, the opposite of a strong material is a weak one, with a low breaking stress; and the opposite of a stiff material is a flexible one, needing only a small stress to produce a large extension. Now for the third descriptor mentioned above: toughness. **Toughness** has to do with the amount of energy a material can absorb or disperse, in the form of plastic deformation, before it fractures from the imposed stress. Put another way, toughness is a measure of the energy required to crack a material. This is especially important for items that suffer impact; in many instances, as with a hammer, car engine, etc., strength without toughness yields a poor product. Although we will revisit this topic in Section 14.8, in which we cover impact testing, it turns out that there is more than one method of calculating toughness. By performing a mathematical integration, one can calculate the area under the stress–strain curve from a tensile test. The resultant value is the property toughness in units of energy per volume. Now consider the shape that a $\sigma(\varepsilon)$ curve may have. Evidently a good combination of strength and ductility produces the toughest materials (considering that high ductility indicates an ability to undergo large plastic deformation before fracture).

14.4 CREEP AND STRESS RELAXATION

Continuing with our discussion of stress and deformation, we are talking now about two phenomena as well as the methods of their characterization. **Creep** is the deformation (i.e., strain) that takes place over time as a result of the imposition of a constant load (i.e., force or stress). A steel cable supporting the constant load of a heavy bucket will lengthen over time. The phenomenon of **stress relaxation** takes place when a constant strain is quickly applied and then maintained. The response, in terms of stress, of the material is measured over time; while the amount of deformation remains constant, the resistance of the material decreases—there is therefore relaxation of the material. Human analogies to these phenomena are easily imagined.

Both creep and stress relaxation are a consequence of *delayed molecular motions*. Stress relaxation is manifested by a reduction in the force (stress) required to maintain a given deformation. Typically, the "protest" of the material is stronger first and weaker later. For example, a rubber band strained to fit around a stack of envelopes responds with a certain stress; over time, while the strain (i.e., length) of the rubber band remains the same, the force exerted by the rubber band on the envelope lessens. Put another way, stress relaxation is the process by which a material relieves the stress caused by a constant strain. The two processes of creep and stress relaxation are thus "twins" of sorts and are presented in a simple way in Figure 14.12.

We know how to measure strain, for instance, in tensile testing. How do we measure stress relaxation? A test setup is described by Figure 14.13. The components illustrated in that figure represent the essential components for stress relaxation testing; these would of

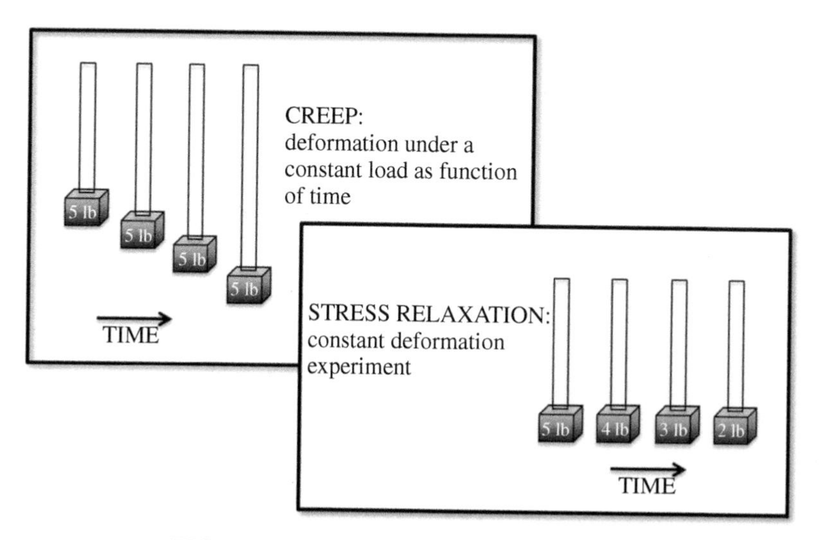

FIGURE 14.12 Creep and stress relaxation.

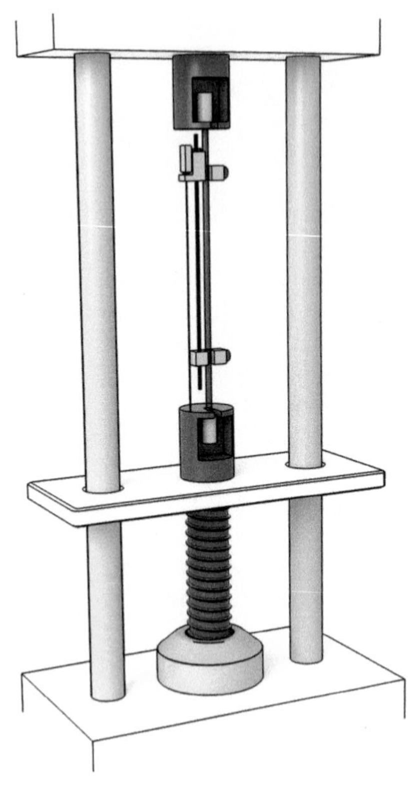

FIGURE 14.13 Schematic of a universal testing apparatus used to measure stress relaxation. A rigid support cylinder is used to determine the extension ratio while the force acting on the sample is measured by displacement of the spring.

course be incorporated in a larger mechanical testing device with a computer interface for monitoring the strain and recording the digital output of the measured force. What kind of results does stress relaxation testing provide? Figure 14.14 provides an example [3] and tells us something important, although not unexpected: the higher the temperature, the lower is the *resistance to deformation*.

MOLECULAR MECHANISMS OF A STRESS RELAXATION IN REAL MATERIALS

We have noted in this text that materials belonging to different classes sometimes behave similarly. This is the case with stress relaxation. We see in Figure 14.14 that the stress decays to a constant value called the **residual stress**. Josef Kubát discovered a very interesting phenomenon [4, 5]: the stress σ versus time t curves for stress relaxation of metals and polymers look practically the same. Kubát has shown that even the slope of the large descending part is virtually independent of the kind of material. Trying to explain this phenomenon, Kubát assumed that the relaxation occurs in *clusters*. That is, neither individual atoms in metals nor the polymer segments relax

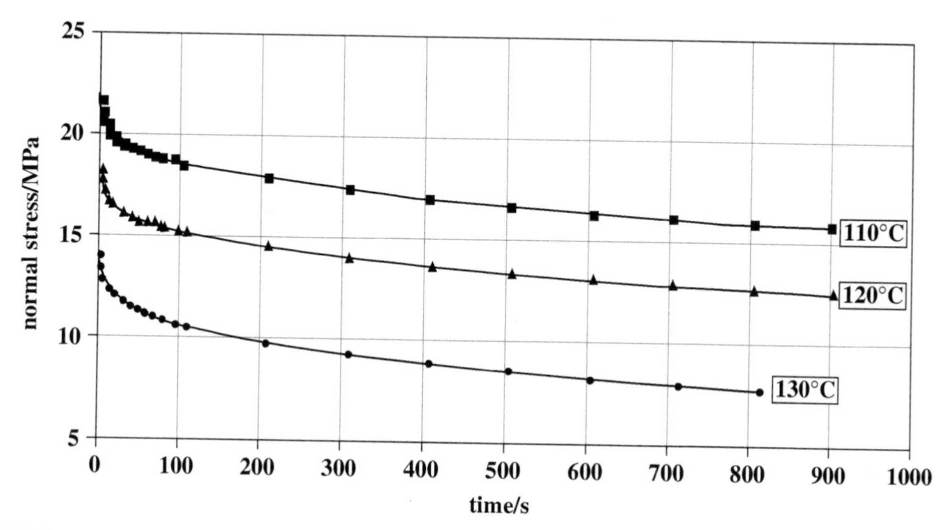

FIGURE 14.14 Isothermal stress relaxation results for a polymer (Bayer Makrolon®, which is a polycarbonate, 3 mm thick sheets) at three temperatures. *Source*: Kubát [3].

individually, but both kinds of units relax collectively. On the basis of this assumption, Kubát and Rigdahl succeeded in developing a general theory of stress relaxation which provides predictions agreeing precisely with experiments [5, 6].

Every theory makes certain assumptions. In Section 9.7 we discussed the molecular dynamics (MD) and Monte Carlo computer simulation methods. Molecular dynamics simulations were used to test the basic assumptions of the Kubát theory [7, 8]. The first results provided the right shape of the curve for stress versus time, but the residual stress was achieved after 1.5 decades of time; in reality for both metals and polymers it takes 4.3 or so decades of time. Those first computer-generated materials had no defects. To make the materials more realistic, some 2 or 3 vol.% of defects such as vacancies were introduced. With that correction, the stress relaxation curves predicted more than 4 decades of time [7, 8]—in agreement with experimental results.

More importantly, MD simulations show that metal atoms and polymer chain segments rarely relax individually. They relax in clusters—just as Kubát assumed (see the shaded box). This is why in the stress relaxation process it does not matter whether the material is a metal or a polymer. Kubát and Rigdahl have also derived an equation for the distribution of cluster sizes [6]. Simulations show that also that that size distribution equation closely matches the behavior of computer-simulated materials [9]. We have here a beautiful example where experiments, theory, and computer simulations all coincide.

Now an obvious practical question: how can we improve mechanical properties? Materials have a tendency to defend themselves against an attack. Furthermore, how can we mitigate the effects of creep and/or of stress relaxation? Refer to Figure 14.15 for a diagram showing creep deformation in a plastic pipe. We see how reinforcement with steel in the form of a plastic composite can substantially reduce the creep. (Read more

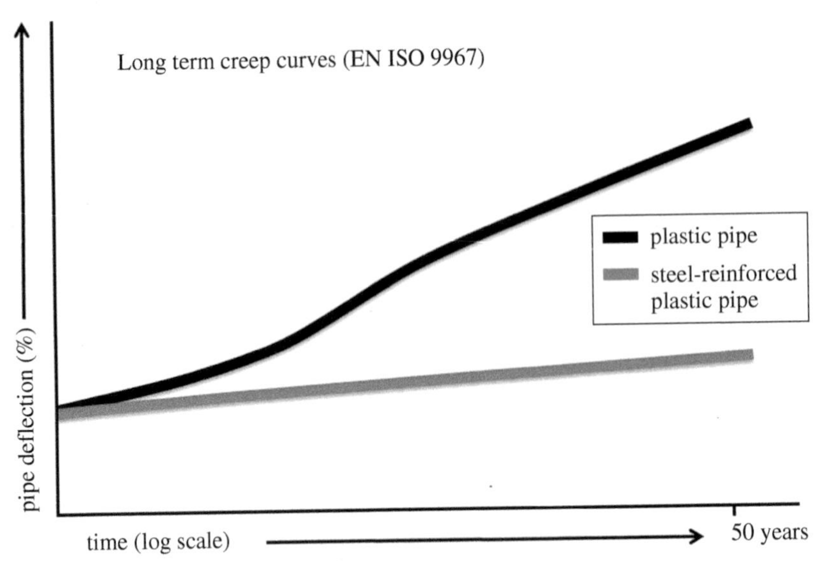

FIGURE 14.15 Creep of a plastic pipe, unreinforced and reinforced with steel. The ISO 9967 Standard was used to determine the creep.

about ISO and other standards for testing in Chapter 21.) Note the time scale. Needless to say, other reinforcements (apart from steel) can also be used for **creep mitigation**.

14.5 VISCOELASTICITY, DYNAMIC MECHANICAL ANALYSIS, AND BRITTLENESS

Some scientists and engineers leave the "safe" field of materials with time-independent properties only reluctantly. However, since all polymers and polymer based materials (PBMs) are viscoelastic (mentioned already in Chapters 9 and 13), we shall now bravely deal with this issue. Let us start with a very simple presentation of viscoelasticity as a phenomenon; see Figure 14.16. The ball in that Figure is a polymer—indeed because a metallic or ceramic ball would not behave so. **Viscoelastic behavior** means that a material shows simultaneously the "face" of an elastic solid and the "face" of a flowing liquid. Recall that we have discussed already viscoelasticity in *liquid* phases in the previous Chapter. Viscoelasticity manifests itself in the solid state, as well. We shall see below some common features of the viscoelastic behavior of melts and solids, particularly since in both cases a sinusoidal force is applied. How can we characterize viscoelasticity in solids quantitatively? The procedure is called dynamic mechanical analysis (DMA), or else dynamic mechanical thermal analysis (DMTA), and is not unlike that used in rheology. There are detailed discussions of DMA by Menard [10, 11].

In contrast to the slow application of force used for the quasi-static tests described in Section 14.2, in DMA one makes use of a dynamic oscillating force as described diagrammatically in Figure 14.17. If the material studied in Figure 14.17 were fully elastic, the maxima and minima of stress and strain would be in-phase and coincide. The more the strain maximum is out of phase from the previous stress maximum, the stronger is the viscous, liquid-like, or loss character of the material.

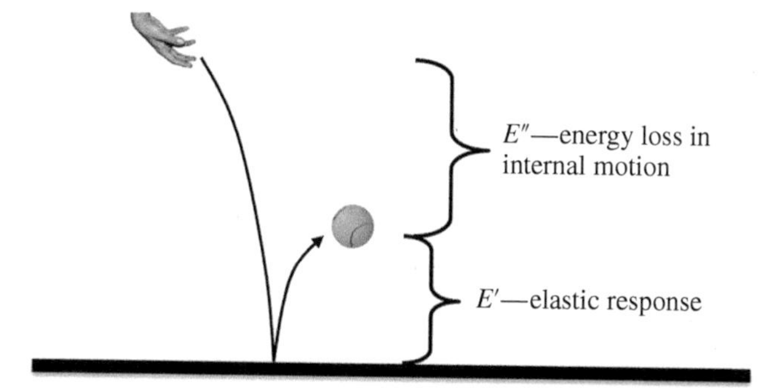

FIGURE 14.16 Bouncing a ball off a floor: There is an elastic response represented by the storage modulus E' and the liquid-like response represented by the loss modulus E''. The latter is so named because it is evident from the illustration that some of the original energy is lost—largely, but not only, by internal motions of the ball. In this scenario the ball and its behavior illustrate a polymer and viscoelasticity.

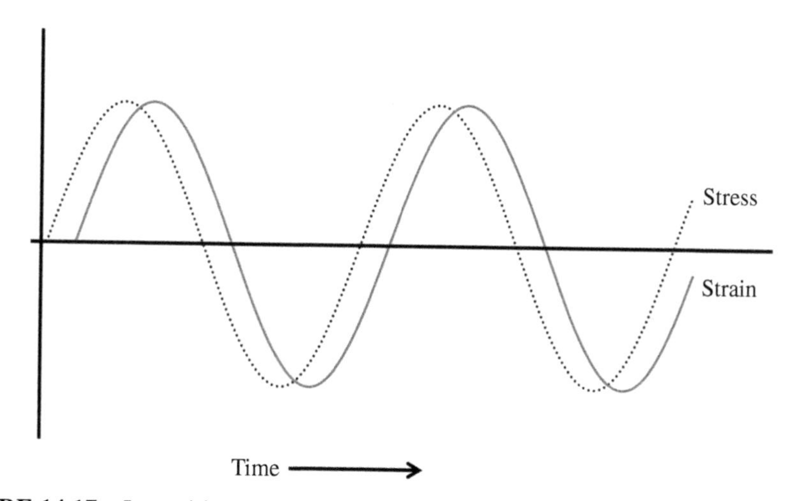

FIGURE 14.17 Imposition of a sinusoidal stress on a viscoelastic solid and the material response in terms of strain.

In DMA experiments a sinusoidal stress σ is applied as a function of time t:

$$\sigma(t) = \sigma_0 \sin(2\pi\nu t) \tag{14.6}$$

Here ν is the frequency in Hertz. Imposing the sinusoidal load yields the following behavior of the strain ε:

$$\varepsilon(t) = \varepsilon_0 \sin(2\pi\nu t - \delta) \tag{14.7}$$

If the material is fully elastic, we have $\delta = 0$. In a viscoelastic material there is always a phase lag between stress and strain as we have seen in Figure 14.17, so $\delta \neq 0$.

For a given deformation mode (such as three-point bending, or tension), we have the following relationship between the stress and strain and the modulae:

$$\frac{\sigma}{\varepsilon} = E^* = E' + iE'' \tag{14.8}$$

Here E^* is the complex modulus; $i=(-1)^{1/2}$. E' is the storage modulus, as mentioned (see again Figure 14.17), a measure of the solid-like (elastic) response of the material. E'' is called the loss modulus and corresponds to the liquid-like (viscous flow) response of the material. Equation (14.8) provides the capability to separate these two kinds of responses in a material. Which of them prevails affects the possible *applications* of a given viscoelastic material in an obvious way.

Since E'' represents energy dissipation, it is also a measure of the *energy converted to heat*. Namely, the heat H created is given by

$$H = \pi E'' \varepsilon_0^2 \tag{14.9}$$

The phase lag δ between stress and strain can be connected to the quantities featured in Eq. (14.8), namely,

$$\tan\delta = \frac{E''}{E'} \tag{14.10}$$

An example of a diagram obtained from DMA [12] is shown in Figure 14.18. Work was done on polystyrene and a styrene/butadiene/styrene (SBS) copolymer as matrices reinforced with the ceramic **Boehmite** [γ-AlO(OH)]. Surfactants were used as the coupling agents to improve adhesion between the polymers and the ceramic filler. The diagram in Figure 14.18 pertains to the polystyrene matrix with boehmite. For each frequency the value of the glass transition temperature T_g manifests itself both in the fall of E' and in the peak of tan δ. In fact, the determination of T_g is more accurate with DMA than with the historically more familiar technique of differential scanning calorimetry (DSC) [10, 11]. It is also evident in Figure 14.18 that at the higher frequency of 1.0 Hz the glass transition shifts to a higher temperature. In general, materials resist deformation. At a lower frequency a material has more time to adapt to a stress wave defined by Eq. (14.6). At a higher frequency there is less time for adaptation, the material shows more resistance to the force applied, and the glass transition region moves to higher temperatures. We have noted this fact already in Section 5.6.

We have also discussed viscoelasticity in Chapter 13 talking about polymer melts. There the objective was improvement of processing. For polymer solids knowledge of their viscoelasticity allows better adaptation to intended applications.

We are now ready to define **brittleness** B of materials [13, 14]:

$$B = \frac{1}{(E'\varepsilon_b)} \tag{14.11}$$

Here ε_b is the tensile elongation at break and E' is the storage modulus determined at 1.0 Hz and the temperature of interest (such as 25°C). The ε_b term in the denominator takes into account potential *large deformations* of a material. On the other hand, the storage modulus

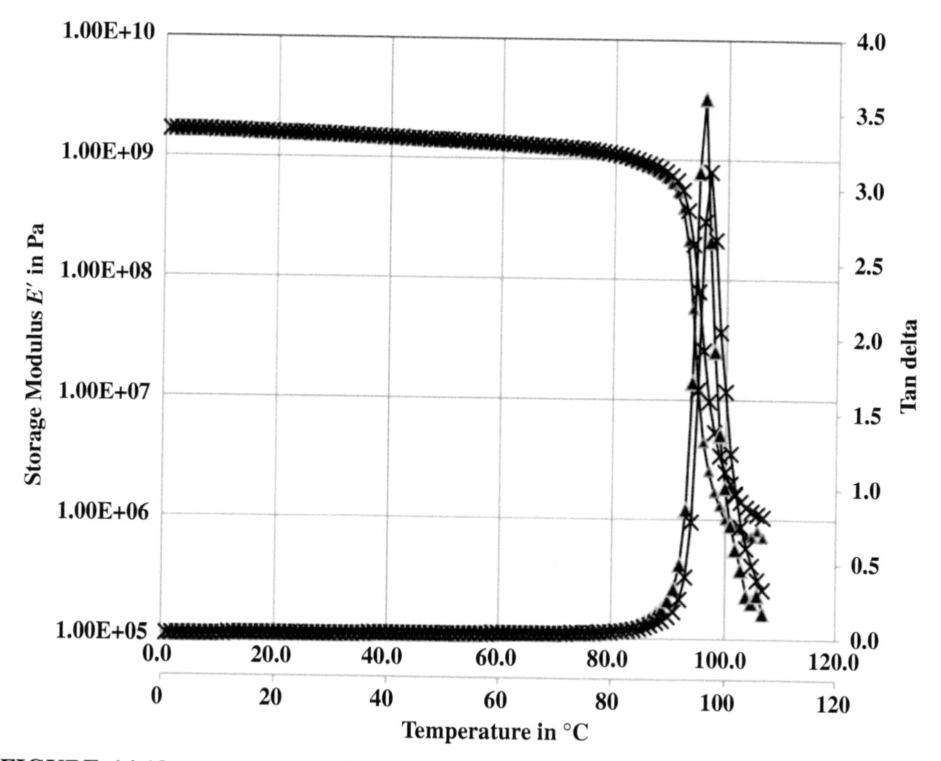

FIGURE 14.18 Dynamic mechanical analysis results for polystyrene + Boehmite. On the left, the curves at the top which at higher temperatures descend are those for E'. The curves which on the left are close to the bottom but form peaks at higher temperatures are for tan δ. Data for 2 frequencies are shown: 0.1 Hz (▲) and 1.0 Hz (x). For the higher frequency, there is a small shift to the right in the major drop in E' and in the peak of tan delta. *Source*: Adhikari *et al.* [12], © 2012. Republished with permission of Maney Publishing; permission conveyed through Copyright Clearance Center, Inc.

accounts for *repetitive loading or fatigue*—so important in service. B values are pertinent, for instance, for laminates [15], polymer blends [16], composites of the polymer + ceramic filler type [17, 18], polymer + carbon nanotubes [19], polymer + metal powder [20], and polymers with reinforcements such as rubber particles or wood flour. [21].

We mentioned that non-polymeric materials exhibit small viscoelastic effects in some properties. Thus, B has been determined, for instance, for copper pastes [22]. Brittleness not only provides information by itself but is also connected to other properties of materials, hence we shall mention B again on various occasions.

14.6 FRACTURE MECHANICS

Fracture is a telling feature of a material's mode of failure. In our earlier discussion of tensile testing, we observed the endpoint of the stress–strain diagram (Figure 14.4) was marked by fracture of the specimen. Fracture may occur by ductile or brittle mechanisms, or by a combination of both. Experts on fracture can tell by just glancing at fractured structures whether the fracture was ductile (Figure 14.19B) or brittle (Figure 14.19C).

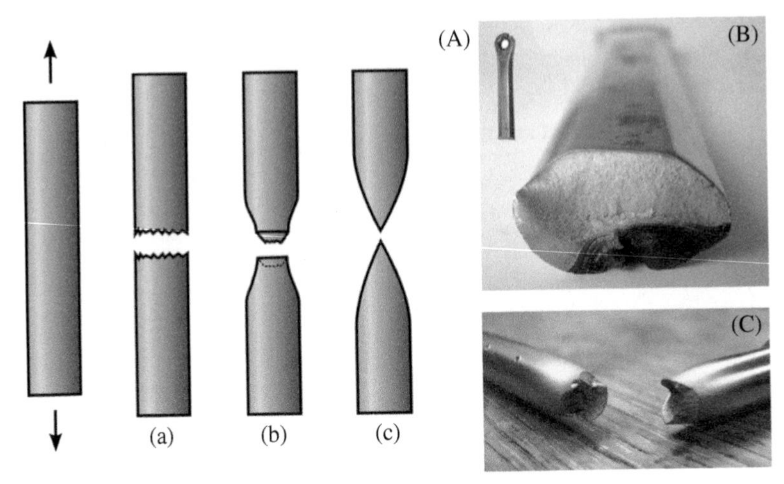

FIGURE 14.19 (A) Schematic appearance of round metal bars after tensile testing. (a) Brittle fracture, (b) Ductile fracture after local necking, and (c) Completely ductile fracture. (B) Stress fracture of a bicycle pedal arm (Kettler aluminum wheel of about 1990). Light: the brittle, forced rupture. Dark: the fatigue fracture with snap lines. (C) Ductile fracture of a metal rod. *Source*: (A): Sigmund, https://commons.wikimedia.org/wiki/File:Ductility.svg. Used under CC-BY-SA-3.0, GFDL. (B): Lokilech, https://commons.wikimedia.org/wiki/File:Pedalarm_Bruch.jpg. Used under CC-BY-SA-3.0, 2.5, 2.0, 1.0, and GFDL. (C): BradleyGrillo, https://commons.wikimedia.org/wiki/File:DuctileFailure.jpg. Used under CC-BY-SA-3.0, GFDL.

In brittle fracture we tend to observe clean fracture surfaces that appear to have been cleaved, while in ductile fracture there are elements of the material that seem to have been "pulled out" or have "adhered to" the surfaces of the fracture.

By intuition, we know that most cases of fracture are preceded by a crack. Thus, looking at **cracks** is the key to fracture mechanics. Near the end of Section 14.2 we mentioned that test specimens are sometimes purposefully notched. So-called **notches** are artificial cracks with well defined geometry we create to study the effects of the presence of cracks in real objects. Most cracks begin as something small and then by some means are propagated. Three modes of crack extension are presented in Figure 14.20: tensile, sliding, and tearing. From this we observe the three ways of applying a force that can lead to crack propagation. In Mode I a tensile force is applied normal to the crack plane. In sliding mode (Mode II) a shear stress acts parallel to the crack plane and perpendicular to the crack front. In Mode III the force also acts parallel to the crack plane but perpendicular to the sides, above and below the crack plane.

We shall now look at the "anatomy" of a crack or notch; see Figure 14.21. Consider first an example from everyday life: tearing a plastic sheet in half without a notch, or alternatively after making a notch with scissors. After a notch is made, tearing is much easier. The lines of the force we apply cannot go through the air; they converge on the notch. The crack or notch has a certain depth (length) h. It also has a radius of curvature L at the tip of the crack (refer again to Figure 14.21). We define the **stress concentration factor** k_i as

$$k_i = 1 + 2\left(\frac{h}{L}\right)^{1/2} \quad (14.12)$$

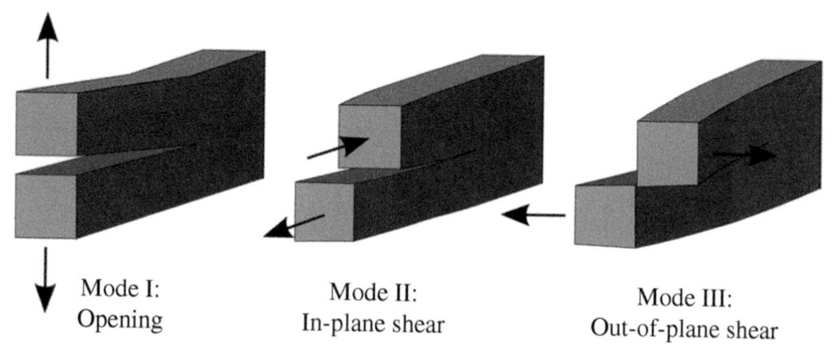

FIGURE 14.20 Three modes of crack extension. Mode I: opening = tensile mode. Mode II: in-plane shear = sliding mode. Mode III: out-of-plane (antiplane) shear = tearing mode.

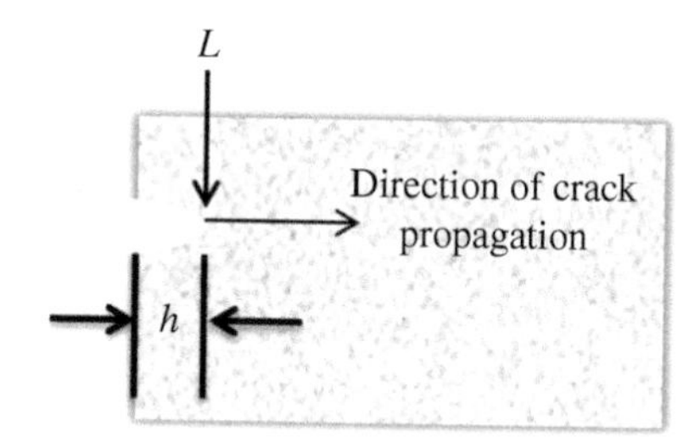

FIGURE 14.21 Geometry of a crack or notch. Depth of the crack is h; radius of curvature at the crack tip is L.

The factor tells us *how many times* larger the stress σ_{max} is at the tip of the crack compared to the nominal (imposed) stress σ. Therefore

$$\sigma_{max} = k_i \cdot \sigma \tag{14.13}$$

These definitions correspond to our intuitive notions of the danger of fracture posed by a crack: the larger h is, the more dangerous the crack; the more blunt the crack tip, the larger L is and the less prone is the crack to propagation.

Although we do not want them, **cracks** appear, sometimes in handling and transport even before service has begun. Clearly crack propagation is an issue of practical concern. There is a spectacular phenomenon of **rapid crack propagation** (RCP) that can occur in metal or plastic pipes filled with a pressurized gas, such as propane. As the name implies, these cracks propagate very quickly. Crack growth speeds of 100–400 m/s have been measured in polyethylene pipes [23], and the cracks may travel (i.e., grow in length) up to several hundred meters. The phenomenon is relatively rare but makes explosion of the gas pressurized inside a possibility.

There are various ways of evaluation of crack propagation and ways of mitigating it. Altstädt and collaborators have performed finite element analysis to study crack initiation and propagation in polymers [24]. The calculated stress distribution was compared with that determined experimentally by a photoelastic technique. The results obtained by the two methods agree for *continuous* crack propagation. Predicting *stick–slip* crack propagation turned out to be more difficult. The Altstädt group developed a method of mitigating fatigue crack growth in epoxy resins by introduction of silica nanoparticles [25]. The method is effective if the nanoparticles exceed the size of the crack tip opening in at least one dimension.

Actually, more insidious than rapid crack propagation is **slow crack propagation** (SCP), where the **crack propagation rate** dh/dt can be *as low as 3 mm/year*. One of the ways of quantifying the potential for crack propagation and fracture under a stress σ is in terms of the **stress intensity factor** as

$$K_{\mathrm{I}} = \alpha^{*} \sigma \sqrt{\pi h} \tag{14.14}$$

Here α^{*} is a geometric factor (determined by the specimen geometry and cracking mode) while h as above is the length of the existing crack. We cannot help the fact that the stress intensity factor (K_{I}) has a similar name to the stress concentration factor (k_{i}); the names have too long a history of use in their related fields and industries.

For polymeric materials, there is experimental evidence that the stress intensity factor is a function of the crack propagation rate dh/dt, where as usual t is time. A relation between the two quantities has been derived [26], namely,

$$\log K_{\mathrm{I}} = \frac{1}{2}\log\left(\alpha^{*2} 2\Gamma E\right) + \frac{1}{2}\log\left[1 + \frac{1}{\left(\beta h_{\mathrm{cr}}\right)}\frac{dh}{dt}\right] \tag{14.15}$$

In Eq. (14.15), Γ is the surface energy per unit area; E is the elastic modulus—tensile if the stress σ is tensile, compressive if σ is compressive, etc. The β term is a time-independent proportionality factor that is characteristic for a material type since it depends on chain relaxation capability (CRC) [27]; the higher that capability, the lower the value of β. The critical length h_{cr} is defined by the fact that a crack will *not* propagate in a given material unless $h > h_{\mathrm{cr}}$.

The validity of Eq. (14.15) has been tested [26, 27]. Hoechst AG in Frankfurt/Main produces a variety of polyethylenes (PEs). Their PEs have various molecular weights and various densities. Uniaxial tension was applied to a number of Hoechst PE specimens at several stress levels and with several initial crack lengths h_0. The stress intensity factor was determined from the data and plotted against the measured crack propagation rate, as shown (for three molecular weights of PE) in Figure 14.22. The solid lines of each curve have been calculated from Eq. (14.15). Points along a given curve pertain to tests of a single molecular weight M; different symbols pertain to different stress levels and to different values of h_0. Actually both σ and h_0 have been taken into account in deriving Eq. (14.15), but as we see they do not appear in the final equation. In agreement with predictions of that equation, points for a PE with a given molecular weight form a single curve, independently of σ and h_0. As expected, the material with the highest molecular weight, here C, has the largest chain relaxation capability and the slowest crack propagation rate.

From Eq. (14.15), we see therefore a connection between crack propagation and energy. In this case surface energy is important. Nevertheless, bear in mind that crack growth in a material facilitates the release of energy from an applied stress. The implications are serious.

Gordon tells of a ship's cook who one day noticed a crack in the steel deck of his galley. His superiors assured him that it was nothing to worry about—the crack was certainly small compared with the vast bulk of the ship—but the cook began painting

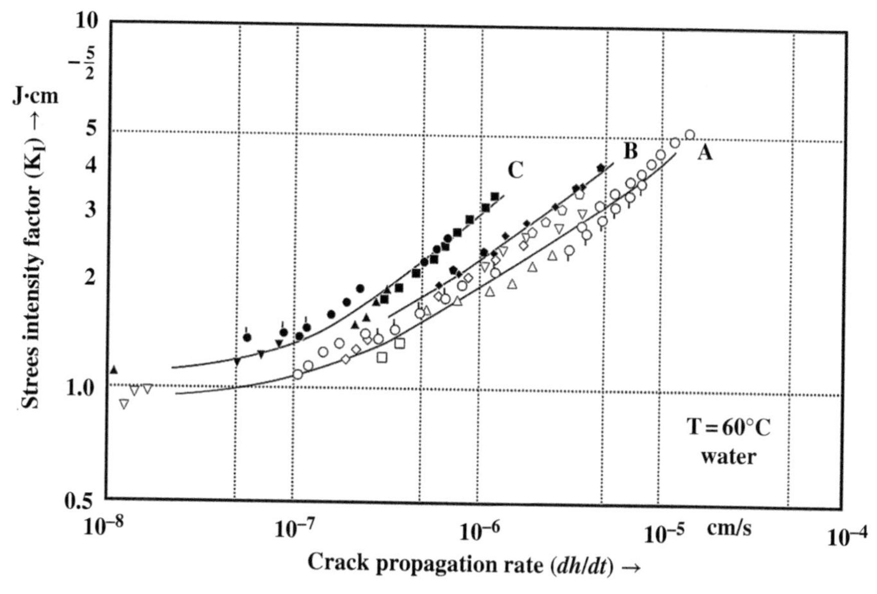

FIGURE 14.22 Crack propagation rates of three polyethylenes with different average molecular masses A, B, and C. *Source*: [26]. Reprinted with permission from Elsevier.

dates on the floor to mark the new length of the crack each time a bout of rough weather would cause it to grow longer. With each advance of the crack, additional decking material was unloaded, and the strain energy formerly contained in it released. But as the amount of energy released grows quadratically with the crack length, eventually enough was available to keep the crack growing even with no further increase in the gross load. When this happened, the ship broke into two pieces; this seems amazing but there are a more than a few such occurrences that are very well documented. As it happened, the part of the ship with the marks showing the crack's growth was salvaged, and this has become one of the very best documented examples of slow crack growth followed by final catastrophic fracture [28] (recounting the story described in [29]).

14.7 IMPACT TESTING

The application of force in impact testing is very fast, in contrast to the slower testing speeds utilized by techniques described in Section 14.2. In quasi-static testing, one observes primarily (but not exclusively) ductile fracture as the mode of specimen failure. This is of course also dependent on the type of material. Impact test conditions, on the other hand, are designed to instigate rapid fracture of the test specimen. The conditions of impact testing include a high strain rate, that is, a rapid rate of deformation. The high strain rate is critical since it affects how much *time* is available for a material to deform—keep in mind that deformation involves the rearrangement of atoms or molecules in a material's structure, a process that takes time. The preparation of a notch in the test specimen, which produces a triaxial stress state, further encourages stress concentration and fracture at the impact site.

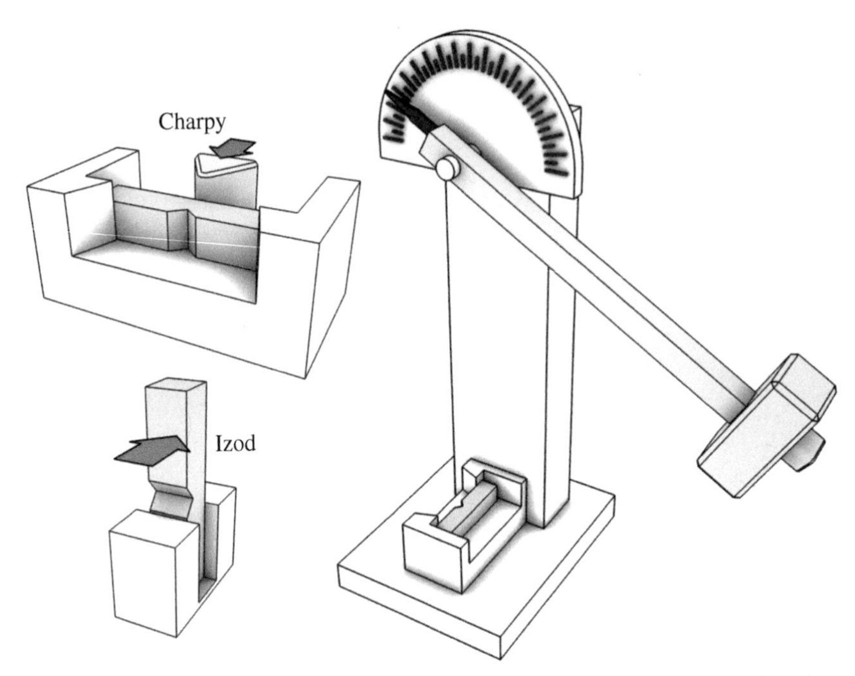

FIGURE 14.23 Impact test of a notched sample. Sample orientation and direction of the hammer strike are indicated for Charpy and Izod test methods.

There are several standard methods for impact testing. The Charpy and Izod tests are the most recognized techniques utilizing a pendulum apparatus. The typical lab model tester can sit on a benchtop; but there are some pendulum impact testers that stand as tall as their human operators. Vertical drop-weight testers offer another mode for impact testing; as the name implies, a specified weight is dropped through a vertical distance before the striker contacts the test specimen. These vertical test units also vary in size and maximum amount of energy that can be delivered to the specimen upon impact. The results of impact testing tell us about the impact *toughness* of a material, although to confuse the issue some report impact strength. But while strength is a measure of force per unit area (typically in Pascals), toughness is a measure of the energy absorbed per unit area (in this case of the crack).

A schematic for the **Charpy test** is shown in Figure 14.23. A centrally notched sample is mounted horizontally against an anvil at the bottom and center of the pendulum's path. The pendulum is released from a specified height h_0, the striker hits the center of the specimen on the side opposite the notch, and the pendulum is allowed to freely swing through to the other side. The height h_1 attained at the end of the swing is used to calculate the energy absorbed by the material during fracture. The **Izod test** is essentially the same as the Charpy; the absorbed energy is calculated the same way, and it differs only in specimen size and orientation. In an Izod test the notch is placed off-center, the specimen is mounted vertically, and the hammer strikes the notched side of the specimen.

A *tough* material requires a lot of energy to break. Frequently this is because the fracture process in tough materials causes a lot of plastic deformation. On the other hand, a *brittle* material, like glass, may be strong but fractures easily once a crack has begun because it has little capacity to absorb energy. Importantly, *toughness* is a temperature-dependent property. Many materials lose toughness and become more brittle as temperature decreases. Impact toughness is also dependent on the strain-rate; therefore one must know the strain rate of testing to properly interpret reported values of toughness.

For most of history, the distinction between brittle and ductile behavior has been defined primarily by inspection of the fracture surfaces such as we have seen in Figure 14.19, with perhaps some little association with other quantitative properties. The explanation of brittleness B defined by Eq. (14.11) provided the first universal quantitative assessment of that quantity [13]. In 1997 an index of brittleness for ceramics was proposed by Quinn and Quinn [30], but as we have seen throughout this book, the elastic properties that characterize most ceramics do not strictly apply to metals and polymers, thus the underlying assumptions used to develop that brittleness index render it invalid for other material types.

In general, specimens used for testing, even from one material, are not all completely identical. Therefore, during a series of impact tests, some specimens might respond in a brittle way and some in a ductile manner. As a consequence of this, the **ductile-brittle impact transition temperature** has been defined as the temperature at which 50% of test specimens undergo ductile fracture. Needless to say, below that temperature the majority of the specimens present with brittle behavior. The University of Cambridge Department of Engineering provides a fun interactive Internet-based resource for exploring the relations between strength and fracture toughness in steels [31].

We noted at the end of Section 14.4 that **brittleness** is related to several other material properties. A diagram of the dependence of **Charpy impact strength** on B is shown in Figure 14.24 for a variety of polymers having a variety of chemical structures [32].

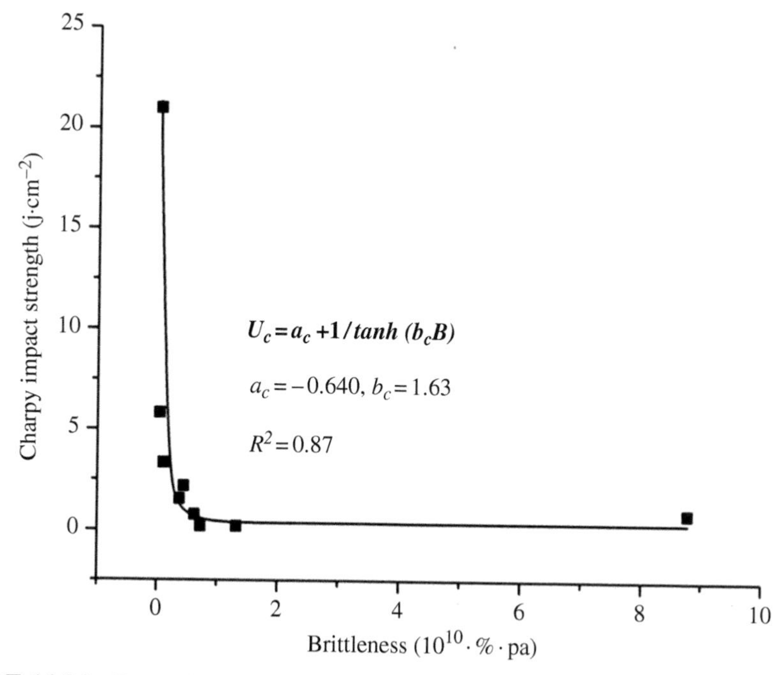

FIGURE 14.24 Dependence of the Charpy impact strength on brittleness for a number of different polymers (each data point represents a different polymer). *Source*: [32]. Reprinted with kind permission from Springer Science and Business Media.

The point at the far right is for polystyrene, a very brittle material. The equation included in the Figure fits well the experimental data. An equation of the same form fits Izod impact results, but the parameters of course are different.

A special kind of impact testing is called impact fatigue testing. One applies a lower force level so that fracture does not occur instantly. Only after some number of impact events does fracture eventually take place. One records the weight of the hammer, the temperature, and the number of the impact event at which the fracture has occurred in order to evaluate the material's susceptibility to fatigue fracture.

Fatigue, as one might suspect, refers not to instantaneous catastrophic cracks and failures but to the kind of damage that occurs over time and with repetitive use and cycling of a product. That cycling may be mechanical in nature, for instance, a repetitive impact of a hammer on another surface or repeated stretching and relaxation of a device component. Thermal cycling between temperature extrema is also possible, and in Chapter 19 we discuss *wear* as a kind of surface fatigue that results from two materials repeatedly sliding over one another.

As noted at the end of Section 14.2, toughness is a measure of the energy required to fracture the material, while one way to obtain toughness τ is to determine the surface area under the tensile stress versus strain curve, that is

$$\tau = \int_0^{\varepsilon_b} \sigma \, d\varepsilon \tag{14.16}$$

The toughness so defined at 25°C is shown in Figure 14.25 as a function of brittleness for a large variety of polymers [33]. Actually, this is only a part of the diagram. Similarly as in Figure 14.24, polystyrene is far to the right. The curve in Figure 14.25 is defined by an equation relating τ to B (which of course includes polystyrene), with steel and aluminum also included [33].

14.8 HARDNESS AND INDENTATION

Hardness, a feature which most of us learn about in infancy, is also important to scientists and engineers. The simplest and most easily understandable definition of hardness was formulated in 1812 by the German geologist and mineralogist Friedrich Mohs. It is based on the ability of a harder mineral to scratch a softer material and is characterized by the ordered list in Table 14.3.

The Table tells us that any mineral can scratch talc while no materials can scratch diamond. When one of us (WB) was nine years old, collecting minerals was an important activity. The parents, seeing this, provided the boy with an atlas of minerals which contained the Mohs scale. Having found out that quartz has the hardness equal seven, the boy took one of his prized possessions, namely, a pink quartz, and tried it on a window pane. Mohs was right, visible scratches appeared on the pane; only *the parents were not amused.* Window glass has the hardness of ≈5.5 on the Mohs scale.

Mohs hardness is not the only scale of hardness in use. The **Vickers test** is relatively easy to use compared to other modern hardness tests since the required calculations are independent of the size of the indenter, and the indenter can be used for all materials irrespective of hardness. The basic principle, as with all common measures of hardness, is to

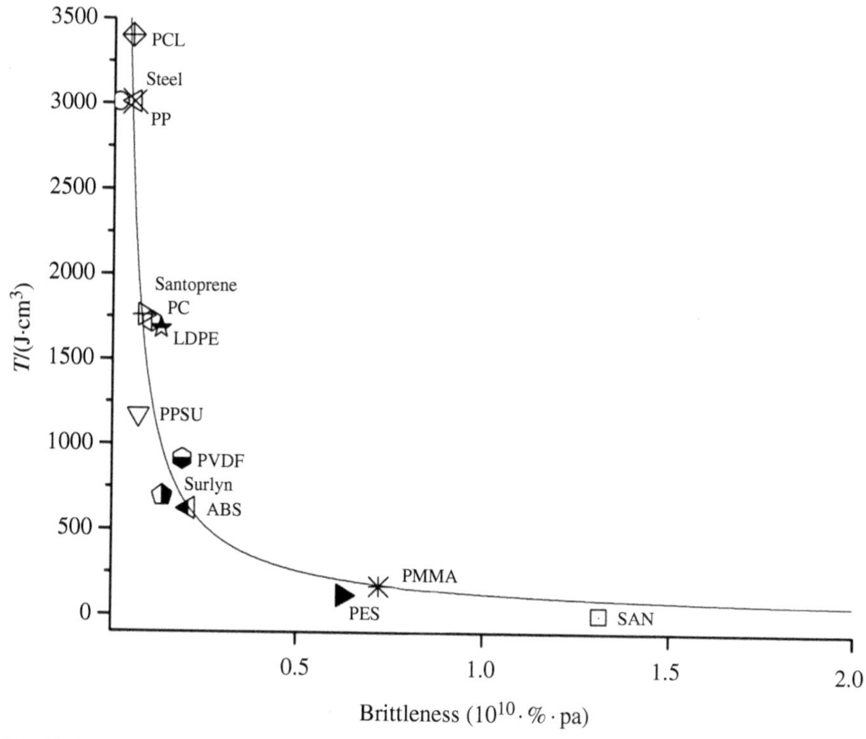

FIGURE 14.25 Toughness τ (the surface area under the tensile stress vs. strain curve) versus brittleness B for a variety of polymers with different chemical structures. *Source*: [33]. Reprinted with permission from Elsevier.

TABLE 14.3 Mohs Scale of Hardness

Material	Chemical Formula	Mohs Relative Hardness
Talc	$Mg_3Si_4O_{10}(OH)_2$	1
Gypsum	$CaSO_4{\cdot}2H_2O$	2
Calcite	$CaCO_3$	3
Fluorite	CaF_2	4
Apatite	$Ca_5(PO_4)_3(OH^-,Cl^-,F^-)$	5
Orthoclase Feldspar	$KAlSi_3O_8$	6
Quartz	SiO_2	7
Topaz	$Al_2SiO_4(OH^-,F^-)_2$	8
Corundum	Al_2O_3	9
Diamond	C	10

observe the material's ability to resist deformation by a standard source. The Vickers test has one of the widest scales among hardness tests. The unit of hardness given by the test is known as the **Vickers Pyramid Number** (h_{Vickers}) or **Diamond Pyramid Hardness** (DPH). The hardness number can be converted into units of Pascals, but should not be confused with a pressure, which also has units of Pa. The hardness number is determined from the *load divided by the whole surface area of the indentation* and not the area normal to the

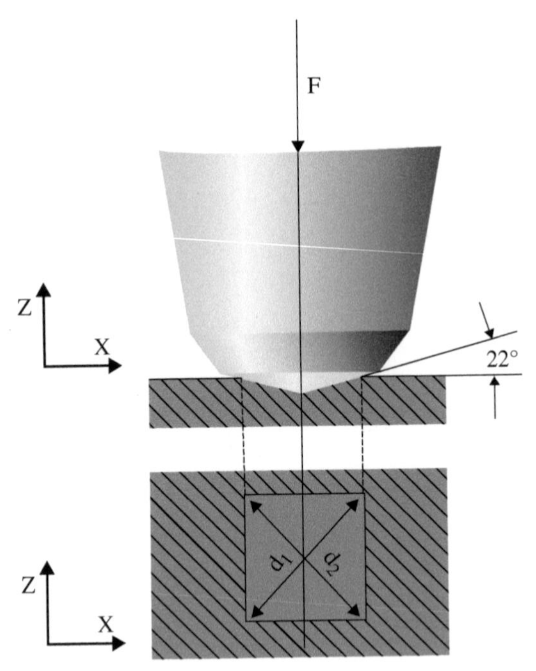

FIGURE 14.26 Vickers hardness determination, a schematic view. *Source*: User A1, https://commons.wikimedia.org/wiki/File:Vickers-path-2.svg. Used under CC-BY-SA-3.0.

force, and for that reason is not a pressure. A schematic of the Vickers hardness test geometry is shown in Figure 14.26.

There exist approximation tables allowing calculation of the Vickers hardness from the Mohs hardness and *vice versa*. There are also a number of other test methods for hardness. Thus, when a hardness value is given, a method by which it was determined has to be provided.

There is a mode of hardness determination called **microindentation** or else **nanoindentation determination**, advanced by Doerner and Nix [34], with some subsequent modifications by Oliver and Pharr [35]. One follows the load versus indenter displacement diagram. One uses a three-sided pyramid invented by Berkovich [36] and known as the **Berkovich hardness** tip. Apparently it is easy to grind such a tip so that it would have a sharp point. The load placed on the indenter tip is increased as the tip penetrates further into the specimen and soon reaches a user-defined maximum value P_{max}. At this point, the load may be held constant for a period or removed. The area of the residual indentation in the sample is measured and the hardness h is defined as

$$h = \frac{P_{max}}{A} \tag{14.17}$$

where A is the residual indentation area.

The indentation test is best suited for small volumes, hence the development of microindentation and nanoindentation versions of the test. By analyzing the loading and unloading displacement curves, one can also try to obtain values of the tensile modulus E, a technique that has more luck with isotropic materials and less with anisotropic ones.

Hydrogels pose special problems in indentation experiments since they are so much softer than metals, ceramics, or even polymers. However, methods of measuring indentation in hydrogels using either spherical or flat ended cylindrical indenters have been developed [37].

Practically speaking, all of the properties discussed in the chapter are extremely important to the process of materials selection. Any time one must select a material to work with for a given purpose, one has to go through the process of selecting an appropriate material based on a host of properties, including mechanical ones. While it is essential to know generally the characteristics of different classes of materials, there are now so many material types that it is impossible to memorize every bit of information. The Department of Engineering at the University of Cambridge has compiled interactive materials selection charts as well as detailed information lists for commonly used materials [38, 39]. The selection charts are also useful for comparing and contrasting properties like toughness and hardness; thus they can be used to ensure one understands the distinctions between these different properties.

14.9 SELF-ASSESSMENT QUESTIONS

1. Discuss the four basic mechanical modulae and the relations between them.
2. Explain effects of necking on the stress versus strain diagrams for non-ferrous metals.
3. Discuss the dependence of elastic modulus on temperature for a crosslinked elastomer.
4. Explain what viscoelasticity is and how it manifests itself in mechanical properties.
5. Complete each sentence. (A) Ductility is the material's ability to _____. (B) Hardness is the material's ability to _____.
6. Why are elasticity and plasticity both important properties? What sort of processes utilize the plasticity of a material to impart deformation on purpose?
7. Complete each sentence. (A) Strength—tensile, compressive, and shear—is the material's ability to _____. (B) Stiffness is the material's ability to _____.
8. What quantitative measurements are associated with the properties of strength and stiffness? (To test how well you can differentiate between the two, complete the fun exercise provided through the MATTER Project at http://schools.matter.org.uk/Content/YoungModulus/stiffnessexercise.html.)
9. Complete the sentence: Toughness is the material's ability to _____.
10. What features of a material would make crack initiation difficult? For instance, why are plastic mailing packages and Tyvek sheets (used to cover unfinished walls during building construction) so difficult to tear but paper is easy to rip? How does the situation change if you try to tear a large stack of paper?
11. Consider earlier discussions from Parts 1 and 2 about defect structure. Using that insight, compare and contrast crack propagation and fracture behavior in metals to that in ceramics?
12. Define brittleness and discuss some uses of this concept.
13. Why is slow crack propagation insidious?

14. You have two specimens of the same material, one with a sharp and the other with a blunt notch. How do you expect they will behave in impact testing?
15. Name three products in which having high toughness is important.
16. Name two components, products, or devices in which a fracture would be catastrophic.
17. Think of (or find) and then describe a real example in which embrittlement due to reduced temperature caused the failure of a product (or device, component, etc.).
18. Which mechanical properties that we discussed (strength, toughness, hardness, etc.) are affected by the strain-rate and how?
19. Think about the motto at the start of this chapter. What kind of stress might a gun barrel be subject to? What kind of stress is the bullet in its brass casing subject to?

REFERENCES

1. P. Tran, S.S. Kandula, P.H. Geubelle, & N.R. Sottos, *J. Phys. D* **2011**, *44*, 034006.
2. S. Kimball, Don't Focus on Stress When Stiffness Is the Problem, *Machine Design Magazine*, July **1999**, http://www.erareplicas.com/misc/stress/machdes.htm, accessed April 3, 2014.
3. J. Kubát, private communication from the Chalmers University of Technology, Gothenburg, September **2013**.
4. J. Kubat, *Nature* **1965**, *204*, 378.
5. J. Kubat & M. Rigdahl, *Mater. Sci. & Eng.* **1976**, *24*, 223.
6. J. Kubat & M. Rigdahl, Chapter 4; in *Failure of Plastics*, ed. W. Brostow, Hanser: Munich/Vienna/New York **1986**.
7. W. Brostow & J. Kubat, *Phys. Rev. B* **1993**, *47*, 7659.
8. S. Blonski, W. Brostow, & J. Kubat, *Phys. Rev. B* **1994**, *49*, 6494.
9. W. Brostow, J. Kubat, & M.J. Kubat, *Mater. Res. Soc. Symp.* **1994**, *321*, 99.
10. K.P. Menard, Chapter 8; in *Performance of Plastics*, ed. W. Brostow, Hanser: Munich/ Cincinnati, OH **2000**.
11. K.P. Menard, *Dynamic Mechanical Analysis: A Practical Introduction*, 2nd edition, CRC Press: Boca Raton, FL **2008**.
12. R. Adhikari, W. Brostow, T. Datashvili, S. Henning, B. Menard, K.P. Menard, & G.H. Michler, *Mater. Res. Innovat.* **2012**, *16*, 19.
13. W. Brostow, H.E. Hagg Lobland, & M. Narkis, *J. Mater. Res.* **2006**, *21*, 2422.
14. W. Brostow, H.E. Hagg Lobland, & M. Narkis, *Polymer Bull.* **2011**, *67*, 1697.
15. J. Chen, M. Wang, J. Li, S. Guo, S. Xu, Y. Zhang, T. Li, & M. Wen, *Eur. Polymer J.* **2009**, *45*, 3269.
16. A. Dorigato, A. Pegoretti, L. Fambri, C. Lonardi, M. Slouf, & J. Kolarik, *Express Polymer Lett.* **2011**, *5*, 23.
17. W. Brostow, T. Datashvili, J. Geodakyan, & J. Lou, *J. Mater. Sci.* **2011**, *46*, 2445.
18. R. Baskaran, M. Sarojadevi, & C.T. Vijayakumar, *J. Mater. Sci.* **2011**, *46*, 4864.
19. A. Szymczyk, Z. Roslaniec, M. Zenker, M.C. Garcia-Gutierrez, J.J. Hernandez, D.R. Rueda, A. Nogales, & T.A. Ezquerra, *Express Polymer Lett.* **2011**, *5*, 977.

20. W. Brostow, M. Brozynski, T. Datashvili, & O. Olea-Mejia, *Polymer Bull.* **2011**, *59*, 1671; W. Brostow, T. Datashvili, P. Jiang, & H. Miller, *Eur. Polymer J.* **2016**, *76*, 28.
21. Y. Deng, G. Gao, Z. Liu, C. Cao, & H. Zhang, *Ind. & Eng. Chem. Res.* **2013**, *52*, 5079.
22. W. Brostow, T. Datashvili, R. McCarty, & J. White, *Mater. Chem. & Phys.* **2010**, *124*, 371.
23. W. Brostow & W.F. Müller, *Polymer* **1986**, *27*, 76.
24. M. Maier, V. Altstädt, D. Vinckier, & K. Thoma, *Polymer Testing* **1994**, *13*, 55.
25. M.H. Kothmann, G. Bakis, R. Zeiler, M. Ziadeh, J. Breu, & V. Altstädt, *Seventh International Symposium on Engineering Plastics*, Xining, August 18–21, **2015**.
26. W. Brostow, M. Fleissner, & W.F. Müller, *Polymer* **1991**, *32*, 419.
27. W. Brostow, *Pure & Appl. Chem.* **2009**, *81*, 417.
28. D. Roylance, *Introduction to Fracture Mechanics*, June 14, **2001**, http://ocw.mit.edu/courses/materials-science-and-engineering/3-11-mechanics-of-materials-fall-1999/modules/frac.pdf, accessed April 3, 2014.
29. J.E. Gordon, *The New Science of Strong Materials, or Why You Don't Fall Through the Floor*, Princeton University Press: Princeton, NJ **1976**.
30. J.B. Quinn & G.D. Quinn, *J. Mater. Res.* **1997**, *32*, 4331.
31. University of Cambridge Department of Engineering, *Strength-Toughness*, http://www-materials.eng.cam.ac.uk/mpsite/interactive_charts/strength-toughness/NS6Chart.html, accessed March 13, **2016**.
32. W. Brostow & H.E. Hagg Lobland, *J. Mater. Sci.* **2010**, *45*, 242.
33. W. Brostow, H.E. Hagg Lobland, & S. Khoja, *Mater. Letters* **2015**, *159*, 478.
34. M.F. Doerner & W.D. Nix, *J. Mater. Res.* **1986**, *1*, 601.
35. W.C. Oliver & G.M. Pharr, *J. Mater. Res.* **1992**, *7*, 1564.
36. E.S. Berkovich, *Zavodskaya Lab* **1950**, *13*, 345.
37. M. Czerner, L.S. Fellay, M.P. Suárez, P.M. Frontini, & L.A. Fasce, *Procedia Mater. Sci.* **2015**, *8*, 287.
38. University of Cambridge Department of Engineering, *Material Selection and Processing*, http://www-materials.eng.cam.ac.uk/mpsite/default.html, accessed April 8, **2014**.
39. University of Cambridge Department of Engineering, *Material Information*, http://www-materials.eng.cam.ac.uk/mpsite/materials.html, accessed April 8, **2014**.

15

THERMOPHYSICAL PROPERTIES

Question: State the relations existing between the pressure, temperature, and density of a given gas. How is it proved that when a gas expands its temperature is diminished?
Answer: Now the answer to the first part of this question is, that the square root of the pressure increases, the square root of the density decreases, and the absolute temperature remains about the same; but as to the last part of the question about a gas expanding when its temperature is diminished, I expect I am intended to say I don't believe a word of it, for a bladder in front of a fire expands, but its temperature is not at all diminished.

—19th Century Schoolboy Blunders. Genuine student answer* to an Acoustics, Light and Heat paper (1880), Science and Art Department, South Kensington, London, collected by Prof. Oliver Lodge. Quoted in Henry B. Wheatley, *Literary Blunders* (1893), 175, Question 1. (*From a collection in which Answers are not given *verbatim et literatim*, and some instances may combine several students' blunders.)

15.1 INTRODUCTION

Don't thermodynamic and thermophysical mean the same thing? The answer is yes. We already have Chapter 3 on thermodynamics, why another one? In Chapter 3 we discussed important concepts, including the stability criteria, fundamental principles of heat and energy, and the meaning of equilibrium states. We have not discussed in detail how thermophysical properties are measured or how we use the numerical values obtained. These issues shall be dealt with in the present Chapter. Dynamic mechanical analysis

Materials: Introduction and Applications, First Edition. Witold Brostow and Haley E. Hagg Lobland.

(DMA) is frequently operated in function of temperature; it provides both mechanical and thermophysical characterization (and is discussed in Section 14.5).

There is also another connection between the present chapter and the previous one. In Chapter 14 we saw that a mechanical force can destroy a material. However, we can disintegrate a multiphase material or a composite *without the use of a force*: in many circumstances it suffices that we *heat* a material. Where a material's constituents have different thermal expansivities, the integrity of the structure may be destroyed by heat. The precise meaning of thermal expansivity will be discussed in the next Section. Temperature, and its variation, which is the constant theme of this Chapter, can through various mechanisms impose thermal stresses on a material and lead to material failure or fracture.

A simple way to think about **thermophysical properties** is that they vary with temperature without altering the material's chemical identity. More specifically, these properties have to do with the transfer and storage of heat. They vary with the state variables of volume, temperature, pressure, and composition (in mixtures) without, as we said, altering the material's chemical identity. Therefore thermophysical properties include: thermal conductivity and diffusivity, heat capacity, thermal expansion, thermal radiative properties, diffusion coefficients (viscosity, mass, and thermal), speed of sound, surface and interfacial tension in fluids. A few of these properties are discussed at greater length in other chapters, but here we will touch on most of the properties just named.

15.2 VOLUMETRIC PROPERTIES AND EQUATIONS OF STATE

Two-dimensional or **surface density** can be used for instance to describe the number of people per km^2 in a given space. Again and again we note in this book similarities between behavior of materials and behavior of humans. In both cases certain *physical parameters determine their behavior.*

We now need some definitions for our description of materials. First of all, there are two kinds of densities. **Mass density** ρ is defined by the relation $\rho = m/V$, where m is the mass and V the volume. There is also **number density**, defined as $\rho_n = N/V$, where N is the total number of particles under consideration such as atoms, molecules, ions, or polymer chain segments. The reciprocal of mass density is the **specific volume** $v_{sp} = V/m$. Recall our discussion in Section 14.2 of how the elastic modulus of polymers changes with temperature. We observed the transition from glassy to leathery to elastic with increasing temperature. We find that those same transitions can be determined from the specific volume by plotting v_{sp} as a function of T.

Next we need to define the **isobaric expansivity:**

$$\alpha = V^{-1}\left(\frac{\partial V}{\partial T}\right)_P \tag{15.1}$$

We realize that without the V^{-1} front factor the expansivity would be an extensive quantity (depending on the *amount* of material present) while with the present definition it is an intensive quantity (see Section 3.2 if you do not remember what these are). We also have **linear isobaric expansivity** α_L, which is

$$\alpha_L = L^{-1}\left(\frac{\partial L}{\partial T}\right)_P \tag{15.2}$$

where L is the length, or in practice height, of the object. One determines α_L in an apparatus for **thermal mechanical analysis** (TMA), which measures the change in height of a specimen as a function of varying temperature. Thermal mechanical analysis apparatus can also be used for determination of α. If the material is isotropic, one measurement is enough. If not, three experiments along three Cartesian coordinates x, y, and z are needed.

One has to be quite careful with the terminology. Frequently the term **coefficient of thermal expansion (CTE)** is used. This can refer to either of the quantities defined by Eqs. (15.1) and (15.2). The importance of isobaric expansivity may be readily understood, but a few aspects deserve extra emphasis. In any application involving a large or rapid temperature change, thermal expansivity needs to be considered. If one has a device component fitted with a very tight tolerance, imagine what might happen if the isobaric expansivity of the component is too high and the device is heated. A second scenario where isobaric expansivity is important is in the case of laminated and coated objects. One frequently hears the phrase "thermal expansivity mismatch" used to describe the situation in which adjacent materials have poorly match CTEs such that they expand at different rates with increasing temperature. (We speak of *expansion*, but obviously the principle applies also to the reverse case of *contraction* upon temperature decrease.)

In Section 2.3 we were talking about the fact that freezing of liquid water into ice results in lowering of density—in contrast to most other materials. We discussed consequences of this fact for aquatic life. Liquid water heated from its freezing/melting point up to +4°C undergoes a density *increase*, that is, negative isobaric expansivity, again an anomaly when compared with other materials. However, water is not the only such exception. In 2010 Greve and coworkers [1] found that cubic scandium trifluoride, ScF_3, has negative α in the solid state in a very wide temperature range: from 10 to about 1100 K. A group at Caltech and the Oak Ridge National Laboratory explains this unusual phenomenon in terms of large vibrational entropy from large-amplitude fluorine atom motions and also from phonon anharmonicity [2]. The harmonic model of thermodynamic properties of crystals created by Albert Einstein is explained in Section 15.3.

As we deal here with volume, temperature, and pressure, we need to define two more quantities. The first is

$$\kappa_T = -V^{-1}\left(\frac{\partial V}{\partial P}\right)_T \tag{15.3}$$

which is called the **isothermal compressibility**. There is also **isoentropic compressibility**, sometimes called **adiabatic compressibility**

$$\kappa_S = -V^{-1}\left(\frac{\partial V}{\partial P}\right)_S \tag{15.4}$$

where S is entropy. In solid phases the difference between κ_T and κ_S can be neglected. The latter can be determined experimentally from the speed of sound. Recall that in an adiabatic system no heat or matter is transferred between the system and surroundings. Now let us go back to Eq. (14.4). If the bulk modulus k_b does not change with pressure P, we have a simple relationship:

$$k_b = \frac{1}{\kappa_T} \tag{15.5}$$

How do we obtain values of the volumetric properties when we need them? We actually need both experimental data and an **equation of state**. By definition, an equation of state is *any* equation which determines completely the state of the system. Thus, for a closed system, an equation for enthalpy of the form $H = H(S,P)$ (see Eq. 3.11) or else $U = U(S,V)$ are equations of state. However, most often the term equation of state is used in reference to relations of the form $V = V(T,P)$.

To discuss equations of state for different kinds of materials, let us begin with gases. **Johannes D. van der Waals** already worked on this and created such an equation. Then **Heike Kamerlingh Onnes** worked on it at the University of Leiden in a laboratory that now bears his name. Here is the **virial equation of state for gases,** sometimes called the **Leiden equation**:

$$PV = NkT\left(1+\frac{B}{V}+\frac{C}{V^2}+\cdots\right) \tag{15.6}$$

Here B is called the second pressure virial coefficient, C the third pressure virial coefficient, etc.; k is the Boltzmann constant.

Equation (15.6) can be inverted; the result is

$$PV = NkT + BP + \left(\left(c - B^2\right)P^2\right)/NkT + \cdots \tag{15.7}$$

We see that neglecting the second and further pressure virial coefficients, we obtain $PV = NkT$, the equation of state for the **ideal gas**. Such a gas does not exist; it could only exist if there were no intermolecular forces in the gas phase—but it provides a useful zeroth order approximation.

Throughout history, materials, which under normal conditions (room temperature, $P = 1\,\text{atm}$) are gases, have been tempting objects for **liquefaction**. On April 5, 1883, Karol Olszewski and Zygmunt Wróblewski at the Jagiellonian University in Cracow liquefied oxygen for the first time [3]. They applied the **cascade method** of gas liquefaction, a multistage method. Lowering the temperature of a gas results in liquefaction; see for instance Figure 3.5. The condensation is accompanied by heat going *out* of the material, the process inverse to vaporization. Olszewski and Wróblewski as a cooling material used ethylene, with its boiling temperature in vacuum about −136°C. Then one lowers the pressure while increasing the volume, so the material becomes a gas again, without adding heat into the system; in fact, taking a larger volume requires energy, thus temperature goes further down. This is repeated several times. Wróblewski and Olszewski reporting the liquefaction of oxygen noted the fact that the liquid so obtained is colorless [3]. Their publication put an end to an earlier belief that there are "permanent gases" which cannot be liquefied—while providing proof for the kinetic theory of gases of the British physicist James C. Maxwell claiming the opposite. On April 13, 1883, they liquefied nitrogen, and later on carbon dioxide and methanol. On August 13, 1894, Lord Rayleigh and Sir William Ramsay at the University College London isolated argon; from a sample of clean air, they removed oxygen, carbon dioxide, water, and nitrogen [4]. Next year Ramsay provided Karol Olszewski with a sample of argon, and the latter liquefied it too [5].

Now moving back to Kamerlingh Onnes: on July 10, 1908, the latter liquefied *helium*. On April 8, 1911, Kamerlingh Onnes found that at 4.2 K the electric resistance in a solid mercury wire immersed in liquid He vanished suddenly. This is how **superconductivity** was discovered; we shall discuss the phenomenon in Section 17.5. In 1926 **Willem Hendrik Keesom** obtained *solid He*.

How about equations of state for liquids and solids? There is a large variety of such equations. We shall include here one developed by Hartmann and Haque [6] which provides good results for polymer solids and liquids [7]

$$P_r V_r^5 = T_r^{3/2} - \ln V_r \tag{15.8a}$$

with the reduced quantities defined as follows:

$$V_r = \frac{V}{V^*};\quad P_r = \frac{P}{P^*};\quad T_r = \frac{T}{T^*} \tag{15.8b}$$

The reducing quantities V^*, P^*, and T^* are obtained from experimental data. Thus, for n experimental points one solves an over-determined system of n equations (15.8a) in three unknowns. The term $T_r^{3/2}$ has a theoretical justification provided by Morton Litt [8]. The idea of using reduced quantities in such a manner has been considered long before by Johannes van der Waals.

Now we have a better framework for interpreting the description of thermophysical properties presented in the first Section. To reiterate, they vary with the state variables of volume, temperature, pressure, and composition (in mixtures) without, as we said, altering the material's chemical identity. Consider the cases of liquefaction just described: the elemental gases maintained their chemical identity throughout changes in the state variables. But one expects that the thermal conductivity and heat capacity under those different conditions will not be the same.

15.3 DIFFERENTIAL SCANNING CALORIMETRY (DSC) AND DIFFERENTIAL THERMAL ANALYSIS (DTA)

A number of thermophysical properties have some relation to phase transitions. The general technique of **calorimetry** has a long history of use. The most popular method of locating phase transitions is **differential scanning calorimetry** (DSC) [9]. In DSC, one uses two identical cells, one empty (the reference) and one containing the object (material specimen) under investigation. Using an absolute method one has to apply a number of mathematical and other corrections. In differential scanning calorimetry, there are two furnaces, and the same temperature is maintained in both cells. Therefore it is possible to determine the *energy* input into the reference cell. This technique is also called power compensation DSC, heat flow DSC, or true DSC; the energy input in mW is recorded. In a less accurate version, often called **heat-flux DSC** or **differential thermal analysis** (DTA), there is one furnace and the difference in *temperature* is recorded. An example of a true DSC diagram is shown in Figure 15.1.

In Figure 15.1 the upwards arrow of the y-axis represents energy absorbed by the material. The left vertical dotted line marks the glass transition region. The horizontal dotted line spanning the valley indicates the temperature range of **cold crystallization**. The two rightmost solid arrows indicate the melting region, with the dotted line pointing to

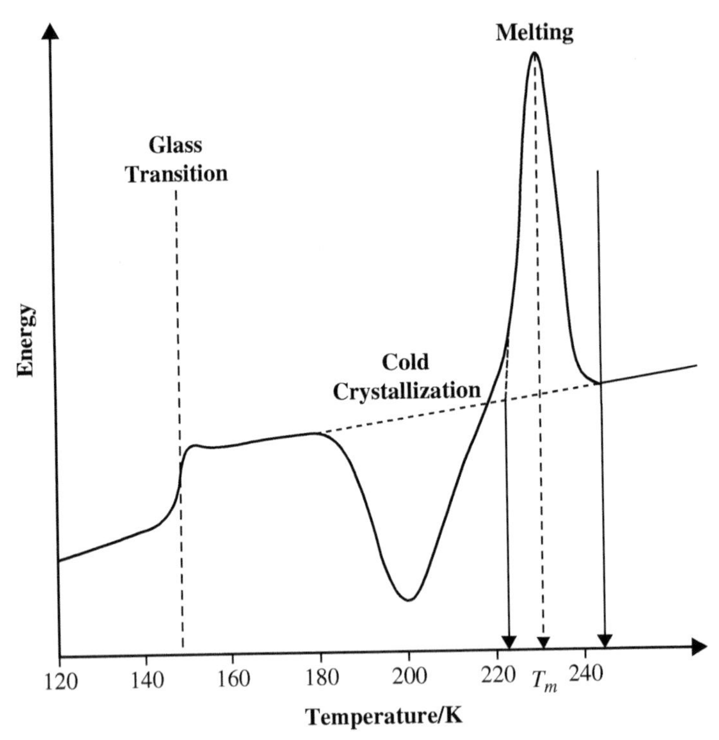

FIGURE 15.1 A DSC diagram for poly(dimethyl siloxane) (PDMS).

the named melting temperature. Cold crystallization was already mentioned in Section 12.2. In a semi-crystalline polymer at temperatures above the glass transition region, there exist both liquid (which used to be glass) and crystalline agglomerates swimming in that liquid. Depending on the thermal history of the material, it is sometimes energetically favorable for some liquid polymeric chains to *join* the crystalline ones even as the material is heated further; this is precisely the process known as cold crystallization. The process is unexpected: we are increasing the temperature and getting *more* of the crystalline phase. However, as we also see in Figure 15.1, the process is finite and is reversed at still higher temperatures below the melting transition.

There are further variations of the DSC technique such as **temperature modulated DSC** (TMDSC) in which a non-linear temperature signal is applied to the furnace. There is **pressure DSC** where pressures other than the atmospheric are applied. There is also **photocalorimetry** [10] so that UV light with a known frequency and intensity is applied to the DSC sample. It has been utilized, for instance, with lightly cured acrylics [8].

Integration within a certain temperature range of the area under the DSC curve, say from T_1 to T_2, provides the amount of energy needed to bring the sample from T_1 to T_2. The amount of energy needed to change the temperature by one degree is called the **heat capacity**, sometimes also called **specific heat**. Actually, we have two such quantities:

$$C_V = \left(\frac{\partial U}{\partial T}\right)_V; \quad C_P = \left(\frac{\partial H}{\partial T}\right)_P \tag{15.9}$$

C_V (heat capacity at constant volume) is more convenient to work with for gases, C_P (heat capacity at constant pressure) for condensed phases. The latter can be a decisive parameter for the decision to launch—or otherwise—an industrial process. If the process requires a temperature change, the heat capacity value will tell whether the process is economically viable given the associated energy costs.

We see that both C_V and C_P as defined in Eq. (15.9) are extensive quantities. More convenient to work with, of course, are intensive quantities such as **molar** C_P or else **specific** heat capacities such as C_P per 1 g. Thermodynamics allows us to calculate the difference between the two heat capacities:

$$C_P - C_V = T\left(\frac{\partial P}{\partial T}\right)_{V,N}\left(\frac{\partial V}{\partial T}\right)_{P,N} = \frac{VT\alpha^2}{\kappa_T} \tag{15.10}$$

thus relating that difference to the volumetric properties defined by Eqs. (15.1) and (15.3). Measurements of C_P are easier, while C_V appears more often in theories. Hence a theory might be tested by using its predictions for C_V, measuring isobaric expansivity and isothermal compressibility, obtaining the C_P–C_V difference using the last member of Eq. (15.10), and comparing with experimental C_P values.

We can also get from thermodynamics the following formula:

$$\kappa_S = \kappa_T - \alpha^2 T/\rho C_P \tag{15.11}$$

where ρ as before is the mass density. Some theoretical models provide us with κ_S, and then we use Eq. (15.11) for confrontation with experimental data.

Heat capacity is an intrinsic property of matter. As said, it can tell us the amount of energy required to heat a material by a given number of degrees. This, in a manner, is also an indication of heat storage capacity. If you have ever burnt your tongue on pizza fresh from the oven, you will recognize that it was most likely the tomatoes, with their high heat capacity, not the crust (with many air bubbles and low heat capacity) or even the cheese topping (which cools quickly due to surface radiation loss) that accomplished the burn. (One can explore more about the chemical and thermophysical properties of food and cooking by looking for resources on Molecular Gastronomy.)

The story of the development of our understanding of heat capacity of crystals is somewhat instructive. Until 1907 people believed in the so-called **Dulong and Petit law** according to which the heat capacity of crystals was independent of temperature. Experiments were telling otherwise.... In 1907 **Albert Einstein** developed a theory of thermodynamics of crystals [11] based on two assumptions: movements of each atom in the lattice can be represented by superposition of three linear harmonic oscillators (each along one Cartesian coordinate) and all oscillators have the same frequency. These assumptions were clearly simplifications but they allowed one to obtain numerical values—in fact pretty close to experimental ones. An improvement of the Einstein theory was provided by **Peter Debye** who assumed a certain distribution of oscillator frequencies instead of a single one. A further improvement was provided by **Eduard Grüneisen** who considered the effect of a changing material volume (such as during heating) on the oscillator behavior. Now we typically work with the so called Grüneisen parameter, where $k_S = 1/\kappa_S$:

$$\gamma = V\left(\frac{\partial P}{\partial U}\right)_V = \frac{\alpha k_S}{\rho C_P} = \frac{\alpha k_b}{\rho C_V} \tag{15.12}$$

The significance of the original model of Einstein was that it demonstrated that the law of Dulong and Petit is valid only as a limit at high temperatures. For solids at low temperatures, the specific heats drop as the effects of quantum processes become more predominant. Deviations from harmonicity have been studied by Guggenheim and McGlashan in solid and gaseous argon, seen there in spite of the spherical force fields [12]. Anharmonicity manifests itself also for instance in cold rubidium atoms confined in an elongated magnetic trap [13].

15.4 THERMOGRAVIMETRIC ANALYSIS

We now consider thermogravimetric analysis (TGA). The basis of the experimental design is very simple: with a sensitive analytical balance, one determines the loss of weight of a specimen while heating the sample at a constant rate. The results are sometimes decisive for materials selection. If a material does not survive in testing the high temperatures to which it will be exposed in service, its wonderful properties at room temperature will not help.

Figure 15.2 provides an example of a TGA diagram. We see how mixing poly(vinyl chloride) with cross-linked polyethylene affects thermal stability. On the y-axis is the weight percentage; in this case as temperature is increased (x-axis), the weight of each material decreases, though not in the same fashion. Information about the thermal stability of different components as well as about various phase transitions can be determined from different aspects of the curves. In an example not shown here, nickel nanoparticles were added to a thermoplastic elastomer; effects of that addition on thermal stability—aside from other properties also measured—were determined [14] by TGA. In fact, TGA can be

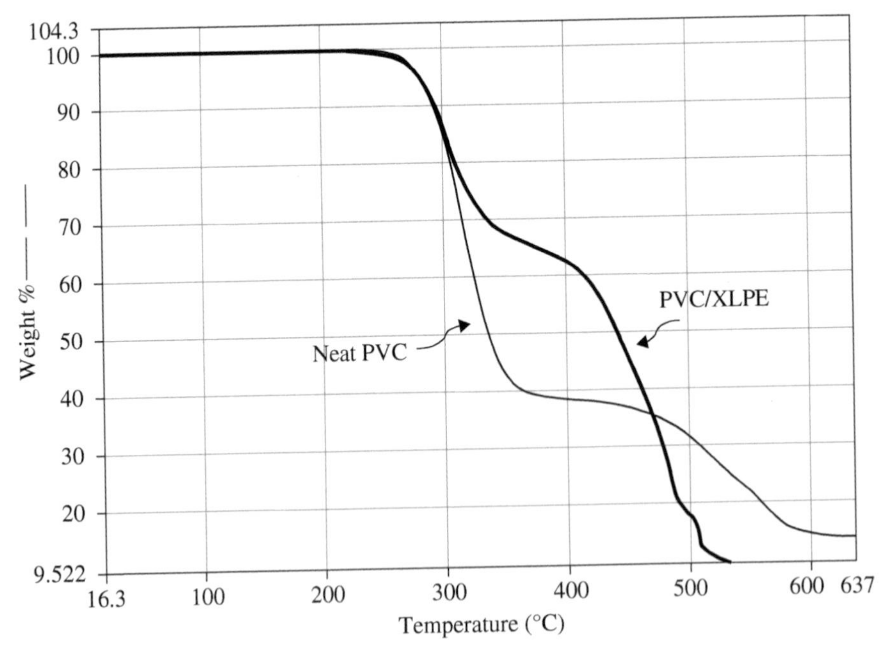

FIGURE 15.2 Thermogravimetric analysis of poly(vinyl chloride) (PVC): curves shown for neat PVC and for 50 wt.% each of PVC and crosslinked polyethylene (XLPE).

applied to all material types. For instance, TGA of a metal may show a weight increase due to oxidation or a decrease from sublimation. Clearly the data produced can be used for a variety of purposes. For example, in laminated materials designed for heat protection, one can evaluate whether the outer protective layer effectively prevents deterioration of the underlying material.

15.5 THERMAL CONDUCTIVITY

Thermal conductivity can be defined as the ability of a material to transfer heat across itself [15]. Some people work with the reciprocal property called **thermal resistivity**. As expected, thermal conductivity is temperature dependent. The appropriate unit is W/m·K. Heat transfer through a material requires a temperature gradient as energy is transferred from a hot region to a cooler region. There is also a dependence on the local geometry, namely, the area of the surface perpendicular to the direction of heat transfer, for instance the face of cylindrical plate and the thickness of the material.

Thus, **thermal conductivity** z is the time rate of steady state heat flow through a *unit area* of a homogeneous material induced by a unit temperature gradient in a direction perpendicular to that unit area. The value of z (W/m·K) can be determined experimentally and is given as

$$z = q\frac{L}{\Delta T} \tag{15.13}$$

Here q = heat flow rate (W/m^2), L = thickness of the specimen (m), while as usual T = temperature (K). This brings us to the **thermal conductance** C, which is the quantity of heat passing in unit time through a plate of a *particular area and thickness* induced by a unit temperature difference between opposite faces. Thus C has the units W·K and is given by the expression

$$C = \frac{zA}{L} \tag{15.14}$$

Therefore, the value of thermal conductance can be easily calculated from the measured thermal conductivity. Now **thermal resistance** R is the reciprocal of thermal conductance, is expressed in the units K·m^2/W, and is given as

$$R = \frac{L}{zA} \tag{15.15}$$

If we consider the rate of steady state heat flow through a *unit area*—rather than another particular area—of a material induced by a unit temperature difference between the object surfaces, we get the **heat transfer coefficient**, equal to z/L and therefore measure in W/m^2·K. This coefficient is also referred to as **thermal admittance**.

The z-value, R-value, and C-value have to do with heat transfer by **conduction**. The property known as **thermal transmittance** quantifies the thermal conductance *plus* the heat transfer due to convection and radiation. Frequently reported as a U-value, thermal transmittance is sometimes used for comparing the insulating capacity of building materials such as windows.

In electronics systems as well as other applications, **heat sink materials** are used to dissipate heat for the cooling of devices. Thus, materials with high thermal conductivity are widely used for this purpose. The thermal resistance is used to determine the suitability of a heat sink material for a given application. Using Eqs. (15.13) and (15.15), R can be expressed in terms of the temperature difference and the heat flow (in watts). Therefore, if one knows the resistance of a component and its heat output, one can calculate the maximum allowed thermal resistance of the heat sink for a specified ΔT.

Underlying all of these aspects of thermal conductivity are the principles laid out in Chapters 2 and 3 regarding molecular interactions and thermodynamics. From those we understand that heat is related to the kinetic energy of molecules and that molecules have more or less ability to move owing to structural and thermodynamic properties such as molecular density, bond structure, entropy, etc. Revisiting the MSE triangle of Figure 1.1, we see that the properties of thermal conductivity are directly connected to the structure and interactions in a material. Therefore, when a material undergoes a phase change from solid to liquid or from liquid to gas, the thermal conductivity will change. Ice has thermal conductivity of 2.18 W/m·K at 0°C; while, after melting into liquid water at the same temperature, the thermal conductivity is 0.56 W/m·K.

The thermal conductivity of pure crystalline substances may vary along different crystal axes due to differences in phonon behavior (phonons of lattice vibrations, as in the Einstein model discussed in Section 15.3) along those axes. A phonon is an elastic quasi-particle, a term representing collective excitation of atoms and molecules in solids and in some liquids. While such particles do not really exist, considering them and their motions is a convenient way to deal with thermal conductivity, and for that matter with electric conductivity as well. Consider sapphire, which at 25°C has thermal conductivity of 35 W/m·K along the c-axis and 32 W/m·K along the a-axis. Phonons, however, are not the only means of transmitting thermal energy. Individual free electrons also contribute to thermal conductivity. The relative contributions of phonons versus freely moving valence electrons to thermal conductivity depend on the material type. Consequently the effects of temperature on thermal conductivity are also dependent on material type.

For metals, we have the Wiedemann-Franz Law according to which *thermal conductivity* is proportional to the *absolute temperature* (in K) multiplied by *electrical conductivity*. In pure metals the electrical conductivity decreases with increasing temperature (see Chapter 17); the (mathematical) product of the two quantities, which is thermal conductivity, remains approximately constant at higher temperatures. See the diagram for copper in Figure 15.3. In alloys the change in electrical conductivity with temperature is usually smaller and thus thermal conductivity increases with temperature, often proportional to temperature.

As expected, the contribution of valence electrons to thermal conductivity in polymers and ceramics is much less than it is in metals. In fact, because of this, most ceramic and polymeric materials are thermal insulators. See Figure 15.4 for polymers and note the comparatively small values for conductivity.

It is not surprising then that thermal conductivity in *ceramics* is generally lower than in metals since phonon scattering by lattice imperfections yields less efficient heat transfer. Because scattering increases with rising temperature, we tend to observe a corresponding decrease in thermal conductivity. However, at still higher temperatures, an

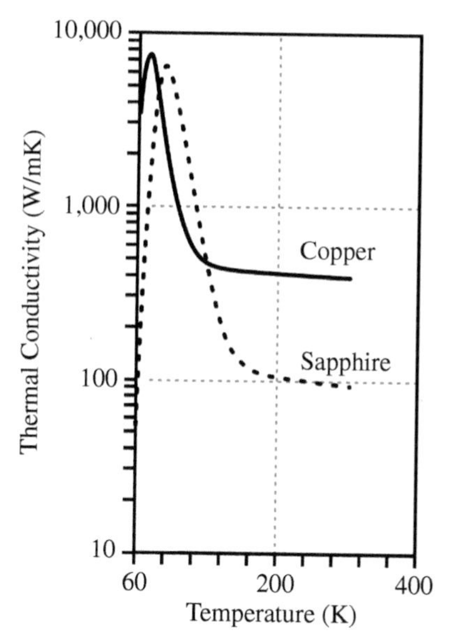

FIGURE 15.3 Thermal conductivity of copper and sapphire as a function of temperature.

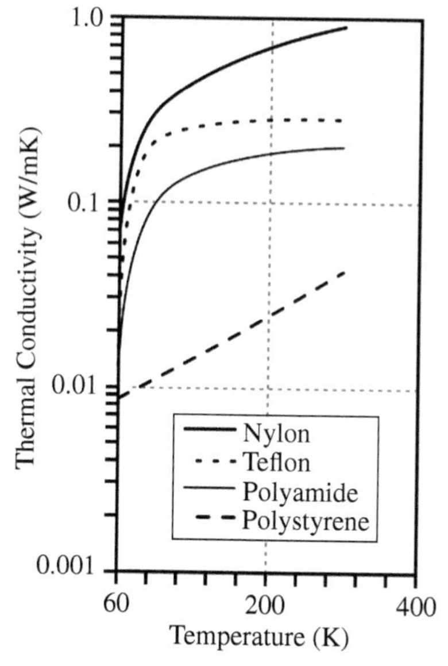

FIGURE 15.4 Thermal conductivities of polymeric materials as a function of temperature.

increase in (usually infrared) radiative heat transfer produces a subsequent increase in thermal conductivity in many ceramics. The absence of a regular lattice structure in amorphous ceramics means that they generally have lower conductivity than crystalline ones. It is to be noted that porosity has a significant impact on thermal conductivity. Heat transfer through internal pores is typically ineffective and slow; thus ceramics used as thermal insulators are frequently porous.

In the absence of convection (which requires movement of the gas), air and other gases are usually good insulators. For this reason the gas-filled pockets of porous ceramics, polymeric foams (e.g., Styrofoam), and aerogels make such materials good thermal insulators. The fur and feathers on animals serve a similar purpose by inhibiting the convection of air or water near an animal's skin. Moreover, light gases (e.g., hydrogen and helium) generally have higher thermal conductivity than dense gases (e.g., argon, frequently used in double paned windows for added insulation).

Thermal conductivity in polymeric materials, as with metals and ceramics, is dependent on temperature, crystallinity, phonons, and orientation. The effects of orientation are more pronounced at higher temperatures, and in general thermal conductivity of polymers increases with temperature. Crystallinity typically results in higher conductivity. The entanglement of molecular chains, which leads to phonon scattering, plays a causative role in the overall low thermal conductivity of polymers (refer again to Figure 15.4). A common approach to increase thermal conductivity of polymers is introduction of a secondary high thermally conductive phase into the polymer matrix. However, high thermal resistance between the two phases somewhat limits the extent of conductivity enhancement. Scale is also important: a single polymer chain can have a very high thermal conductivity, behaving as a one dimensional conductor [16]. On the other hand, because of their very low thermal conductivity, three-dimensional polymeric materials are useful in cryogenics applications.

We have not yet discussed methods to determine thermal conductivity. There are several such methods; we shall mention here one based on DSC [9]. By this approach one melts a known standard together with and on top of a sample specimen of known size and density. The heat transfer due to presence of the second material causes a change in slope of the leading edge of the melting curve (see again Figure 15.1). The thermal conductivity can be calculated from that change in slope. Naturally, the use of this method requires the existence of standards with sharp (very defined) melting curves [9].

Finally, let us revisit the topic of food. Thermal conductivity plays an especially important role in cooking as well as in food storage. A tray of cookies baked on an aluminum sheet will turn out differently than ones baked on a stone (i.e., ceramic) sheet. Silicone can be used as gelatin molds and will remain flexible at room temperature even when just taken from the freezer. Silicone also appears in rubber spatulas that can be used to stir hot foods. A corkboard mat can protect a finished table surface from being scorched by a hot dish.

15.6 NEGATIVE TEMPERATURES

Let us now explore the topic of **negative temperatures**. First we need to define a parameter τ related to the absolute temperature T:

$$\tau = -\frac{1}{kT} \tag{15.16}$$

The Boltzmann constant k appears here because of the "wrong" choice of temperature scales and units proposed by the Polish engineer and physicist of Dutch and German extraction **D. Gabriel Fahrenheit** and the Swedish astronomer **Anders Celsius**. Both Celsius and Fahrenheit scales require the k factor, although temperature scales in which this factor is absorbed in the temperature unit and therefore not visible have been proposed. When we use the absolute temperature starting at 0 and the Celsius unit of one degree, that is the **Lord Kelvin temperature scale**, thermodynamics "forces us" often to multiply T by k. Now, when we are at $T = 0\,\mathrm{K}$, all atoms in a crystal are in their ground state. Since there is only one state, we know what that state is; Eq. (3.2) tells us that entropy $S = 0$. While in some cases (where many atoms are involved) the discrete energy levels of electrons can be considered as forming continuous bands, for the present purpose we need to consider them individually. States with lower energies have higher probabilities of being occupied. This can be expressed by an equation as follows:

$$\frac{N_i}{N_j} = \exp\left[\frac{-(\varepsilon_i - \varepsilon_j)}{kT}\right] \tag{15.17}$$

where N_i is the number of electrons with the energy ε_i, and likewise for the index j. We see that also here the Boltzmann constant k "sticks like a glue" to the temperature T. At high temperatures we have $[(\varepsilon_i - \varepsilon_j)/kT] \ll 1$, so that $N_i/N_j \approx 1$ for any pair of levels i and j. We have, therefore, two competing factors. On the one hand, thermal agitation produces randomization and thus the tendency to populate the energy states equally, giving $N_i/N_j \approx 1$. On the other hand, we cannot forget the stability criterion (3.17), that is, the tendency of a material to go into the state with the lowest possible value of energy.

Consider a system having a maximum allowed energy ε_{max}. Now go back to the absolute zero temperature. We have

$$T = 0; \quad \tau = -\infty; \quad N_1 = N; \quad N_{i\neq 1} = 0 \tag{15.18}$$

where N_1 is the number of particles in the ground state and N the total number of particles. Now consider pumping energy into the system, that is, increasing T, but first so little that only one particle will jump from the ground level 1 into the next state 2, having slightly higher energy. Then

$$N_1 = N - 1; \quad N_2 = 1; \quad N_{i\neq 1,2} = 0 \tag{15.19}$$

We continue to put energy into the system fearlessly, with the higher energy states getting populated more and more. Being fearless is important here, since we eventually arrive at $T = \infty$. Equation (15.17) now tells us that all states are equally populated; we thus have

$$T = +\infty; \quad \tau = 0; \quad N_1 = N_2 = N_3 = \cdots = N_{max} \tag{15.20}$$

with N_{max} being the number of particles in the maximum allowed energy state.

Now, however, consider the strange situation where the thermodynamic temperature is *minus infinity*. Using again Eq. (15.17), we have

$$T = -\infty; \quad \tau = 0; \quad N_1 = N_2 = N_3 = \cdots = N_{max} \tag{15.21}$$

The populations of quantum states are the same whether we have $T = -\infty$ or $T = +\infty$. Clearly the parameter τ defined by Eq. (15.16) is a better indicator of what is going on than T. We have *one* state of the material, the same population of quantum states, *one* value of τ, while T performs outlandish jumps between $-\infty$ and ∞. In other words: $T = -\infty$ and $T = \infty$ correspond to a single state of the material.

We have already demonstrated how brave we are, but we need even more bravery: starting from this state where all quantum states are equally populated, we *now* add a very small amount of energy so that one particle will jump from the ground level 1 to level 2. Now, therefore

$$N_2 > N_1 \tag{15.22}$$

If we use once more Eq. (15.17) and assume $T>0$, we shall find a contradiction: $\varepsilon_1 > \varepsilon_2$. We do have a way out: assume that $T<0$. We now have

$$T < 0; \quad \tau > 0; \quad \text{and} \quad \varepsilon_1 < \varepsilon_2 \tag{15.23}$$

as it should be. We have just discovered something amazing: all temperatures $T<0$ are above or "hotter", by the measure of τ, than temperatures $T>0$.

Bravery was needed in this operation so far, but our resources of bravery cannot be exhausted yet! We are now outside of positive temperatures, that is, in the range of negative temperatures; we keep on pumping energy into the material and eventually get to the state when all particles have the highest allowed values of energy ε_{max}. Now

$$\lim_{T \to 0^-}\left(-\frac{1}{kT}\right) = +\infty; \quad \text{hence } \tau = +\infty; \; N_{max} = N; \; N_{i \neq max} = 0 \tag{15.24}$$

where the subscript $T \to 0^-$ tells us that we are approaching zero from negative values. We thus went through the entire range of τ values, from $-\infty$ to $+\infty$. The function τ behaved "decently" all the time, while the temperature did its wild jump in the middle of the range; also in (15.21) we had to use a limit of a function. We have also seen that matter at negative temperatures is more energetic and this is why it is conceptualized as being "hotter."

Can negative temperatures be produced in real materials? Yes, this is why we discuss them. This was achieved first in 1951 in lithium fluoride by two Americans, Purcell and Pound [17]. $T<0$ can be maintained inside LiF for some minutes. In 1952 Edward M. Purcell received the Nobel Prize in Physics, shared with Felix Bloch. Negative temperatures can be also created inside lasers. Lasers will be discussed more in Section 16.6. See also the illustration in Figure 15.5 which provides a visualization of the Boltzmann distributions discussed in relation to negative temperatures, while the work of Braun *et al.* [18, 19] can be referenced for further insight on the capacity to achieve negative temperatures in real materials.

FIGURE 15.5 Artistic illustration of five different thermal Boltzmann distributions. The first container on the left shows a gas at a very small positive temperature, close to absolute zero. Most atoms are close to the lowest energy state, which is given by the lower energy bound, indicated by the cover at the bottom. The second container also shows a gas at positive temperature, but at a much higher temperature. Some atoms also occupy high energy states. In the center container, the gas is at positive or negative infinite temperature, which are physically the same. All energies are equally likely. In this case, both a lower and an upper energy bound is required. In the fourth container, the gas is at negative temperature, at a large negative value. For negative temperatures, an upper energy bound is required, but not necessarily a lower. The gas in the container on the right is at negative temperature, at a very small negative value. Most atoms are close to the maximum energy. *Source*: Image courtesy of Simon Braun, LMU (Ludwig-Maxmilians-Universität) and MPQ (Max-Planck-Institut für Quantenoptik) Munich, Germany, and pertaining to the original work S. Braun *et al.*, *Science* (2013), *339*, 52. http://www.quantum-munich.de/media/negative-absolute-temperature/.

15.7 SELF-ASSESSMENT QUESTIONS

1. If removed from the freezer, the typical glass jar containing a frozen liquid will crack if immediately placed under very hot water. Conversely, heated glass or stoneware (e.g., pottery) will crack if exposed to cold water while still hot. On the other hand, hot Corningware® or Pyrex® bakeware will typically not crack if placed immediately under cold water. Likewise, hot metal may be cooled in water without risk of fracture. What thermal property is at play in these scenarios and why does each type of material behave as described?
2. What is DSC and what is it used for?
3. What is the heat capacity of a material? Why is it important?
4. What was the significance of the Einstein model of thermodynamics of crystals as it relates to heat capacity?

5. What information is available from thermogravimetric analysis and for what can it be used?
6. Discuss uses of materials with low thermal conductivity.
7. What do you expect is the thermal conductivity of diamond and why? After you answer these questions, look up the answer. If you answered incorrectly, explain how you erred in your assessment.
8. How does a change of temperature affect energy states of electrons?

REFERENCES

1. B.K. Greve, K.L. Martin, P.L. Lee, P.J. Chupas, K.W. Chapman & A.P. Wilkinson, *J. Am. Chem. Soc.* **2010**, *132*, 15496.
2. C.W. Li, X. Tang, J.A. Muñoz, J.B. Keith, S.J. Tracy, D.L. Abernathy & B. Fultz, *Phys. Rev. Letters* **2011**, *107*, 195504.
3. Z.F. Wróblewski & K. Olszewski, *Comptes Rendus* **1883**, *96*, 1140. The full title of this journal was then: *Comptes rendus hebdomadaires des séances de l'Académie des Sciences*; one notices that the location of this particular Academy of Sciences, namely Paris, was considered obvious and therefore not included in the title.
4. Lord Rayleigh & W. Ramsay, *Proc. Royal Soc. London* **1894**, *57*, 265.
5. K. Olszewski, *Philos. Trans. Royal Soc. London*, **1895**, *186*, 253.
6. B. Hartmann & M.A. Haque, *J. Appl. Phys.* **1985**, *58*, 2831.
7. W. Brostow, *Pure & Appl. Chem.* **2009**, *81*, 417.
8. M.H. Litt, *Trans. Soc. Rheol.* **1976**, *20*, 47.
9. K.P. Menard, Chapter 8, in *Performance of Plastics*, ed. W. Brostow, Hanser: Munich/ Cincinnati, OH **2000**.
10. W. Chonkaew, P. Dehkordi, K.P. Menard, W. Brostow, N. Menard & O. Gencel, *Mater. Res. Innovat.* **2013**, *17*, 263.
11. A. Einstein, *Ann. Physik* **1907**, *22*, 180.
12. E.A. Guggenheim & M.L. McGlashan, *Proc. Royal Soc. A* **1960**, *255*, 456.
13. L. Llorente Garcia, B. Darquié, C.D.J. Sinclair, E.A. Curtis, M. Tachikawa, J.J. Hudson & E.A. Hinds, *Phys. Rev. A* **2013**, *88*, 043406.
14. W. Brostow, M. Brozynski, T. Datashvili & O. Olea-Mejia, *Polymer Bull.* **2011**, *59*, 1671.
15. C.L. Choy, *Polymer*, **1977**, *18*, 984.
16. A. Henry & G. Chen, *Phys. Rev. Lett.* **2008**, *101*, 235502.
17. E.M. Purcell & R.V. Pound, *Phys. Rev. A* **1951**, *81*, 279.
18. Quantum Optics Group, *Quantum Many Body Systems Division*, http://www.quantum-munich.de/research/negative-absolute-temperature/, accessed February 16, **2016**.
19. S. Braun, J.P. Ronzheimer, M. Schreiber, S.S. Hodgman, T. Rom, I. Bloch & U. Schneider, *Science* **2013**, *339*, 52.

16

COLOR AND OPTICAL PROPERTIES

One of the first rational thoughts of early man must have been to understand the importance of the sun to provide light and heat and in essence, enable life on earth. This began a quest for understanding the nature of what would become known as radiation, leading to our present understanding of that most important elementary particle, the Photon.

—Raymond H. Pahler, 2013.

16.1 INTRODUCTION

Color in materials may be important purely for aesthetic reasons or it can be critical to a material's function. This notwithstanding, the optical properties of matter are not limited to features involving only visible light. Materials interact in various ways with light, including the ultraviolet (UV) and infrared (IR) regions of the electromagnetic spectrum, which we cannot see with the naked eye. Many of the connections between structure and optical properties (recall the MSE triangle, Figure 1.1) are well understood. For instance, the size of particles and quantity of matter interacting with light play a large role in determining the observed color. Interestingly, there are often direct correlations between the optical properties of a material and its electrical properties, as we shall see in this and the subsequent chapter.

16.2 ATOMIC ORIGINS OF COLOR

The property known as **color** arises from interactions between light waves and atoms, particularly with their electrons. Thus, the various aspects of color, including hue, saturation, brightness, and intensity, are all expressions of the structure of matter. **Light** is an electromagnetic (EM) wave.

Materials: Introduction and Applications, First Edition. Witold Brostow and Haley E. Hagg Lobland.

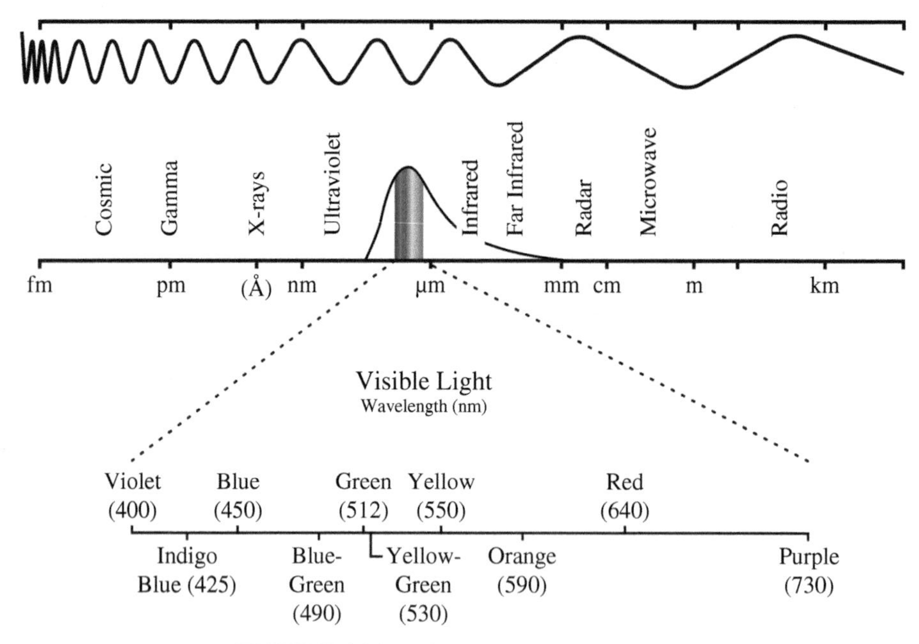

FIGURE 16.1 Electromagnetic spectrum.

Recall the electromagnetic spectrum, which for reference is shown here in Figure 16.1. Recall also the de Broglie Eq. (5.1) defining the wave-particle duality. Following the same approach, we find that light is associated with an energy E, and that is related to the wavelength λ and frequency ν of the light wave according to the expression

$$E = h\nu = \frac{hc}{\lambda} \tag{16.1}$$

where h is Planck's constant and c is the speed of light. The shorter the wavelength, the higher is the energy per photon of light.

Frequently, color is the result of a change in the state of an electron. The transition of an electron from one state to another involves energy absorption or emission. Thus, color in a material can arise from **absorption** or **emission** of light. Color is also observed due to **transmission** and **reflection**; see Figure 16.2. To understand this, consider what happens when an opaque material is exposed to **white light** (containing all the colors). If the color is derived primarily by absorption, then certain colors from the white light are absorbed by the material, and the light reflected and scattered will be lacking the absorbed color(s). In consequence, the color perceived is the **complementary color** to the one absorbed (see Figure 16.3 left). On the other hand, if the material is transparent, certain wavelengths of light may be transmitted through the material. The color seen by looking through the material will be complementary to whatever color or colors were absorbed by the material from the ambient light. If the surface behaves similarly to the bulk, the colors seen by transmission and reflection will be the same. However, a varied dependence on wavelength of the surface reflectivity may cause transmitted and reflected colors to differ; see the table on the r.h.s. of Figure 16.3.

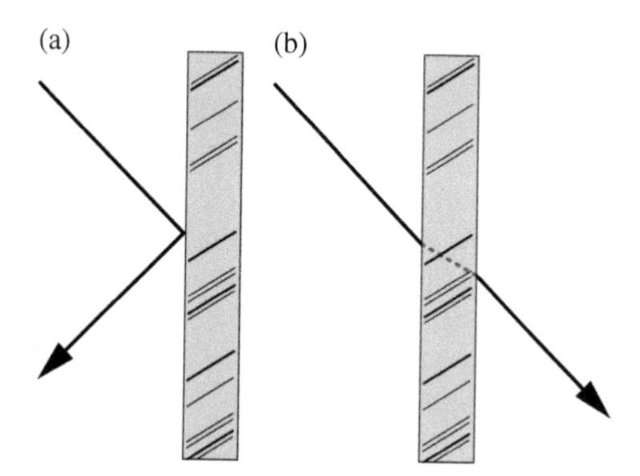

FIGURE 16.2 Some of the incident light is reflected (a), some is transmitted (b).

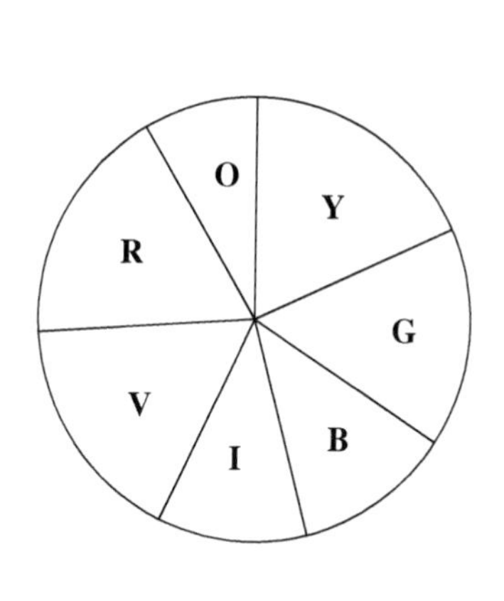

Absorbed light		
Wavelength (nm)	Corresponding color	Observed (transmitted) color
400	Violet	Yellow-Green
425	Indigo Blue	Yellow
450	Blue	Orange
490	Blue-Green	Red
512	Green	Purple
530	Yellow-Green	Violet
550	Yellow	Indigo Blue
590	Orange	Blue
640	Red	Blue-Green
730	Purple	Green

FIGURE 16.3 Left: Newton's color wheel, in which colors opposite to each other are complementary. The table at right lists the complementary colors with greater specification. (R, red; O, orange; Y, yellow; G, green; B, blue; I, indigo; V, violet.)

Now let us return to the subject of electronic transitions. From an energy standpoint, electronic states are further apart than vibrational states, which are likewise separated more than rotational states. A useful tool here is the **work function** which is the minimum energy required to move an electron from a solid to vacuum—meaning the electron is now located immediately outside the solid surface. The quantity of energy separating the electronic states often corresponds to the energy of visible light. When an electron that has been excited to higher energy state returns to its ground state, there can be an emission of visible light that is observable as color. A familiar example is the appearance of a yellow color when sodium is heated in a flame. Other examples in which light emission occurs from electronic transitions between high and low energy states: neon signs, mercury lamps, the Northern lights (*Aurora borealis*), and in fireworks from elements such as Sr (strontium) and Ba (barium).

"Black body" radiation is the well-known term referring to the idealized material that absorbs all wavelengths of light (without reflection or transmission) and perfectly emits all wavelengths of light. The spectrum of emitted light depends only on the temperature. The concept has significant ramifications in leading to the knowledge that at high temperatures, materials emit radiation with a characteristic range and intensity of wavelengths. Comparing the emission spectra (emittance as a function of wavelength) of gases to condensed matter (solids and liquids), gases have much sharper emission (and absorption) lines, while the emission spectra of condensed matter usually span a wide range of wavelengths.

Vibrational transitions usually correspond to the infrared region, which is lower in energy than the region of visible light. Note therefore the relationship between IR-testing and vibrational states, which have characteristic values dependent on the atoms involved.

In rare instances, however, vibrational transitions are responsible for observable color. In solid and liquid H_2O, the absorption of red-orange light can excite molecules to a higher vibrational state. This absorption leaves the complementary green-blue color visible in the reflected light. Because the absorption coefficient for this transition is low, a detectable green-blue color is only visible when there is a very long path length for the light to travel through. That is why we do not see color in a glass of ice water, but we do observe the blue-green color in a large body of water such as the ocean or through ice in a glacier; see a photograph of glacier in Figure 16.4.

Staying with the theme of absorption and emission, color in some crystalline solids may involve light absorption by ions within the solid. We have seen that valence electrons

FIGURE 16.4 Antarctica: The blue ice covering Lake Fryxell, in the Transantarctic Mountains, comes from glacial meltwater from the Canada Glacier and other smaller glaciers. The freshwater stays on top of the lake and freezes, sealing in briny water below. *Source*: Joe Mastroianni, National Science Foundation, https://commons.wikimedia.org/wiki/File:Fryxellsee_Opt.jpg. Public Domain.

in gaseous phase atoms can be thermally excited, leading to visible emissions; by contrast the lower ground state energy of atoms chemically bonded (i.e., in the solid state) necessitates a greater energy to promote electrons to excited states. In this situation, excitation is associated with the UV region. However, for transition metals with incomplete d-shells, the excitation energy for valence electrons can be in the visible spectrum. Therefore, compounds containing impurity ions of these metals can likewise produce their beautiful colors. (Refer to Figure 6.1: transition metals are identified there in the periodic table.) The electric field in a crystal, i.e. a **crystal field**, can alter the energy levels of such an ion located in the crystal and subject to the crystal field (in comparison to a free ion of the same element). Thus, the crystal field can give rise to observable colors by modulating the electronic transitions and light absorption of impurity ions. At the same time, we can say these crystal defects modify optical, as well as other, properties. We noted—in Chapter 6—that the concentrations of impurity atoms causing an appearance of a color or a change of color are quite low. We also saw in Chapter 6 how defects alter mechanical properties; in Chapter 17 we shall see how the very function of photovoltaic solar cells depends upon such defects.

Scientists also recognize extra charge centers (electrons or holes), known as **color centers** or **F-centers**, as sources of color. The phenomenon involved is somewhat similar to that in crystal field colors, but in this case the electrons are unattached to a particular ion and typically merely associated with a defect. The excitation energy of these electrons falls within the range of visible light, yielding observable color. Color centers are not uncommon; holes resulting from the presence of Fe^{3+} in SiO_2 are responsible for the violet color of **amethyst**. In consequence of the hole, an alteration of energy levels causes absorption of yellow light, yielding the complementary violet color characteristic of amethyst.

The appearance of color in covalently bonded molecules (e.g., organic carbon-containing compounds) offers yet another scenario for the origination of color. Charges and electronic states are important in gases, crystalline solids, and others. Normally, it requires light in the UV range to excite electrons that are paired in a *chemical bond*. As a consequence, such materials are not colored. However, a particular phenomenon that arises in greatly conjugated molecular systems (e.g., alternating single and double carbon-carbon bonds) is the delocalization of electrons over several chemical bonds. In such a system electrons can be excited from the highest occupied molecular orbital (HOMO) to the lowest unoccupied molecular orbital (LUMO). For some materials, the energy of the HOMO-LUMO transition is in the visible range of the spectrum, and the absorption of visible light causes the material to be colored. This **charge delocalization** to molecular orbitals is the operating mechanism responsible for the color in the thousands of known organic dye molecules. Charge delocalization as a source of color occurs in other types of materials as well. The color in some solid pigments and gemstones comes from charge delocalization and charge transfer. The technology associated with carbonless copy paper is an interesting application of principles associated with color and charge delocalization. An article by Mary Ann White describes the technology in more detail [1].

16.3 COLOR AND ENERGY DIAGRAMS

The color phenomena discussed in the previous Section were applicable primarily to insulating materials and involved mainly single atom electron transitions or small groups of atoms—as in molecules and crystal field effects. With metals and semiconductors, both color and luster are noted; and the appearance of color involves large numbers of atoms.

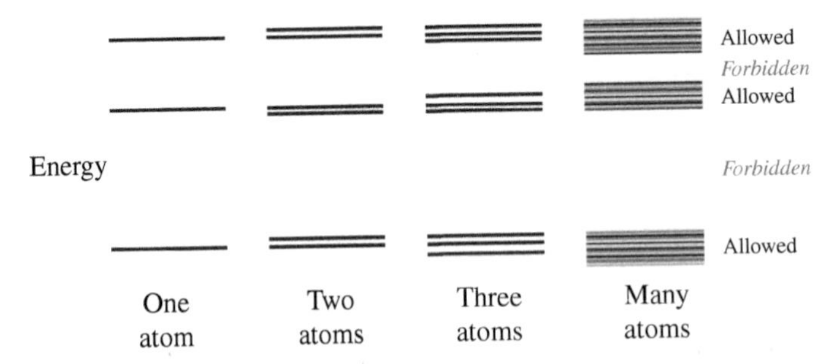

FIGURE 16.5 Increased number of lines represent density of states.

In a pure metal, electrons have the maximum possible delocalization. The free conduction electrons are practically equivalent, although owing to the **exclusion principle** of **Wolfgang Pauli**, a maximum of two electrons can have exactly the same energy but they have to have different spins. The outcome is that in a metal there is a virtual continuum of energy levels that we commonly describe as a band; this is shown schematically in Figure 16.5. The probability of a state with energy E being occupied is based on the **Fermi energy** E_F and the Fermi-Dirac distribution function, Eq. (16.2). The probability that a state with energy E will be occupied by an electron is $p(E)$ defined as

$$p(E) = \frac{1}{e^{(E-E_F/kT)} + 1} \tag{16.2}$$

The Fermi energy is the energy at which $p(E) = ½$; k is the Boltzmann constant, and T is the temperature. This is explained more thoroughly by the diagram in Figures 16.6 and 16.7. At $T = 0$ K, the energy levels are filled up to E_F; at higher temperatures, some energy levels are occupied above the Fermi energy, as indicated in these two Figures. Owing to the near continuum of energy levels above E_F, metals are good light absorbers, able to absorb many wavelengths of light, including visible light. Metals are also reflective, rapidly re-radiating absorbed light; thus metals are not black, from absorbing all light, but in fact are shiny due to the emission of light from their surfaces. The different colors of metals arise from their having different numbers of energy states available above the Fermi level. Interestingly, color depends not only on electron energy levels but also on surface smoothness: the luster and hue of metals (and wood and other materials too) can be altered by surface polishing.

In a metal, the electrons that are above the Fermi level are considered conducting electrons. They participate in the thermal and electrical conductivity of the material. We can better picture the electron distribution in metals by an energy diagram (Figure 16.8) that is essentially the probability distribution of Figure 16.6 turned on its side. Now we can see in Figure 16.8 how energy states above and below the Fermi energy are occupied by electrons.

In contrast to metals, semiconductors are not good conductors of electricity. This can be explained by an energy gap E_g, or **band gap**, between the valence and conduction bands (energy levels). There are no energy levels in the band gap (see Figure 16.8), much as there are no available states between molecular bonding and anti-bonding orbitals. The band gap increases with increasing bond interaction and with shorter, stronger bonds. Thus for the following elements, the band gap increases in the order Sn < Ge < Si < C. For a given element, increased pressure or decreased temperature can increase the width of the band gap.

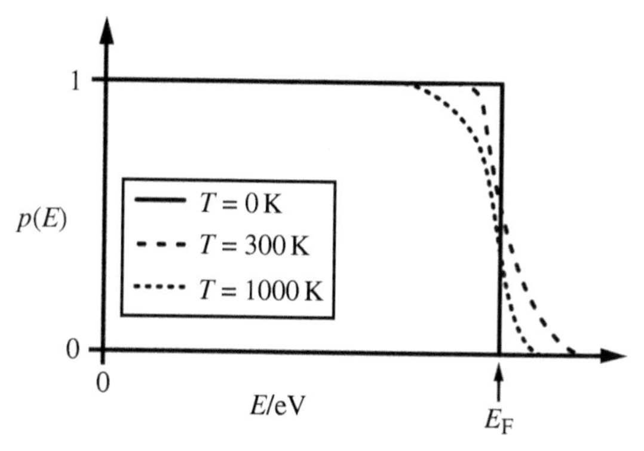

FIGURE 16.6 The probability of an electron in a metal occupying a certain energy level is $p(E)$, defined by Eq. (16.2). At $T=0\,K$, the Fermi energy, denoted E_F, separates occupied from unoccupied levels. At higher temperatures, the function is no longer rectangular.

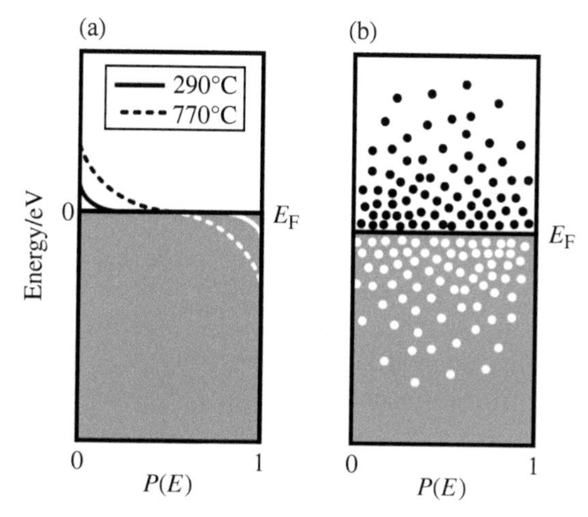

FIGURE 16.7 Alternate presentation of the Fermi-Dirac distribution of electronic energy states. Curves in (a) indicate the changes in occupied levels as temperature increases. The image in (b) pictorially shows, for $T>0\,K$, as dots the conducting electrons above the Fermi energy E_F and as open circles the holes (electron vacancies).

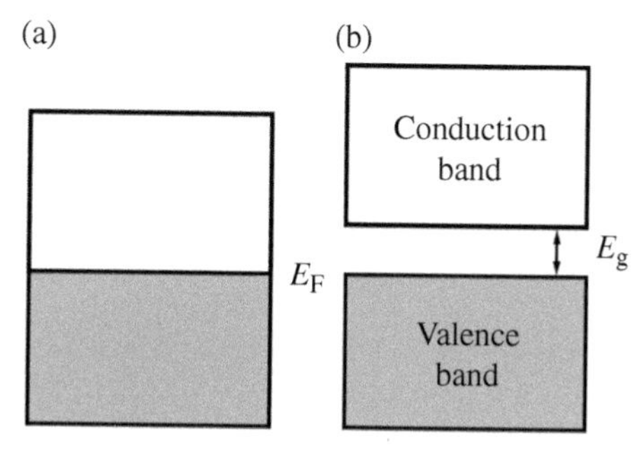

FIGURE 16.8 The energy gap E_g is illustrated by a modified schematic of the electron probability distribution shown in Figure 16.6. Energy bands are shown for (a) a metal and (b) a semiconductor. E_F is the Fermi energy.

So then, how do the electrons in a semiconductor get promoted to the conduction band? Thermal energy can achieve this, but generally only at very high temperatures. Visible light, however, can provide sufficient energy to promote electrons from the valence band to the conducting band with resultant color imparted to the semiconductor material. The exact color depends on the value of E_g: pure semiconductors are colorless if E_g is very large and black if E_g is small. The presence of impurities (often added on purpose by doping) introduces color. Donor impurities donate electrons to the conduction band resulting in **n-type semiconductors** while acceptor impurities produce electron vacancies in the valence band resulting in **p-type semiconductors**. In consequence to the presence of impurities, energy levels can be added between the valence and conduction bands. Obviously, the added energy levels will impact the color of the material. A familiar example is diamond, which is yellow when doped with nitrogen and blue when doped with boron. Some doped semiconductors can act as **phosphors**, that is solid materials that emit light. Still other features can be obtained from placing a p-type semiconductor next to an n-type semiconductor to form a **p,n-junction** (discussed at greater length in Chapter 17). Of note here is that the p,n-junction is key in the operation of **light emitting diodes** (familiarly known as **LEDs**). While LEDs are largely made from inorganic semiconductors, conjugated polymers can be used too; see Section 17.6.

SOLID-STATE LIGHTING

LEDs—light emitting diodes—are a common source of light in electronic devices; they are also used in traffic lights, decorative lighting, and even general purpose light bulbs. In LEDs light can be produced electronically—rather than by heating a thin metal filament (incandescent bulbs) or by fluorescence.

Diodes are explained further in Chapter 17, but basically LEDs are constructed from p-type and n-type semiconductors. Recombination of electrons and holes in the *active region* of the device causes light to be emitted. This solid state process is termed **electroluminescence**. By adjusting the materials involved, specified colors, each with its own hue, can be emitted. Furthermore, one can obtain more control over the spectrum of light emitted by putting differently colored LEDs together in an array and selectively activating them.

Light output in a LED is linearly proportional to the current; light output can therefore be modulated to send an undistorted signal through a fiber optic cable. An LED is also a directional light source, with its maximum power emission in the direction perpendicular to the emitting surface. Plastic lenses are sometimes used to diffuse the light for a wider angle of visibility.

The semiconducting materials in LEDs are mainly of a mixture of Group III and Group V elements with dopants such as tellurium and magnesium for adjusting the wavelength of colored light emitted [2]. Bright white light has been a challenge with LEDs.

One way to generate white light is to combine the output of red, green, and blue LEDs. However, this is difficult to do efficiently with good uniformity and control. A second route is to have an LED photon excite a phosphor. This may involve yellow, blue, red, green, and UV phosphors, packaged in various ways. The process, which is similar to that in fluorescent tube lighting, is simpler than mixing three LED colors but quite inefficient. Ultimately, applications will determine which route is used.

Organic LE devices (**OLEDs**) have been developed and are successfully used in many low-power applications. It is important to distinguish OLEDs from LEDs. The former are area sources, not blinding, and suitable for large-area applications. Traditional LEDs, on the other hand, are point sources, blindingly bright, and suitable for imaging and optics applications. The luminance of OLEDs is about four orders of magnitude lower than for LEDs. However, OLEDs can be fabricated by low cost reel-to-reel manufacturing methods while production of LEDs by epitaxial growth is expensive. A benefit from both types of solid state lighting is reduced energy consumption during operation compared to traditional light sources.

LASER DIODES

Laser diodes, although seemingly similar to LEDs, are a different beast intended for different purposes [2]. Laser diodes are comprised of semiconductors sandwiched between mirrors, forming a resonator cavity. Electrical current induces emission of photons. The photons, of the same wavelength, bounce between the mirrors and continuously increase in intensity. Light that escapes forms a narrow column of pure, bright light at a single wavelength, what is known as **coherent** photons or **coherent light**.

By contrast, LEDs produce scattered light, not like the well-defined beam of laser. The difference is principally owing to the mirrors, which form the resonator cavity. The entirety of photons produced by LEDs may not be at one exact wavelength, but they are close enough for the human eye to perceive it as one color.

Nonlinear optics (NLO) is the branch of optics that describes the behavior of light in nonlinear media, that is, media in which the dielectric polarization P responds nonlinearly to the electric field E of the light. This nonlinearity is typically only observed at very high light intensities (values of the electric field comparable to interatomic electric fields, typically 108 V/m) such as those provided by pulsed lasers.

So-called **photonic materials** are used for light detection, generation, and manipulation. These activities can be performed via light (that is photon) emission, transmission, modulation, switching, amplification, and sensing (detection). In principle we are dealing with the entire light spectrum, but most current applications are in the visible or infra-red region. Photonic materials are used primarily in light detection (obviously) and also in information processing, telecommunications, holography, laser material processing, visual arts, robotics, agriculture, and medicine. In this last area there are a variety of applications: vision correction, health monitoring, endoscopy (an endoscope is used to examine an interior of a cavity or a hollow organ inside the body), and surgery.

Photonic crystals are structures, periodic at the nanoscale, that affect the motions of photons similarly to the way ions in ionic lattices affect the motions of electrons. They consist of regions of low and high dielectric constants. Actually the word "crystals" in the name is misleading since these materials do not have crystal unit cells repeated throughout the material. Photonic materials are created in one, two, or three dimensions. 1D photonic crystals can be made by layer deposition, 2D by photolithography, and 3D by stacking 2D layers on top of each other or else by laser writing.

We talked about opals in Section 5.2; see Figure 5.2. Opals are in fact 3D photonic crystals made by Nature. Since photonic crystals are not crystalline, according to a more accurate terminology, opals are mineraloids (see Section 5.2). The high and low dielectric constant regions are provided by the spherical colloidal silica particles and by voids, respectively. The result is diffraction of light entering the mineraloid, varying with location in the material and with the viewing direction. Opals can display many colors, from colorless and white through yellow, orange, red, purple, blue, green, brown, and gray all the way to black.

The impression that only inanimate objects can diffract light is unfounded. An aqueous creature called a **sea mouse** (*Aphrodita aculeata*) has colorful spines, the result of diffraction—and an entity in existence long before humans developed photonic engineering. The spines warn potential predators since they appear to have a deep red luster. In the animal world, red or orange color is often used as a signal: "you can eat me since I do not have strong teeth or claws; but then you will regret it since I am toxic". However, light shining on the spines perpendicularly provides blue or green color instead of red.

16.4 LIGHT AND BULK MATTER

We have progressed in our discussion of color from phenomenon involving one or a few atoms to many atoms in metal and semiconductor materials. Now we consider interactions of light with bulk matter, involving the order of Avogadro's number of atoms. In this Section we shall touch on some key properties of light and color involving bulk matter:

- Refraction
- Interference
- Light Scattering
- Diffraction Gratings

Refraction refers to the change in velocity of light passing from one medium to another. Simply put, light rays take the path of least time. Materials are compared by the **refractive index** n, which is the ratio of the speed of light in a vacuum (ν) to the speed of light in the selected medium (ν'):

$$n = \frac{\nu}{\nu'} \tag{16.3}$$

Importantly, n depends on temperature and on the wavelength of light; higher energy light is more refracted. Thus, white light is dispersed by a prism. Many other common optical phenomena are related to refraction. Mirages are one example.

Recall that light is an electromagnetic wave, and waves have with them an associated phase. **Interference** has to do with the interaction of waves. If two waves come together exactly in phase, there is constructive interference; if the two waves are exactly out of phase, there is destructive interference. As it relates to color in materials, interference especially comes into play in thin films. Familiar instances include the colors seen in a bubble and in soap films and oil films on water. Mary Anne White provides a description of the light interference associated with soap films [3]. Hologram images also utilize the effects of light interference to achieve their 3-dimensional appearance.

Light scattering occurs when light waves are bent (or scattered) as they encounter the edge of an opaque object. The effect is minor with large objects, but significant with small ones. The largest scattering effect occurs as the object size nears the wavelength of the incident light. Particles in Earth's atmosphere scatter sunlight, resulting in a blue sky (because the shorter wavelength light is more scattered). At sunset, a longer path to travel means that the blue light is scattered away and what remains for us to see is mainly scattered red light. It may come as a surprise that **chameleons** and **Hercules beetles** achieve their color changes through light scattering; more or less light is scattered in response to alterations of the creatures' surface skin layers.

Diffraction gratings are more an object than a property. Having one or more slits through which light is allowed to pass, they produce colors based on the combined effects of interference and light scattering. The phenomena associated with diffraction gratings are treated at length in many resources, so we will not elaborate further here. It is common to think of a diffraction grating only as a 2-dimensional sheet with small slits, but it is important to realize that a regular lattice array—such as in a crystalline material—can operate as a diffraction grating, reflecting light constructively and destructively to produce colors, which may vary with the viewing angle. The colors of liquid crystals arise through a similar mechanism; the molecular alignment and regularity of stacking in liquid crystals allows them to function like a diffraction grating. The spacing may change with temperature, leading to thermochromism, which was explained previously in Chapter 12.

16.5 OPTICAL PROPERTIES AND TESTING METHODS

The optical properties of materials entail more than just color and the interactions of matter with visible light. Characteristic information obtained from optical testing methods includes:

- Morphology, size (e.g., grain growth by reflective optical microscopy)
- Transparency, opacity
- Color: reflected and transmitted
- Refractive indices
- Polarization, dichroism, pleochroism
- Crystal systems
- Birefringence
- Luminescence, fluorescence (UV, V, IR)
- Melting point, polymorphism, eutectics (using a hot stage)
- Chemical composition (e.g., from infrared spectroscopy)

The lenses used in optical microscopes and in many other testing apparatuses have their operation and fabrication based on the knowledge outlined in the preceding sections of this Chapter. The aspects of magnification, contrast, and illumination are left for a course on instrumentation and characterization techniques. There is however a surprising extent of information that can be extracted even from simple optical testing equipment. For example, a simple spherical spectrophotometer, but having the capacity to detect transmission of 1,340 colors, can be used to assess minor defects, structural changes, or chemical decomposition in a material (such as yellowing of polypropylene after gamma irradiation).

We have used the word spectroscopy several times without really defining it. Generally speaking, **spectroscopy** is the study of the interaction between radiation and matter as a function of wavelength or frequency. Spectroscopic techniques used frequently for materials analysis include Fourier transform infrared (FTIR) spectroscopy (utilizing the Michelson interferometer), Raman spectroscopy, nuclear magnetic resonance (NMR) spectroscopy, and Auger spectroscopy. Light scattering is associated also with FTIR and Raman spectroscopy. Bragg's law describes the fundamental relationship between the wavelength of incident light and diffraction angle. X-ray diffraction techniques, for example, employ the scattering of x-rays to identify crystal structures and determine the degree of crystallinity in polymers. Somewhat related are electron and neutron scattering techniques, which also provide topographical and structural information. Practitioners of neutron scattering claim their results are much more accurate than those from the x-ray diffractometry.

Light polarization is mentioned in the list at the beginning of this Section (and also in Chapter 12); let us examine the topic in more detail. Light, an electromagnetic wave, travels in a plane. We have **plane-polarized light** when all the light from a source is in the same plane. A **polarizer** is a material that acts like a filter to permit only light oscillating in one orientation to pass through it. This feature of light is utilized in Polarized Light Microscopy, which allows one to observe numerous crystallographic features of small crystals. It has been said that "It is much easier to demonstrate polarized light than to clearly describe what it is" [4]. Therefore, consider the following illustrative examples. Certain liquid crystalline helicene molecules have been shown to rotate the plane of plane polarized light [5]. The glare outdoors on a sunny day arises from light partially polarized in a horizontal plane. Sunglasses with a vertical polarization axis block out the glare. Polaroid camera filters block out polarized light from the sky to increase contrast between clouds and sky. Two polarizers with polarization axes aligned at right angles to each other will completely block the passage of light.

Birefringence is an important related property. We are talking here about optically anisotropic materials, also called **birefringent** or **birefractive**. Recall that liquid crystals—discussed in Chapter 12—are by nature anisotropic. Owing to that anisotropy, liquid crystals are birefringent, that is they exhibit double refraction (having two indices of refraction). In liquid crystals, light that is polarized parallel to the director (refer to Section 12.2) has a different refractive index than light polarized perpendicular to the director. When light enters a birefringent material (e.g., nematic liquid crystals or helicenes mentioned above [5]), it is broken up into fast (ordinary ray) and slow (extraordinary ray) components. The two components travel at different velocities, thus the waves get out of phase; see Figure 16.9. When the waves recombine upon exiting the birefringent material, the polarization state will have changed because of the phase difference. Therefore, the length of the sample is also an important parameter because phase shift accumulates as long as the light is propagated through the material. Other birefringent materials are tourmaline, calcite, quartz, and rutile (TiO_2). The birefringence is often quantified as the maximum difference between refractive indices exhibited by the material. Crystals with asymmetric crystal structures are often birefringent, as are some polymers under mechanical stress.

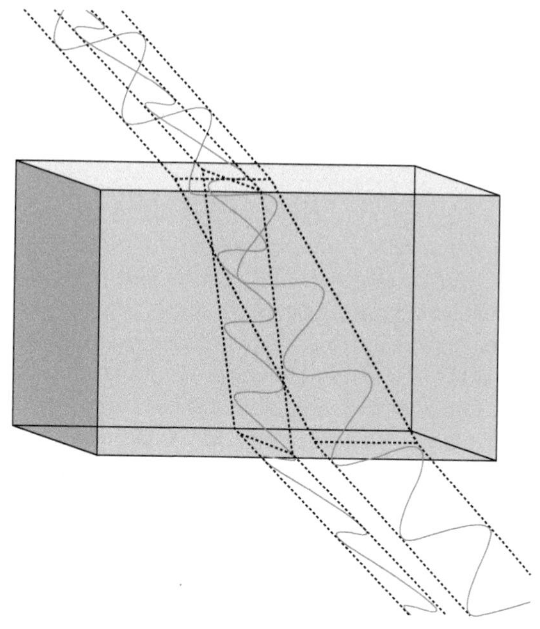

FIGURE 16.9 Birefringence (explanation in text). Shown here: two rays with parallel directions but perpendicular polarization passing through a birefringent material. *Source*: Mikael Häggström, https://commons.wikimedia.org/wiki/File:Positively_birefringent_material.svg. Public Domain.

Now consider a liquid crystal sample placed between crossed polarizers whose transmission axes are aligned at some angle between the fast and slow directions of the material. Owing to birefringence of the LC, incoming polarized light becomes elliptically polarized; a component of this light can pass through the second polarizer, making a bright region. On the other hand, if the transmission axis of the first polarizer is parallel to the ordinary or extraordinary directions, then the light is not broken into components and there is no change in polarization. In consequence, no component is transmitted and the region appears dark. Because birefringence in liquid crystals is typically not uniform, one observes light and dark regions as seen in Figure 12.1.

Thus, in birefringent materials a ray of light is split by polarization into two rays taking slightly different paths; see again Figure 16.7. This effect was first seen and described already in 1669 by the Danish scientist **Rasmus Bartholin** in **calcite**, a crystal having one of the strongest birefringences. Explanation in terms of polarization was provided only in the early 19th century by **Augustin-Jean Fresnel**, a French engineer and physicist.

We have discussed light polarization in relation to liquid crystal displays in Section 12.2. The phenomenon of birefringence also has several applications. **Light modulators** modulate the intensity of light through electrically induced birefringence of polarized light—followed by a polarizer. **Wave plates** are thin birefringent sheets used in certain optical equipment for modification of the polarization state of light passing through it. Birefringence also has applications in medicine, for instance, for location of anomalies such as gouty joints or monitoring of glaucoma.

16.6 LASERS

Lasers have been mentioned in Section 15.6 when we discussed negative temperatures. **Laser** = *Light Amplification by Stimulated Emission of Radiation*. Fundamentally, operation of lasers is based on the stability criterion (3.17): energy of a material *decreases* in a natural process. Electrons can be excited to higher energy states either by an outside light source or by an electrical field which supplies energy for atoms to absorb and be transformed into their excited states. Now referring back to Figure 16.7, let us once again personify inanimate particles; electrons "do not like" the situation in Figure 16.7b. The stability criterion (3.17) tells the electrons to decrease their energy; they do so rapidly, getting rid of excess energy fast—in the form of photons—and in large quantities. If there is $T<0$ inside of a laser, we know from the discussion in Section 15.6 that the material is even more willing to get into a $T>0$ state by rapidly radiating the excess energy out.

We know that the stability criteria discussed in Chapter 3 are general; they apply to solids, liquids, gases, and plasma. Therefore, laser materials can be in any state. In contrast to other sources of light, lasers emit light coherently, allowing a laser to be focused on a small spot, enabling applications such as laser cutting and lithography. We already talked above in the second track about laser diodes sending out coherent photons. Spatial coherence also allows a laser beam to stay narrow over long distances (collimation), enabling applications such as laser pointers. Lasers can have high coherence—what allows them to provide a very narrow spectrum, observable as a single color of light. Temporal coherence—in time—can be used to produce **pulses of laser light**, quite short ones (e.g., a femtosecond).

Lasers have still further important applications in common consumer devices: DVD players, laser printers, and also barcode scanners. In medicine, lasers are used for laser surgery and various skin treatments, in industry for cutting and welding materials, and in laboratory research as well. Furthermore, lasers are used in military and law enforcement devices for marking targets and measuring range and speed. Laser lighting displays use laser light as an entertainment medium.

16.7 ELECTRO-OPTICAL EFFECTS AND LUMINESCENCE

Electro-optical modulators (EOMs) are optical devices in which a signal-controlled element exhibiting the electro-optical effect is used to modulate a beam of light. The modulation may be imposed on the phase, frequency, amplitude, or else polarization of the beam. Modulation bandwidths extending into the gigahertz range are possible with the use of laser-controlled modulators—one more application of lasers.

The **electro-optical effect** is the change in the refractive index of a material resulting from the application of a DC or low-frequency electric field. This is caused by forces that distort the position, orientation, or shape of the molecules constituting the material. Generally, a nonlinear optical material with an incident static or low frequency optical field will see a modulation of its refractive index.

The simplest kind of EOM consists of a crystal, such as lithium niobate, whose refractive index is a function of the strength of the local electric field. When lithium niobate is exposed to an electric field, light will travel more slowly through it, since the refractive index varies with the electric field. Note that the phase of light leaving the crystal is directly proportional to the length of time it takes that light to pass through it. Therefore, the phase of a laser light exiting an EOM can be controlled by modulating the electric field in the crystal. The electric field can be created by placing a parallel plate capacitor across the crystal. Since the field inside a parallel plate capacitor depends linearly on the potential, the index of refraction depends linearly on the field, and the phase depends linearly on the index of refraction; thus, the phase modulation must depend linearly on the potential applied to the EOM. Therefore, EOMs can be used as fast optical switches. Organic polymers have very fast response rates, so they are often used in such applications.

Light-emitting diodes (LEDs), discussed in a second-track feature earlier in this Chapter, are an example of **luminescent** materials, that is, they emit light not as a result of heat. Typical LEDs are composed of semiconductor materials, however there are now also organic light-emitting diodes (OLEDs) and polymer light-emitting diodes (PLEDs) that have organic constituents providing their functionality. Importantly, there are a variety of routes to modify the luminescent properties of organic and polymeric LEDs via structure control of the materials and device configurations [6–10]. Polymer light-emitting diodes can be made from poly(*p*-phenylenevinylene) (PPV) layer-by-layer with dodecylbenzene sulfonic acid (DBS). Creation of a poly(*o*-methoxyaniline) cushion layer hinders any changes in molecular conformation of PPV and DBS [10]. Therefore, in line with our discussion of defects in Section 16.2, the presence of the cushion results in a decrease in the number of structural defects.

ORGANIC MATERIALS THAT CHANGE COLOR

Light is manipulated by materials in a variety of ways through: transmission, amplification, detection, modulation, and integration. Light is also a driving force in the function or design of numerous products and applications such as: photovoltaics, lenses, mirrors, LEDs, lasers, and heat management. The refractive index of a material is a critical determining factor in how it interacts with light. Generally inorganic materials have much higher indices of refraction than organic ones; however novel hybrid and nanocomposite materials are being developed with tunable properties, which allow achievement of refractive indices intermediate between the extremes. There are limitations with the design of nanocomposites, but hybrid materials promise higher capacity for modulation of photo-related properties.

Chromic responses of materials are utilized in a variety of applications; several of these were mentioned in Chapter 12. Functional structures that change color are desirable for use in products ranging from billboards to fashion accessories to security devices. Light, temperature, and pH are common stimuli. Hybrids prepared by Stingelin and her colleagues [11] from titanium oxide hydrates in poly(vinyl alcohol) (PVA) display chromic features in response to light (photochromism) and temperature (thermochromism). Their use of titanium oxide hydrates, in contrast to more common systems based on (nano-) crystalline TiO_2, results in finer variability of chromic effects.

Note several interesting features of the titanium oxide hydrate/PVA hybrids. The color change to blue upon irradiation with UV-light is reversible upon exposure to humid, oxygen-rich environment. On the other hand, exposure to heat leads to a permanent red coloration of the resultant structure. These characteristics along with good solution-processability and high transparency yield an inorganic/organic hybrid system characterized by "a desirable and highly tunable property set that can be readily manipulated by adjusting the hybrids' composition and selection of the post-treatment protocols" [11].

We can consult [6–10] to find out how high precision in creation of small scale structures is needed to achieve desired optical effects. Another such example is nanocrystalline films of indium tin oxide (often called ITO) combined with 30–50 nm ZnO crystallites. Kozytskiy and his colleagues in Kyiv and in Chemnitz [12] sensitized these films to visible light by $Cd_xZn_{1-x}S$ nanocrystals deposited using the method of successive ionic layer adsorption and reaction (SILAR).

Since world reserves of indium are limited, other materials for LED applications are being developed. Thus, ZnO is used as the active material in LEDs, in OLEDs, and also in solar cells. A study by Du, Shepherd, and their coworkers [13] shows how the properties of zinc oxide can be modified by argon sputter cleaning and oxygen plasma treatment. The effects are evaluated in terms of values of the work function; the plasma makes the surface electronegative and the work function increases. Combinations of experimental and theoretical work are rare, while here density functional theory (DFT) calculations were confronted with experimental ones; in agreement with the experiments, DFT predicts an in increase in the work function as a consequence of plasma cleaning.

Additionally, LEDs can serve as important tools for analyzing materials that possess certain optical and electronic properties. For instance, the molecular solid perylenetetracarboxylic dianhydride (PTCDA) tends to form polycrystalline films wherein the molecular planes are oriented in fixed directions with respect to the substrate; this behavior gives rise to interesting electronic transport and optical properties [13]. One observes in PTCDA anisotropy in the transport direction of the charge carriers; namely, electrons are transported parallel to the molecular planes while holes are transported in the perpendicular direction [14]. We have discussed anisotropy in earlier chapters, especially in regard to its relationship with mechanical properties. Now we have seen that there may be anisotropy also in optical and electrical properties, with more of the latter discussed in Chapter 17.

Luminescence, as already defined, is a form of cold body radiation. As such, the stimulus for light emission can vary. Light emitting diodes provided an example of electroluminescence, wherein light was emitted as a result of an electric current. However, there are many types of luminescence, including:

- Photoluminescence—resulting from absorption of photons
  - Fluorescence, phosphorescence; in both cases we have re-emission of absorbed radiation. However, fluorescence is a rapid and thus short-lived process. By contrast, phosphorescence can last several hours after radiation absorption. Organic light-emitting diodes exhibiting electrophosphorescence have been made such that variation in the concentration of the platinum dopant provides as the emitted color blue-green, or yellow, or else orange-red [15].
- Mechanoluminescence—resulting from a mechanical action on a solid
  - Triboluminescence (scratching or rubbing), fractoluminescence (crystalline bond fracture), piezoluminescence (pressure), sonoluminescence (sound wave excitation in a liquid).
- Electroluminescence—resulting from an electric current passed through a substance
  - Cathodoluminescence.
- Chemiluminescence—resulting from a chemical reaction
  - Bioluminescence (biochemical reaction in a living organism), electrochemiluminescence (electrochemical reaction).
- Radioluminescence—resulting from bombardment by ionizing radiation.

Availability of techniques for deposition of thin films is being used to advantage in creation of photoluminescence materials. Scharf, Shepherd, and collaborators [16] used pulsed laser deposition in oxygen to create 150 nm thick films of zinc oxide. It turned out that photoluminescence observed is strongly dependent on Zn interstitials. Recall our discussion in the beginning of Chapter 4 about structure-sensitive properties, that is, about properties that can be largely affected by components present at very low concentration.

16.8 PHOTOINDUCTION

Light photons, as we have seen in this Chapter, are associated with more than just the color of matter. Photons can induce a variety of changes in a range of materials; research on such photo-induced changes is abundant. There are reports on photoinduced ordering in polymers, ceramics, liquid crystals, and others. Such ordering can take on a variety of forms. There are photoinduced phase transitions that lead to ferromagnetic order in magnetic semiconductor heterostructures, and there may be photo-induced charge ordering in a variety of material types. Photoinduced structural and spatial ordering are also achievable in a range of materials. Photoinduced chemical reactions can be exploited for various purposes, for instance, to create long-range order in polymeric blends. Photoinduced polymerization reactions were mentioned in Chapter 9.

Chalcogenide materials are especially responsive to light. Popescu takes more than three pages to list all the photo induced effects possible in chalcogenide materials [17]. A few of these are: photo-darkening effect, photo-plastic effect, photo-induced fluidity effect, optome-chanical effect, photo-induced amorphization effect, photo-elastic birefringence effect. Some applications of these photo-induced effects (not limited only to chalcogenide materials) are in integrated optics (volume expansion), CD-RW drives (amorphization/devitrification), micro-electromechanical systems (MEMS) actuators (plasticity), lithography (chemical properties).

Let us now mention a relatively simple example of **photomechanics**. A film of a liquid crystal network containing an azobenzene chromophore can be repeatedly and precisely *bent along any chosen direction* by using linearly polarized light [17]. A photoselective volume contraction takes place. It is claimed that such a process may be useful in the development of high-speed actuators for microscale or nanoscale applications, such as for microrobots in medicine or for optical microtweezers [18].

Let us note also that the human eye is a wonderful instrument to observe colors but not an infallible one. In Figures 16.10 and 16.11 are examples of optical illusions.

FIGURE 16.10 An optical illusion: the blue and green spirals are actually the same color.

FIGURE 16.11 An optical illusion from effects of lighting and reflection. The reflection of light on the lower tile makes it seem lighter than the upper tile. However, if one covers the region where the dark gray and white tiles come together, it becomes obvious that both tiles are the same shade of gray.

In 1922 (with the date of 1921) Albert Einstein was awarded the Nobel Prize in Physics for his theory of the **photoelectric effect** (electrons are emitted from solids, liquids, or gases when they absorb energy from light; such electrons are called **photoelectrons**). He was nominated for this Prize for several preceding years, year after year, but during those years, the Prize always went to somebody with lesser achievements. The theory of relativity that Einstein created in 1905 caused strong protests; the very idea that the results of an observation depend on the observer was considered absurd. In fact the theory of relativity is clearly explained to all including laymen in a book by Einstein and Infeld [19]. A simple example from [19] should suffice. Consider a moving train with glass walls. In one of the train cars, a girl is playing with a ball, bouncing it off the floor. In the coordinate system of the girl, the ball is moving up and down. But there is an observer standing in a meadow outside of the train. To him the ball follows an approximately sinusoidal trajectory. The Einstein harmonic oscillator theory of crystals (see Section 15.3) was also too difficult for the Nobel Committee to absorb. He received the Prize for the photoelectric effect because this was something that the Committee was able to understand and agree with.

16.9 INVISIBILITY

A trick used in science fiction films is invisibility; a person or persons disappear, gradually or abruptly. Humans are always looking for what superheroes and magicians in those films or in fairy tales have already been using for a long time. Several magic cloaks providing invisibility are more or less well known: the Laurin cloak in the *Thidreksaga*, the elves cloak Frodo received from Galadriel in the *Lord of the Rings*, and the cloak of Harry Potter in *Harry Potter and the Goblet of Fire*. Abrupt disappearance in movies is in fact easy to create; just tell the people to go out of the visible area of the camera and film the remaining background. Is there any connection to reality? Actually, yes.

Serious scientific work in the area started in the late 1990s [20], first for military applications. The inspiration comes from nature. We already discussed in Sections 16.3 and 16.4 the "talents" of sea mice, chameleons, and Hercules beetles. Cephalopods, which include the squid *Doryteuthis (Loligo) pealeii* and the cuttlefish *Sepia officinalis*, can imitate their environment by changing their skin in terms of pattern, color, shape, and texture. The familiar color tone called sepia is of course is related to the scientific name of the cuttlefish. Roger Hanlon and his coworkers [21] at the Woods Hole Marine Biological Laboratory have studied extensively the behavior of cephalopods. They have found that pigmented chromatophore organs are controlled by a non-synaptic neural system. Chromatophores in the cephalopod skins expand or shrink, thus filtering light. There are also iridophores that act as reflectors and leucophores diffusing light. Hanlon and his colleagues say:

> Animals can be extraordinarily colourful. There are innumerable examples in both the vertebrate and invertebrate worlds, and, while in some animals these colours function to camouflage their owners, in others they have clear functions in signalling, and some may even perform both functions simultaneously. [21]

Listening to a talk by Hanlon, Alon Gorodetsky found inspiration for a new project in his laboratory at the University of California, Irvine, namely, focusing on camouflage that can be varied by external stimuli [22]. In cephalopods the camouflage is largely provided by reflectins, a class of proteins. Gorodetsky, as a materials scientist, had the idea to treat reflectins as ordinary polymers. His aim was "to translate the principles underlying cephalopod adaptive coloration to infrared camouflage systems" [22]. With that mission, Gorodetsky and colleagues have created transparent, flexible adhesive substrates with biomimetic camouflage coatings. Reportedly, the substrates can be deployed on almost any kind of surface since they are a soft, conformable adhesive. Reflectance can be reversibly modulated from the visible to the near IR regions of the electromagnetic spectrum by applying a mechanical stimulus, namely, strain. These camouflage stickers enable the disguise of objects with varied geometries and surface roughness from IR visualization; see Figure 16.12.

Another attempt at invisibility is based on creation of **metamaterials** [23]. These are artificial materials—typically made of metals and polymers—shaped at a small scale so they can affect electromagnetic or acoustic waves in a way that is not possible by the bulk material. Shape, geometry, orientation, and size are the essential parameters. Proper adjustment can even produce a negative refractive index [23, 24]. Invisibility can be accomplished by guiding the electromagnetic radiation (light) *around* the object to be hidden. The technique has some hurdles yet to overcome, however in the visible range, objects could be successfully cloaked invisible in a scattering medium like murky water.

(a)

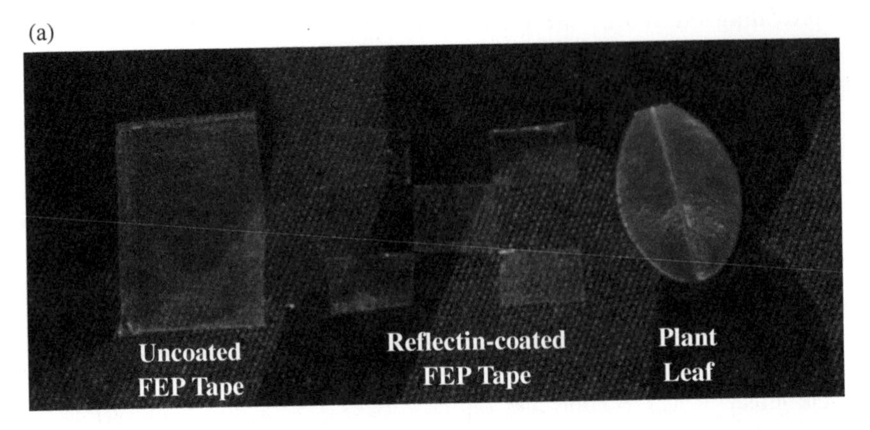

(b)

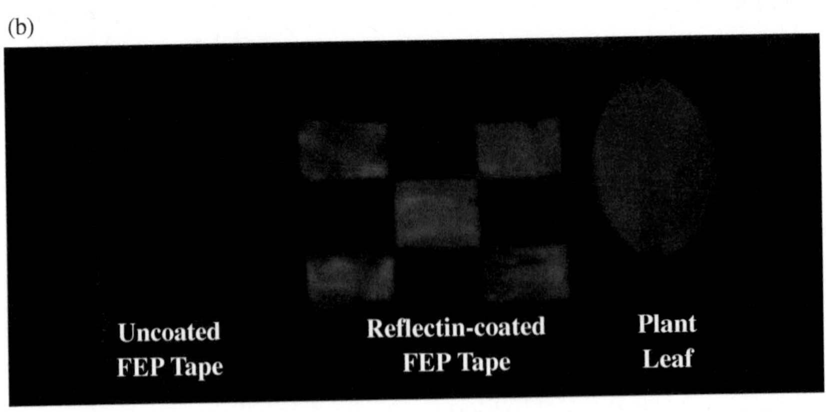

FIGURE 16.12 An optical camera image of a camouflage fatigue; (a) top left: no modification; top middle: reflectin coated tape squares; top right: a plant leaf; (b) bottom images correspond to the top ones but under IR illumination. *Source*: Reproduced from Ref. [22] with permission from The Royal Society of Chemistry.

There is also the so-called **monitor technique**. One creates on the surface of an object (e.g., a car) the image of what is behind the car in a way that it perfectly fits in with the background [25]. This can be done by projection or LCDs on the object, using it as a video screen for the background. The technique has been used for impressive advertising effects by a certain car brand.

The final “invisibility” technique to be named here is **refractive index matching**. One immerses a colorless transparent material in an environment of identical refractive index. No light scattering is observed and a refractive index detector in liquid chromatography shows no response. This can be applied, for instance, in glass fiber reinforced polymer composites—by assuring that the refractive index is the same in both constituents.

Now a very practical comment to conclude this chapter. Understanding what color actually is and how you can manipulate it is important for *selling* a product—especially in a world where people so often require constant stimulation. If an industrial company tries to make a commercially popular product, *but does not make it appealing to the eye*, poor sales and possibly even a failure of the product might ensue.

16.10 SELF-ASSESSMENT QUESTIONS

1. Compare and contrast color phenomena in ionic versus covalent materials.
2. What is the role of defects in determining the color of matter?
3. Give the meanings of the following terms: absorption, emission, transmission, reflection. Why are surfaces important in considering optical properties?
4. What is the Fermi energy?
5. Describe the features that result in color in metals.
6. What is the relationship between color and band gap in pure and in doped semiconductors?
7. Why is the sky blue in daytime but looks red at sunset?
8. Describe examples of practical importance for the following: polarizers, birefringent materials.
9. What are the distinguishing features of laser light? How does a laser device work?
10. What are luminescent materials?
11. Contrast photoinduction with luminescence.

REFERENCES

1. M.A. White, *J. Chem. Ed.* **1998**, *75*, 1119.
2. M.G. Craford, N. Holonyak Jr., & F.A. Kish Jr., *Scientific American*, February **2001**, p. 62.
3. M.A. White *Physical Properties of Materials*, CRC Press: Boca Raton, FL **2012**, pp. 77–81.
4. G.H. Needham, *The Practical Use ofThe Microscope*, Charles C. Thomas: Springfield, IL **1958**.
5. A.J. Lovinger, C. Nuckolls, & T.J. Katz, *J. Am. Chem. Soc.* **1998**, *120*, 264.
6. A. Marletta, D. Gonçalves, O.N. Oliveira Jr., R.M. Faria, & F.E.G. Guimaraes, *Adv. Mater.* **2000**, *12*, 69.
7. S.Y. Kim, S.Y. Ryu, J.M. Choi, S.J. Kang, S.P. Park, S. Im, C.N. Whang, & D.S. Choi, *Thin Solid Films* **2001**, *398–399*, 78.
8. A. Marletta, E. Piovesan, N.O. Dantas, N.C. de Souza, C.A. Olivati, D.T. Balogh, R.M. Faria, & O.N. Oliveira Jr., *J. Appl. Phys.* **2003**, *94*, 5592.
9. E.M. Therézio, E. Piovesan, M. Anni, R.A. Silva, O.N. Oliveira Jr., & A. Marletta, *J. Appl. Phys.* **2011**, *110*, 044504.
10. A. Marletta, S. de Fatima Curcino da Silva, E. Piovesan, K.R. Campos, H. Santos Silva, N.C. de Souza, M.L. Vega, M. Raposo, C.J.L. Constantino, R.A. Silva, & O.N. Oliveira Jr., *J. Appl. Phys.* **2013**, *113*, 144509.
11. M. Russo, S.E.J. Rigby, W. Caseri, & N. Stingelin, *Adv. Mater.* **2012**, *24*, 3015.
12. A.V. Kozytskiy, O.L. Stroyuk, S.Ya. Kuchmiy, V.D. Dzhagan, D.R.T. Zahn, M.A. Skoryk, & V.O. Moskalyuk, *J. Mater. Sci.* **2013**, *48*, 7764.
13. F.-L. Kuo, Y. Li, M. Solomon, J. Du, & N.D. Shepherd, *J. Phys. D* **2012**, *45*, 065301.

14. J.R. Ostrick, A. Dodabalapur, L. Torsi, A.J. Lovinger, E.W. Kwock, T.M. Miller, M. Galvin, M. Berggren, & H.E. Katz, *J. Appl. Phys.* **1997**, *81*, 6804.
15. M. Li, W.-H. Chen, M.-T. Lin, M.A. Omary, & N.D. Shepherd, *Org. Electronics* **2009**, *10*, 863.
16. F.L. Kuo, M.-T. Lin, B.A. Mensah, T.W. Scharf, & N.D. Shepherd, *Phys. Status Solidi A* **2010**, *207*, 2487.
17. M. Popescu, *J. Opt. Adv. Mater.* **2005**, *7*, 2189.
18. Y. Yu, M. Nakano, & T. Ikeda, *Nature* **2003**, *425*, 145.
19. A. Einstein & L. Infeld, *The Meaning of Relativity*, Princeton University Press: Princeton, NJ **1950**.
20. P.H. Halloway, M. Davidson, N. Shepherd, A. Kale, W. Glass, B. Harrison, T. Foley, J. Reynolds, K. Schanze, J. Boncella, S. Sinnott, & D. Norton, Near infrared display materials. *Proceedings of Conference 5080 on Cockpit Displays X*, Orlando, FL. SPIE: Bellingham, WA, April 21–25, **2003**, p. 340.
21. L.M. Mäthger, E.J. Denton, N.J. Marshall, & R. Hanlon, *Interface Focus* **2009**, *6* Suppl. 2, 1.
22. L. Phan, D.D. Ordinario, E. Karshalev, W.G. Walkup IV, M.A. Shenk, & A.A. Gorodetsky, *J. Mater. Chem. C* **2015**, *3*, 6493.
23. R. Schittny, M. Kadic, T. Bückmann, & M. Wegener, *Science* **2014**, *345*, 427.
24. M. Hess, Interphases and interfacial properties of glass fiber epoxide composites. *Seventh International Symposium on Engineering Plastics*, Xining August 18–21, **2015**.
25. T. Imhof, Wie ein unsichtbares Auto durch Hamburg fährt. *Die Welt*, Mercedes-Benz, http://www.welt.de/motor/article139087 81/Wie-ein-unsichtbares-Auto-durch-Hamburg-faehrt.html, accessed March 30, **2016**.

17

ELECTRONIC PROPERTIES

I used to wonder how it comes about that the electron is negative. Negative-positive—these are perfectly symmetric in physics. There is no reason whatever to prefer one to the other. Then why is the electron negative? I thought about this for a long time and at last all I could think was "It won the fight!"

—Albert Einstein, quoted in George Wald, *The Origin of Optical Activity* (1957), 352–368.

17.1 INTRODUCTION

Electrical conductivity is not a new topic. We first visited the concept in our discussion of bonding and intermolecular forces in Chapter 2; it was mentioned again briefly as we elucidated the nature of crystalline materials in Chapter 4; and of course it is an important topic in connection with our discussions of metals, semiconductors, smart materials, and optical properties. In this module we shall fill in the gaps that exist in the framework laid thus far.

As for the citation from Einstein, it seems we should take it as a tongue-in-cheek comment—for which he was well known. Atom nuclei are charged positively. To preserve electroneutrality, electrons have to be charged negatively. **Positrons** (positively charged electrons) would not do this job. Ordinary negatively charged electrons, called also **negatrons**, are needed.

Before we delve in too deeply, simply consider what happens—in general—when an electric field is applied to a material. What is the most responsive entity? The most responsive matter is electrons; for this reason the properties of materials subjected to external electric fields—as well as to magnetic fields and to electromagnetic radiation—are commonly referred to as electronic properties. Thus one can also refer to the *electronic behavior of materials*. Since the electronic behavior of materials may vary drastically, it is an important

Materials: Introduction and Applications, First Edition. Witold Brostow and Haley E. Hagg Lobland.

distinguishing feature of materials. A critical determinant is how tightly the outer electrons are bound to the "parent" atoms. At one extreme the valence electrons are held tight, on a short leash so to speak, and not allowed to wander around. At the other extreme, the valence electrons are essentially independent of the parent atoms, which is the case for the conduction electrons in metals. Of course, then there are cases intermediate between the extremes.

Given this situation, the response of a material to an electric field depends on the strength of the leash, or equivalently on the strength of the interactions between the nuclei and outer electrons. Coulombic forces resultant from the field make electrons and ions want to move in the direction of those forces. **Electric polarization** results if the electrons or ions *cannot* move large distances, for instance, in the case of ionic materials at low temperatures. The outcome: positive ions are displaced slightly in the direction of the applied field; electrons and negative ions are displaced slightly in the opposite direction; and both categories of particles return to their original positions when the field is removed. A frequent consequence of electric polarization is an imposed mechanical strain. The reverse phenomenon is responsible for piezoelectric behavior (described in Section 12.5), exhibited, for example by ionic crystals including quartz, tourmaline, and Rochelle salt. An example at the other end of the spectrum is the recent development of solid-state ionics; the function of these materials entails the phenomenon of ion migration in solids as the basis for varied electrochemical applications such as power generators and chemical sensors.

17.2 CONDUCTIVITY, RESISTIVITY, AND BAND THEORY

In descriptions of the flow of electric current in a material, **resistivity** ρ and **conductivity** σ are key parameters, and these depend in part on the dimensions (length L and cross-sectional area A) of the material. The **resistance** R to the flow of electric current is

$$R = \rho\left(\frac{L}{A}\right) \tag{17.1}$$

Resistance is related to current by Ohm's Law, a fundamental principle of electronics, defined by

$$V = iR \tag{17.2}$$

where V = voltage and i = current. Current is measured in Amperes (denoted A). R is given in the units ohms (denoted Ω) and ρ in the unit Ωm. The electrical conductivity of a material is simply the reciprocal of the resistivity:

$$\sigma = \frac{1}{\rho} \tag{17.3}$$

The units of σ are therefore $\Omega^{-1}\,\text{m}^{-1}$, or equivalently S m^{-1}, where S is the unit Siemens and $1\,\text{S} = 1\,\Omega^{-1}$. In the absence of external forces other than the imposed **electric field** E, conductivity can be expressed in terms of the **current density** j:

$$\sigma = \frac{j}{E} \tag{17.4}$$

Units of the current density j are $A\,m^{-2}$ (where A is amperes); units of the electric field E are $V\,m^{-1}$ (where V is volts). Therefore the units for conductivity can be expressed as $A\,m^{-1}V^{-1} \equiv \Omega^{-1}m^{-1} \equiv S\,m^{-1}$. Note that the units are named after scientists who laid the groundwork for our current understanding of electrical behavior: Georg Simon Ohm, Werner von Siemens, Alessandro Volta, and André-Marie Ampère (in Lyon there is a painting on a windowless wall of a six floor building showing Ampère standing on a balcony, with Antoine de Saint-Exupéry and other famous sons of Lyon on other painted balconies). The current density, drift velocity, and field strength are, strictly speaking, vector properties; but for simplicity we can consider the above equations without vector notation.

How are the variables defined in Eqs. (17.1)–(17.4) related to electrons? Based on the concept of free electrons, if one were to calculate the resistivity of a material from the number of valence electrons per unit volume, the mean free path between collisions, the electron mass, and the average electron velocity, the result would not match (by an order of magnitude) an experimentally determined value of resistivity. A bit of quantum mechanics is therefore required to better explain the energetic behavior of electrons.

Several factors are at play. The movement of particles is affected by their proximity to one another; this was shown in Figure 2.5, a plot of potential energy versus interparticle distance. Also at work is the **Wolfgang Pauli exclusion principle**, which tells us that no two electrons can occupy the same quantum state simultaneously. (For instance, two 3s electrons should have spins antiparallel.) What happens then when a mole of sodium atoms come together in a crystal lattice? The valence electrons are now considered as belonging to the whole crystal rather than to parent atoms. One energy level with two spin states cannot accommodate all the 3s valence electrons. Thus, that single energy level gives rise to a large number of closely spaced energy levels, so many and so close that we can consider them together as an **energy band**. The rise of energy bands was discussed also in the previous Chapter in connection with the optical properties of materials. This was explained in Section 16.3, and the reader is encouraged to review that short section.

Consider next the energy distribution of electron states. Without laboring through the specifics, one can draw from the thermodynamic principles outlined in Chapter 3 to determine the probability $p(E)$ of an energy level E being occupied by an electron. That probability is known as the **Fermi function**, given already by Eq. (15.2) but shown again here, for reference:

$$p(E) = \frac{1}{e^{(E-E_F/kT)}+1} \qquad (17.5)$$

The Fermi energy E_F is a cutoff energy; that is at $T=0\,K$, $p(E)=1$ for $E<E_F$ and $p(E)=0$ for $E>E_F$. As we see in Figure 17.1 (the same as Figure 16.6), the Fermi function is rectangular at $T=0\,K$ and $p(E_F)=½$. For higher temperatures there is a rounding of the function at the two right angles. The schematic in Figure 17.2 (what we saw already in Figure 16.9) portrays the valence and conduction bands, with shaded regions indicating occupied levels. The band gap E_g in semiconductors is also shown in Figure 17.2. Insulators differ primarily in that E_g is much larger.

Let us look more closely at the topic of permitted energy bands and forbidden gaps. Here we shall approach the problem from a different perspective, considering the wave character of electrons. Remember the wave-particle duality and the de Broglie equation, Eq. (5.1). Imagine one free electron moving in a one-dimensional system. The electron has momentum

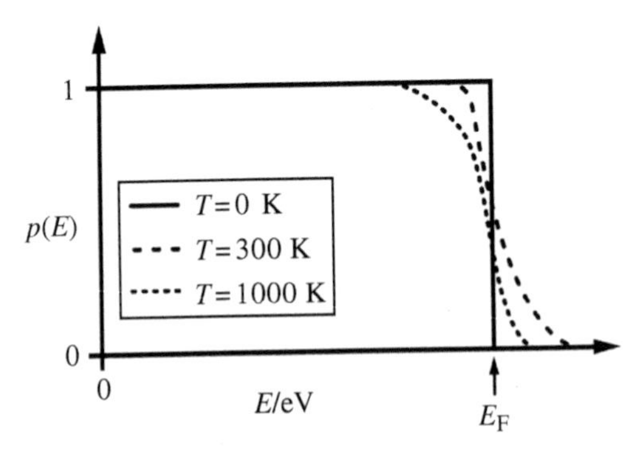

FIGURE 17.1 The probability of an electron in a metal occupying a certain energy level is $P(E)$, defined by Eq. (16.2). At $T=0$ K, the Fermi energy, denoted E_F, separates occupied from unoccupied levels. At higher temperatures, the function is no longer rectangular.

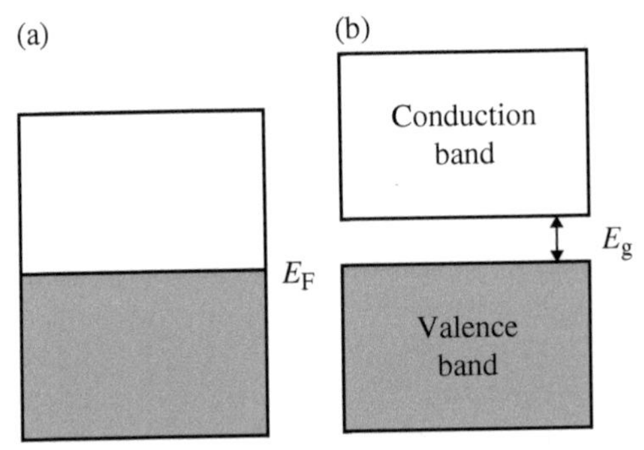

FIGURE 17.2 The energy gap E_g is illustrated by a modified schematic of the electron probability distribution shown in Figure 16.6. Energy bands are shown for (a) a metal and (b) a semiconductor. E_F is the Fermi energy.

and therefore kinetic energy due to this momentum. The relationship between the kinetic energy U^M (M for momentum) and the wave vector $\boldsymbol{k}$ of the electron is parabolic (see Figure 17.3). Although $\boldsymbol{k}$ is not strictly a continuous variable, we can treat it as a quasi-continuous variable because the permitted values of wave vectors are so numerous and close together.

The magnitude of the wave vector is inversely proportional to the wavelength λ; ordinarily, the direction of the vector is the direction of wave propagation. Now, consider the situation in which our electron is placed in a solid crystal. The electron necessarily interacts with the crystal, especially via Coulombic interactions with positive ions. The potential energy depends on the kind of material: the acting forces are relatively weak in a metal but strong in covalent solids. As the electron travels along a crystal axis, its potential energy is perturbed to a small degree in metals but to a larger degree in covalent solids. This variation is represented by changes in amplitude, according to the illustration in Figure 17.4.

We gather from Figure 17.4 that the kinetic motions of the electron are confined within a periodic crystal lattice; and since electrons are waves, they are diffracted by atoms or ions of the lattice. Using Bragg's law Eq. (5.2), the key relation of diffractometry, one can determine that there are forbidden wave vectors for the electron. Thus, the simple parabolic relationship between the kinetic energy and wave vector of the free electron seen in Figure 17.3 must be modified in a crystal to account for the forbidden zones; see Figure 17.5.

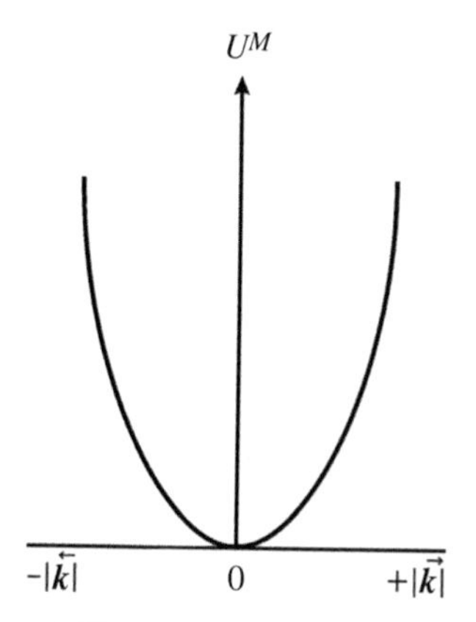

FIGURE 17.3 Kinetic energy U^M as a function of the wave vector $\boldsymbol{k}$ for a free electron.

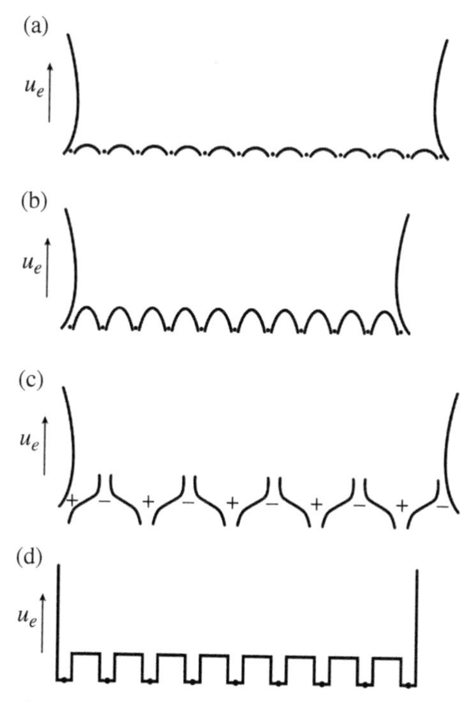

FIGURE 17.4 Schematic representation of potential energy u_e of electrons in different types of solid crystals: (a) metal; (b) tightly bound covalent crystal; (c) ionic solid; (d) a simplified rectangular well model (also called the Kronig-Penney model). *Source*: W. Brostow, *Science of Materials*, Wiley-Interscience: New York 1979. Reprinted with permission from Wiley.

The *permitted* regions are known as **Brillouin zones**. By expanding a one dimensional Voronoi diagram description in turns to two and then three dimensions, the Brillouin zones become volumes bounded by surfaces. In consequence, the Brillouin zones can overlap, eliminating some forbidden zones.

Constant-energy contours can be constructed for any crystal lattice. For example, the contours for the first Brillouin zone in a two-dimensional square lattice are shown in Figure 17.6.

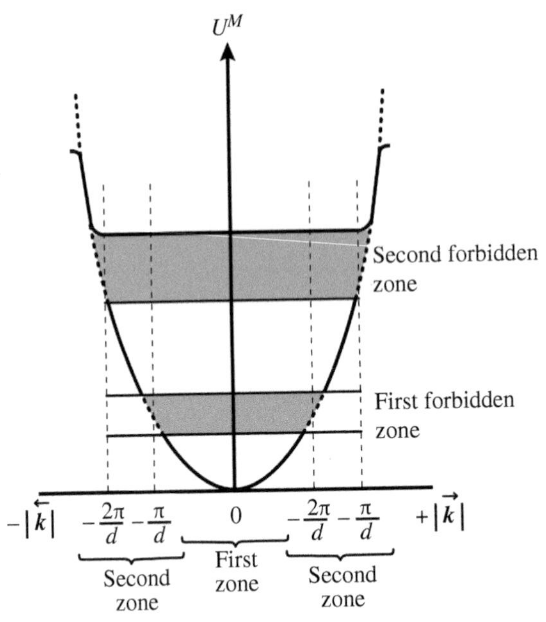

FIGURE 17.5 Relation between the kinetic energy U^M of an electron in the crystal lattice and the wave vector $\boldsymbol{k}$.

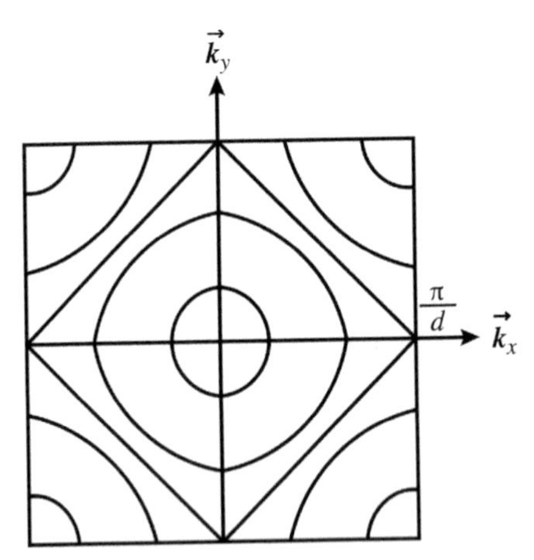

FIGURE 17.6 Equal energy contours in the first Brillouin zone of a two-dimensional square lattice.

In three dimensions, the contours are spherical for free electrons near the center of a zone, though the situation for bound electrons is much more complicated. With the use of further quantum mechanics, it is possible to calculate the number of quantum states available to electrons in each zone. Just as we have discussed before, the lowest energy states are filled first with higher energy levels filled successively afterwards until all valence electrons are accommodated. The Fermi energy, as we already know, corresponds to the highest level occupied; and the corresponding surface in $\boldsymbol{k}$ space is referred to as the **Fermi surface**. Moreover, there can be zone overlap, such that before the first zone is completely filled the second Brillouin zone begins to fill. Thus we arrive at the same qualitative and quantitative description of permitted and forbidden energy regions as by the band approach. The existence of these energy regions forms the foundation for explanations of the behavior of conductors, semiconductors, and insulators.

One final thought regarding the conductivity of electrons: the mean free path of an electron traveling in a solid decreases as the temperature increases. This is of course owing to the increasing amplitude of atomic vibrations and resultant increase in collisions of electrons. Thus, electrical resistivity ρ is expected to *increase* (linearly) with T. Such behavior is generally observed in metals. However, we shall soon see that for semiconductors there are factors other than the mean free path that contribute markedly to the temperature dependence of ρ.

17.3 CONDUCTIVITY IN METALS, SEMICONDUCTORS, AND INSULATORS

Metals, semiconductors, and insulators can be distinguished by their values of conductivity. Typically, the ranges for σ are delineated as follows:

Metals: $\sigma > 10^4\,\Omega^{-1}\,m^{-1}$
Semiconductors: $10^{-3}\,\Omega^{-1}\,m^{-1} < \sigma < 10^4\,\Omega^{-1}\,m^{-1}$
Insulators: $\sigma < 10^{-3}\,\Omega^{-1}\,m^{-1}$

There are some interesting non-metallic materials that have conductivities in the range of metals. For instance, compounds of sulfur and nitrogen denoted $(SN)_x$ have just such high conductivity. Likewise the conductivity of graphite inclusion compounds and conductive organic crystals may lie above $10^4\,\Omega^{-1}\,m^{-1}$. A little lower in conductivity is doped polyacetylene, a polymer with conductivity in the semiconducting range. Graphene, which consists of monoatomic layers of graphite, each layer separated by its environment from the other layers, is essentially a semiconductor. Phthalocyanines are also named among non-metallic semiconductors; these materials are discussed further in the following Section on semiconductors. Non-doped (neat) polyacetylene is an insulator, like most macromolecules, including familiar compounds such as organic dyes, polyethylene, polystyrene, Teflon, and others. There are also non-conducting materials that are made conductive by addition of fillers. Thus, silver particles are added to poly(aniline-*co*-*p*-phenylenediamine) [1], or graphite nanoplatelets plus multi-walled carbon nanotubes plus boron nitride particles are put into an epoxy [2].

Given the prior understanding of energy bands, it can be seen that for a metal the valence band is not completely full, and there are thus unoccupied energy levels immediately above the highest occupied levels. With no forbidden gap above the filled levels, excitation of electrons at temperatures above 0 K is easy. Thereby the conduction of electrons is facilitated in metals.

Refer now to Figure 17.7; the energy difference between the Fermi energy and vacuum level is the **work function** ϕ, which represents the minimum amount of energy needed to remove an electron from the metal. In metals, the work function and **ionization energy** are the same. In practice, the work function of a surface is strongly affected by contamination or chemical reactions that alter the condition of the surface. A contact potential is formed when dissimilar metals having different work functions are placed in contact with one another; there is an equalization the Fermi energies and an electric field generated as a result of the

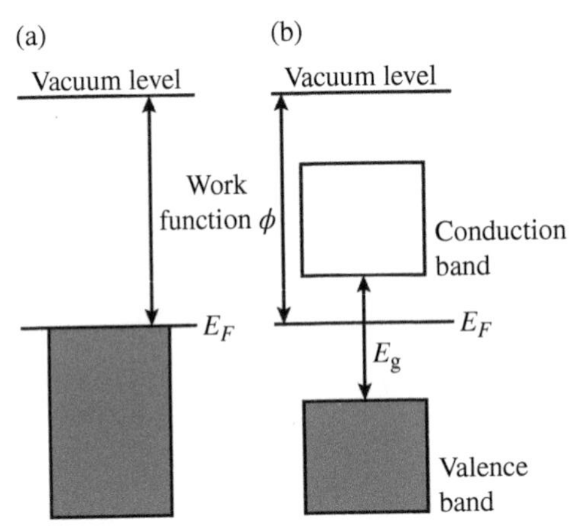

FIGURE 17.7 The work function ϕ, showing the minimum energy required to remove an electron to a vacuum away from the atom, is shown for (a) a metal and (b) a semiconductor.

contact potential. The work function and contact potentials are important factors in fabrication and operation of electrical devices, especially in sensitive instrumentation.

As stated earlier, an insulator has a very large gap between its valence and conduction bands. The amount of thermal energy required to excite electrons to the conduction band is so large that electrical conductivity is negligible in these materials—they would melt before they conduct electricity! They are used then for their capacity to *insulate* other materials from electrical current.

The electronic behavior of semiconductors is intermediate between that of insulators and metals. At $T=0\,\mathrm{K}$, all the electrons are in the valence band; but above 0 K, some electrons are thermally excited into the conduction band, imparting slight electrical conductivity. This situation is depicted for us in Figure 17.8. As the temperature increases, more electrons acquire the thermal energy needed to be excited into the conduction band. The outcome is observed as a *decrease* in the electrical resistivity of pure semiconductors with increasing temperature. Naturally, low band-gap semiconductors (e.g., silicon) require less energy than high band-gap materials to promote electrons to the conduction band. Interestingly, depending on their precise structure, carbon nanotubes can exhibit either metallic or semiconducting electrical behavior.

17.4 SEMICONDUCTORS: TYPES AND ELECTRONIC BEHAVIOR

Pure semiconductors, composed of a single element, are frequently referred to as **intrinsic semiconductors**. Examples are Si and Ge. **Extrinsic semiconductors**, on the other hand, have impurities added purposely to alter their electronic energy levels. Extrinsic semiconductors are further classified as n-type or p-type; we shall explain these soon. The conductivity mechanisms differ for each type, based on their structures.

Consider Ge, with its tetrahedral bonding and full valence shell (in the bonded state). The conductivity mechanism is described by Figure 17.9. The mechanism for Si is the same (but with $E_g = 1.1\,\mathrm{eV}$ for intrinsic Si).

What happens if there are defects in a semiconductor? In connection with optical properties, we saw in Chapter 16 that defects result in the addition of electronic energy

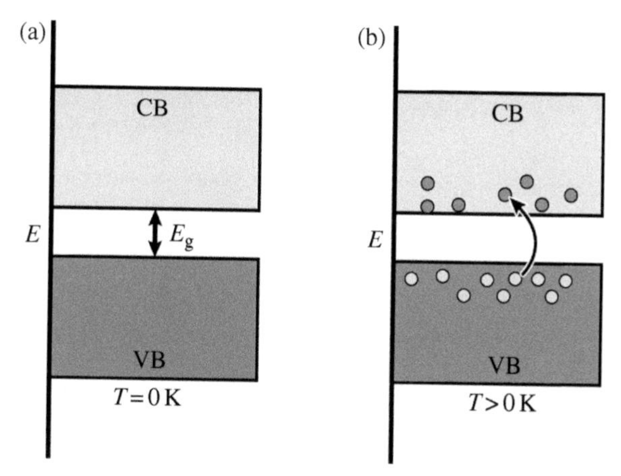

FIGURE 17.8 Energy bands of a semiconductor. (a) Valence band (VB) is full, and there is no conductivity at $T=0$ K. (b) At temperatures above absolute zero, electrons can acquire enough energy to move to the conduction band (CB). Dark dots represent electrons in the conduction band, and white dots represent the corresponding holes in the valence band.

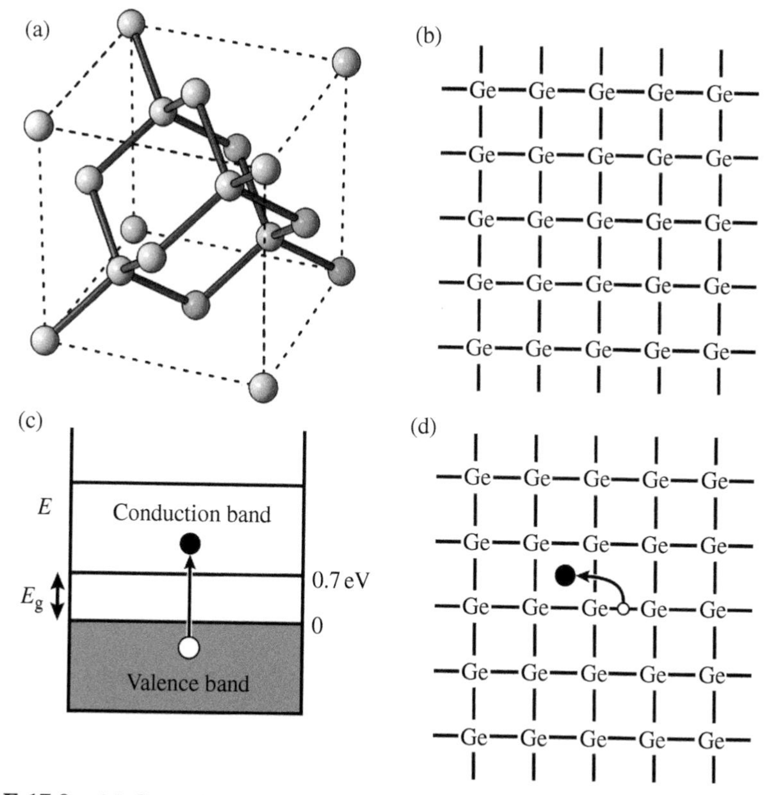

FIGURE 17.9 (a) Crystal structure of Ge. (b) Schematic of Ge lattice. (c) The band gap for Ge is 0.7 eV. The mechanism of conductivity for an intrinsic semiconductor involves the moving of an electron from the Ge lattice into the conduction band, leaving a hole behind, as shown here and in (d).

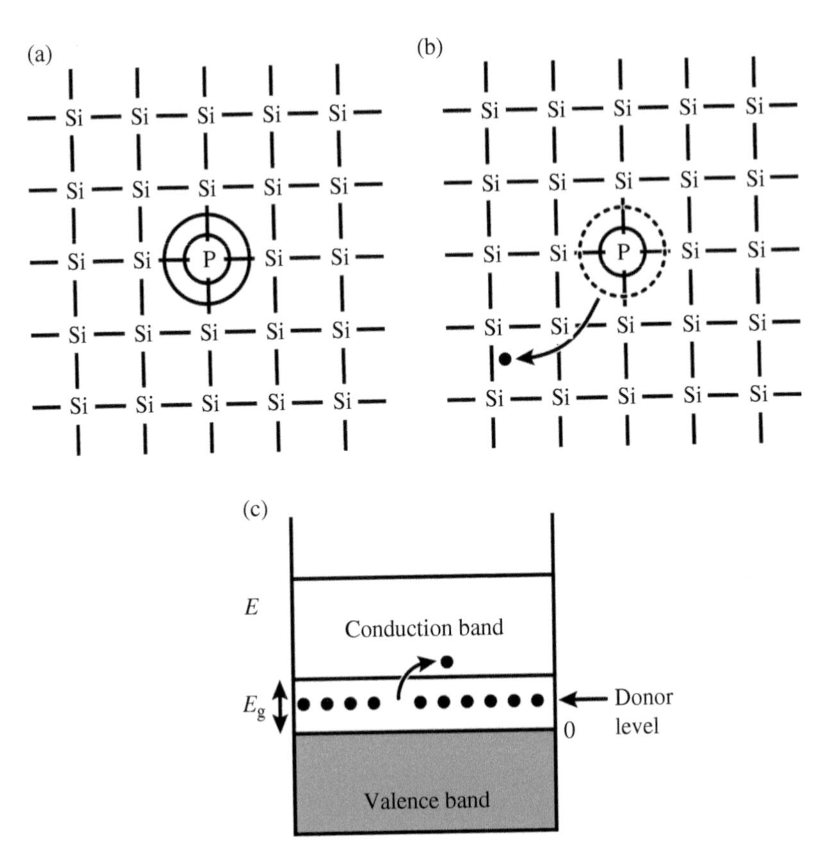

FIGURE 17.10 An extrinsic semiconductor. (a) Two-dimensional representation of *P* inserted into Si lattice. (b) *P* donates an extra electron (compared to Si) to the conduction band. (c) A donor level is created in the band gap, from which electrons can be promoted to the conduction band.

states. Thus extrinsic semiconductors have defects imparted from the insertion of impurities, a process known as **doping**. If Si, with its four valence electrons, is doped with *P*, having 5 valence electrons, there is now an extra electron in the lattice. A localized energy level, called a donor level, is created within the band gap as a result of the electron donated by phosphorus. Electrons can be promoted to the conduction band from the donor level; and since the current is carried by negatively charged particles, this type of semiconductor is referred to as **n-type**. The mechanism is illustrated by a schematic diagram in Figure 17.10.

When holes are the charge carriers, we have **p-type** semiconductors, so called because of the positive charge of holes. In Si this can be achieved by doping with Al, for instance, which has three valence electrons. In this case, a localized energy state referred to as an acceptor level is generated within the band gap and can accept electrons from Si. The mechanism is depicted in Figure 17.11.

When p-type and n-type semiconductors are utilized together, a multitude of electrical devices can be made. The term **p,n-junction** is used for the junction of a p-type and n-type semiconductor. A variety of behaviors can be obtained by applying an electric field across

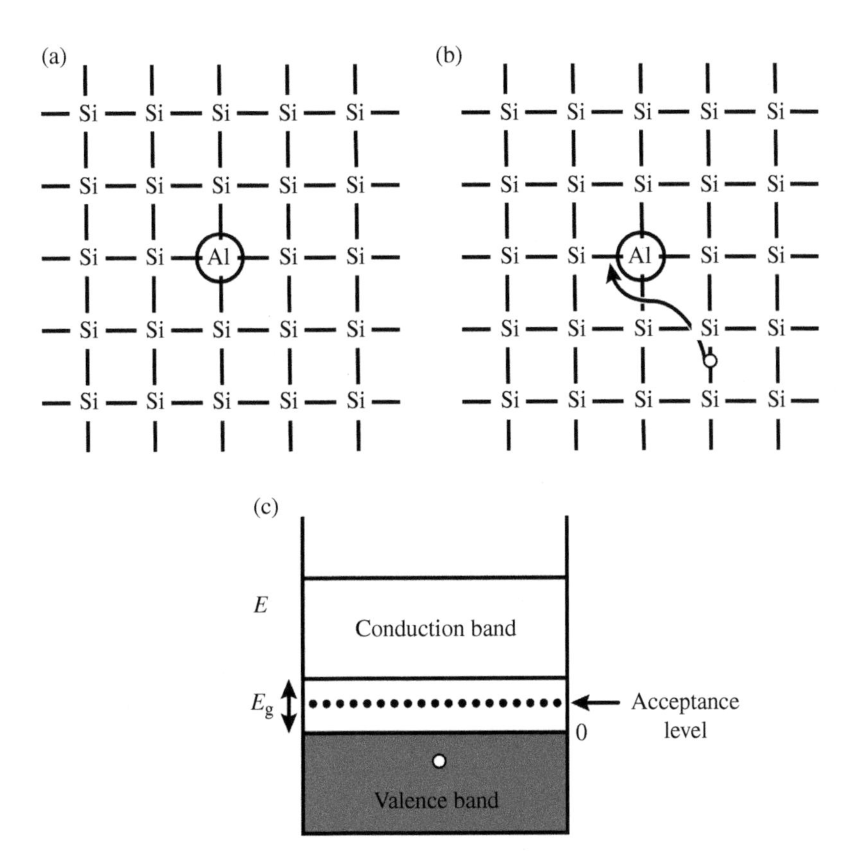

FIGURE 17.11 A p-type semiconductor: Si doped with Al. (a) The lattice. (b) Movement of an electron within the structure. (c) The presence of a hole in the valence band drives the movement of electrons from Si to the acceptor band.

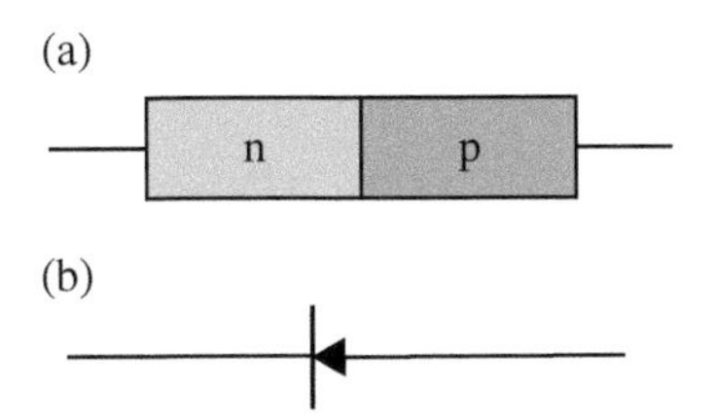

FIGURE 17.12 (a) Representation of a diode and (b) the electrical symbol for a diode.

the junction. Electrical devices commonly known as **diodes** are essentially p,n-junctions in operation (see Figure 17.12). "Sandwiched" junctions of the type p,n,p-junction or n,p, n-junction are used to make **transistors**.

In Section 4.4 we discuss the strong influence of structure defects in crystals on some of the crystal properties. This applies to electrical properties of semiconductors. As discussed by Zawadzki [3], electrons are scattered in such materials by impurities (charged or neutral), dislocations, optical phonons (interactions of both nonpolar and polar types), and acoustic phonons—as well as by other electrons. The results are nonspherical energy bands in Ge and Si as well as nonparabolicity of the bands in narrow-gap semiconductors.

A TALE OF ELECTRONIC TAILS

The phenomenon known as Urbach tailing is a good illustration of the influence of defects on electronic properties. In 1953, Urbach [4] identified exponential tails at band edges in semiconducting ionic crystals, the presence of which resulted in unexpected hybrid excitonic transitions. Three decades later, N. F. Mott [5–7] expressed the state-of-the-art of investigations on this intriguing phenomenon in his Nobel Lecture [5]: "*...This problem (Urbach tailing) is going to be quite a challenge for the theoreticians, but up till now we depend on experiments for the answer ...*". In other words, for over 30 years though experiments continued to prove the existence of Urbach tailing in various types of semiconductors, the phenomenon still defied explanation.

Urbach tailing is commonly defined as the broadening of the exciton absorption band, as depicted in Figure 17.13 [8], which occurs when charged defects in the crystal induce phonon-induced microelectric fields. Urbach tailing inside semiconductor crystals alters drastically the electrical conductivity and has been attributed to atomic-scale topological disorder, unexpected hybrid excitonic transitions, or topological filament occurrence. Another more realistic explanation has been provided by Dow and Redfield [9]. Their explanation claims that Urbach tailing is due to the broadening of the exciton absorption band which occurs when charged impurities in the lattice induce phonon-induced microelectric fields. In recent works on the Lattice Compatibility Theory [10–21], the Urbach tailing phenomenon has been described for II–VI compound semiconductors as a reduction of Coulombic interactions of holes-electrons, which results in more free excitons and consequent resistivity changes. Many other causes have been proposed in the relevant literature [22–25].

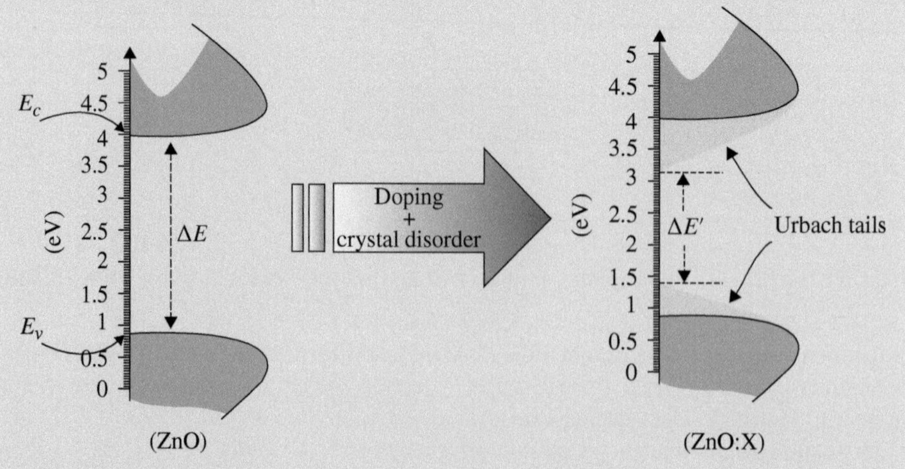

FIGURE 17.13 Urbach tails scheme. Source: Boukhachem *et al.* [8], © 2012. Used under CC-BY-3.0.

Thermoelectric devices are derived from semiconductors assembled in a fashion to utilize the Peltier effect or Seebeck effect. As an industry, thermoelectric devices are growing in number and importance. These devices make use of the fact that a temperature gradient across certain combinations of dissimilar materials results in the generation of an electronic voltage—or the reverse that application of a voltage across such materials generates a temperature change. We shall discuss these devices more in Section 20.5. *Thermoelectric generators*, which can convert heat to electrical energy, are a notable application of semiconductor materials. The familiar *integrated circuit* also operates on the principles and properties just described. It consists of several metal oxide semiconductor field-effect transistors (MOSFETs) mounted on one p-type semiconductor (e.g., a silicon wafer).

Another distinction among semiconductors is between those with a direct band-gap and indirect band-gap. The distinction goes back to the concept of wave vectors mentioned in Section 17.2. In a **direct band-gap** semiconductor, the $\boldsymbol{k}$-vectors are the same for the maximum-energy state of the valence band and the minimum energy state of the conduction band. If the $\boldsymbol{k}$-vectors are different, the material has an **indirect band-gap**. This distinction has implications for photon absorption and impacts **photovoltaic** capacity of semiconductor materials.

Electric current (voltage) is created in a **photovoltaic** (PV) material upon exposure to light. If the radiation is sunlight, the photovoltaic devices are called **solar cells**. Polymer-based photovoltaic devices offer the promise of low cost and lightweight solar energy conversion systems [26, 27]. We discuss them from a different perspective also in Section 20.4. Appropriate balance of constituents in PV devices has to be provided. Mitzi and coworkers [28] have studied $Cu(In, Ga)(Se, S)_2$ photovoltaic materials from the point of view of structure and performance. Defects, namely, insufficient concentration of Se, have little effect on the material grain structure, crystal orientation, or bulk composition. However, there is a significant negative effect of those defects on solar cell photovoltaic performance.

ORGANIC SEMICONDUCTORS FOR ELECTRONICS

Optoelectronics applications include photovoltaic devices (discussed in the present chapter and in Chapter 20), light-emitting diodes (discussed in Chapter 16), and **field effect transistors** (FETs) (mentioned in Section 17.6 of the present chapter). For commercial success of organic semiconducting matter, we still need better understanding of how chain conformation, molecular packing, miscibility, and other physicochemical aspects affect semiconducting phenomena from nanoscale to macroscale. To understand the behavior of organic semiconductors, Natalie Stingelin is developing ways to manipulate the respective phases [29, 30]. Strategies to use ternary blends as the constituent of the polymer bulk heterojunction of organic PV devices are being developed (recall ternary phase diagrams discussed in Chapter 3) [29, 30].

While we talk later in Section 17.6 about **3D** electronics, there are also one-dimensional organic semi-conducting materials. Among these are **conjugated polymers**, which consist of a backbone chain of alternating double- and single-bonds. Their semiconducting functionality results from a process of **mesomerization** in which one kind of bond becomes similar to the other type.

Somewhat similarly, in a cyclic molecule of benzene there are alternating single and double bonds continuously interchanging their positions (yielding the **August Kekulé model** so named after a German chemist). Conjugated polymers for electronics applications have been pioneered by Alan J. Heeger, Alan G. MacDiarmid, and Hideki Shirakawa [31] who were awarded the 2000 Nobel prize in Chemistry for their work. A very simple conjugated polymer is polyacetylene:

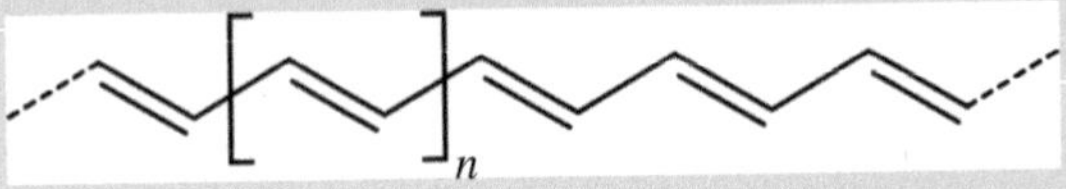

Other conjugated polymers are polyphenylvinylene and polythiophenes; the latter have the following backbone structure:

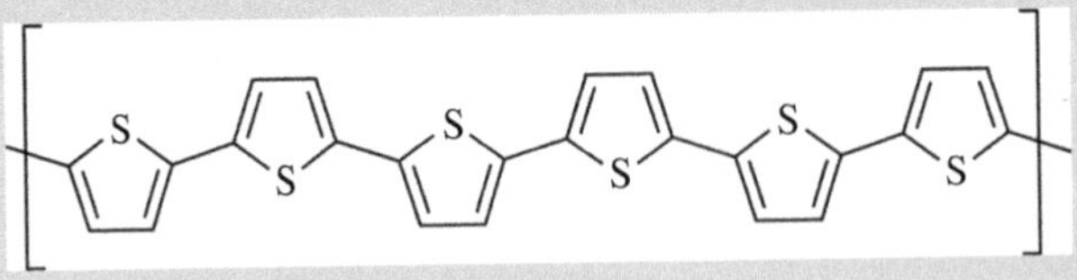

There is, in fact, a definable relationship between disorder, aggregation, and charge transport in conjugated polymers [32].

As said, devices made from inorganic semiconductors can be made also from conjugated polymers. Recall once again Figure 1.1: structure and interactions determine properties. The structures of conjugated polymers are fairly complicated, including amorphous phases as well as ordered phases with defects. High electric conductivity is *not* expected in such systems, and yet it exists. This puzzle was solved by a combined Stanford–ETH Zurich–Imperial College team [29]. They showed that the polymeric chains interact weakly with each other and such short range interactions are sufficient for long-range transport of electric charges. Thus it is possible to make LEDs and PV devices from organic materials.

While organic solar cells sound attractive, there are problems involved in their implementation. Christian Müller and his colleagues studied a polymer liquid crystal based on fluorine and containing a low molecular weight fullerene [33]. On the one hand, photovoltaic performance increases with increasing molecular mass of the polymer, eventually reaching an asymptote around $M = 10{,}000$. On the other hand, higher M causes increased viscosity, thus processing difficulties, and in turn difficulties with molecular ordering that can be achieved [33]. Müller soberly states [34]:

> A challenging perspective is provided with regard to the thermal stability of the blend nanostructure vs. the mechanical robustness and ductility of the active layer material. Conflicting demands on the blend glass transition temperature, i.e., higher vs. lower than the processing and operating temperature, require a satisfactory compromise that must be achieved before truly flexible polymer solar cells with a high light-harvesting efficiency can be realized.

There is also an option based on two-dimensional phthalocyanine covalent organic frameworks (COFs) created by Dichtel and coworkers [35]. As an example, there is a COF having a square lattice of phthalocyanine macrocycles joined by phenylene bis(boronic acid) linkers. This COF exhibits broad absorbance over the solar spectrum while its good thermal stability and potential for efficient charge transport through the stacked phthalocyanines show promise for applications in organic PV devices.

17.5 SUPERCONDUCTIVITY

Superconductivity is an interesting phenomenon in which there is no resistance to the flow of electrical current ($\rho = 0$). The phenomenon was discovered in 1911 by the Dutch scientist Heike Kamerlingh Onnes in Leiden [36]. There is now a laboratory named after him at the University of Leiden. The data collected for Hg at low temperature showed a sudden drop to zero in resistivity at 4.2 K, a behavior unpredicted by any previously proposed theory. The same phenomenon was then confirmed also for lead and tin; at some critical temperature T_c the electrical resistance dropped to zero. The lack of resistance means that superconductors can carry electric current almost indefinitely because there is no Ohmic heat loss.

The practical use of superconductors is limited by their low transition temperatures. For 75 years since the discovery of superconductivity, the search for more materials exhibiting this phenomenon produced only metal alloys, and those with insignificant results. However, in 1986 J. Georg Bednorz and K. Alex Müller at the IBM Zurich Research Laboratory pushed the critical temperature up to 30 K with their new ceramic superconductive material, which was basically an oxide consisting of Ba+La+Cu+O [37]. That original material consisted of three phases. In such a combination, electrons originating from copper atoms are more mobile, as there are both Cu^{+2} and Cu^{+3} ions; and though oxygen is present, the material is oxygen deficient in its stoichiometry. A year later Bednorz and Müller were awarded the Nobel prize in Physics. Suddenly more and more laboratories started working on new superconducting materials.

Superconducting magnets are used in magnetic resonance imaging (MRI) and nuclear magnetic resonance (NMR) to generate high magnetic fields. Superconductors are also characterized by the **Meissner effect**: they completely expel a magnetic field. This feature has implications, for example, for developing magnetically-levitated trains. Nb+Ti and Nb_3Sn magnets are used for MRI and NMR and also in the Large Hadron Collider at CERN in Geneva. They operate between 2 and 6 K. Superconductors that have found industrial use include MgB_2, with a T_c of 39 K, and Nb_3Ge, with a T_c of 23 K. Cuprate-based superconductors, including cuprates of La, Y, and Hg, boast a T_c of 92 K or higher. Current density and electric field strength affect superconductivity—meanwhile the search for better such materials continues [38–41]. In Section 18.7 we shall discuss the use of superconductors in magnetic levitation (MagLev) trains.

17.6 PHENOMENA OF DIELECTRICAL POLARIZATION

In simple terms, the distinction between electrical properties and dielectric properties is the following: with the former electric charges *move*, with the latter they *sit*. Many insulators are also **dielectrics**, which have the property of reducing the Coulombic force between two charges. Another way to think about this is as the capacity to *store* electric charge. If, however, an applied electric field exceeds a certain limit, **dielectric breakdown** occurs and charges will begin to move.

Dielectric properties arise from polarization. Consider the simple case of an electrical capacitor with a dielectric material between the two plates. The larger the **dielectric constant** ε (also known as **relative permittivity**), the more charge can be stored and the higher the capacitance. The dielectric constant is therefore related to the capacitance with (C) and without (C_0) the dielectric material:

$$\varepsilon = \frac{C}{C_0} \tag{17.6}$$

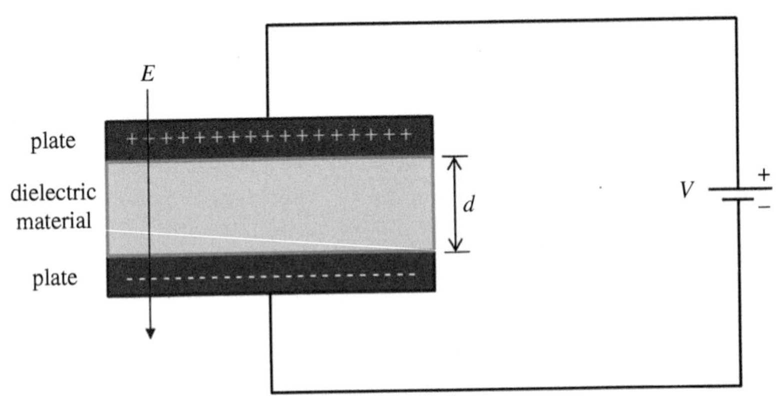

FIGURE 17.14 A parallel-plate capacitor with a distance d between the plates, a dielectric material between the plates, an electric field E across the plates, and a voltage V.

Placed in the electric field of the capacitor, the positive and negative charge centers of the material are separated, inducing a dipole moment. This process is called **polarization**. The dipoles align, generating an electric field in opposition to the field of the capacitor: in effect reducing the electric field. The process is shown schematically in Figure 17.14.

A typical dielectric is un-polarized in the absence of an applied electric field. Note that various mechanisms contribute to polarization; at least four of these contributions are:

- Electronic (or electrical), resulting from displacement of electrons with respect the nuclei.
- Ionic, from deformation and displacement of charged ions with respect to other ions.
- Orientational or dipolar, in consequence of a change in orientation when a field is applied to molecules already possessing permanent electric dipole moments.
- Interfacial, which occurs in heterogeneous materials due to the buildup of charges at structural interfaces.

Ferroelectrics are a special class of materials in which the polarization does *not* depend linearly on the field strength. Instead, we observe the behavior shown in Figure 17.15. Using that figure as a guide, consider a ferroelectric material in an electric field $E=0$; hence the polarization P of the material is also zero. As E is increased, the charges in the material gradually become oriented in the field. Eventually we reach the top horizontal asymptotic part of the curve. This is the saturation polarization; further increase of the field produces no more polarization since all electric charges are already oriented. If the electric field is then reduced to zero, the figure shows the polarization does not simultaneously return to zero. In fact, P is reduced only slightly; the value is referred to as the **remanent polarization** and indicates that a part of the previously imposed orientation remains. The polarization can be removed by imposing the opposite field, referred to as the **coercive field**, thus returning the material to a state where $P=0$ again. By increasing the negative field still further, the material eventually reaches the full "negative" polarization. Reducing that field to zero leaves the material again with a remanent polarization, a polarization which can be reduced to zero according to the measured coercive field (in reverse orientation). Continued increase of the electric field brings the material back to the full (saturated) positive polarization. In the hysteresis curve of Figure 17.15, we see that the lower part is the mirror image of the top part. It is evident that with continued manipulation

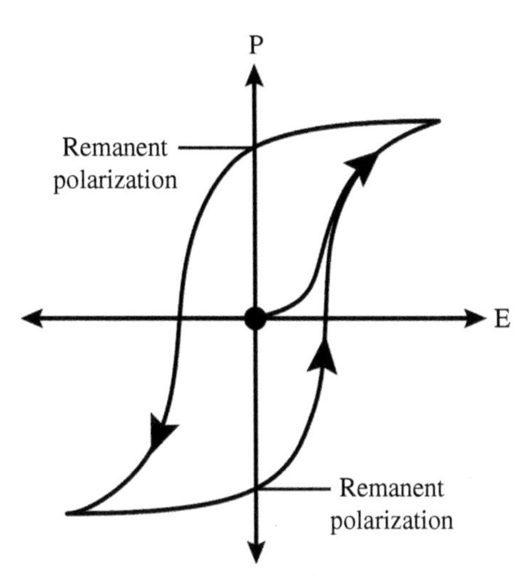

FIGURE 17.15 Hysteresis curve for polarization P versus imposed electric field E for a ferroelectric material.

of E, the polarization values can only follow along the hysteresis loop. Though the process began at $E=0$ and $P=0$, we cannot get back to that point. $BaTiO_3$ is one example of a ferroelectric material.

It is important to note that the ferroelectric behavior just described disappears above a critical temperature (the ferroelectric Curie temperature). Thus, it is only below the T_C that we observe the behavior presented in Figure 17.15.

What are the uses of ferroelectrics? They serve to make capacitors with tunable capacitance. A ferroelectric capacitor consists essentially of a pair of electrodes with a ferroelectric material between them. The permittivity has typically a very high value; thus, such capacitors are much smaller than so-called dielectric (not-tunable) capacitors having comparable capacitance. The hysteresis effect in ferroelectrics makes them useful in memory devices, such as for the RAM (random access memory) in computers or radiofrequency identification devices.

There are other useful phenomena related to polarization. Among these is **piezoelectricity**, namely, the dielectric polarization of a material as a result of mechanical stress. Thus, there is a Mechanical Energy-to-Electrical Energy connection; see the cartoon in Figure 17.16. A simple application is in making spark igniters, often from lead zirconium

FIGURE 17.16 An illustration of piezoelectricity. The active people induce some mechanical stress to the material, and the resultant dielectric polarization completes the circuit, turning on the light.

titanate. Microphones also operate from the phenomenon of piezoelectricity. Further discussion of piezoelectric materials appears in Section 12.5. As stated there, a number of ferroelectric materials exhibit this property. Applications involving piezoelectricity are numerous.

There is also the phenomenon of **inverse piezoelectricity**: application of electric voltage causes strain. This is illustrated in Figure 12.14. Inverse piezoelectricity is used in the generation of ultrasound waves, in ultrasonic cleaners or toothbrushes, and also in actuators.

A material is **pyroelectric** if its spontaneous polarization changes in response to temperature. Several examples and their behavior are described in Section 12.5. Above a certain phase transition temperature (the critical temperature T_c), a pyroelectric is no longer spontaneously polarized. **Ferroelasticity** describes a phenomenon wherein a material exhibits a spontaneous strain; it is the mechanical equivalent to ferroelectricity and ferromagnetism (discussed in the next Chapter). As with ferroelectrics, there is a critical temperature with an associated phase change at which the permanent polarization or strain is "erased". There are also **antiferroelectrics**, which possess permanent dipoles but not all aligned in the same direction, as in ferroelectrics. The result in antiferroelectrics is a net zero spontaneous polarization. The property of antiferroelectricity disappears at the anti-ferroelectric Curie temperature.

Field effect transistors (FETs) use the electric field to control the conductivity of a channel of one type of charge carrier in a semiconductor material. They have high input resistance, $100\,\mathrm{M\Omega}$ or more, and low output resistance; the voltage controls the device. A typical use of FETs is in amplifiers.

Three-dimensional (3D) printing has more and more applications. One can design a structure by computer aided drafting (CAD) and almost instantly begin its creation by the guided laying down of successive layers. Objects so made can have a large variety of shapes and compositions. Owing to the ease of design and rapidity of processing, the time-to-market is dramatically reduced for low-volume products. Within the field of 3D printing, an important aspect is providing functionality to the products—for instance, insertion of electrical interconnects. A variety of embedding processes are in use. Direct-write wire embedding enables the printing of **3D structural electronics** [42]. A crucial aspect associated with insertion of the electromagnetic components is retaining the mechanical strength of the product or device.

A different area in which functionality has to be provided while mechanical and other properties have to be preserved is **X-ray shielding**. Three-dimensional-printed shield structures have been made by inserting tungsten into a polycarbonate matrix [43]. Impact testing is used to verify that tungsten addition does not negatively affect mechanical behavior. Electromagnetic characterization has been achieved by **terahertz time-domain spectroscopy** (THz-TDS) in which short pulses of terahertz radiation are applied to the specimen. The signal detected depends on the material nature and structure. It turns out that the addition of tungsten affects the electromagnetic characteristics of polycarbonate only insignificantly [43] while X-ray shielding is achieved.

Terahertz time-domain spectroscopy has in fact a variety of other applications. Terahertz radiation is safe for biological tissues since it is non-ionizing—in contrast to X-rays. Many materials are transparent to terahertz radiation. For those that are not, sample thickness, density, and defect location can be determined—useful for materials such as foams difficult to be studied by other techniques. The fact that certain materials can be studied across visually opaque layers, such as thick coatings, clothing, or packaging, is the basis of crime detection applications of THz-TDS. One can detect hidden explosives, active pharmaceutical ingredients in medications, as well as narcotics.

Finally, we shall note **electron spin resonance** (ESR). It is often compared to nuclear magnetic resonance (NMR) which we discuss in Section 18.7. Many chemical and biological materials do not have unpaired electrons, which constitutes an obstacle in ESR application. There is an advantage, however [12]. The electron has a much larger magnetic moment than a nucleus, so ESR is more sensitive per spin. Applications of ESR in the study of membranes and proteins are known, based on spin labeling of certain molecules with a small spin-bearing moiety [44]. A **moiety** is a part or functional group of a molecule.

17.7 SELF-ASSESSMENT QUESTIONS

1. Describe temperature dependence of conductivity for metals and for semiconductors. Explain the reasons for the differences.
2. Radiative recombination occurs in direct band-gap materials when holes and electrons combine, leading to emission of light. Compare and contrast this phenomenon with light emissions from magnesium burning in a flame and from a crystalline solid containing a transition metal impurity.
3. Compare n type and p type semiconductors.
4. In contrast to conventional electronic devices, which use electron charges to move information, the field of spintronics takes advantage of electron spin for the same purpose. Semiconductors can function as spin polarizers and spin valves in order to carry out spin transport and spin injection. Imagine how a material with large magnetoresistance might be used for injection of spin in spintronic materials—describe your ideas.
5. Explain the phenomena of piezoelectricity, inverse piezoelectricity, and pyroelectricity.
6. Consider the origins of magnetism in a ferroelectric metal. How might the rolling process of cold working affect magnetization of the material? What is the effect of the field strength H on the magnetization B?
7. Compare benzene to polyacetylene, describing similarities and differences in composition, structure, properties, and function.

REFERENCES

1. R. Moucka, M. Mrlik, M. Ilcikova, Z. Spitalsky, N. Kazantseva, P. Bober & J. Stejskal, *Chem. Papers* **2013**, *67*, 1012.
2. Yu.S. Perets, L.Yu. Matzui, L.L. Vovchenko, Yu.I. Prilutskyy, P. Scharf & U. Ritter, *J. Mater. Sci.* **2014**, *49*, 2098.
3. W. Zawadzki, *Mechanisms of Electron Scattering in Semiconductors*, Ossolineum: Wroclaw **1979**.
4. F. Urbach, *Phys. Rev.* **1953**, *92*, 1324.
5. N.F. Mott, Nobel Lecture, Electrons in glass, in *Nobel Lectures, Physics 1971–1980*, Ed. S. Lundqvist, World Scientific Publishing Co.: Singapore **1992**.

6. N.F. Mott, *Conduction in Non-Crystalline Materials*, 2nd edn., Oxford: Clarendon Press **1993**.
7. N.F. Mott, *Metal-Insulator Transitions*, London: Taylor & Francis **1974**.
8. A. Boukhachem, B. Ouni, A. Bouzidi, A. Amlouk, K. Boubaker, M. Bouhafs & M. Amlouk, "Quantum Effects of Indium/Ytterbium Doping on ZnO-Like Nano-Condensed Matter in terms of Urbach-Martienssen and Wemple-DiDomenico Single-Oscillator Models Parameters," *ISRN Condensed Matter Physics* **2012**, Article ID 738023; 10.5402/2012/738023.
9. J.D. Dow & D. Redfield, *Phys. Rev. B* **1972**, *5*, 594.
10. P. Petkova & K. Boubaker, *J. Alloys & Compounds* **2013**, *546*, 176.
11. D. Gherouel, S. Dabbous, K. Boubaker & M. Amlouk, *Mater. Sci. in Semicond. Proc.* **2013**, *16*, 1434.
12. K. Ben Messaoud, A. Gantassi, H. Essaidi, J. Ouerfelli, A. Colantoni, K. Boubaker & M. Amlouk, *Adv. Mater. Sci. & Eng.* **2014**, Article ID 534307; 10.1155/2014/534307.
13. M. Haj Lakhdar, B. Ouni, R. Boughalmi, T. Larbi, A. Boukhachem, A. Colantoni, K. Boubaker & M. Amlouk, *Current Appl. Phys.* **2014**, *14*, 1078–1082.
14. K. Boubaker, *ISRN Nanomaterials* **2012**, 4; 10.5402/2012/173198.
15. K. Boubaker, *J. Ceramics* **2013**, *6*, 2013.
16. K. Boubaker, M. Amlouk, Y. Louartassi & H. Labiadh, *J. Austral. Ceram. Soc.* **2013**, *49*, 115.
17. K. Boubaker & M. Amlouk; *Internat. J. Appl. Ceram. Technol.* **2014**, *11*, 530.
18. K. Boubaker, A. Colantoni & P. Petkova, *Internat. J. Chem. Phys.* **2013**, *2013*, Article ID 728040; 10.1155/2013/728040.
19. K. Boubaker & P. Petkova, *J. Molec. Struct.* **2013**, *1049*, 233.
20. K. Boubaker, *Adv. Phys. Chem.* **2013**, *2013*, 5.
21. K. Boubaker, *Mendeleev Commun.* **2013**, *23*, 160.
22. J. Zhou, Z. Zou, A.K. Ray & X.S. Zhao, *Ind. Eng. Chem. Res.* **2007**, *46*, 745.
23. P. Lacovara, H. Chai, C.A. Wang, R.L. Aggarwal & T.Y. Fan, *Opt. Lett.* **1991**, *16*, 1089.
24. S.E. Lin, Y.L. Kuo, C.H. Chou & W.C.J. Wei, *Thin Solid Films* **2010**, *518*, 7229.
25. S. Tekeli & U. Demir, *Ceram. Internat.* **2005**, *31*, 973.
26. C.E.Z. de Souza, A.P. Ibaldo, D.J. Coutinho, E. Valaski, O.N. Oliveira Jr. & R.M. Faria, *Appl. Phys. A* **2012**, *106*, 983.
27. D.J. Coutinho & R.M. Faria, *Appl. Phys. Letters* **2013**, *103*, 223304.
28. Q. Cao, O. Gunawan, M. Coppel, K.B. Reuter, S.J. Chey, V.R. Deline & D.B. Mitzi, *Adv. Energy Mater.* **2011**, *1*, 845.
29. R. Noriega, J. Rivnay, K. Vandewal, F.P.V. Koch, N. Stingelin, P. Smith, M.F. Toney & A. Salleo, *Nature Mater.* **2013**, *12*, 1038.
30. N. Stingelin, *Polymer Internat.* **2012**, *61*, 886.
31. H. Shirakawa, E.J. Louis, A.G. MacDiarmid, C.K. Chiang & A.L. Heeger, *J. Chem. Soc. Commun.* **1977**, 578.
32. F. Goubard & G. Wantz, *Polymer Internat.* **2014**, *63*, 1562.
33. C. Müller, E. Wang, L.M. Andersson, K. Tvingstedt, Y. Zhou, M.R. Andersson & O. Inganäs, *Adv. Funct. Mater.* **2010**, *20*, 2124.

34. C. Müller, *Chem. Mater.* **2015**, *27*, 2740.
35. E.L. Spitler & W.R. Dichtel, *Nature Chem.* **2010**, *2*, 672.
36. H. Kamerlingh-Onnes, *Comm. Phys. Lab. Univ. Leiden* **1911**, Nos. 120b, 122b.
37. J.G. Bednorz & K.A. Müller, *Z. Physik B* **1986**, *64*, 189.
38. M.A. White, *Physical Properties of Materials*, CRC Press, Boca Raton, FL **2012**.
39. D. van Delft & P. Kes, "The Discovery of Superconductivity", *Physics Today*, September **2010**, pp. 38–43.
40. D. Larbalestier & P.C. Canfield, "Superconductivity at 100—Where we've been and Where we're Going", *MRS Bulletin* **2011**, *36*, 590–593.
41. Emergent Universe *Superconductivity: Resistance is Futile* http://www.emergentuniverse.org/#/superconductivity, accessed March 15, **2016**.
42. R.B. Wicker & E.W. MacDonald, *Virtual Phys. Prototyp.* **2012**, *7*, 181.
43. C.M. Shemelya, A. Rivera, A. Torrado Perez, C. Rocha, M. Liang, X. Yu, C. Kief, D. Alexander, J. Stegeman, H. Xin, R.B. Wicker, E. MacDonald & D.R. Roberson, *J. Electron. Mater.* **2015**, *44*, 2598.
44. P.P. Borbat, A.J. Costa-Filho, K.A. Earle, J.K. Moscicki & J.H. Freed, *Science* **2001**, *291*, 266.

18

MAGNETIC PROPERTIES

"Have you asked some good questions in school today, Isidor?"

—Sheindel Rabi, as quoted by her son Isidor Isaac Rabi. He invented the atomic and molecular beam magnetic resonance method of measuring magnetic properties of atoms, molecules, and atomic nuclei. His work led to the invention of the laser, the atomic clock, and to diagnostic uses in medicine of nuclear magnetic resonance (NMR). Rabi was born in a little town of Rymanów, then in the Kingdom of Galicia. At the age of 73 he returned to his native Rymanów (city charter in 1376, then and now in south-eastern Poland) for a triumphant visit. Rabi was then a Columbia University Professor (the first with such a broad title), the 1944 Nobel laureate in Physics, and also the Science Advisor to President Dwight Eisenhower. Inevitably, he was asked: how has he achieved so much? Rabi replied that it was his mother who made him what he is. In Rymanów and elsewhere, Rabi attributed his success to the way his mother used to greet him most days when he came home from school—with the words quoted above.

18.1 MAGNETIC FIELDS AND THEIR CREATION

There are many analogous features between electrical and magnetic properties; there are also certain analogies between dielectric and magnetic properties. As we examine magnetic properties of materials, we shall see the similarities as well as differences among these three categories of properties. We shall also find various ways in which they are related and interact. To begin, let us be clear that with electric charges, we have separate positive and negative entities; thus each entity is a **monopole**. No corresponding magnetic monopole has been found; magnets always appear as **dipoles**.

Materials: Introduction and Applications, First Edition. Witold Brostow and Haley E. Hagg Lobland.

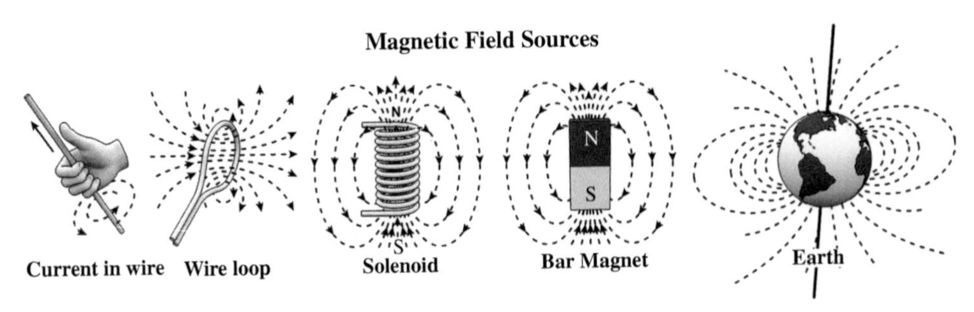

FIGURE 18.1 Magnetic field sources.

Like electric charges that generate an electric field, magnetic entities generate a **magnetic field**. Several ways of creating magnetic fields are shown in Figure 18.1. The five scenarios shown in Figure 18.1 really fall under two simple categories. A magnetic field can be generated either *by permanent magnets* or *by conducting materials* carrying electric current. In the former, microscopic currents associated with electrons in atomic orbits are at play; in the latter, macroscopic currents.

Bar magnets remain "permanent" until something happens to change the alignment of "atomic magnets" in the bar of iron, nickel, cobalt, etc. Neodymium permanent magnets are a type of rare-earth magnet actually composed of an alloy of neodymium, iron, and boron; they are presently the strongest commercially available magnets. The magnetic field of Earth seemingly comes from a giant bar magnet with its south magnetic pole located near Earth's north pole and with field lines as shown in Figure 18.1. On the other hand, the "geodynamo" theory (still not proven) suggests that the rotation of the earth somehow contributes to development of a current loop that in turn generates the magnetic field [1]. The magnetic field of Earth also helps to protect its inhabitants from the dangers posed by energetic protons of cosmic radiation.

The connection between electric current and magnetic fields is important. The effect of the former on the latter is plainly evident in the changing direction of a compass needle when a magnetic compass is near a wire carrying electrical current. As electric charges gave rise to electric fields, electric current gives rise to magnetic fields. The direction of the magnetic field lines around a portion of wire can be ascertained simply by using the mathematical **right-hand** rule for cross products of vectors. If one points the thumb in the direction of electric current flow, then the magnetic field lines follow along the path of the fingers as they bend at the knuckles and begin to curve around. This technique is illustrated in part (a) of Figure 18.1.

The **magnetic field** describes a region in space surrounding a magnetic body and in which a magnetic field can be induced into other bodies. The magnetic field is schematically shown by the familiar **field lines** or **flux lines** of Figure 18.2. Flux, of course, refers to the flow of a physical property through space; and the magnetic flux through a material can be calculated. By treating a surface as a collection of smaller elements, in which each has a constant magnetic field, the total flux can be obtained by summation over these elements. Most scientists and engineers working with magnetism aim to study the influence of a magnetic field on some object; typically this involves the use of conducting materials (as illustrated in Figure 18.1) to generate what is referred to as an *applied magnetic field*.

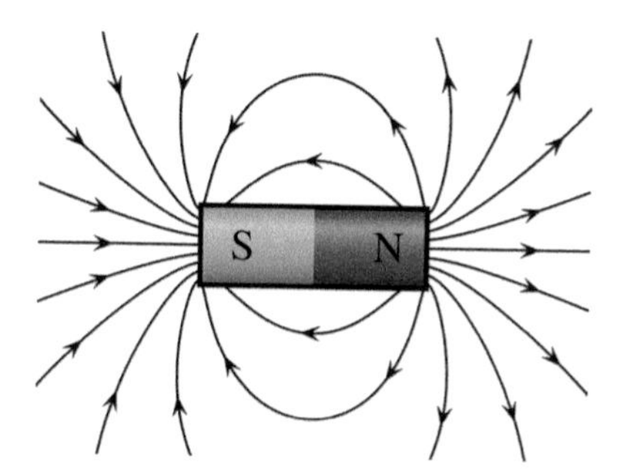

FIGURE 18.2 Magnetic field lines. *Source*: "VFPt cylindrical magnet thumb" by Geek3 https://commons.wikimedia.org/wiki/File:VFPt_cylindrical_magnet_thumb.svg#/media/File:VFPt_cylindrical_magnet_thumb.svg. Used under CC BY-SA 3.0 via Wikimedia Commons.

There are two key properties here. An applied magnetic field is characterized by its **magnetic field strength** $\overline{H}$. Because an applied magnetic field can induce magnetic polarization in an object within its field, we have also the **magnetic induction** $\overline{B}$, sometimes referred to as the **magnetic flux density**. In a vacuum, the relationship between them is

$$\overline{B} = \mu_0 \overline{H} \tag{18.1}$$

where μ_0 is the magnetic **permeability of a vacuum**, which is a universal constant. For a material with a **magnetization** $\overline{M}$, the contribution of the material to $\overline{B}$ must be considered; thus we have

$$\overline{B} = \mu_0 \left(\overline{H} + \overline{M}\right) \tag{18.2}$$

The magnetization corresponds to the density of net magnetic dipole moments in the material. For a given material, there is also a characteristic permeability μ_{m} of the material; the relationship between $\overline{B}$ and $\overline{H}$ is:

$$\overline{B} = \mu_{\mathrm{m}} \overline{H} \tag{18.3}$$

The relative permeability μ_{r} is therefore:

$$\mu_{\mathrm{r}} = \frac{\mu_{\mathrm{m}}}{\mu_0} \tag{18.4}$$

In Figure 18.3 we illustrate the experiment of removal of a small cylinder from inside a larger one. Extra current passing around the cavity walls would be needed to restore the relative permittivity value to the situation before the removal.

To determine how the relative permeability differs from zero, the quantity **magnetic susceptibility** χ is defined:

$$\chi = \mu_{\mathrm{r}} - 1 \tag{18.5}$$

From Eq. (18.5) it is evident that χ will be negative for $\mu_m \le \mu_0$ and positive for $\mu_m > \mu_0$. As a consequence, different classes of magnetic materials can be distinguished by magnetic susceptibility (see Table 18.1).

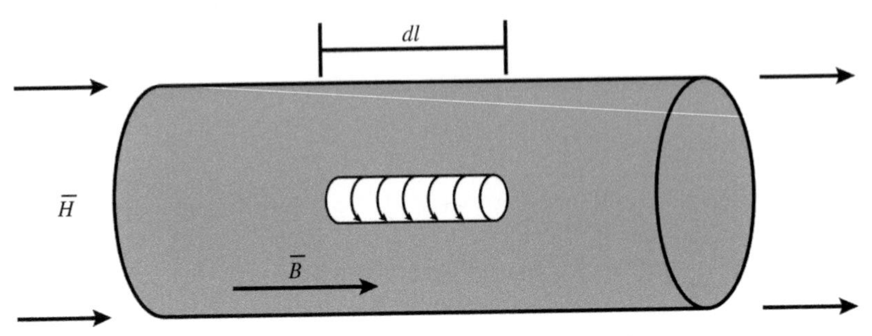

FIGURE 18.3 Removal of a small cylinder of length *dl* from inside a larger cylinder having relative permeability μ_r. A current, indicated by the circular arrows passing around the cavity walls, restores the magnetic induction to the value before the removal.

TABLE 18.1 Magnetic Susceptibility Values

MATERIAL	$10^5\chi$
Diamagnetic Materials	
Au	−3.44
Hg	−2.84
Te	−2.43
Ag	−2.38
Diamond	−2.18
Al_2O_3	−1.81
SiO_2	−1.63
Graphite	−1.6
Zn	−1.58
NaCl	−1.41
Cu	−0.96
Water	−0.91
ZrO_2	−0.83
Ge	−0.79
Si	−0.37
Paramagnetic Materials	
K	0.57
Na	0.85
Al	2.11
Ca	2.17
Ti	18.2
Y_2O_3	22.6
Nitinol (Ni + Ti, 50% each)	24.5–32.0
V	38.4
Pd	80.8
Stainless steel (nonmagnetic, austenitic)	350–670

18.2 CLASSES OF MAGNETIC MATERIALS

The five major groups describing the magnetic behavior of materials are as follows:

- **Diamagnetic**: no permanent dipoles, this in contrast to all other classes of materials. Opposition to the applied field is the only reaction that takes place. χ is always negative.
- **Paramagnetic**: random orientation of dipoles. Partial orientation of dipoles takes place upon imposition of the field. χ is small and positive. Magnetization is proportional to the applied magnetic field in which the material is placed.
- **Ferromagnetic**: there are strong interactions between dipoles resulting in parallel or antiparallel alignment of dipoles. Net magnetization is large even in the absence of an external magnetic field.
- **Antiferromagnetic**: neighboring magnetic spins (or alternate layers of magnetic spins) are equal but antiparallel; the net moment is zero. χ is small and positive.
- **Ferrimagnetic**: complex magnetic ordering, typical in ionic solids. Magnetic sublattices within the structure are characterized by different magnetic spins and electron interactions, resulting in a net magnetic moment. χ is large and positive.

Examples of diamagnetic and paramagnetic materials are listed in Table 18.1 along with corresponding values for the magnetic susceptibility at 293 K. One has to be careful here since mass, molar, and volume magnetic susceptibilities are all in use. In the Table we list volume susceptibilities; as Eqs. (18.4) and (18.5) tell us, the values are dimensionless.

Really all matter is magnetic, some much more so than others. Generally speaking, detectable magnetism depends on whether there is a strong collective interaction of atomic magnetic moments. Diamagnetic, paramagnetic, and antiferromagnetic materials are essentially non-magnetic. Materials that we typically think of as magnetic, such as Fe, Co, and magnetite are either ferromagnetic or ferrimagnetic. The behavior of the latter two categories is similar, but the underlying mechanism differs. Figure 18.4 gives a schematic to help describe the distribution of magnetic moments characterizing each category. We shall now discuss briefly each of these classes of magnetic materials.

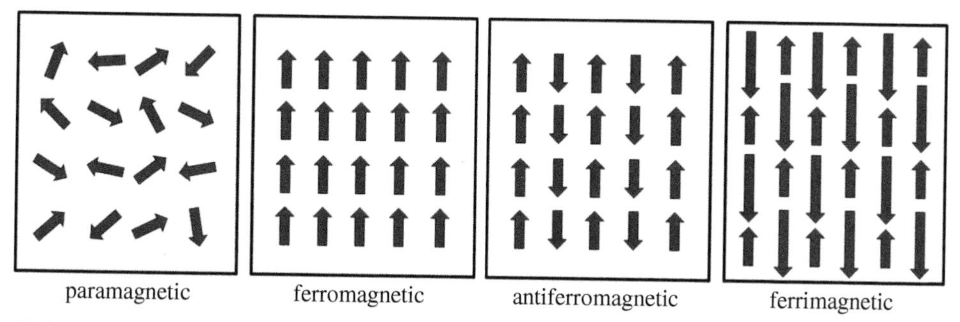

FIGURE 18.4 Magnitude and orientation of magnetic spins in materials. Arrows indicate spin direction, while different sizes of arrows reflect unequal spin size.

18.3 DIAMAGNETIC MATERIALS

In **diamagnetic** materials, electrons are paired, thus the (quantum) spins are also paired, resulting in no net magnetic moments. Diamagnetic substances respond to an external magnetic field by repulsing the field; thus negative magnetization is produced. This has already been mentioned and we have seen in Table 18.1 that χ is always negative for diamagnetic materials. While χ is independent of $\bar{H}$, the magnetization $\bar{M}$ *is* dependent on $\bar{H}$. Therefore, the weak magnetism of diamagnetic materials persists only as long as an external field is being applied.

18.4 PARAMAGNETIC MATERIALS

If there are unpaired electrons in a material, the net spin is non-zero. This is the case for **paramagnetic** materials; however, in the absence of a magnetic field, the magnetic moments do not interact and align, so the magnetization is zero. In the presence of a field, there is a partial orientation in the direction of the field—observed as a slight attraction to the field. There is a net positive magnetization and χ is positive.

Many transition metals and compounds containing oxygen (e.g., clays, silicates, carbonates) exhibit paramagnetism. It is important to note that for paramagnetic materials χ is dependent on temperature but independent of the applied field.

18.5 FERROMAGNETIC AND ANTIFERROMAGNETIC MATERIALS

Ferromagnetic materials are very attracted to a magnetic field; χ is very large and *dependent* on $\bar{H}$. This behavior results from the collective interaction of the magnetic moments of individual atoms. The class gets its name from Fe, the classic example of ferromagnetism. Even in the absence of an external magnetic field, the magnetic dipoles tend to align within microscopic regions called **domains**. All domains are not necessarily oriented in the same direction—unless a **coercive field** is applied. They exhibit hysteresis (see Figure 18.5), qualitatively similar to that of ferroelectrics. The spin alignments (parallel and antiparallel) associated with ferromagnetic behavior are closely related to the electronic band structure—which we know is an effect of interactions involving many atoms. The density of unoccupied states near the Fermi level plays an important role in the resultant magnetism of ferromagnetic materials. Thus, although electrical and magnetic properties are distinct, there is also a clear correspondence between them.

Temperature dependence of magnetization has already been hinted. The susceptibility χ of normal paramagnetic materials obeys the Curie law:

$$\chi = \frac{C}{T} \tag{18.6}$$

where C is the Curie constant.

Effects of increasing thermal energy and randomization of magnetic moments on magnetic susceptibility are evident. For ferromagnetic materials, the characteristic behavior is displayed only below the material's **Curie temperature** T_C; above T_C behavior is paramagnetic; see Figure 18.6.

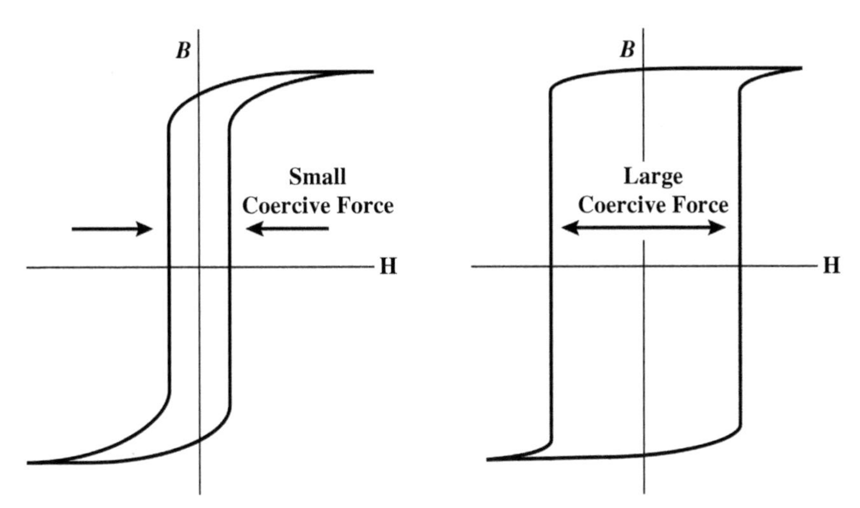

FIGURE 18.5 The magnetic hysteresis loop in the magnetic induction B versus magnetic field strength H coordinates. The distinction between "soft" and "hard" material is somewhat subjective, it relies on the perception that the loop is "lean" or "fat".

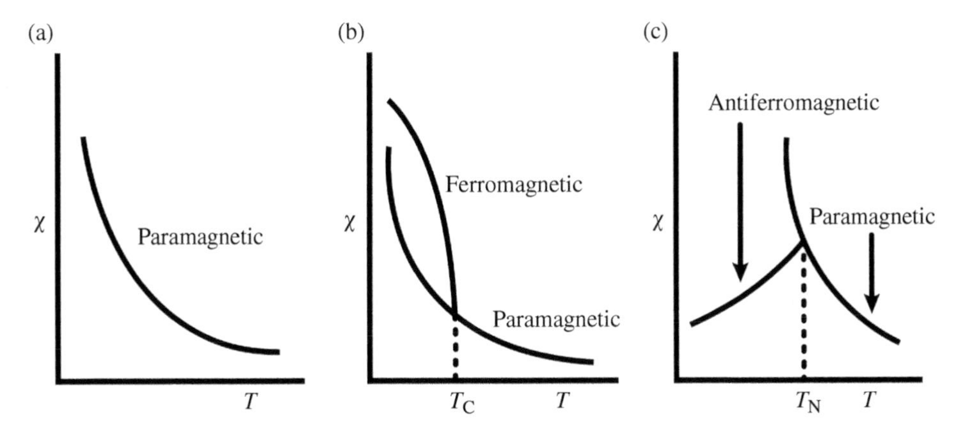

FIGURE 18.6 Temperature dependence of magnetic susceptibility for: (a) paramagnetic material; (b) ferromagnetic material with transition into paramagnetic shown; and (c) antiferromagnetic material, also with transition into paramagnetic behavior shown. T_C is the Curie temperature; T_N is the Néel temperature.

The Curie law for ferromagnetic materials is therefore:

$$\chi = \frac{C}{T - T_C}, \quad T < T_C \tag{18.7}$$

In Figure 18.6c we see the **Néel temperature** T_N at which the transition from the antiferromagnetic into paramagnetic occurs, so named after the French physicist Louis Néel. It is sometimes also called the **antiferromagnetic Curie point**. In contrast to the situation with ferromagnetics (in Eq. 18.7), the temperature dependence of magnetic susceptibility of antiferromagnetic materials is related to the Néel temperature as follows:

$$\chi = \frac{C}{T + T_N}, \quad T < T_N \tag{18.8}$$

Recall that the defining feature of antiferromagnetic materials is that neighboring magnetic spins (or alternate layers of magnetic spins) are equal but antiparallel; the net moment is zero.

Actually, not that many materials are ferromagnetic. The reason this fact is not obvious is that ferromagnetics include iron, cobalt, nickel, and necessarily alloys of these metals, the alloys of which are common and somewhat numerous.

There is an ongoing search for **tunable ferromagnetic materials**. Thus, Moodera and coworkers [2] have studied chromium doped In_2O_3, varying the concentration of the dopant. They have found that both electrical and magnetic behaviors—ranging from ferromagnetic metal-like to ferromagnetic semiconducting to paramagnetic insulating—can be controllably tuned by the defect concentration.

18.6 FERRIMAGNETIC MATERIALS

Ferrimagnetic materials have antiparallel spins of unequal size; see Figure 18.4. Ferrimagnetism may seem to be only a special case of antiferromagnetism, in which neighboring spins are equal and opposite. Ferrimagnets also share some properties with ferromagnets. However, there remain several important distinguishing features of ferrimagnetics: the materials that comprise them, their set of magnetic properties, and their temperature dependence.

Ferrimagnetic materials are typically ceramic and oxide materials that tend to be electric insulators. The magnetic moments are aligned antiparallel (as with antiferromagnets), but due to unpaired electrons there is spontaneous magnetization and a net non-zero magnetic moment. Again, look at Figure 18.4, which shows the ordering of magnetic spins in a ferrimagnetic lattice. Arrows pointing in opposite directions indicate antiparallel alignment; longer arrows pointing down indicate unpaired spin magnetic moments, resulting in a net magnetic moment of down spin moments.

There is a Curie temperature for ferrimagnetism at which the magnetic moments get randomized and the materials begin to behave paramagnetically. Below T_C, the magnetization decreases until reaching zero at the Curie temperature. Typically lower values for T_C are seen for ferrimagnets than for ferromagnets. Ferrimagnetism is exhibited by magnetic garnets, including the so-called yttrium iron garnet (YIG). The mineral **magnetite** Fe_3O_4, known as Lodestone, is a ferrimagnet and the only mineral that acts as a natural magnet. We note again the contrast between ferromagnetic and ferrimagnetic materials: the former form microscopic domains; the latter do not.

18.7 APPLICATIONS OF MAGNETISM

This concludes a fundamental explanation of magnetic properties. One can find tables of Curie and Néel temperatures, of the properties of hard and soft ferromagnets, and of susceptibilities of a large host of compounds in other reference works. Likewise one can read elsewhere in more detail about hysteresis curves and how they help identify materials for particular applications. The mechanisms and phenomena of magnetism are indeed quite complex and extensive. Here we want to close by mentioning a few interesting magnetic properties; the reader is encouraged to explore these topics further.

Let us look a little closer at **magnetic resonance**. Recall the motto of the present Chapter. Isidor Rabi first predicted that nuclear magnetic resonance (NMR) should exist and then demonstrated experimentally that it does [3]. Atomic nuclei in a magnetic field absorb and then re-emit electromagnetic radiation. The emission depends on two factors: the strength of the applied field and the local atomic structure of the material—either solid or liquid. Thus NMR is a powerful technique enabling determination of the structural arrangement of atoms in molecules. The ramifications are significant. Recall the general principle that materials—similarly as humans—"do not like" perturbations. Electrons moving around a given atom create a small magnetic field which opposes the externally applied field.

Mary Ann White [4] says that "NMR is arguably the most important structural technique in all of chemistry today". It is important also to geochemistry and soil chemistry [5]. However, as noted already under the motto of this chapter, NMR has applications in a variety of ostensibly unconnected fields. Laymen usually have at least heard about NMR applied in medical diagnosis—usually in the form of magnetic resonance imaging (MRI); diseases alter the magnetic behavior of atoms and molecules in the human body. A lucid presentation of the NMR technique, covering applications in chemistry as well as in medicine, is provided in a book by Blümich, Haber-Pohlmeier, and Zia [6].

Another interesting property of some materials is **magnetoresistance**, which refers to a change in electrical conductivity in response to an applied external magnetic field. Peter Grünberg and Albert Fert shared a Nobel Prize in Physics for their work in magnetoresistance [7]. As an example, Moodera and coworkers [8] observed a change of 24% in junction resistance in $CoFe/Al_2O_3/Co$ junctions upon exposure to a magnetic field at 4.2 K. The respective change of 11.8% at 295 K is still significant.

Magnetocaloric effects are also being investigated. Samanta, Dubenko, Stadler, and coworkers [9] studied the $MnNiGe_{1-x}Al_x$ system by magnetization and differential scanning calorimetry. First-order magnetostructural transitions (MSTs) from hexagonal ferromagnetic to orthorhombic antiferromagnetic phases have been found: for $x=0.085$ at 193 K and for $x=0.09$ at 186 K. Enthalpy and entropy changes at transitions have been determined, the former relatively large ones. The authors discuss possibilities of moving MST values towards the room temperature, thus towards magnetic refrigeration devices. Similar opportunities based on MSTs exist also for ternary or quaternary alloys containing Ni, Mn, In, Ga, Al or Ge [10].

Magnetic refrigeration takes place in adiabatic (thermally isolated) materials by decreasing the strength of an externally applied magnetic field. Phonons moving in the material then cause randomization of magnetic domains—a process that takes up energy and thus lowers the temperature. Recall that magnetic domain formation and randomization result in the magnetic hysteresis loops seen in Figure 18.5 (a function of magnetization and demagnetization processes). Magnetic refrigeration is also known under the name **adiabatic demagnetization**.

There exist also efforts towards the design of **molecular magnets**. The goal of such efforts is to build single molecules (or one-dimensional chains of molecules) with controlled magnetic properties [11]. Achieving magnetic performance at ambient or higher temperatures is also an aim in the development of this relatively new class of materials. Potential applications of molecular magnets include information storage and information processing.

Finally, we shall discuss **magnetic levitation (maglev) trains**. The simple fact that two identical magnetic monopoles (both either north or south) repulse each other is used here. A vehicle is levitated only, say, 10 mm above the rails—thus eliminating solid state friction and allowing much higher speed. In such a construction, the very high speed then makes *air drag* rather than solid state friction the main problem. The result of this technology created a paradigm shift that has started to change in some parts of the world the way people travel.

For the operation of maglev trains, the bottom part of the train wraps around the track. This is shown in Figure 18.7. A set of electromagnets on either side of the train pushes the train upwards so it floats above the track. Another set exerts a sideways force to keep the train in the correct position. The electromagnets must also generate forward motion of the train. How this is done is shown in Figure 18.8. The magnetization on the

FIGURE 18.7 A magnetic levitation train car. *Source*: Állatka, "Transrapid series 09 vehicle at the Emsland Test Facility, northern Germany", https://commons.wikimedia.org/wiki/File:Transrapid-emsland.jpg. Public Domain.

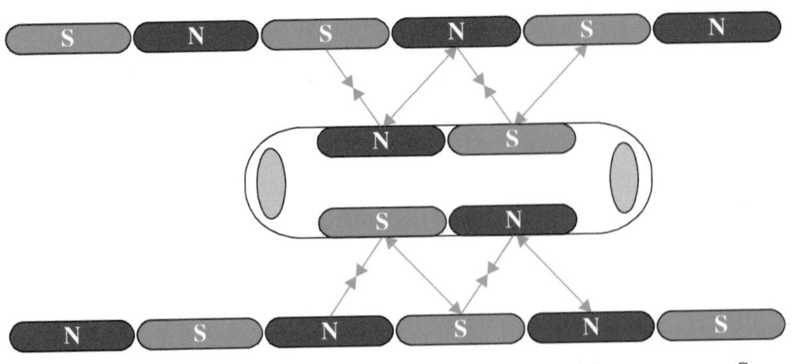

FIGURE 18.8 Illustration of how magnetic levitation propulsion operates. *Source*: Cool Cat, https://commons.wikimedia.org/wiki/File:Maglev_Propulsion.svg. Used under CC-BY-SA-3.0, GFDL-en.

train car remains constant while magnetic polarity in sections of the track can be altered. The switching magnetism of the track is indicated by the double arrows in Figure 18.8. Due to the alternating attraction and repulsion the train is continually pulled into the next section of track, with electromagnets running the entire length of the guideway.

The power needed for levitation is usually not a particularly large percentage of the overall consumption of the power needed for the moving train. As already noted, most of the power used is needed to overcome air drag, as it is for that matter with any other high speed train. Deutsche Bahn, the German national rail company, had their original patent for maglev trains issued as far back as August 14, 1934, the main inventor being Hermann Kemper. Japan has maglev trains in operation, including the famous Chūō Shinkansen. Speeds exceeding 600 km/h have been achieved in Japan. There are also magnetic levitation trains in China (Shanghai) and in Korea (serving the Seoul Incheon Airport). Acceleration is such that 580 km/h can be achieved after covering less than 9 km.

18.8 SELF-ASSESSMENT QUESTIONS

1. Describe ways to create a magnetic field.
2. Consider the origins of magnetism in a ferroelectric metal. How might the rolling process of cold working affect magnetization of the material? What is the effect of the field strength $\bar{H}$ on the magnetic induction $\bar{B}$?
3. Explain the principle of operation of magnetic levitation trains.
4. What do you think a ferrimagnetic material could be used for? After providing an answer, find out from other resources some uses of ferrimagnetism and describe them.

REFERENCES

1. R. Hollerbach, *Physics of the Earth and Planetary Interiors*, **1996**, *98*, 163.
2. J. Philip, A. Punnoose, B.I. Kim, K.M. Reddy, S. Layne, J.O. Holmes, B. Satpati, P.R. LeClair, T.S. Santos & J.S. Moodera, *Nature Mater.* **2006**, *5*, 298.
3. I.I. Rabi, J.R. Zacharias, S. Millman & P. Kusch, *Phys. Rev.* **1938**, *53*, 318.
4. M.A. White, *Physical Properties of Materials*, CRC Press, Boca Raton, FL **2012**.
5. M.A. Wilson, *NMR Techniques and Applications in Geochemistry and Soil Chemistry*, Pergamon Press, Oxford **1987**.
6. B. Blümich, S. Haber-Pohlmeier & W. Zia, *Compact NMR*, Walter de Gruyter, Berlin/Boston **2014**.
7. P.A. Grünberg, *Rev. Modern Phys.* **2008**, *80*, 1531.
8. J.S. Moodera, L.R. Kinder, T.M. Wong & R. Meservey, *Phys. Rev. Letters* **1995**, *74*, 3273.
9. T. Samanta, I. Dubenko, A. Quetz, S. Temple, S. Stadler & N. Ali, *Appl. Phys. Letters* **2012**, *100*, 052404.
10. I. Dubenko, T. Samanta, A. Kumar Pathak, A. Kazakov, V. Prudnikov, S. Stadler, A. Granovsky, A. Zhukov & N. Ali, *J. Magn. Magnet. Mater.* **2012**, *324*, 3530.
11. L. Bogani & W. Wernsdorfer, *Nature Mater.* **2008**, *7*, 179.

19

SURFACE BEHAVIOR AND TRIBOLOGY

Patina,
august,
mellow green,
flowering on church spires,
and on rooftops
billowing through the centuries.

Necklace, virescent,
as eucalyptus leaves,
patiently burnished,
lustrous,
warm
on the skin of a woman.

And lit by the desert sun
gray-green columns
of eastern temples
recalling the ancient Persians.

—Malachite, by Janina Brzostowska; translated from the Polish by Neil S. Snider.
The chemical formula of malachite and of patina is $CuCO_3{\cdot}Cu(OH)_2$.

19.1 INTRODUCTION AND HISTORY

The term *tribology* is derived from the Greek word *tribos* meaning "to rub", hence "the science of rubbing". It entails the familiar study areas of friction, wear, and lubrication, but not only these. **Tribology** refers to the science and technology of interacting surfaces in

Materials: Introduction and Applications, First Edition. Witold Brostow and Haley E. Hagg Lobland.

relative motion [1]. It encompasses the art of addressing issues of reliability, maintenance, and wear of equipment ranging from common household appliances to advanced aircraft—thus tribology deals with problems that significantly affect economies. The interfacial interactions in tribological contacts are highly complex—not surprising based on the information in Chapter 2 and others in this book—therefore understanding these interactions requires knowledge from various disciplines including physics, chemistry, applied mathematics, solid mechanics, fluid mechanics, thermodynamics, heat transfer, materials science, rheology, lubrication, machine design, performance, and reliability [1].

Tribology is new in name only. The term came to popular use after being coined in the 1966 report by H. Peter Jost, which detailed the extent of economic loss due to wear and the potential for monetary savings for industry in the United Kingdom by better application of tribological principles and practices [2]. Prior to the Jost report, David Tabor along with Philip Bowden established through their research a great deal of the scientific understanding of tribology (refer to their many individual and joint articles and books). Several hundred years earlier, in the 15th Century AD, Leonardo da Vinci explained the ratio of friction force to normal load for blocks sliding over a flat surface. Other noteworthy individuals who have contributed (during the 15th to 20th Centuries) to the knowledge of tribological principles include Guillaume Amontons, Charles Augustin Coulomb, Robert Hooke, Sir Isaac Newton, Beauchamp Tower, Osborne Reynolds, N.P. Petroff, Ragnar Holm [1]. The rapid development of machinery during the Industrial Revolution (ca. AD 1750–1850) and the expansion of the petroleum industry (beginning around 1850) spurred on the demand for better knowledge in all areas of tribology.

The great and useful contributions of the above-named scientists notwithstanding, there is sufficient evidence indicating that human interest in the constituent parts of tribology is as old or older than recorded history [3]. Consider the long history of wheels—for transportation, pottery, grinding, and otherwise—and that a number of these were designed to use some sort of bearings in their operation. Ancient records, such as from Egypt ca. 1880 BC, indicate the use of lubricants to facilitate the transportation of heavy objects on sledges. A chariot from an Egyptian tomb dated several thousand years BC was found to have some remaining animal-fat lubricant in its wheel bearings [1]. An insightful article entitled "The Ten Greatest Events in Tribology History" provides a fascinating account of tribological discoveries throughout the course of history [4]. In that article, Mark Riggs is quoted as saying:

> The singular event in tribology was the guy 3,200 years ago who thought to slap some animal fat on the axle of his wagon [4].

Another illuminating statement is: "If there were no wheels, there would be no machines. Wheels are closely connected to bearings, and bearings are the main subject of tribological research" [4]. We shall see that tribology deals with both plane sliding and rolling contacts, but certainly the author of that statement understands the importance of tribology to wheels and bearings, which are abundant in modern society.

It is easy to think of many instances and operations in which we desire to minimize friction and wear. Unproductive friction and wear occurs in bearings and seals, in the knees of denim pants, on the surfaces of non-stick cookware, and in damaged or artificial body joints. There are also cases in which friction and wear are *productive*. For example, the presence of *friction* is useful in brakes, clutches, bolts, and nuts. Activities such as writing with a pencil, machining, polishing, and shaving provide examples of *productive wear*.

Tribology is at play in individual components, in assemblies and products, in manufacturing processes, in construction and exploration, and in natural phenomena. Tribology is also at play on multiple levels, from the macroscale to the nanoscale, and from bulk surfaces to atoms in contact with one another. Virtually every area of engineering and industry must contend with some aspect of tribology; from agriculture to cosmetics, dental implants to food processing, shoe manufacturers to sports equipment companies, each of these entities must have an appropriate knowledge of tribology as it relates to their own unique procedures and processes. In every situation, however, there is a common feature: surfaces in contact and in relative motion—sliding or rolling.

The scope of this chapter is not limited to tribology only; we discuss also general properties of surfaces, including those not in contact or motion. In fact, that is where we shall go now.

19.2 SURFACES: TOPOGRAPHY AND INTERACTIONS

Let's take a close look at a solid surface. We already know from earlier discussions that the properties of the surface differ from those of the bulk. This feature of matter also affects optical, electrical, and thermal properties and influences the appearance of an object and even its ability to be painted.

First, there is **surface roughness**. No matter how a solid material has been prepared, its surface will always contain irregularities. These may range in scale from deviations in shape to deviations on the order of interatomic distances. A molecularly flat surface cannot be produced in conventional materials by even the most precise method of machining. Nor does even crystal cleavage yield a molecularly flat surface. There always exist irregularities whose height exceeds that of the interatomic distances. These peaks or irregularities are called asperities; and these asperities have mechanical and chemical properties. Tribological forces are dictated by the interaction of asperities. Furthermore, the geometry and distribution of asperities is a result of past history including the manufacturing method, handling, and prior rubbing.

A solid surface also contains layers or *zones having different physicochemical properties* (see Figure 19.1). The forming processes of metals produce work-hardened or deformed layers that are characterized not only by chemical reactivity but also by mechanical behavior that is different from that of the bulk. On top of these deformed layers is a microcrystalline or amorphous region known as the Beilby layer. There is also some deformation in the surface layers of polymeric and ceramic materials, but owing to differences in their structure and processing, the deformation is not as definite as it is for metals.

The physical and chemical aspects of surfaces affect mechanical, optical, and other properties of materials. Some of these effects are described in the left side of Figure 19.2. In turn, there are various techniques, as shown in the right side of Figure 19.2, used to characterize those aspects.

Many material surfaces are chemically reactive. Most metals and alloys as well as many non-metals react with oxygen and possibly also carbon dioxide in air, thus forming surface layers; see again the motto of this chapter. In other environments layers such as nitrides, sulfides, and chlorides may be formed, yielding a chemical corrosion film on the surface. Moreover, there may be adsorbed films that result from the physisorption or chemisorption of elements such as oxygen, water vapor, and hydrocarbons from the environment. Such films form on metallic and non-metallic surfaces. A film that is even a fraction of a monolayer can significantly alter the surface interaction with another solid,

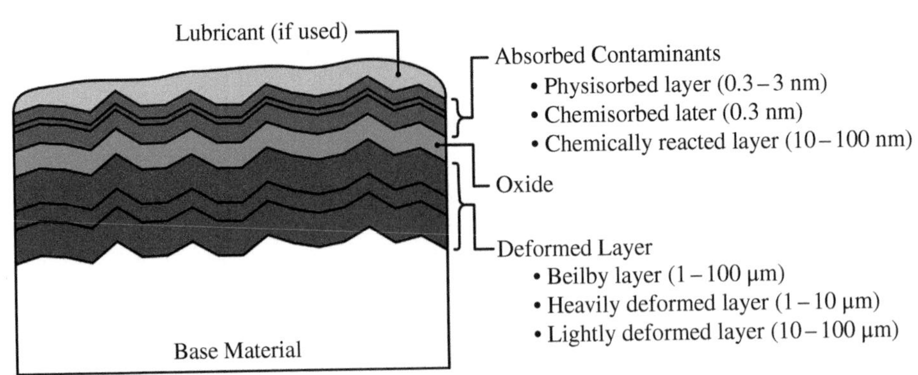

FIGURE 19.1 Schematic of a solid surface and typical surface layers. Adsorbed contaminants may result from physisorption, chemisorption, and chemical reactions. The oxide layer formed on many types of materials constitutes a chemically reacted layer. The deformed layer is subdivided into 3 separate layers. In some metals and alloys a microcrystalline or amorphous structure known as the Beilby layer is present. Underneath that are more and less heavily deformed layers, resulting from processing and prior history of the material part. Finally, underneath all the surface layers is the base material from which are derived the bulk material properties. All surfaces have a roughness, determined by prior history and the nature and composition of the layer.

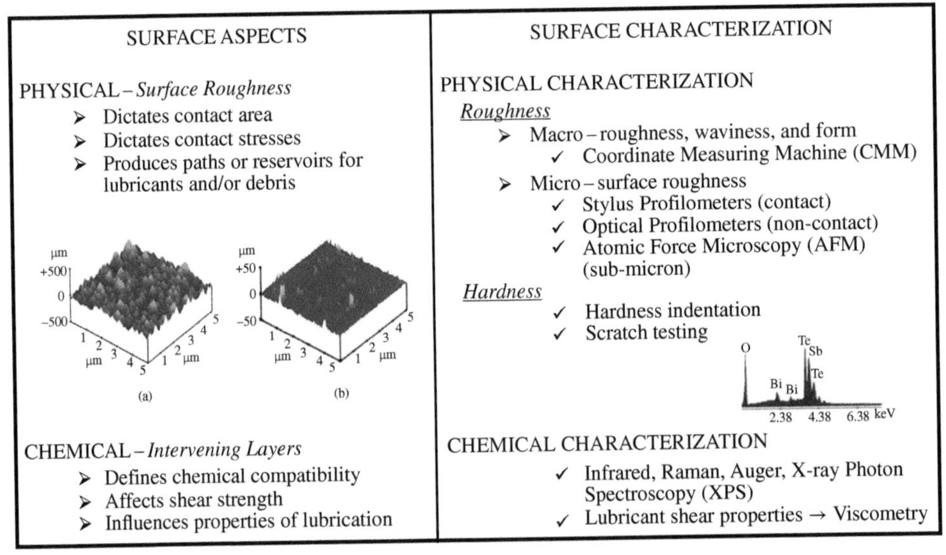

FIGURE 19.2 Physicochemical properties, effects, and testing methods in solid surface layers.

liquid, or gas. In some cases, these surface films may be worn out in the initial period of operation and thereafter exert no influence. However, a greasy or soapy film derived from the environment may be more persistent and markedly alter surface interactions.

The energetic properties of a surface differ from those of the bulk material; because work is required to form a surface, there is an associated energy. The **surface energy** is quantified as the work per unit area for surface formation and is related to the disruption of intermolecular bonds needed to accomplish the process. As a general trend (since there are a number of exceptions) ionic solids have surface energy $<1\,J{\cdot}m^{-2}$, metals $\approx 1\,J{\cdot}m^{-2}$, and covalent solids $>1\,J{\cdot}m^{-2}$. Considering the physics of solids, it makes sense that surfaces should be intrinsically less energetically favorable than the bulk of a material, otherwise there would be a driving force for more surfaces to be created, thereby removing the bulk of the material! Therefore, one can think of surface energy as the excess energy of the surface material compared to the bulk.

Consider that in the bulk, atoms are evenly surrounded and the cohesive forces tend to balance. At the surface, except for a small number of atoms in the gas phase, most atoms are attracted from primarily one side, which yields a net inward cohesive force that favors minimization of surface area. This force is called **surface tension**. In liquids the surface tension and surface energy are identical. As temperature increases the atoms in a solid vibrate more, reducing the cohesive force binding the atoms. Since the surface energy depends directly on the net inward cohesive force, the surface energy decreases with increasing temperature. Glass at high temperature has a tendency to form into a sphere. The transformation of a cylindrical specimen into a sphere, which has a smaller surface area, results in energy being released. Surface energy can also be lowered by contaminant molecules adhering to the surface. How do you suppose such contaminants affect the net inward force? We shall return to the story of surface tension in Section 19.5.

When we cut a solid body of material, bonds are disrupted and new surface is created; thus we know that energy is consumed in the process. If the process were reversible, the law of conservation of energy implies that the energy consumed will be equal to the energy inherent in the newly created surfaces. All other things being equal, the unit surface energy of two new faces would be half the energy of cohesion. The simplicity of this implied "cleaved bond" model is not realized, though it may be possible for a surface freshly prepared in a vacuum. In practice, surfaces are dynamic regions that readily rearrange or react; processes such as passivation and adsorption often reduce the surface energy.

Surface energy for solids is frequently measured by fracture and indentation methods. Surface tension at solid surfaces can be calculated by mathematical methods from the measurement of the contact angles of liquid droplets on the surface [5–7]. If working with liquids only, it is more common to report the values of surface tension, since it refers to the same dimensional quantity as surface energy. The size of a liquid bubble is determined by the balance between excess pressure and surface tension.

Surfaces are important also because they mediate many types of reactions and processes. The mechanism of some catalysts is to speed up chemical reactions by facilitating the reaction on their particulate surfaces. Surfaces can be modified by chemical or physical means to have different properties than the bulk interior. For example, plasma treatment of hydrophobic polymers can render the surface hydrophilic and biologically compatible [8] without compromising the bulk properties needed to maintain appropriate mechanical performance. Alternatively, an unreactive surface might be plasma treated to increase its chemical reactivity for the purpose of adhering a coating to it.

19.3 OXIDATION

The chemical reactions of oxidation and corrosion affect the vast majority of materials. Because of their unique situation, surfaces are more susceptible to these processes.

In the case of steels, the oxidation of iron to rust leads ultimately to destruction of the original material. In other cases, such as for aluminum, an oxidized surface layer actually serves a barrier that protects the underlying bulk. Oxidation, the reaction of a material with O_2, can happen in wet and dry environments. For example, the reaction at the silicon surface during dry oxidation is:

$$Si + O_2 \rightarrow SiO_2 \quad (19.1)$$

For wet oxidation, the reaction is:

$$Si + 2H_2O \rightarrow SiO_2 + 2H_2 \quad (19.2)$$

The dry oxidation of Si yields a higher quality oxide film but at a lower growth rate than by wet oxidation. Thus, this method is preferable for the preparation of thin silica films (e.g., screen oxide, pad oxide, gate oxide). The higher oxide density resultant from dry oxidation imparts a higher breakdown voltage to the material. During wet oxidation, hydroxide formed by the dissociation of water at high temperatures can diffuse more quickly through silicon than can molecular oxygen. The outcome is a notably higher oxidation rate that enables the formation of thick oxides (e.g., masking oxide, blanket field oxide). These two situations describe oxide materials intentionally produced for use in electronic components.

As already implied in the previous discussion, the process of oxidation for a given material is typically temperature dependent. Thermoelectric materials such as the compounds Bi_2Te_3, SiGe, and PbTe are relatively stable at room temperature but suffer from oxidative degradation at elevated temperatures in oxygen-containing atmosphere—more on this subject in the next Chapter. In general, high temperature speeds up the oxidation rate for most material types—making oxidation one of the most significant high-temperature corrosion reactions. However, oxidation can take place whenever the oxygen content and oxygen activity are suitable and is therefore possible even in reducing atmospheres for certain cases.

Now let's take a closer look at what happens to a single crystal of pure metal (M) in an oxidizing environment. First, we note that oxidation takes place at the surface; we have empirically $M + O \rightarrow MO$. Once the surface is completely oxidized, there will only be further oxidation if (1) metal atoms diffuse through the oxide layer to the surface and react with oxygen at the air/oxide interface *or* (2) oxygen diffuses through the oxide layer and reacts with the metal at the metal/oxide interface. Therefore, it is evident that oxidation is diffusion controlled. The rate of oxidation is thus defined quantitatively in terms of the thickness of the oxide layer and a calculated rate constant. Already in 1933 Wagner established widely used theories on oxide growth kinetics [9]. However, his prediction of parabolic oxide growth rarely occurs in real systems because one or more of the underlying assumptions are not met. For instance, the assumption that the migration of ions or electrons across the oxide is the rate controlling process is frequently not true in real systems. In fact, the self-diffusion of metal or oxygen atoms through the oxide layer ought to be the rate limiting step since the migration of one or the other component is required for the reaction to proceed. Moreover, calculations that take into account self-diffusion predict rates that are several orders of magnitude too low. Such underestimates are attributed to the presence of structural defects in the oxide layer. To understand this, consider how diffusion might be affected by dislocations, grain boundaries, and impurities in the oxide layer. It turns out that diffusion is higher in the vicinity of these defects. Not surprisingly then, the oxidation process becomes more complex when alloying elements are present in a metal.

The nature of the oxide layer that forms in a binary alloy of metal A and B depends on several factors including the reactivity of each metal and the concentration of B in A.

Thus, for example, one may end up with an oxide layer consisting primarily of one metal oxide, say, BO, while the underlying bulk has its interfacial layer slightly deficient in B. Specimen size has an impact on the oxidation kinetics, while diffusivity of elements and scale also play important roles.

In nickel-based superalloys and many other alloys as well, the surface oxide layer serves to protect the underlying bulk from further oxidation and degradation. However, certain conditions can reduce the protectiveness of an outer oxide layer. The evaporation of volatile oxides from the layer will cause thinning of the barrier. Mechanical damage to the oxide such as by erosion, cracking, or spalling will lower the protectiveness. Likewise stresses during formation or service can reduce the overall protectiveness of an oxide layer on the surface of an alloy. Therefore, additional alloying elements are sometimes incorporated to help mitigate these points of weakness. This is often the situation with large-scale structures and components. At the same time, there is ongoing research to better understand the oxidation process at the atomic level.

Our discussion of oxidation has centered on inorganic materials; what about organic raw materials and polymers? In this case the mechanisms and outcomes of oxidation are quite different than with inorganic materials. We know from experience that the reaction of many carbon-containing materials with oxygen is associated with burning. Of course one can also set metal on fire (causing it to oxidize), but this requires thin strands and a lot of surface area; bulk metal objects typically conduct heat away too quickly to attain the necessary temperature for burning.

The majority of hydrocarbons discussed in Chapter 8 react with O_2 forming CO_2 and H_2O accompanied by an energy release from breaking the hydrocarbon bonds. Thus we are familiar with the oxidation of raw organic materials for power generation. Perhaps less familiar are chemical oxidation methods used for environmental remediation in the event of hydrocarbon and petroleum product spills or contamination. Chemical oxidation methods are used to quickly initiate decomposition of petroleum contaminants by instigating their breakdown. Among the oxidants used are the following: hydrogen peroxide, Fenton's Reagent, permanganate, and ozone. Apart from some inherent risks with the chemicals themselves, there are also some serious health risks posed by the breakdown products (e.g., benzene, polycyclic aromatic hydrocarbons (PAHs), and methyl tert-butyl ether (MTBE)) associated with chemical oxidation methods [10].

There is of course another kind of organic raw material somewhat less utilized by practitioners of MSE: that is the organic material of soil, vegetable matter, animal manure, and the like. As with petroleum hydrocarbons, this matter undergoes oxidation. The process is more commonly referred to as aerobic decomposition, and heat is generally a byproduct. Recall our earlier discussion of the relation between diffusion and oxidation kinetics. It is not surprising therefore that grinding organic matter into smaller particle sizes, thereby increasing the total surface area, improves the rate of oxidation. Living biological organisms also play a role in the oxidative degradation of organic matter. In the absence of oxygen, we get decomposition of organic matter by fermentation, a somewhat more smelly process.

Now on to the issue of oxidation in polymer based materials, including engineering plastics. The rate of diffusion of oxygen will tend to be faster through amorphous regions of a polymer than through closely packed crystalline regions. Oxidative reactions in polymers may proceed by mechanisms operating within the main macromolecular chain or within the branched groups. In polyolefins, a greater number of branches may increase the diffusion of oxygen to the main chain, accelerating oxidation. However, side groups can affect the access of oxygen and thus affect the oxidation rate. Polypropylene (PP), as compared to PE, is more sensitive to oxygen and degrades more rapidly.

Molecular oxygen reacts easily with other free radicals. The oxidation of polyolefins, rubber, and polyamides involves hydroperoxide and peroxy radical intermediates. For halogenated polymers, the first step is dehydrohalogenation followed by thermal oxidation. Thermal oxidation is simply a thermally initiated reaction with molecular oxygen. Typically that process also involves radicals. As expected, the absorption of oxygen is proportional to the amount of free surface. Chemical constituents as well as structure affect susceptibility to oxygen.

Photo-oxidation in polymers results from the susceptibility of certain chemical groups to degradation by solar energy. Incident photons create excited molecular species, including molecular oxygen. The initiation is followed by a typical cascade of reactions and is frequently accompanied by various photo-physical processes such as fluorescence, internal energy conversion, and phosphorescence. Depending on the nature of the substrate, environmental conditions, and temperature, excited molecular oxygen may be involved in many of the reactions owing to its own high reactivity.

There exists also so-called radio-oxidation, or oxidation incited by the impingement of ionizing radiation. It is well known that either scission or crosslinking, or both, may occur in polymers due to irradiation. The free radicals present at the initiating centers for scission and/or crosslinking open up sites to modification of the molecular chain. The scission-to-crosslinking ratio is higher for a polymer irradiated in air (oxygen) due to the peroxy radicals that in turn lead to alcohols, aldehydes, ketones, acids, and so on.

Physicochemical processes underlying degradation of polymers are still rather poorly understood. Because polymers are macromolecules, their solubility is naturally quite low. However, their tendency to swell in the presence of liquids can indeed compromise their morphology and mechanical properties since the small molecules of the liquid can permeate the structure. As noted above, degradation in plastics is also achieved by breaking the polymer chains, that is, chain scission. This can occur by ionizing radiation (e.g., UV light)—significant for DNA—as well as by free radicals and oxidizing agents—including not only oxygen but also ozone or chlorine. Degradation due to UV light is known as photodegradation [11] (of which photo-oxidation would be a subset). A well-known problem of oxidation is the ozone cracking that affects natural rubber tubing. As indicated earlier, elevated temperatures may accelerate the breakdown and chemical reactivity of polymers. Furthermore, as with all the aspects of degradation or corrosion discussed in this section, material *surfaces* provide the points of vulnerability.

To combat the degradation of polymers due to oxidative processes, various methods are employed. These include structural modification, improved processing and forming techniques, and the incorporation of small amounts of chemical substances designed to hinder and/or suppress the degradation processes. The last technique is referred to as **polymer stabilization**; the chemical substances are called **stabilizers**. We saw mention of these in Chapter 9. To protect against oxidation, one needs a suitable **antioxidant**, typically with the following properties:

- Good compatibility with the polymer matrix
- Good stabilizing effects for the polymer in the specified process and use conditions
- Low volatility
- Low water extractability

- Low or no toxicity level; odorless
- Reasonable cost

In addition to these requirements, there are other requirements specific to the predominant method of oxidation. For instance, light stabilizers used to protect against photo-oxidation must additionally satisfy the following conditions:

- Good UV optical absorption and good energy dissipation
- UV stability
- Low optical absorption in the visible spectrum

Antioxidants used to retard thermo-oxidation in polymers function by breaking the kinetic oxidation chain and/or decomposing the generated hydroperoxides into non-radical products. Stabilization of polymers against ionizing radiation is achieved mainly by antioxidant additives that are **radical scavengers** or **energy scavengers**. The first category operates by a mechanism of controlling the free radical reactions. The mechanism of the second involves deactivation of polymer excited states during irradiation.

As we have seen, not all oxides are bad. However, in any material type where there is unwanted oxidation, the integrity and various properties of the material may be compromised. Therefore, whether or not oxidation is desirable, it is important to understand the underlying processes.

19.4 CORROSION

Corrosion is the term used to describe the deterioration and destruction of a material due to exposure to and interaction with the environment in which it resides. Oxidation is not the only chemical reaction that causes corrosion, but it is the primary mechanism of corrosion in metals. In general, *ceramic materials* are quite unreactive and almost entirely immune to corrosion.

Galvanic corrosion occurs when dissimilar metals immersed in a common electrolyte are in physical or electronic contact. The more reactive anode metal corrodes at an accelerated rate. This is a great concern for marine applications in which metal components are exposed to salt water.

Pitting corrosion occurs over a limited section of a metal's surface but leaves behind holes much larger in depth than width. As the pitting progresses, the thinning metal is more susceptible to fatigue and stress cracks begin to form at the base of the corrosion pits. In Guadalajara, Jalisco, Mexico, a single pit that had formed in a gas line placed over a sewer line caused the death of more than 200 people when the leaking gas got into the sewer line, igniting the fumes.

Other modes of corrosion in metals are: stress corrosion cracking, corrosion fatigue, intergranular corrosion, crevice corrosion, filiform corrosion, erosion corrosion, fretting corrosion. Some of these produce very fine cracks that are difficult to see. In several cases, the underlying corrosive mechanisms operate by the combined effects of a corrosive environment and applied stress or mechanical fatigue. **Filiform corrosion** occurs specifically under thin films that serve as coatings on metal surfaces, especially on steel, magnesium, and aluminum. The corrosion appears as threadlike filaments which are really tunnels consisting of corrosion products that lie underneath the bulged and cracked coating. To avoid catastrophe, it is often necessary to have in place means of detection and monitoring of the various types of corrosion.

Concrete is also subject to corrosion, though typically the process begins with the corrosion of the reinforcing steel rods. Although the alkaline environment provided by concrete is generally conducive to formation of a *protective* layer on steel, various conditions can introduce chemicals or abrasion that break down the protective layer and lead to corrosion. For instance, poor construction of the concrete or exposure to freeze-thaw cycles can allow de-icing salts, chemicals, and other water admixtures to seep in. Chloride absorbed from such admixtures facilitates corrosion, and as steel corrodes, it expands. The expansion eventually causes the concrete to break, which allows even more oxygen, water, and chlorides to enter, thus accelerating corrosion, weakening the structure, and hastening the time to failure. On the other hand, corrosion is completely (or nearly so) prevented in concrete structures protected by air conditioning or wholly submerged in water. It is the fluctuation of temperature and moisture that provide ideal conditions for initiation of corrosion.

Not surprisingly, there are extensive efforts to develop new methods to mitigate corrosion. For example, Scharf and coworkers [12] deposited nanocrystalline zinc titanate on a steel using atomic layer deposition (ALD). They studied corrosion rates in saline solution and in simulated body fluid (SBF). In both cases corrosion rates decreased by a whole order of magnitude.

In some situations polymers can be deposited on surfaces as protective coatings. For instance, composites based on an epoxy and containing a fluoropolymer plus in turn Ni, Al, Zn, or Ag were deposited on a mild steel [13]. After curing the epoxy at 30°C, dynamic friction and wear decreased significantly as a result of phase separation between the epoxy and the fluoropolymer. By contrast, both these properties increased after epoxy curing at 80°C. (Thus, the epoxy curing temperature was also an important aspect of the application.)

Perhaps less familiar is the process of **microbial corrosion** of metallic and nonmetallic materials that can occur in the presence or absence of oxygen. In anaerobic conditions, sulfate-reducing bacteria produce hydrogen sulfide and initiate sulfide stress cracking. Under aerobic conditions, oxidizing microbes instigate degradative oxidation that can lead to the various types of corrosion named earlier. These modes of microbial corrosion are typically associated with metals. Similar processes in plastics are more often referred to as biodegradation. Many plastics are quite resistant to such microbial corrosion; thus it turns out that when we want a plastic to degrade more quickly, we in fact try to build it with a greater capacity for biodegradation. (Note that the term biodegradation does not refer exclusively to degradation by microbes. This is a general term that includes also other environmental factors that instigate the chemical degradation of materials. See more on this topic in Chapters 11 and 20.)

19.5 ADHESION

A very primitive description of adhesion is: if surfaces of two materials stay together, we have adhesion; otherwise, we do not. Apart from all else, adhesion entails an interface. The International Union of Pure & Applied Chemistry (IUPAC) [14] defines **interfacial adhesion** as: "Adhesion in which interfaces between phases or components are maintained by intermolecular forces, chain entanglements, or both, across the interface". When we use an epoxy glue, we have two components: epoxy plus a curing agent. After curing, the epoxy network provides strong intermolecular forces that hold the two glued surfaces together. People use adhesives for a gamut of reasons; from simply taping something on a piece of paper to repairing a broken item to constructing buildings and automobiles; adhesives are ubiquitous. Without adhesion, many objects well known to us would simply not exist. Importantly, adhesives function at various size scales. The Dreamliner airplane would not be possible without adhesion of the reinforcing fibers to the matrix in the composite materials used in its airframe and primary structure.

The question at hand then is, how do we achieve adhesion? In fact, there are multiple ways. One kind of adhesion involves the formation of primary chemical bonds. This most often occurs in the mixing of reagents—for example, compatibilized silica in an organic polymer matrix—to prepare blends and composite materials. For adhering somewhat larger surfaces together, however, other mechanisms predominate. Otherwise adhesion is caused by hydrogen bonding, van der Waals forces, and/or electrostatic dipole interactions. It turns out that geckos, which are known to climb on walls, can walk on vertical surfaces due to adhesive forces of the van der Waals type between fibrils on their feet and the wall surface.

Mechanical bonding is another mechanism by which adhesives bond to substrates. Except for very highly polished surfaces, all surfaces have some pores and roughness. When an adhesive flows into these pores and then hardens, a mechanical bond is created. There is also the kind of mechanical bonding seen with Velcro, where protrusions on the opposing surfaces essentially interlock with one another to hold the two surfaces together. This kind of phenomenon is seen also as a smaller scale.

The IUPAC statement also mentions entanglements—for a reason. Consider two polymers in contact. With time, there will be diffusion of parts of chains from one material into the other one, with such occurring in both directions across the interface. As we discussed in Chapter 9 with respect to thermoplastic elastomers, the entanglements make separating the two materials difficult.

Adhesion has to be distinguished from **cohesion**; the latter pertains to a single phase and is sometimes called "internal cohesion". For instance, we have rain droplets more often than a fine mist because there is a strong cohesion between water molecules. Smog appears in the form of mist since molecules other than water destroy that cohesion.

To obtain good adhesion between a liquid and solid, the adhesive must **wet** the substrate, that is to say, it must spread itself out in a thin film over the substrate surface. The so-called property of wetting is illustrated in Figure 19.3. The smaller the angle θ, the better is the adhesion. "Good wetting" typically refers to $\theta < 90°$. Thus "not wetting" is

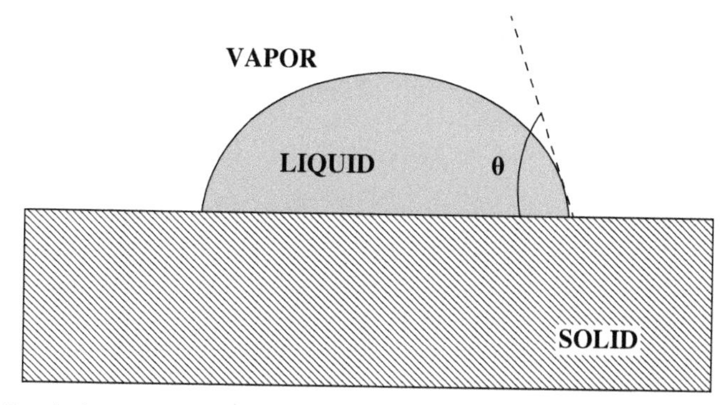

FIGURE 19.3 A drop of a liquid on a solid surface, with the wetting angle θ between the two phases indicated.

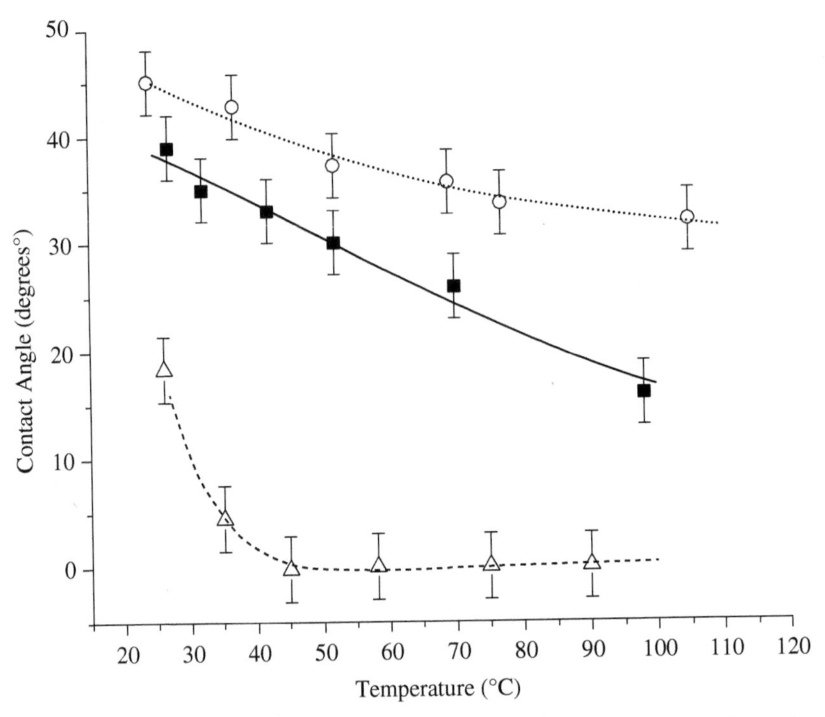

FIGURE 19.4 The wetting angles of several liquid polymers on the surface of a semiconductor, before curing them, as a function of temperature. Lower wetting angles prove to be an indicator of good adhesion.

characterized by $\theta > 90°$ while the term "spreading" may be used when θ is close to zero. "Not wetting" means also that cohesion prevails over adhesion.

Somebody might say at this juncture: you were talking above about an epoxy adhesive, I need to put two *solid* surfaces together. Well, before you cure the epoxy, it is a liquid, so its wetting angle to one of the solids can be measured, often as a function of temperature [15]. In the cases when curing is used to achieve long term adhesion, the wetting angle before curing is a good indicator of how successful the adhesive will be; see an example in Figure 19.4.

Quantitatively adhesion is characterized in terms of the surface energy. We briefly discussed surface tension in Section 19.2. **Surface tension** γ is defined as the work W_a needed to increase the area of a surface by one unit, such as by 1 cm^2. For a solid and a liquid, such as in Figure 19.3, we have

$$W_a = \gamma_A + \gamma_B - \gamma_{AB} \tag{19.3}$$

where

γ_A is the surface tension of material A,
γ_B is the surface tension of material B, and
γ_{AB} is the interfacial surface tension.

Considering the angle θ described in Figure 19.3, we can obtain the interfacial tension between the solid and liquid γ_{SL} as follows:

$$\gamma_{SL} = \gamma_{SL} + \gamma_{LV} \cos\theta \qquad (19.4)$$

where the index V pertains to the vapor phase.

Let us stay with our example of epoxy curing. Once the curing is completed, we wish to know how good the adhesion is. With the adhesive applied and cured, a popular method consists in measuring the critical load at which the adhered layer is peeled off the substrate—the so-called peel test.

NATURAL ADHESIVES

As in other areas of MSE, nature has done the job first, and we are often imitating it with more or less success. An adhesive protein called Mefp-1 that is found in blue mussels (*Mytilus edulis*) is an impressive one. It can adhere to almost any type of surface (e.g., glass, stainless steel, wood, concrete, and Teflon), and it is very strong and durable even in water [16]. At least nine additional proteins (Mefp-2, Mefp-3, and so on) in the mussel are implicated in adhesion. The full mechanism of adhesion and the role of each protein are still not completely understood. We know that oxidized forms of tyrosine are essential in the adhesion process since they form complex with metal ions (Fe^{3+}, Al^{3+}) and semimetals; hydrogen bonds are also formed. Other constituents believed to be important in the adhesion process are lysine and glycine. Lysine forms ionic bonds to negatively charged surfaces (such as collagen and acidic polysaccharides) and intermolecular crosslinking with quinones (to which belong hydroxylated forms of Mefp-1). It appears that glycine in the protein structure contributes to the open, extended conformation of the peptide chain, facilitating intermolecular bonding and therefore adhesion. Each variant of the Mefp adhesive proteins contributes to the total adhesive strength of the blue mussel, with their roles depending on peptide composition and how they are distributed along the adhesive thread released by the mussel.

Numata and Baker [17] tried to prepare one of these adhesive foot proteins, Mefp-5, by multiple chemo-enzymatic synthesis using catalysts. Mefp-5 can be found on the surface of the mussel adhesive plaque. The synthesis was successful, and the adhesive shear strength tested in different pH solutions. Strong adhesives were formed at pH = 10 and 12. However, the pH of seawater is between 7.5 and 8.5, not enough to provide strong adhesive bonding reported. We conclude that mimicking nature is a difficult task...

One might get a false impression that we always want as high adhesion as possible; this is not true. The very simple example of Post-It notes—which adhere lightly to paper and other surfaces—provides a prime example of a situation in which we want the adhesive to adhere weakly and be easily removed. Furthermore, considering adhesiveness as a surface property, there are many instances in which little or no adhesion to other surfaces is desirable.

The goal for **self-cleaning surfaces** is to achieve as low adhesion to those surfaces as possible. Typically such surfaces consist of glass substrates covered with a coating. Self-cleaning glass is expected to stay free of dirt particles and grime. The coatings are of two kinds, hydrophobic and hydrophilic. Somewhat unexpectedly, both kinds are cleaned with water. The hydrophobic surfaces can be made by plasma etching the glass or else by coating the glass with a polymer or wax to achieve high water wetting angles, such as $\theta > 160°$, needed for water droplets to roll-off the surface together with dirt. On hydrophilic surfaces with titania-based coatings, **photocatalysis** contributes to the cleaning: chemical reactivity of the coating with sunlight causes chemical decomposition of the dirt, making the dirt flow easily away with water. Though the concept is intriguing, self-cleaning glass is not in widespread use as there are still performance problems and challenges remaining with the technology.

19.6 FRICTION

Now that we have explored several important aspects of material surfaces, we begin to consider phenomena of surfaces in contact and *in relative motion* with one another. Conceptually, we know that **friction** is the *resistance* to relative motion between two bodies in contact. Where does the resistance come from? Two places really. Between the two bodies in contact there are microscopic forces of **molecular adhesion** (including electrostatic, van der Waals, metallic bonds) and microscopic forces of mechanical **abrasion** (including elastic and plastic deformation) that oppose motion. The surface "contaminants" described earlier—oxides, adsorbed films, adsorbed gases, foreign or "domestic" particles—contribute to friction. Fluid friction, or the resistance of flow in liquids, is measured as viscosity. This was a topic of Chapter 13 and is discussed in detail there.

The first theory of friction arose from da Vinci's experiments sliding rectangular blocks of materials on varied surfaces. Based on his observations, he proposed (1) the area of contact has no effect on friction, that is to say, friction is a function of the weight of the body—whether we slide a given brick on its narrow or broad side, it will require the same amount of force—and (2) if the load of an object is doubled, its friction will be doubled—a relation from which we get the direct proportionality between the normal load and the friction force. The constant of this proportionality μ is known as the coefficient of friction, or better yet, simply friction. Da Vinci's observations were later established as Amontons' Law. The essential relationship so defined between the friction force F and the normal force N (the load) is the following:

$$F = \mu N \tag{19.5}$$

It is apparently simple to determine μ for a given material. There are only two measurable parameters involved, F and N. So then, what factors affect the value of μ?

- Surface roughness
- Lubrication
- Surface chemistry
- Contact stress
- Contact geometry
- Environment

- Temperature
- Sliding speed
- etc. ...

Now it seems not so simple. What we need to realize is that friction is not a material property; it is a ***system property***. There is no such thing as the *friction of polyethylene* or the *friction of steel*. The friction of a material is only defined for a particular system. Therefore, when assessing a material's friction behavior, we must consider conditions and aspects including: the opposing contact material, temperature, coatings, lubricants, contact area, geometry, stress, roughness, sliding speed, sliding mode, duty cycle (continuous or intermittent contact), humidity, atmosphere. One will find tables of friction values in various resources; it is essential to take note of the test conditions used and consider how they compare to one's intended operational conditions.

Presently we recognize several so-called laws of friction. The **first law of friction**, defined by Eq. (19.5), tells us the friction force (that resists motion between two bodies, and that must be overcome to move a stationary object) is proportional to the load (N). This law, posited by Amontons, was confirmed by Coulomb. As already hinted, one should use with caution tables of friction coefficients, because, for instance, the value of the proportionality constant μ will be lower with an oil present than on a clean, dry surface. In the case of aluminum oxide paired with steel, surface cleanliness will have a large impact on the observed friction.

Consider problems 1 and 2 shown in the shaded box. In Example 19.1, which deals with a stationary object, the value of the **static friction** tells us about the force that must be overcome to get the body in motion. The force required to get a body in motion is typically larger than that required to keep it moving at a constant velocity thereafter. Thus static friction is typically larger than the **kinetic** (also called **dynamic**) **friction**, which is the resistance to motion of surfaces already sliding across one another.

If in Example 19.2 we measure the force required to keep the aluminum oxide body constantly moving across the steel surface, then we have calculated a value for the kinetic friction; if only the force to get it in motion, then we have the static friction. One sometimes hears the word "stiction" used to describe the force that must be overcome to initiate the relative motion of stationary objects in contact. The term is a portmanteau of the words static and friction, but also conveys the notion of the verb "stick", which is applicable to the concept of friction.

There are some materials for which friction is *not* proportional to the load. Some very hard materials like diamond and very soft materials like DuPont's Teflon do not obey the first law. For these materials, the friction is proportional to some reduced value of the load.

EXAMPLE 19.1

A 400 kg landscaping rock is on a concrete driveway. It must be dragged 15 m to be placed in a garden bed. The friction between the rock and concrete is 0.3. If a rope is wrapped around the rock to drag it, what will be the force in the rope?

The load is 400 kg × 9.8 m s^{-1}, which equals 3920 N.

$$F = \mu N = (0.3)(3920\,\text{N})$$

$$F = 1176\,\text{N} \approx 265\,\text{pound-force (lbf)}$$

EXAMPLE 19.2

We want to know the friction for aluminum oxide sliding on steel. Say we use a spring balance to pull the aluminum oxide along the steel surface. If the former weighs 150 g, and the force required to pull it along the steel surface is 30 g (both quantities normalized by acceleration due to gravity), then what is μ?

$$\mu = \frac{F}{N} = \frac{30}{150} = 0.2$$

The **second law of friction**, confirmed by Coulomb after observations by da Vinci and Amontons, states that friction is independent of the contact area between two surfaces. If we drag a sheet of plywood across the floor, the friction force is the same whether we slide it on the 1-inch by 96-inch edge or on the flat side that is 48×96 inches. It is not less by a factor of 48. The reason: there is a difference between the **apparent area of contact** and the **actual (real) area of contact** between the plywood sheet and floor. Remember that most surfaces are not flat; even at the microscopic level they are wavy and bumpy, exhibiting some extent of surface roughness. Thus two adjacent surfaces are never in total contact; the upper body is supported by the peaks of the irregularities of the lower body—known as asperities, already noted in Section 19.2. The asperities bend and deform until the load is fully supported, but even at that point typically less than 1 part in 10,000 of the apparent area is usually in contact. Therefore friction is effectively independent of the apparent area of contact. Exceptions to this law occur in the case of extremely clean and highly polished surfaces in which the real area of contact is a significant portion of the apparent area. As for the real area of contact, friction is directly proportional to it.

Consider again kinetic friction: the **third law of friction** is that the resisting force between moving bodies is independent of velocity. Again, if we are dragging the sheet of plywood, the force is the same if we slide it slowly or quickly. The presence of a lubricant, of course, would change the friction force. The independence of friction on velocity is, however, not always true [18]. At very high speeds, the friction tends to decrease as the speed increases. At very low speeds, one observes a gradual rise in friction with decreasing sliding velocity. In oriented (i.e., anisotropic) materials, the sliding direction is important; tribological properties depend on it [19].

A relatively simple and common laboratory technique for measuring friction is the **sled test**. One uses a standard mechanical testing machine according to the setup shown in Figure 19.5. This type of friction test fixture consists of a horizontal surface, referred to as the table (i.e., the supporting base), and a moveable sled attached to a load cell by a cord. The cord is guided by a pulley during the test. As the sled is pulled across the table, the force to start the sled (static friction) and to keep the sled moving (dynamic friction) are measured by the load cell and recorded by the acquisition system.

A second widely used apparatus for measuring friction is a **tribometer**. A pin-on- disc tribometer is useful for providing simple wear and dynamic friction data for bulk materials as well as for specimens with coatings; refer to the illustration in Figure 19.6. In this setup (rotative mode), the test specimen is mounted on a rotating platform. A pin of known weight (and known material type) is loaded on the specimen as it is rotated at a specified rate (rpm), creating a circular wear track. Deflection of the direct load cell during the test

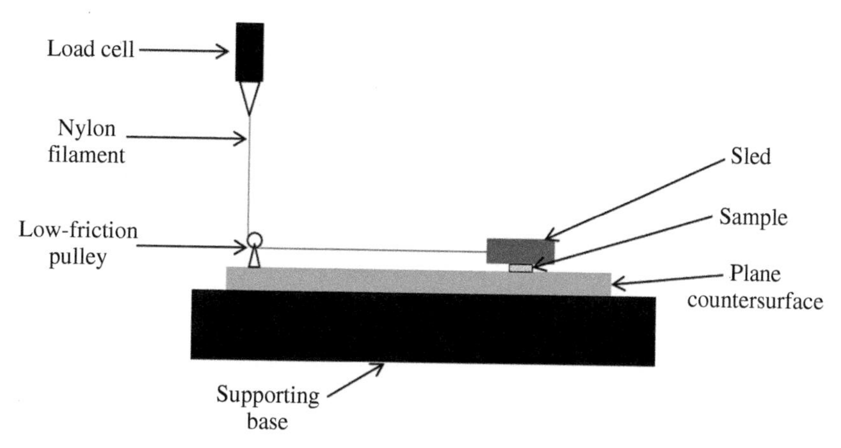

FIGURE 19.5 Test setup for determination of static and dynamic friction using a mechanical testing machine. The standard load cell is fitted with a nylon filament that passes through a low-friction pulley. The material (steel, Teflon, etc.) of the plane countersurface is selected for the desired system conditions. The sample is mounted underneath the sled, which is pulled by the nylon filament.

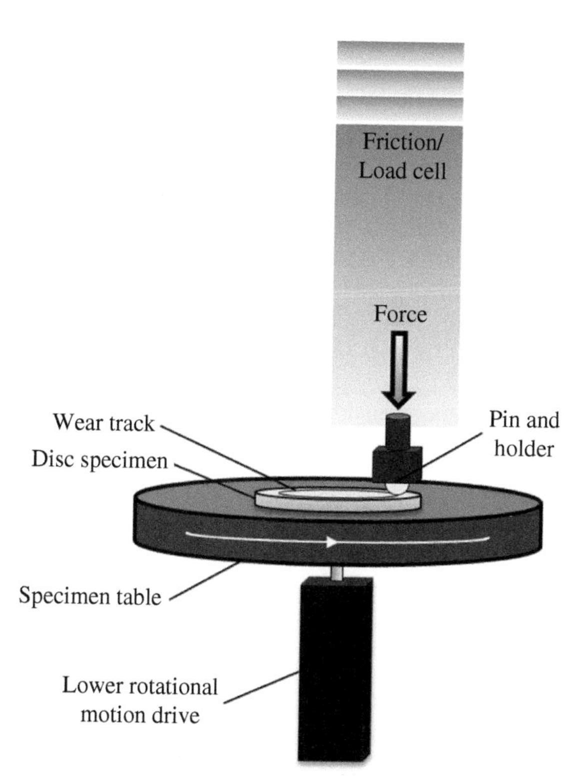

FIGURE 19.6 Schematic of the main components of a pin-on-disc tribometer.

is measured and used to determine kinetic friction. Wear rates for the pin and material disc can also be calculated from measurements of the lost material volume. There are tribometers that operate in linear mode—analogous to the sled test—and for both pin-on-disc and linear models there may be attachments that allow for testing at high temperatures, with controlled humidity or with lubricants. Since we talk about wear in Section 19.8, we shall discuss wear results obtained by tribometry there.

Friction is experienced by materials in **plane sliding** (linear or rotative, as just discussed) as well as when objects roll over one another. Ball bearings and wheels provide prime examples of **rolling friction**. Consider a car tire rolling on the road. To keep the tire from slipping, some amount of friction is needed at the point of contact of the tire with the road surface. Static friction between the tire and road determines the *traction*. The force of friction once the wheel is locked and sliding is kinetic friction, usually a smaller quantity than the static friction. If a tire is rolling without slipping, theoretically the surface friction does no *work* against the wheel motion so there is no energy lost. For any real tire, however, there is some energy loss and deceleration due to friction; this is known as rolling friction—which arises partly from friction at the axle and partly from flexing of the tire, which dissipates energy. For car tires, the rolling friction is typically less than the static friction between the tire and the road. For all rolling contacts, energy is dissipated due to friction at the contact interface, due to elastic properties of the materials, and due to roughness of the rolling surface. Again, we are reminded that friction is a *system property*; the nature of the surface affects the friction of the material sliding or rolling upon it. Some materials are more resistant to rolling. For ease of riding a bicycle, how does sand compare to pavement as a surface on which to ride? Given two flat surfaces, one softer than the other, which imposes more resistance to rolling?

Friction can be a useful phenomenon or a hindrance to a desired operation. Friction is necessary to create music from the action of taking a bow along violin strings; putting on rosin increases friction in this scenario. Friction in a door hinge, however, yields an unwelcome noise that one tries to eliminate by applying oil on the hinges. It turns out that static friction is not constant but depends on the duration of static contact. The increase in static friction during the first 1/10 of a second can be very large, followed by a more gradual increase after that until the body is in motion. If a system is sufficiently elastic, and given static friction greater than kinetic friction, a situation of intermittent sliding known as **stick-slip** may occur. The squealing of automobile breaks as a car stops is one example; the squeaking door hinges are another. Stick-slip can be a particular problem in slow moving mechanisms because of the long duration of static contact.

For two solid surfaces in contact and moving relative to one another, friction has the effect of diminishing the kinetic energy and increasing the thermal energy of the bodies in contact. As a consequence of the opposition of the friction force to the relative motion of solid surfaces, mechanical energy is transferred and the ensuing change in thermal energy of the two bodies may result in a temperature difference. The presence of a temperature difference can in turn generate heat. If one is seeking to make primitive fire by rubbing two sticks together, heat is welcomed. By contrast, the heat generated by friction between moving parts in a car engine needs to be dissipated in order to avoid causing engine damage by overheating. This can be achieved by recirculating a lubricating oil, that undergoes cooling, through the engine. There is an apparent paradox, which has been addressed by Sherwood and Bernard [20], associated with the traditional analysis of work and energy in conjunction with friction. Some textbook explanations and calculations do not account for changes in internal energy associated with friction; Sherwood and Bernard present a model that explains the observations of real experiences [20]. John Williams in his textbook *Engineering Tribology* provides a good treatment of the subject of thermal conditions in sliding contacts [21].

What makes one material have higher or lower friction than another? Bowden and Tabor presented the adhesion theory of friction for metals, based on the fact that the surfaces make contact only at the tips of the asperities [22]. According to that adhesion theory, for two surfaces brought into contact under an applied load forcing them together, junctions are formed where asperities come into contact. The total contact area of the junctions is a function of load and penetration hardness of the *softer* material. The relationship is approximated by the following equation:

$$A = \frac{N}{h} \tag{19.6}$$

where A is the real area of contact, N is the load, and h is the hardness. The theory was consistent with earlier observations that friction is independent of the apparent area of contact and that friction is proportional to the load. However, it does not account for loose particles generated and assumes the contribution of plowing is always small. In the case of polymeric materials, it is essential to utilize a model of friction that accounts for both adhesion and plowing contributions (where plowing, like its use in reference to farming, refers to the digging into a softer material by the harder). Irwin Singer points out what was hinted at earlier in saying that "any credible model of friction should account for the work done; friction is merely a special mechanism(s) for dissipating work" [23]. To actually describe and predict the friction for a given material, one needs to know something about its elasticity, plasticity, viscoelasticity, if present, and rheological response behaviors. To make *a priori* predictions of the friction of polymers requires prior understanding of the rheological responses "over an extensive range of stresses, strain rates, temperatures, hydrostatic stresses, and so on" [23]; and that is something very difficult to attain. In practice we get by with somewhat less than this full spectrum of knowledge.

How can we modify friction? In metal contacts liquid lubricants can be applied; see more about lubrication in Section 19.9 below. This does not work for polymers since they are likely to absorb the liquid and swell. Other options to reduce friction exist or are in development. Consider the scenario of an ordinary commercial epoxy in contact with steel. By preparing the epoxy with a small amount of fluoropolymer additive [24], friction between the epoxy and steel can be lowered; dynamic friction (obtained from the sled test) for this scenario is presented in Figure 19.7. The static friction data are similar. There is also an interesting warning in this diagram: if we wish to accelerate the epoxy curing by raising the temperature to 70°C, we get *higher* friction instead of lower.

The two opposite effects at two temperatures seen in Figure 19.7 require an explanation. See now Figure 19.8 with SEM results for the same epoxy plus 12F-PEK fluoropolymer blend of Figure 19.7. There is an obvious difference between images (a) and (b) in Figure 19.8, both containing 5% of fluoropolymer in epoxy. In the image at left, of a sample cured at 24°C, we observe that the fluoropolymer, that is, the minority component appearing white in the micrograph, has preferentially migrated to the surface. In the image at right, of a sample cured at 70°C, we observe a surface consisting essentially of pure epoxy. This is

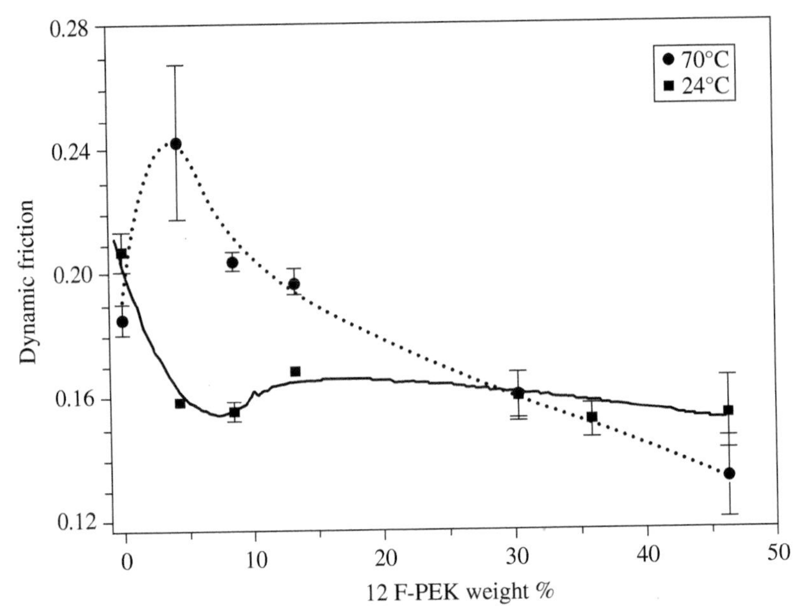

FIGURE 19.7 Dynamic friction of a commercial epoxy as a function of concentration of a fluoropolymer (12F-PEK) additive. *Source*: [24]. Reprinted with permission from Elsevier.

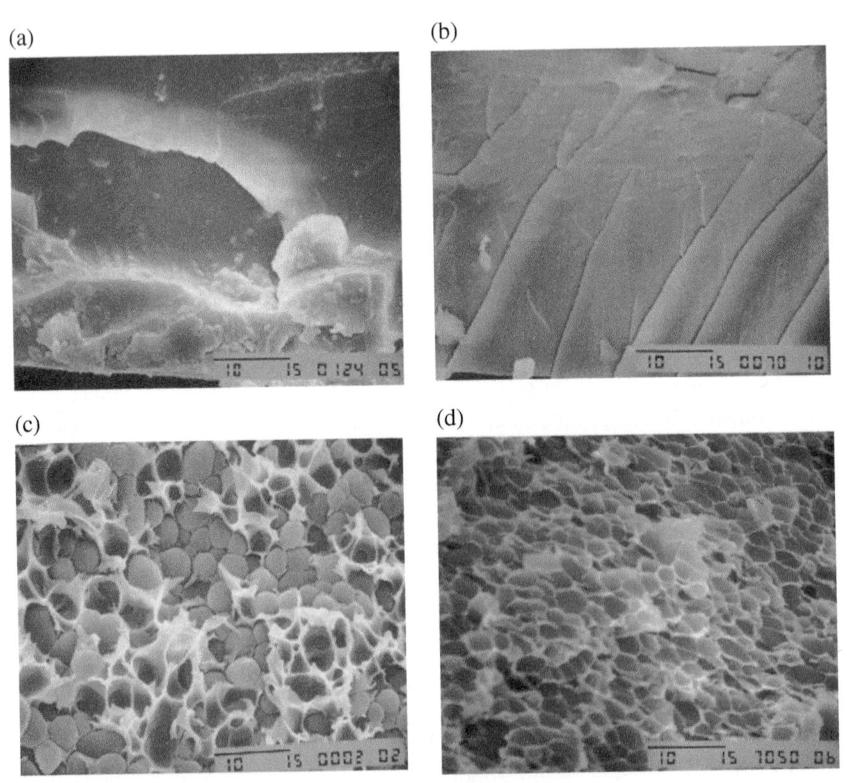

FIGURE 19.8 Scanning electron micrographs of blends of epoxy + 12F-PEK (a fluoropolymer). The top images show epoxy + 5% 12F-PEK cured at (a) 24°C and at (b) 70°C. The bottom two images show epoxy + 30% 12F-PEK cured at (c) 24°C and at (d) 70°C. *Source*: After [24]. Reprinted with permission from Elsevier.

the reason for higher friction in the latter. Faster curing at 70°C prevents migration of the low-friction fluoropolymer to the surface. The situation changes somewhat, however, at higher concentrations of 12F-PEK. At 30% fluoropolymer in epoxy, the fluoropolymer forms a continuous phase around the epoxy; see Figure 19.8c and d. In spite of different curing temperatures, the morphology is similar, hence the similarity in friction values at that concentration.

Let us take a closer look at a few more cases of friction in materials. Most people are familiar with the very low friction polymer polytetrafluoroethylene (PTFE), with DuPont's Teflon® being a recognizable brand. Polytetrafluoroethylene is utilized in plumber's tape, as non-stick coatings on cookware, as a solid lubricant, and more. Polytetrafluoroethylene is a polycrystalline polymer whose crystalline layers slide easily over one another. In most scenarios involving a polymer sliding over a hard counterface, the formation of a transfer film is observed; the nature of the transfer film impacts tribology of the polymer. The molecular structure of PTFE renders a unique mode of film transfer. The low shear stress for exfoliation of crystalline "slices" seems to facilitate the deformation of PTFE by a series of discrete lamina [25]. The continuous deposition of laminae during sliding across a harder counterface results in very low friction but also very high wear of PTFE.

As argued by Jonna Lind and Åsa Kassman Rudolphi [26], there is ongoing gradual replacement in the construction industry of metal elements by polymer-based ones. An obvious motivation is lower density, hence lower weight of the components. An also obvious drawback is lower mechanical strength. Therefore, fillers such as glass fibers or carbon fibers are used to reinforce engineering polymers such as poly(phenylene sulfide), polyetheretherketone, Polyamide 66, and their blends containing also PTFE. The two researchers in Uppsala have found that the use of silicon oil as an internal lubricant provides a dramatic decrease of dynamic friction, while graphite also lowers friction.

Since adhesion and abrasion contribute to friction, surface roughness naturally plays a role in the friction between two material faces. A smoother surface, being less abrasive, tends to have lower friction, especially if the material is unreactive with the counterfacing material. In some scenarios, adhesion may facilitate the film transfer (e.g., contacts involving a polymeric component); in other cases, adhesion due to asperities can increase frictional resistance during sliding. Because like materials tend to experience high adhesion at the interface, it is generally unadvisable to design a sliding contact from two like materials. Furthermore, deformation of surfaces (typically associated with abrasion) increases friction in sliding contacts. These principles apply to all types of materials; and thinking back to the effect of load on friction, one can ascertain how an increasing load might affect adhesion and abrasion insofar as they contribute to friction.

Finally, while in so many cases we wish lower friction, there are cases when *high friction* is a blessing. Namely, when Kevlar yarns and fabrics were treated with multi-walled carbon nanotubes (MWCNTs) and subjected to sonication in solution, the result was higher friction—along with the desired higher penetration resistance to ballistic impact [27].

19.7 SCRATCH RESISTANCE

The phenomena associated with scratching of a surface are treated separately from friction. We spoke of an abrasive component of friction; a consequence may be the appearance of scratches in the surface of the softer contact. **Scratch resistance** is a

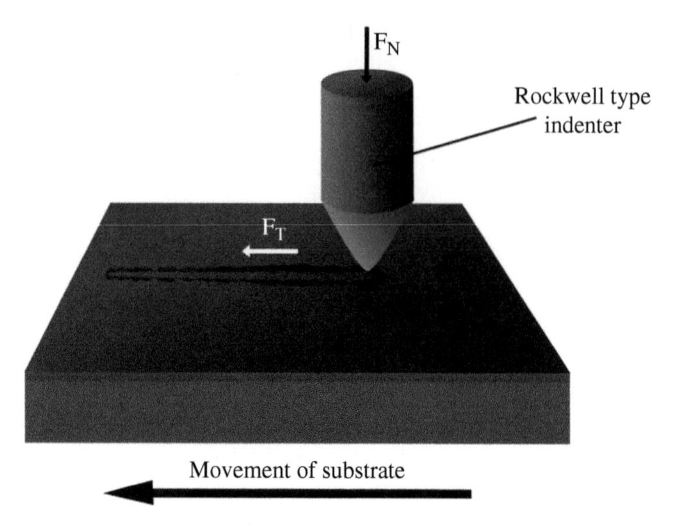

FIGURE 19.9 Schematic illustration of micro-scratch test operation for scratch resistance determination. *Source*: After Djhé, https://commons.wikimedia.org/wiki/File:Scratch_test.png. Used under CC-BY-SA-3.0, 2.5, 2.0, 1.0, GFDL.

critical service parameter. Historically, the scratch test was designed to measure adhesion of thin hard films; further developments of the technique ensued to better explain the phenomena.

In scratch testing the indenter, often a diamond, moves at a constant speed across a surface; see Figure 19.9. We are familiar with scratch grooves created when one material deforms a softer one, for instance, when a child has dragged his fork across a wooden dining table, leaving behind a perhaps unwelcome deformation. During a scratch test, the applied load may be held constant, continuously increased, or increased stepwise. The critical load L_c is said to be the smallest load at which a coating on the test specimen *is damaged*, which of course is a rather subjective definition. A better approach consists in recording the load under which the indenter went across the entire material and reached the substrate; this provides a measure of adhesion. Often a scratch resistance tester provides also an acoustic emission signal along the scratching path. There is a strong signal when the indenter moves from one phase to a different one.

A scratch test provides instantaneous measurement of the **penetration depth R_p** of the tip into the material as the indenter moves along a straight line; see Figure 19.10. The value of R_p depends on the elasticity, plasticity, and in the case of polymers viscoelasticity of the material. Owing to viscoelasticity, the scratch groove in polymers is expected to heal or recover somewhat, with the bottom of the groove going up and settling finally at a level called the **residual depth R_h**. This depth too can be measured. For the case of a load linearly increasing during the movement, we can plot both depths; see Figure 19.10. Detailed descriptions of a micro-scratch testing are provided in the *Journal of Materials Education* [28, 29]. From the difference between R_h and R_p one can calculate the percentage of viscoelastic recovery f for scratching

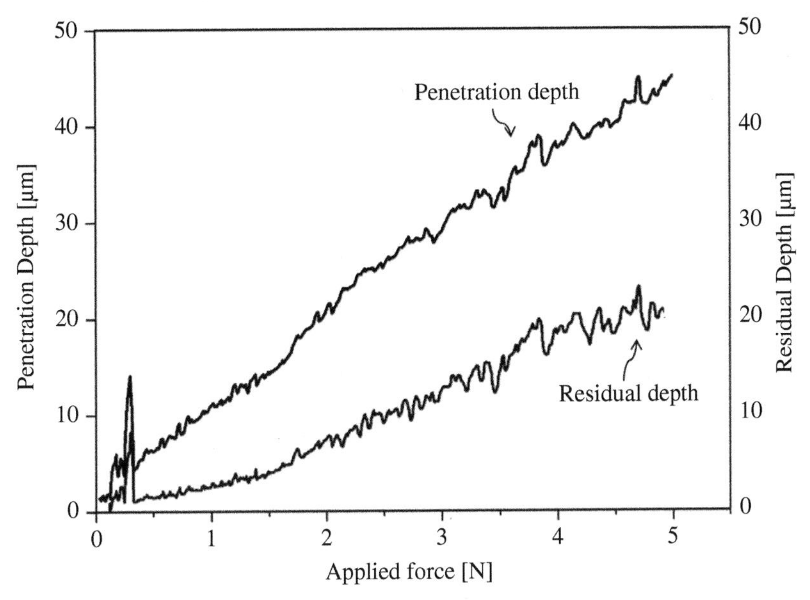

FIGURE 19.10 Penetration and recovery depths as a function of increasing load on a polymer specimen.

$$f = \frac{R_p - R_h}{R_p} \tag{19.7}$$

An equation relating f to brittleness B, valid for polymers with different chemical structures and for some composites, has been obtained [30] (refer also to Section 14.5 and Eq. (14.11) for a definition of B). After scratch testing, wear tracks remain; some examples are displayed in Figure 19.11. Note especially the debris generated at higher test loads. This is comparable to the plowing that occurs from asperities during sliding contact, as discussed above in the section on friction.

High scratch resistance and low friction often do not go hand in hand. For instance, PTFE (Teflon) has low friction but at the same time quite poor scratch resistance. An epoxy modified with a fluorinated polyetherketone, however, has been shown to exhibit both low friction (see again Figure 19.7) and good scratch resistance (Figure 19.12) [24, 31]. This work apparently ended earlier speculations based on the PTFE case that low friction and high scratch resistance are incompatible. We see in Figure 19.12 that the minimum of either R_p or R_h is at ≈5 wt.% fluoropolymer, the same at which we have the minima of static and dynamic friction. By looking at the difference in scale between the penetration and residual depth curves, we also see how large is the viscoelastic recovery defined by Eq. (19.7).

We have noted above that surface tension is an important property—and it necessarily has to be related to other tribological properties. This has been demonstrated for the same epoxy plus fluoropolymer system we have been discussing [6]. In Figure 19.13, the values are calculated from Eq. (19.3), and the minimum of the

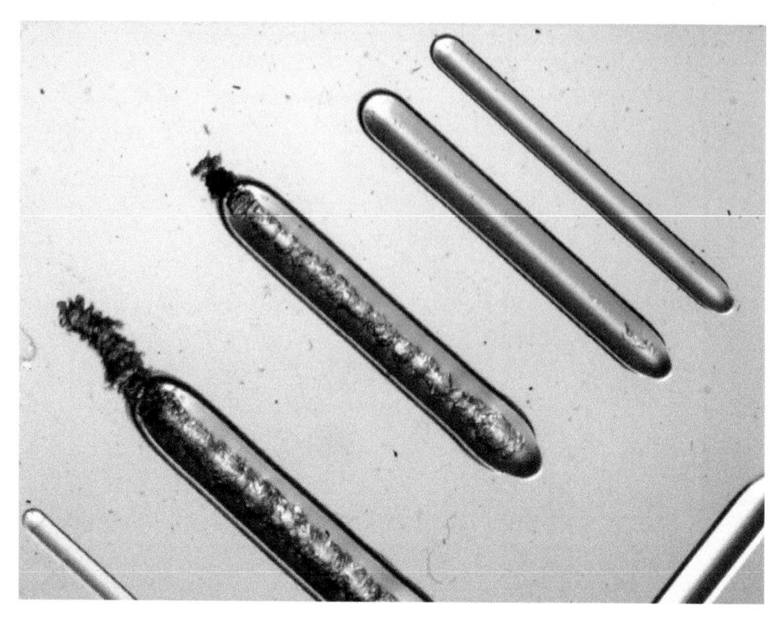

FIGURE 19.11 Grooves resultant from scratch testing a polymer under a constant load. The loads decrease from right to left.

surface tension is at ≈5% fluoropolymer, the same as for the static friction, dynamic friction, penetration depth, and recovery depth.

Human teeth and their protection deserve a separate short discussion. Nearly 95% of the whole human population suffers or has suffered at some stage of their lives from tooth decay or some other disease related to the oral environment. A naked eye view of a tooth suggests that we are dealing with a smooth relatively flat surface. This is not true; see Figure 19.14 with results of scratch testing, showing again penetration and recovery depths [32]. Neither the words "smooth" nor "flat" apply. Figure 19.14 shows also that teeth are viscoelastic; a result not entirely expected—given the hardness of tooth enamel.

Synthetic dental materials need to closely match the properties of the native tooth so as not to damage the enamel by scratching it nor be damaged itself from being scratched by the teeth. Apart from materials already commercialized, there are a variety of experimental obturation materials; many based on polymer + ceramic nanopowder composites [33–35] are under development.

A microscratch tester allows also another kind of test option: multiple scratching along the same groove, resulting in the determination of the **sliding wear** [36]. This is a more accurate method of wear determination than abrasion wear tests, in which one simply measures the weight of material lost after abrading a material's surface, though the kind of results differ. As with other scratch tests, in sliding wear one determines two basic quantities, the penetration depth R_p and the recovery depth R_h; in the case of sliding wear, however, the depths are determined as a function of the number scratches (or passes of the indenter) along the same groove. Equation (19.7) is applicable here as well.

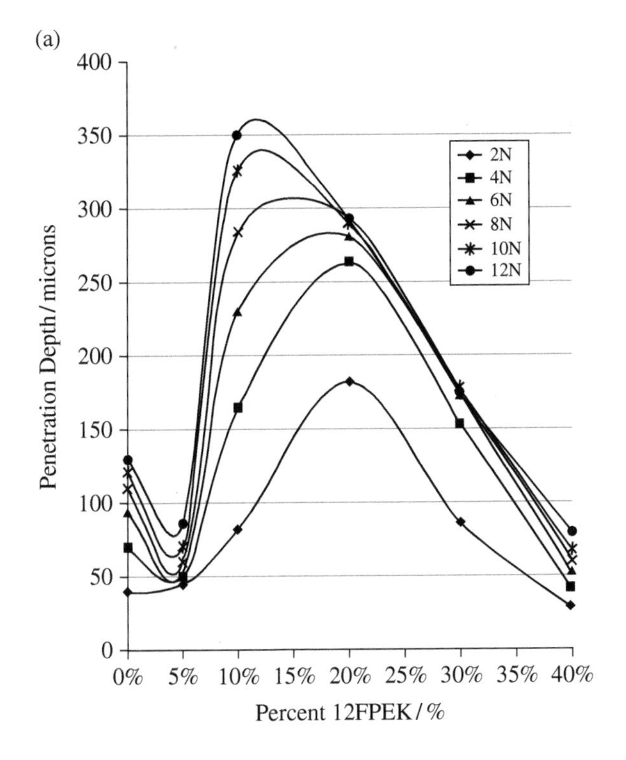

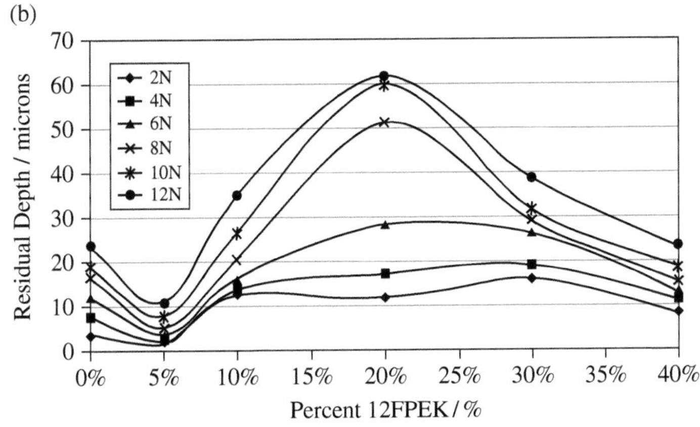

FIGURE 19.12 (a) Penetration depth and (b) residual depth of a commercial epoxy with fluoropolymer additive, as a function of concentration of the 12F-PEK fluoropolymer. The epoxy curing was performed at 24°C. *Source*: [31]. Reprinted with kind permission from Springer Science and Business Media. © Springer-Verlag 2002.

Results of such testing for several different polymeric materials are shown in Figure 19.15; the phenomenon of **strain hardening in sliding wear** [36] is manifested there. If we look at what happens to the depth with each subsequent scratch, we find that the depths of the grooves increase over the first few passes. Eventually, further passages of the indenter provide very small or negligible effects. The asymptotic behavior seen in Figure 19.15 for both penetration and residual depths naturally also applies to the

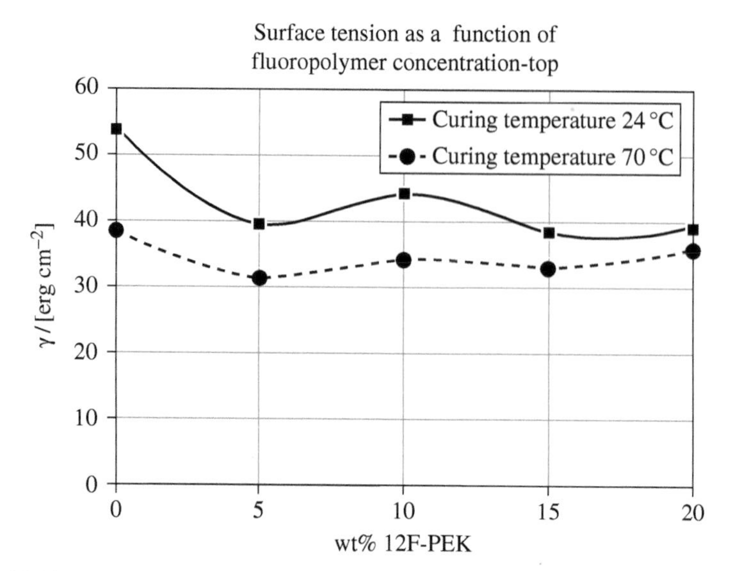

FIGURE 19.13 Surface tension of epoxy + 12F-PEK fluoropolymer as a function of the fluoropolymer concentration. Surface tension total values at 25°C calculated from the van Oss–Good method. Measurements taken on the top surface. *Source*: After [6]. © 2003 Society of Chemical Industry. Reprinted with permission from John Wiley & Sons.

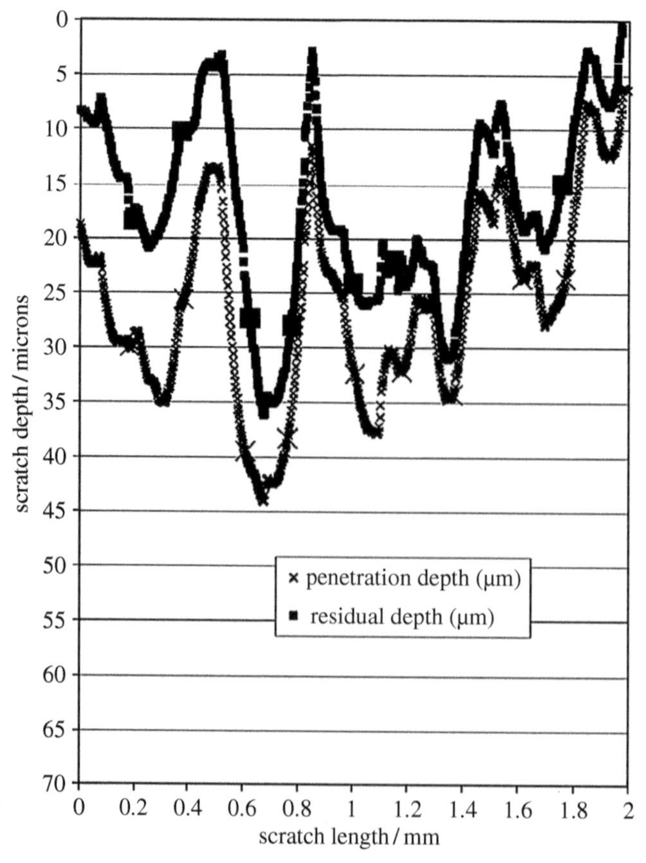

FIGURE 19.14 Results of scratch testing of an uncoated tooth (extracted from a volunteer). *Source*: de la Isla *et al.* [32]. With kind permission from Springer Science and Business Media. © Springer-Verlag 2003.

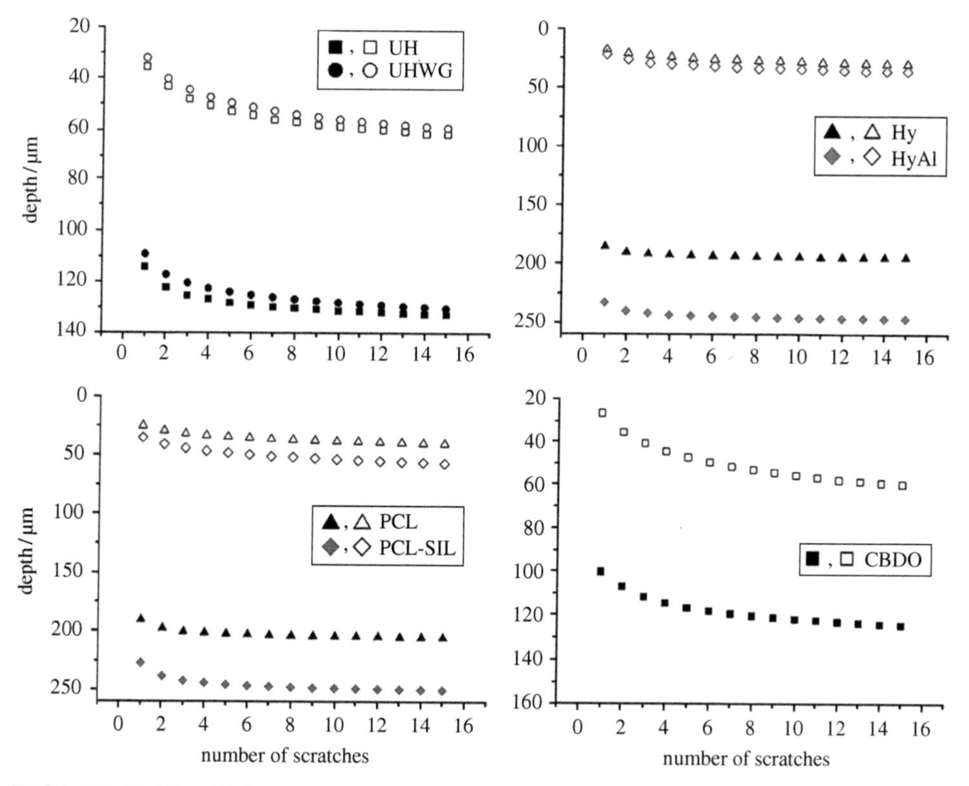

FIGURE 19.15 Sliding wear determination (i.e., multiple scratching along the same groove) for polymers and polymer composites. Solid symbols are used for penetration depth (R_p); open symbols are used for residual depth (R_h). UH, Hy, PCL, and CBDO are neat polymers. UHWG, HyAl, and PCL-SIL are composites. The force applied during testing was 10N.

viscoelastic recovery *f*, defined by Eq. (19.7). The main mechanism consists in conformational changes of polymeric chains inside the groove resulting in material densification [37, 38]. Subsequent studies have shown that this phenomenon occurs in more polymers but not in all. The fact that polystyrene does not exhibit this behavior was actually the motivation for the definition of brittleness *B* provided in Section 14.5. A relationship between the asymptotic viscoelastic recovery *f* (Eq. 19.7 again) and *B* has been demonstrated for a variety of polymers [30] and also polymer-based composites; the relation [30] is shown in Figure 19.16. Recall also the relationship of *B* to impact strength [39].

Elsewhere in this book we have talked about structural orientation, isotropy, and anisotropy. These had effects, for example, on tensile strength of materials. Similarly, scratch resistance of a material can be influenced by orientation. The morphology of polymer liquid crystals can be adjusted by imposition of a magnetic field (refer to Chapter 12); not surprisingly the observed scratch resistance varies with scratch direction (in reference to the crystal orientation) [40]. Such features can be used to advantage in applications of PLCs.

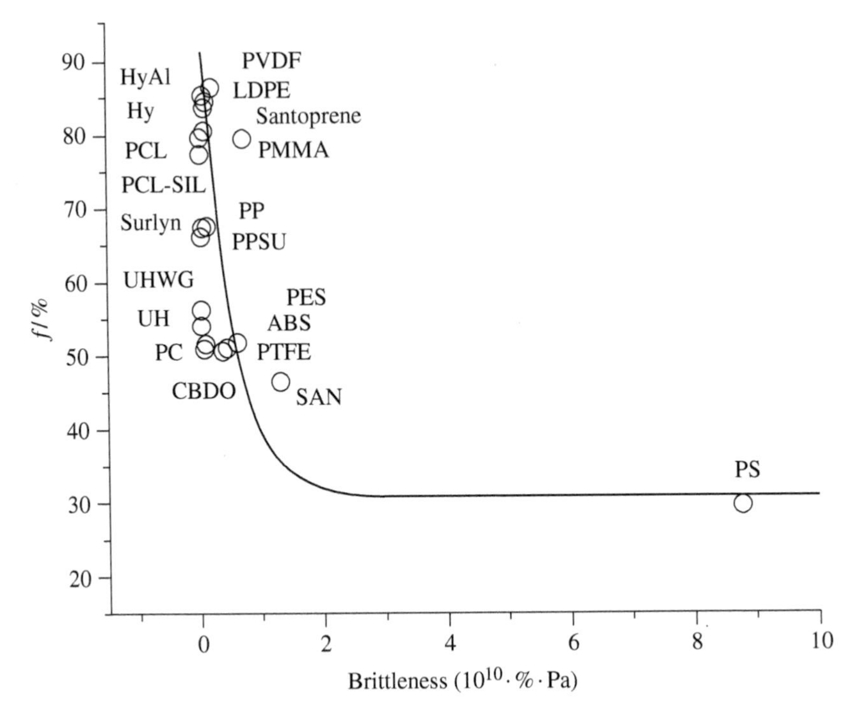

FIGURE 19.16 A relationship between the viscoelastic recovery *f* in sliding wear determination (horizontal asymptotic behavior) and brittleness for a variety of polymers and composites. (Abbreviations for materials are defined in [39].) Original Caption: The percentage of viscoelastic recovery as a function of brittleness for all materials (excluding metals). The solid line represents an exponentially decaying function defined by Eq. (19.3) fit to the experimental data points. *Source*: [39]. With kind permission from Springer Science and Business Media. © Springer Science+Business Media, LLC 2009.

As stated at the beginning of this Section, the surface phenomena associated with scratching may have a significant impact on performance. For instance, a scratch groove may provide a convenient host site for bacterial growth in a dental filling; this could invite further tooth decay. Scratches in coatings may compromise the intended functions of those coatings, altering the friction or facilitating corrosion. In optical materials, scratches reduce transparency and clarity.

19.8 WEAR

We read earlier in this chapter about Peter Jost's 1966 report on the economic costs due to poor the tribological behaviors of materials in service. Decades later, there are still economic concerns associated with tribology. Thus, Rabinowicz says [41]:

> At the time the Jost report appeared it was widely felt that the Report greatly exaggerated the savings that might result from improved tribological expertise. It has now become clear that, on the contrary, the Jost report greatly underestimated the

financial importance of tribology. The Report paid little attention to wear, which happens to be (from the economical point of view) the most significant tribological phenomenon.

The statement of Rabinowicz has been supported in the *Materials Research Society Bulletin* [42]:

> According to surveys carried out in different countries, the economic losses resulting from friction and wear are estimated to total 1–2% of gross domestic product (GDP). Using the World Bank's estimate of a $54 trillion global GDP (2007), it can be estimated that friction and wear cost the better part of $1 trillion dollars per annum. Since nearly all machinery involves sliding or rotating contact, the potential applications for a superhard, low friction coating are vast, spanning such diverse components as pump components, timing chains, cutting tools, abrasive water jet nozzles, and materials handling systems.

Wear is the unwanted loss of material from a solid surface by the mechanical action of another solid. Wear is a function of the applied load F and the total sliding distance D and can be determined, for instance, in a pin-on-disc tribometer (see again Figure 19.6). The rate of wear u_{sp} is defined quantitatively in terms of the volumetric loss V_{loss} of material as follows:

$$u_{sp} = \frac{V_{loss}}{F \cdot D} \tag{19.8}$$

This general definition does *not* take into account various factors which affect V_{loss}.

There are various kinds of wear, including adhesive wear, delamination wear, fretting wear, abrasive wear, erosive wear, corrosive wear, and more. Myshkin, Petrokovets, and Kovalev [43] discuss friction and wear in terms of three main factors: (1) interfacial bonds, in particular their type and strength; (2) real contact area; and (3) shear and related phenomena at and around contact points. In the next section we shall see how the factors considered important by Myshkin and his colleagues, plus more factors, play a role also when we are dealing with nanometric dimensions.

There is another side to wear, noted in particular by Zambelli and Vincent [44], also quite important for the economic well-being of the industry. Not only do otherwise good moving parts of machinery have to be replaced periodically, but while these parts are in motion in service, the *energy needed to keep them moving* is much higher than it would have been in the absence of friction as well as in the absence of wear.

19.9 LUBRICATION AND NANOSCALE TRIBOLOGY

In the broadest sense, the role of a lubricant is to reduce friction, prevent or minimize wear, transport debris away from the interface, and provide cooling. Lubricants are typically viscous liquids, but there exist a number of solid lubricants as well. In liquid lubrication, the various regimes can be classified according to fluid film thickness.

We recall that liquid lubricants are not usable for polymers or polymer-based composites—except for low amounts of internal lubricants. There is an intensive activity aimed at development of **solid state lubricants** [45–51], also for applications in aerospace, aviation, nuclear power, materials forming and tooling industries. The traditional solid state lubricants are graphite and molybdenum disulfide. However, graphite is not usable in vacuum or dry environments; it needs some moisture to terminate dangling chemical bonds which otherwise have adhesive tendencies. By contrast, MoS_2 cannot tolerate the presence of water since it then undergoes oxidation—while at room temperature it exhibits adhesive properties. Neither of the two is usable at high temperatures such as those exceeding 600°C while their wear resistance is insufficient for demanding applications. This situation has led Muratore and Voevodin [47] to the idea of multilayer or multiphase and multifunctional solid lubricants. They have found carbon nanotubes + MoS_2 composites usable as lubricants up to 300°C [48], and molybdate ($Ag_2Mo_2O_7$) and silver tungstate (Ag_2WO_4) thin films usable at 500°C [49]. The behavior of silver tantalite is particularly interesting; at 750°C or so, $AgTaO_3$ converts into a mechanically mixed system containing also Ta_2O_5 and silver nanoparticles [51]. This modified system exhibits low friction (0.6–0.15) against Si_3N_4 counterfaces. The melting temperature of $AgTaO_3$ is 1172°C, another advantage compared to a number of metals used as solid lubricants.

Laser surface treatments are an alternative approach in use for improvement of tribological properties. Voevodin and Zabinski use such a treatment on ceramic surfaces, combined with application of self-lubricating films [52, 53]. First micro-reservoirs are machined by a focused UV laser beam on the surface of a hard coating such as TiCN [53]. Such reservoirs can have the diameter of 10 μm, spaced on the surface at 50 μm intervals, resulting in surface coverage of some 10%, apparently an optimum. MoS_2 and/or graphite are then applied as self-lubricating films. The service life of the solid lubricant is an order of magnitude longer than on the ordinary unmodified TiCN surface [53].

Another type of laser surface treatment known as **laser surface texturing** (LST) can be applied to steel surfaces [54, 55]. The technique imparts textural features, described as dimples, bulges, etc. to the metal surface. Etsion, Rapoport, and their colleagues have found that long service life of MoS_2 can be obtained from smearing of the solid lubricant around the dimples, thus providing a supply of solid lubricant that preserves its thin solid film around the bulges.

Microelectromechanical systems (**MEMS**) deserve a few words at this point. The components of MEMS have sizes ranging from 0.001 mm to 0.1 mm, with the entire MEMS devices ranging in size from only 0.02 mm to 1 mm. Among the parts there is usually a central processor that along with other components interacts with the surroundings, for instance, with microsensors. Microelectromechanical systems are used in inkjet printers (control of deposition of ink on paper), in accelerometers in cars (for airbag deployment), and also in radio controlled flying machines such as airplanes, helicopters, and drones.

A relevant question at present is whether solid lubricants used in larger scale devices are also usable in MEMS. As discussed by Hall and his colleagues [56], in MEMS devices, friction and adhesion forces can overwhelm actuation forces; the result is lock-up and eventually failure. These forces are affected by surface roughness and the real contact areas, a topic called **contact mechanics** that is discussed in detail by Persson [57]. Small roughness values such as 20 nm are preferred. Actually, as demonstrated by Scharf and his colleagues [45], tungsten sulfide is useful in MEMS applications. Smooth transfer films of WS_2 are formed in dry nitrogen atmosphere, resulting in long term low friction as well as low wear. There is a technique called **LIGA** for creating high-aspect-ratio microstructures; LIGA is an acronym based on German words Lithographie, Galvanoformung, Abformung

(Lithography, Electroplating, Molding), which thus define the three consecutive stages of the process. The technique was developed in Karlsruhe in 1982 by Erwin Becker, Wolfgang Ehrfeld, and coworkers [58] and is now increasingly more used [59]. Structures having high aspect ratios such as 100:1 can be achieved through the use of X-rays originating from a synchrotron. Somewhat lower ratios are obtained using more accessible UV radiation. The radiation is applied to a photoresist through a radiation-absorbing mask prepared according to the pattern for the intended product. Afterwards one chemically removes the exposed—or else unexposed—photoresist and obtains a three-dimensional structure. The structure can be filled with a metal by electrodeposition. The produced metallic mold can be used for making parts from polymers or ceramics by injection molding. Other variations of this general process are also possible. Detailed images of the LIGA process are accessible on the website of the Karlsruhe Institute of Technology [60].

The capability to perform atomic layer deposition (ALD) is also important in creation of MEMS devices [61]. Atomic layer deposition was already mentioned in Section 19.4. The technique allows creation of films or coatings with thicknesses down to the nanometer range, and the procedure can be applied to yield films of a variety of material types: metals, oxides, nitrides, sulfides, and more. Furthermore, one can take hard coatings such as metal nitrides and cover them with diamond-like carbon (DLC, amorphous carbon) overcoatings [59, 62]. In fact, such overcoatings can be deposited on metals, ceramics, and polymers to protect them from wear resultant from rubbing. For instance, beneficial transfer films have been observed for metal-carbide reinforced amorphous carbon coatings (i.e., metal-doped diamond-like carbon coatings) against sapphire and steel hemispheres in tribometry testing (similar to pin-on-disc tribometry, with sapphire and steel as probes) [62]. After an initial run-in time for film formation, the films on the hemispherical sapphire or steel surfaces lowered friction as well as enhanced wear resistance.

With the capacity to create nanometer scale surface structures, it is necessary to be able to determine their properties. An appropriate technique is **atomic force microscopy (AFM)**. Its principle of operation is displayed in Figure 19.17. The cantilever interacts in various ways (depending on the operation mode) with the sample surface. A laser beam directed at the cantilever is reflected towards a photodiode that records displacement of the light as the cantilever encounters varied surface features. Atomic force microscopy is widely used to study materials surfaces. As an example of the kind of data that can be extracted, consider a study of effects of helium plasma on multi-layer graphene [63]. It was established by AFM that He plasma removed carbon atoms one layer at a time, without reduction of lateral dimensions and without formation of pits.

19.10 FINAL COMMENTS

Real life situations sometimes involve conflicts between various requirements. Consider an electric train, tramway, or an electric bus. A **pantograph** is mounted on the roof; it provides electricity from a cable spanned above the route of the vehicle. For low friction we wish to have a low contact area between the cable and the pantograph. At the same time, for good electricity transmission we would like to have a large contact area. There is one more factor: the part of the pantograph in contact with the cable undergoes wear. The current solution is to use a pantograph material such that the pantograph is worn more than the line is, while the wear of both should still be as low as possible [64]. This is

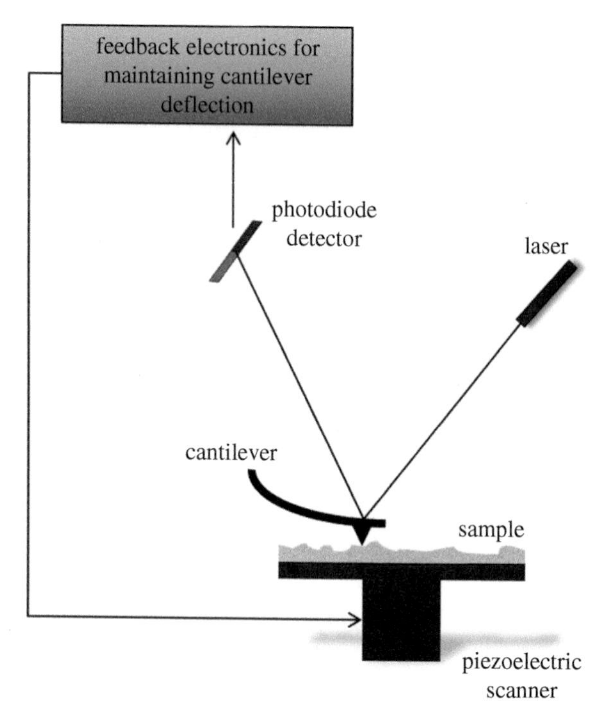

FIGURE 19.17 Schematic of operation of atomic force microscopy (AFM) with a piezoelectric (PZT) scanner.

because it is easier to replace the pantograph rather than to replace sections of the cable along a lengthy route.

19.11 SELF-ASSESSMENT QUESTIONS

1. Discuss the importance of tribology from a technical and economic point of view.
2. Explain how friction is determined by the sled test on a universal mechanical testing machine. How could you use a simple spring scale to determine friction?
3. Explain in molecular terms viscoelastic recovery seen in scratch resistance determination.
4. Explain strain hardening in sliding wear determination.
5. Discuss some methods of improvement of tribological properties of polymers.
6. What is surface tension and how does it differ between solids and liquids?
7. What is oxidation? Compare and contrast oxidation in metals, ceramics, and organic materials.
8. Considering what was discussed in Parts 1 and 2 about chemical bonds, energy, and reactivity, why do ceramic materials (typically) not corrode?
9. Name three products or devices that utilize ball bearings. What is the principal type of friction at work in ball bearings: sliding or rolling? Is friction expected to be high or low? How can the friction in ball bearings be modified (higher or lower)?

REFERENCES

1. B. Bhushan, *Introduction to Tribology*, John Wiley & Sons: New York **2002**.
2. H.P. Jost, *Tribology; Education and Research; Report on the Present Position and Industry's Needs*, HM Stationary Office: London **1966**.
3. D. Dowson, *History of Tribology*, Longman Group Ltd: London **1979**.
4. K. Carnes, The Ten Greatest Events in Tribology History, *Tribology and Lubrication Technology* **2005**, *61*, 38.
5. C.J. Van Oss, R.J. Good & M.K. Chaudhury, *Langmuir* **1988**, *4*, 884.
6. W. Brostow, P.E. Cassidy, J. Macossay, D. Pietkiewicz & S. Venumbaka, *Polymer Internat.* **2003**, *52*, 1498.
7. A.F. Stalder, T. Melchior, M. Müller, D. Sage, T. Blu & M. Unser, *Colloids & Surfaces A* **2010**, *364*, 72.
8. A. Bismarck, W. Brostow, R. Chiu, H.E. Hagg Lobland & K.K.C. Ho, *Polymer Eng. & Sci.* **2008**, *48*, 1971.
9. C. Wagner, *Z. phys. Chem.* **1933**, *B21*, 25.
10. Environmental Protection Agency, *Chapter 13: Chemical Oxidation*, https://www.epa.gov/sites/production/files/2014-03/documents/tum_ch13.pdf, accessed April 29, 2016.
11. J.E. Guillet, J. Dhanraj, F.J. Golemba & G.H. Hartley, Fundamental processes in the photodegradation of polymers, in *Stabilization of Polymers and Stabilizer Processes*, American Chemical Society: Washington, DC **1968**, p. 272.
12. V. Ageh, Y.H. Ho, R. Rajamure & T.W. Scharf, *Nanomater. & Energy* **2014**, *3* (2), 47.
13. W. Brostow, M. Dutta & P. Rusek, *Eur. Polymer J.* **2010**, *46*, 2181.
14. W.J. Work, K. Horie, M. Hess & R.F.T. Stepto, Definition of Terms Related to Polymer Blends, Composites, and Multiphase Polymeric Materials, *Pure & Appl. Chem.* **2004**, *76*, 1985.
15. W. Brostow, H.E. Hagg Lobland, S. Sayana, S. Shipley, A.L. Wang & J.B. White, to be published.
16. H.G. Silverman & F.F. Roberto, *Mar. Biotechnol.* **2007**, *9*, 661.
17. K. Numata & P.J. Baker, *Macromolecules* **2014**, *15*, 3206.
18. M.D. Bermudez, W. Brostow, F.J. Carrion-Vilches, J.J. Cervantes, G. Damarla & J.M. Perez, *e-Polymers* **2005**, no. 003.
19. M.-D. Bermudez, W. Brostow, F.J. Carrion-Vilches & J. Sanes, *J. Nanosci. & Nanotech.* **2010**, *10*, 6683.
20. B.A. Sherwood & W.H. Bernard, *Am. J. Phys.* **1984**, *52*, 1001.
21. J. Williams, *Engineering Tribology*, Cambridge University Press: Cambridge **2005**.
22. F.P. Bowden & D. Tabor, *Friction: An Introduction to Tribology*, R.E. Krieger: Malabar, FL **1973**.
23. I.L. Singer & H. Pollock eds., *Fundamentals of Friction*, Springer: New York **1992**.
24. W. Brostow, P.E. Cassidy, H.E. Hagg, M. Jaklewicz & P.E. Montemartini, *Polymer* **2001**, *42*, 7971.
25. G. Stachowiak & A.W. Batchelor, *Engineering Tribology*, Butterworth-Heinemann/Elsevier Science: Oxford/Burlington, MA **2013**, p. 682.
26. J. Lind & Å. Kassman Rudolphi, *Tribologia – Finnish J. Tribol.* **2015**, *33* (2), 20.

27. E.D. LaBarre, X. Calderon-Colon, M. Morris, J. Tiffany, E. Wetzel, A. Merkle & M. Trexler, *J. Mater. Sci.* **2015**, *50*, 5431.
28. W. Brostow, J.-L. Deborde, M. Jaklewicz & P. Olszynski, *J. Mater. Ed.* **2003**, *24*, 119.
29. W. Brostow, V. Kovacevic, D. Vrsaljko & J. Whitworth, *J. Mater. Ed.* **2010**, *32*, 273.
30. W. Brostow, H.E. Hagg Lobland & M. Narkis, *J. Mater. Res.* **2006**, *21*, 2422.
31. W. Brostow, B. Bujard, P.E. Cassidy, H.E. Hagg & P.E. Montemartini, *Mater. Res. Innovat.* **2002**, *6*, 7.
32. A. de la Isla, W. Brostow, B. Bujard, M. Estevez, R. Rodriguez, S. Vargas & V.M. Castaño, *Mater. Res. Innovat.* **2003**, *7*, 110.
33. B. Bilyeu, W. Brostow, L. Chudej, M. Estevez, H.E. Hagg Lobland, J.R. Rodriguez & S. Vargas, *Mater. Res. Innovat.* **2007**, *11*, 181.
34. M. Estevez, S. Vargas, V.M. Castaño, J.R. Rodriguez, H.E. Hagg Lobland & W. Brostow, *Mater. Letters* **2007**, *61*, 3025.
35. M. Estevez, J.R. Rodriguez, S. Vargas, J.A. Guerra, H.E. Hagg Lobland & W. Brostow, *J. Nanosci. & Nanotech.* **2013**, *13*, 4446.
36. W. Brostow, G. Damarla, J. Howe & D. Pietkiewicz, *e-Polymers* **2004**, *4* (1), 255.
37. W. Brostow, W. Chonkaew, L. Rapoport, Y. Soifer & A. Verdyan, *J. Mater. Res.* **2007**, *22*, 2483.
38. W. Brostow, W. Chonkaew, R. Mirshams & A. Srivastava, *Polymer Eng. & Sci.* **2008**, *48*, 2060.
39. W. Brostow & H.E. Hagg Lobland, *J. Mater. Sci.* **2010**, *45*, 242.
40. W. Brostow & M. Jaklewicz, *J. Mater. Res.* **2004**, *19*, 1038.
41. E. Rabinowicz, *Friction and Wear of Materials*, 2nd Edn., John Wiley & Sons: New York **1995**.
42. Technology Advances, Wear-Resistant Boride Nanocomposite Coating Exhibits Low Friction, *MRS Bulletin* November **2009**, *34*, 792.
43. N.K. Myshkin, M.I. Petrokovets & A.V. Kovalev, *Tribology Internat.* **2005**, *38*, 910.
44. G. Zambelli & L. Vincent, *Matériaux et contacts: Une approche tribologique*, Presses Polytechniques et Universitaires Romandes: Lausanne **1998**.
45. T.W. Scharf, S.V. Prasad, M.T. Dugger, P.G. Kotula, R.S. Goeke & R.K. Grubbs, *Acta Mater.* **2006**, *54*, 4731.
46. S.M. Aouadi, Y. Paudel, B. Luster, S. Stadler, P. Kohli, C. Muratore, C. Hager & A.A. Voevodin, *Tribol. Letters* **2008**, *29*, 95.
47. C. Muratore & A.A. Voevodin, *Ann. Rev. Mat. Res.* **2009**, *39*, 297.
48. X. Zhang, B. Luster, A. Church, C. Muratore, A.A. Voevodin, P. Kohli, S. Aouadi & S. Talapatra, *ACS Appl. Mater. & Interfaces* **2009**, *1*, 735.
49. D. Stone, J. Liu, D.P. Singh, C. Muratore, A.A. Voevodin, S. Mishra, C. Rebholz, Q. Ge & S.M. Aouadi, *Scripta Mater.* **2010**, *62*, 735.
50. T.W. Scharf & S.V. Prasad, *J. Mater. Sci.* **2013**, *48*, 511.
51. D.S. Stone, S. Harbin, H. Mohseni, J.-E. Mogonye, T.W. Scharf, C. Muratore, A.A. Voevodin, A. Martini & S.M. Aouadi, *Surface & Coatings Tech.* **2013**, *217*, 140.
52. A.A. Voevodin, J. Bultman & J.S. Zabinski, *Surf. & Coatings Tech.* **1998**, *107*, 12.
53. A.A. Voevodin & J.S. Zabinski, *Wear* **2006**, *261*, 1285.

54. I. Etsion, *J. Tribology* **2005**, *127*, 248.
55. L. Rapoport, A. Moshkovich, V. Perfilyev, I. Lapsker, G. Halperin, Y. Itovich & I. Etsion, *Surf. & Coatings Tech.* **2008**, *202*, 3332.
56. A.A. Hall, M.T. Dugger, S.V. Prasad & T. Christensen, *J. MEMS Systems* **2005**, *14*, 326.
57. B.N.J. Persson, *Surf. Sci. Reports* **2006**, *61*, 201.
58. E.W. Becker, W. Ehrfeld, D. Münchmeyer, H. Betz, A. Heuberger, S. Pongratz, W. Glashauser, H.J. Michel & R. Siemens, *Naturwissenschaften* **1982**, *69*, 520.
59. S.V. Prasad, T.W. Scharf, P.G. Kotula, J.R. Michael & T.R. Christenson, *J. MEMS Systems* **2009**, *18*, 695.
60. Karlsruhe Institute of Technology, Institute of Microstructure Technology, *LIGA Process*, http://www.imt.kit.edu/english/liga.php#LIGA, accessed August 6, **2016**.
61. R.L. Puurunen, J. Saarilahti & H. Kattelus, *ECS Trans.* **2007**, *11* (7), 3.
62. T.W. Scharf & I.L. Singer, *Tribol. Letters* **2009**, *36*, 43.
63. J.D. Jones, R.K. Shah, G.F. Verbeck & J.M. Perez, *Small* **2012**, *8*, 1066.
64. J. Lind, private communication from Uppsala University, February **2016**.

20

MATERIALS RECYCLING AND SUSTAINABILITY

A society grows great when old men plant trees whose shade they know they shall never sit in.

—Greek proverb.

20.1 INTRODUCTION

The first decades of the 21st century have witnessed continued growth in technology along with increasing constraints through policy on technological developments. Thus the need for materials scientists and engineers is not expected to dwindle. For the materials scientist and engineer, it is crucial to understand the fundamental principles that govern structure and interactions of matter (from the micro- to the macro-scale); moreover it is essential to understand the properties and behavior of different materials, the processes by which they are synthesized or fabricated, and the typical modes of application. For these reasons the majority of this text has focused on the just named aspects. This emphasis is further warranted by the fact that for decades cost and performance have been the most important factors in consumer product design. However, as LeSar, Chen, and Apelian [1] state, "Materials science and engineering must also take into account materials sustainability in the context of society and the environment." The need for this consideration is highlighted by the diminishing of certain natural resources. For instance, we are finding reason to assess the reserves of raw materials in light of the need for their use in manufacturing or in energy production. Some individuals are also beginning to assess quantitatively the effects of chemical toxicity—on human and environmental health—from the production, use, and disposal of materials [2]. In fact, the toxic effects of materials on human health deserve more attention. This applies to all materials in the environment and even more to materials

Materials: Introduction and Applications, First Edition. Witold Brostow and Haley E. Hagg Lobland.

intended for direct contact with human bodies. As Katsarava and Gomurashvili state [3], "The limitation for many synthetic biodegradable polymers as biomedical materials is the potential toxicity of the degradation products".

According to the United Nations' Brundtland Commission, **sustainable development** is development that "meets the needs of the present without compromising the ability of future generations to meet their own needs" [4]. Let us think now about MSE operating within the framework of a sustainable society and consider society as a *system*. Again drawing from the ideas of LeSar and coworkers [1], a sustainable society is "one that meets societal needs while maintaining the integrity of the environment and ecosystems, the economy, and the social needs of individuals". Therefore, the demand for sustainable development adds a new constraint to the design processes of materials engineers. Consequently, we shall consider in this chapter the sustainability of materials and the impact of materials on the environment. Materials scientists and engineers have a responsibility to the planet and therefore ought to bear in mind the constraint of sustaining human and environmental health even while exploring new frontiers in science and technology.

20.2 WATER

It deserves reiteration that water is the most valuable material of all. The World Health Organization (WHO) collects the information on water availability. According to WHO, one in nine people worldwide does not have access to clean water. As a result, every year 1.8 million people die from diarrheal diseases; 90% of those are children. While it is primarily thought of as a resource for agricultural and human consumption, water is also tied to energy production. In fact, power plants use very large quantities of water, primarily for the removal of waste heat [5]. This is no less true for nuclear power plants than for those burning fossil fuels. Additionally, water is used in significant volumes in agricultural and mining operations as well as by other industries and in manufacturing processes.

There is an important connection between energy production and water use. Even though utilization of solar or nuclear power may replace petroleum-based power, those processes still require water—and there is no substitute for water. Many regions of the world suffer at times from drought. In recent decades drought conditions have obliged the limiting of nuclear power production. Certain regions of the world frequently suffer from scarcity of water, notably the Middle East and North Africa. In those regions drinkable (i.e., potable) water is typically extracted from seawater by desalination. Kuwait obtains *all* its fresh water from the sea. The two main desalination processes are:

- Distillation: seawater is heated to evaporate off pure water—which is then condensed and collected.
- Reverse osmosis: passing seawater through a semi-permeable membrane that allows the passage of water molecules but blocks salt. To overcome the natural tendency of water to travel from low salt solutions to high salt solutions (osmosis), water is forced through the membrane at a high pressure. The membranes are typically made from 1 μm thick polyamide coated on a microporous polysulfone.

Both processes involve high energy costs: providing pressure for reverse osmosis or for evaporation for distillation.

There are two issues to contend with: (1) reducing contamination of water itself and in consequence of the environment with such "dirty water" and (2) slowing down the

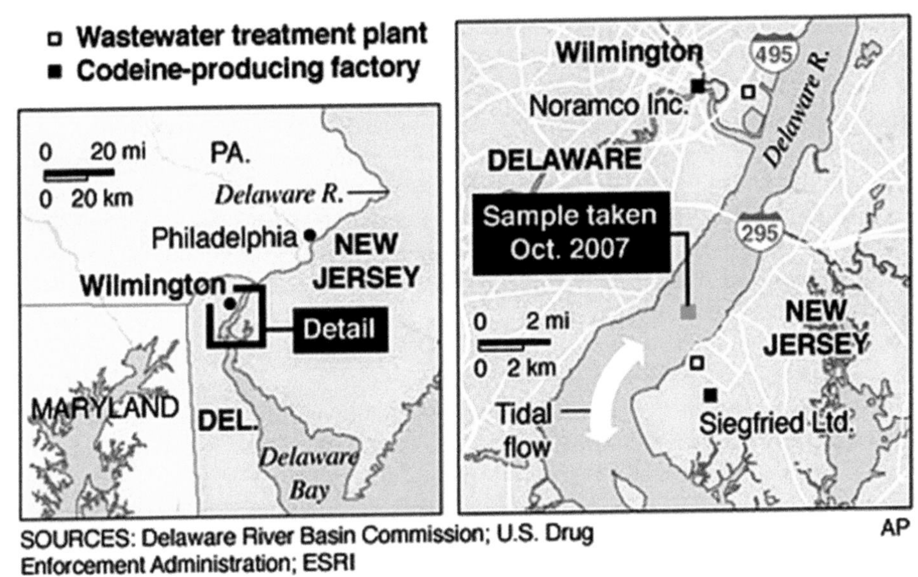

FIGURE 20.1 Unscrupulous industrial companies contaminate water, in this case the Delaware river. *Source*: "Tons of released drugs taint U.S. drinking water", http://www.cleveland.com/nation/index.ssf/2009/04/tons_of_released_drugs_taint_u.html.

depletion of raw water resources. MSE can offer some solutions. First consider the contamination of water: pharmaceutical byproducts, agricultural compounds, and waste products from industrial and mining processes can all contaminate surface and underground water sources, some of which may supply water for human consumption or agricultural irrigation. For example, Figure 20.1 shows how unscrupulous manufacturers of codeine (a painkiller medication) in the United States have been pumping their treated wastewater into the Delaware river. River water samples in the southern reaches of that river were found to contain a concentration of codeine 10 times higher than in the rest of the river. In some cases—especially with larger particulate material, not the small molecules of pharmaceutical drugs—advanced materials can be used to remove contaminants from water allowing its re-use, not necessarily as potable water but at least to serve various industrial needs. We have already discussed at the end of Chapter 13 how **polymeric flocculants** work and what are the advantages of their use.

Of course, in some cases more judicious use of materials can eliminate or reduce the *initial* water contamination. Indeed, elimination of the initial water contamination ought to be the primary goal because even if toxins can be filtered out, the toxins must still somehow be disposed of! It is not a simple matter to dispose of toxic waste extracted from water without re-contaminating some other water supply.

A frequent procedure consists in disinfecting tap water from fungi, bacteria, and viruses by chloride dioxide [5]. However, exposure to this chemical causes polyethylene pipes for water distribution to degrade at inner walls; pipes exposed to chlorine-free water last longer. Phenolic antioxidants are therefore added to PE. They slow down the PE pipe material degradation, but chlorine dioxide is a one-electron oxidant that now attacks the phenolic rings. Antioxidants are a solution: Ulf Gedde and coworkers found significant differences between antioxidant depletion times dependent on their chemical structures [5]. Some antioxidant depletion times are almost twice those of the other antioxidants—for the same initial molar concentration of phenolic groups. Thus, judicious choices of those antioxidants are now possible.

Let us return to **water cooling** in power generation. On one hand, water demand could be reduced by improved efficiency of gas and steam turbines. Materials are important here as such improvements are likely to be achieved by progress in developments of superalloys, ceramic matrix composites, and hydrophobic condenser surfaces. Providing these materials at low cost is essential to render air-cooled and low-water-demand processes economically viable [6]. On the other hand, the supply of water could be expanded to include nontraditional water sources, such as seawater, brackish aquifers, processed municipal wastewater, produced water from hydrocarbon extraction, and more. While this would conserve fresh water for other purposes, nontraditional water sources are likely to foul equipment. Advances in MSE that would provide biofilm resistant coatings and membranes and other mechanical or surface improvements (to improve equipment capacity to function with "subpar" water) might make it possible to use nontraditional water resources for cooling [6]. And think, we have not even mentioned here the fact that water itself can be used for power generation via dams and hydropower plants.

20.3 NUCLEAR ENERGY

Energy is of course a key figure in discussions of sustainability. A collective book edited by David Ginley and David Cahen published in 2011 [7] deals with connections between materials, energy, and sustainability. As in previous chapters, we have here a rather broad topic, in this case well treated in Ref. [7], that we need to cover inside of one chapter.

When one talks to laymen about science and technology, likely the most feared is the operation of nuclear power plants. The instantaneous mental association is with atomic bombs. Unscrupulous politicians (aided possibly unconsciously by some journalists) in certain countries sometimes create hysteria for gains of their political parties. We know how much pollution is created by products of coal combustion. In the United Kingdom the last coal mine in Kellingley in Yorkshire has been closed in December 2015. This while nuclear power plants *reduce* significantly consumption of fossil fuels for energy production; on the other hand they *produce* radioactive waste. As discussed by Hecker, Englert, and Miller [8]:

> Nuclear power holds the promise of a sustainable, affordable, carbon-friendly source of energy for the twenty-first century on a scale that can meet the world's growing need for energy and slow the pace of global climate change. However, a global expansion of nuclear power also poses significant challenges. Nuclear power must be economically competitive, safe and secure; its waste must be safely disposed of; and, most importantly, the expansion of nuclear power should not lead to further proliferation of nuclear weapons.

The authors discuss also related issues such as radioactivity of fluids used in mining uranium and lowering of the ductile-brittle transition temperature (see Section 14.7) of steel pressure vessels by neutron irradiation.

In this context, the story of two nuclear power plants in the now defunct Soviet Union is instructive. One plant was in Chornobyl in then Ukrainian Soviet Socialist Republic (called Chernobyl in Russian), the other in Ignalina in then Lithuanian Soviet Socialist Republic. Both plants were already obsolete by the time they were built. Both were under direct supervision from Moscow. A disaster occurred in Chornobyl on April 26, 1986; reactor vessel rupture and a series of steam explosions led to ignition of the graphite moderator. The main preoccupation of the authorities in Moscow was in keeping the disaster secret. Unusually high levels of radioactivity were discovered in Sweden, and the event became public worldwide when the Swedes defined the direction from which the radioactivity came and the approximate location of the origin. The plant managers in Moscow clearly were neither engineers nor scientists. The local fire fighters who were first on the scene of the accident in Chornobyl were not even told that there is radioactivity; secrecy was the only consideration. The fire fighters leader Lt. Volodymyr Pravik died in pain from radiation exposure on May 9. Other firefighters in his team died the same week. When high levels of radioactivity were discovered on the Chornobyl-Homel highway (Homel is nearby but already in Belarus), the road was covered with new asphalt paving. Within a week the new paving exhibited the same level of radioactivity as the old one. Thus in the beginning and then for some time symptoms but not the source were dealt with.

The story of the identical Ignalina plant was different. In spite of the Chornobyl disaster, the plant remained in operation until December 31, 2009. There was no other option. While Lithuania regained independence in 1990, one third of the electricity for the entire country came from Ignalina. It took many years to gradually develop alternative electricity sources. During those years the Lithuanian Energy Institute in Kaunas was watching over the plant to prevent a possible disaster, with a number of emergency scenarios developed. Since the idea of nuclear energy did and does make sense, a new such plant is being built by the Hitachi Corp. in Visaginas, in fact not far from Ignalina. Declared participants include Governments of Latvia and Estonia, with other countries of the region showing interest as well.

The nuclear power plants in use around the world derive energy from the process of *fission*. Plans for **nuclear *fusion* reactors** are underway but as yet not one has been realized. In this case temperature is also a problem since the fuel temperature is high enough to melt most materials. Mechanical creep due to irradiation is also a concern as it could lead to eventual failure of the vessel walls.

Nuclear power plants are not the only sources of irradiation in the human environment. There are more and more medical procedures that entail exposure to various kinds of irradiation. Detectors and scanners through which people must pass at airports and other institutions incur additional exposure to irradiation and intense sources of electromagnetic frequencies. People are also exposed to various levels of UV radiation indoors depending on the type of lighting used. This is especially important because it has been explained throughout this book that materials respond in various ways to radiation from electromagnetic waves of various frequencies. Apart from the effect of such exposure on naturally occurring bodily materials, what is happening to the synthetic materials in and on human bodies, for instance, to polymer-based hip implants, metal rods inserted into bone, dental obturations, and the like, when they are exposed to these sources of irradiation? Similarly, plastic tubes and other devices are often used internally during medical procedures that involve some exposure to radiation. Frequently biomedical materials are not tested for all the circumstances under which they are used; thus more caution with regard to the possibilities of degradation and release of byproducts is warranted. Realizing the situation, Abby Whittington and colleagues at Virginia Tech have begun to investigate the effects of radiotherapy and imaging on polyurethane (PU) medical devices [9]. For example, catheters inserted into neonates (i.e., infants) are monitored by frequent imaging, thus the catheters are repeatedly exposed—in a warm, saline environment—to small doses of radiation. Whittington's group is therefore assessing the effects of such exposure on the PU materials.

20.4 ENERGY GENERATION FROM SUNLIGHT

As Raymond H. Pahler says in the motto to our Chapter 16, the sun is not only essential to life on Earth but is also a key fixture that has garnered our attention for ages. Energy coming from the sun to the earth is inexhaustible—at least on the human lifespan scale. While windmills are another option for energy creation, wind appears because of different temperatures in different areas on the Earth surface—and those differences are related to activity of the sun.

To get an idea about the amount of energy coming from the sun, consider all organic raw materials (discussed in Chapter 8) available for burning, including natural gas, petroleum, and coal. Their total energy value is quite close to the energy coming from the sun to the earth over the course of 20 sunny days. As expected, the amount of sunlight received varies by location. In the desert one gets more than 6 kWh/(day·m^2), in the city of Boston 3.6 kWh/(day·m^2), while the average for the entire surface of the Earth has been estimated as 4.2 kWh/(day·m^2) [10]. There are multiple ways to utilize the energy of sunlight. Here we describe briefly the predominant ones and their important connections to materials.

Passive solar design of buildings is an ancient strategy used in construction to maximize absorption of the sun's heat during winter and to minimize it during the summer by reflecting or blocking radiation from the sun. The selection of appropriate materials types is important in such construction. From previous chapters we know that some materials are more thermally conductive than others; light transmittance and reflectance also varies with material type and surface finishing. Some modern materials can offer advantages and new possibilities that were not attainable with materials of the past. For instance, consider the possibility of using *electrochromic materials* (described in Section 12.4) to permit sunlight through the windows in winter while reflecting it back in the summer. If such technology could be applied to a large proportion of existing windows, the impact on energy consumption would be considerable. In the United States, residential and commercial buildings account for more than one-third of the total energy use [10], and the unwanted exchange of thermal energy through windows likely accounts for a significant portion of that.

So-called **solar thermal** devices generate electricity *indirectly* from sunlight. They are contrasted with photovoltaic devices which convert sunlight *directly* into electricity. The principles of solar thermal technology are used not only for power generation but also for the direct heating. For a long time, solar thermal technologies were considered the best method for generating hot water from solar power exclusively. Solar thermal water heating systems typically involve special coils of tubing installed in the roof where they absorb heat from the sun. See Figure 20.2 for an early model solar water heater. In some instances, this special equipment is not even needed. One of us lived for a time in a Texas residence with

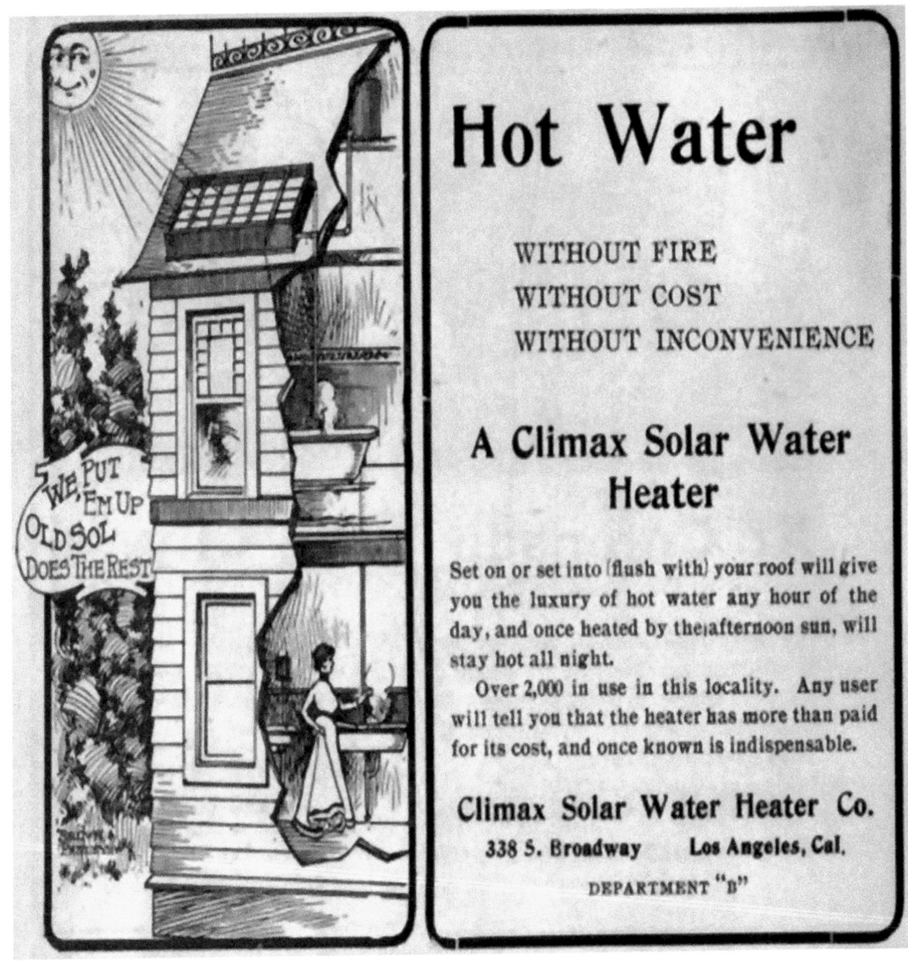

FIGURE 20.2 An advertisement for a solar water heater dating to 1901. *Source*: https://archive.org/stream/outwestland15archrich#page/n17/mode/2up from *The Land of Sunshine: The Magazine of California and the West*, Vol. XV, June, 1901 to December, 1901, Land of Sunshine Publishing Co.: Los Angeles, CA. Public Domain.

a poorly insulated roof; with only a single occupant in the house and having moved in during the summer, it was more than a month before it was discovered that the gas supply for the hot water heater was not connected, and all the hot water used for showers and dishwashing had been possible due to the placement of the hot water tank in the attic space under the roof. Indeed, this simple concept has certainly been used for a long time; but it is a poor method to store hot water as the tank and water in it seek equilibrium with the environment, thus heat from the water is lost more or less quickly depending on the outside temperature. Therefore, advancements have been made from this primitive method. On a much larger scale, such as for laundromats and college dormitories, advanced solar thermal technologies are an effective means of providing large quantities of hot water at lower cost than from the electrical power grid; see Figure 20.3.

FIGURE 20.3 A laundromat in California with panels on the roof providing hot washing water. *Source*: https://commons.wikimedia.org/wiki/File:Laundromat-SolarCell.png. Used under CC-BY-SA-3.0, GFDL.

Solar heat collectors are often comprised of large shallow containers, black inside and covered with glass as seen in Figure 20.3. Pipes or tubing in the box contain a water+ethanol mixture or other specified liquid; the liquid, heated by the sun, runs through pipes into the building and either heats water in a tank (i.e., a solar water heater) or else passes through radiators that heat the air. As strange as it seems, such systems can also provide cooling. With a so-called **absorption chiller**, solar energy heats a refrigerant under pressure. When the pressure is released, the refrigerant expands and the air around it therefore becomes cooler. Recall the story of liquefaction of gases in Section 15.2.

As mentioned above, solar thermal energy can also be used indirectly to generate electricity. Devices sometimes called solar heat collectors, similar in some ways to the direct heating devices just described, operate on the principle of converting the sun's *heat* into electricity. Thus, solar thermal power devices concentrate sunlight to create heat that is then used to run a heat engine, which turns a generator to make electricity. As with the heat collectors just described, a working fluid is needed to hold and transfer the heat. The fluid can be gas or liquid; among those commonly used are water, oil, salts, air, and nitrogen. Engine types include steam engines, gas turbines, Stirling engines, and more. Engines with efficiencies between 30 and 40% have the capacity to produce tens to hundreds of megawatts of power. **Solar thermal concentrating systems** with mirrors and lenses can provide temperatures as high as 3000°C. Molten salts are used for energy storage, so energy can be provided also when the sun is down.

Photovoltaic (PV) devices—in contrast to solar thermal devices—are capable of direct sunlight-to-electricity conversion [11], a phenomenon observed by the French physicist Alexandre-Edmond Becquerel in 1839. Solar cells collect, or concentrate, sunlight. Semiconducting materials housed in the cells undergo electronic excitation from the photons of light, thereby generating electric current (refer to explanations of semiconductors and photon excitation in Chapters 16 and 17). Under the action of an electric field, both positive and negative photo-generated carriers are extracted from the solar cell. Hence, the process involves both physical and electrochemical processes. An improved understanding of the processes involved has been achieved by Coutinho and Mendonça Faria [12] who have succeeded in explaining the photocurrent behavior of a heterojunction photovoltaic device by its having similar mean free paths for electrons and holes.

The photovoltaic semiconducting materials used include various forms of silicon (monocrystalline, polycrystalline, amorphous), cadmium telluride, copper indium gallium selenide, and more. Perovskites are important players as PV materials. In the past, they had been used for PV device applications but exhibited low efficiencies of solar energy absorption. In 2014, Malinkiewicz and her coworkers [13, 14] created devices with higher than usual efficiencies by making sandwich structures, with a $CH_3NH_3PbI_3$ perovskite between an organic electron layer and a hole blocking layer. Critical to the function of such PV structures is high purity of the perovskite layer, which is achieved by sublimation in a high vacuum chamber.

A single cell would only power a telephone for a short time. For practical applications, therefore, cells are connected together electrically into **photovoltaic modules** or **solar panels**. The development of new applications and improvements of photovoltaic technology are ongoing. While there remain certain limitations on use and energy storage, solar panels can boast of the rare advantage of having *no moving parts*. This feature, along with other improvements in the housing of solar cells, increases their durability and renders them very useful as portable energy sources.

The field of photovoltaic materials and solar power generation is a large and active one. Given this situation, with an abundance of books and literature available elsewhere, we shall conclude the topic with a final idea for contemplation. Springing from the success of photovoltaic modules, some researchers are looking for new ways of harvesting electricity from the sun. One of these developments is artificial photosynthesis for solar energy storage, discussed by Rybtchinski and Wasielewski [15] and also by Savage [16]. How do you think such a technology might work?

20.5 ENERGY GENERATION FROM THERMOELECTRICITY

There are two potentially useful thermoelectric (TE) effects, both with some two hundred years of history. In 1821 the Estonian-German physicist Thomas Johann Seebeck discovered in Tallinn an effect that is named after him. The **Seebeck effect** takes place when a temperature difference ΔT between two dissimilar materials in contact with each other produces a voltage V. Seebeck used metals; we now know one can also use semiconductors.

In TE materials there are free electrons; their density (the number of e^- per unit volume) differs from material to material. Therefore, when we put two different materials in contact, their electron gases diffuse into one another. The extent of diffusion is a function of the temperature. If there is a temperature difference across the materials, a potential difference will exist between the junctions and a current will flow. The larger the ΔT, the more efficient the energy conversion.

In 1834 a French physicist Jean Charles Athanase Peltier discovered an effect somewhat the reverse of the Seebeck effect. The generation of a current through the junction of two dissimilar conductors or semiconductors results in the cooling of one junction and heating of the other. This behavior is referred to as the **Peltier effect**. Both phenomena are discussed in Ref. [17].

Consider now opportunities based on these effects, one at a time [17, 18]. **Thermoelectric generators** (TEGs) operating on the principle of the Seebeck effect are capable of converting heat into useful electricity. Therefore, TEGs can take waste heat given off from home heating, automotive exhaust, industrial processes, and the like, and recycle it into a usable power supply. In the case of automotive exhaust, there is a large temperature difference ΔT between the exhausts and the surroundings. Theoretically the large ΔT should result in more electricity created. In practice, in order to achieve such good performance one has to combat the degradation of the TE materials exposed to such high temperatures [18].

On the flip side, **solid state TE coolers** (TECs) operate based on the Peltier effect. These have several advantages over current cooling methods. The coolants such as chlorofluorocarbons (CFCs) (now banned) and hydrofluorocarbons (HFCs) used in refrigerators are known to destroy the ozone layer of the earth. Finding a suitable non-toxic and non-flammable replacement, however, is a challenge. Imagine using instead solid state TE coolers. There would be no need for CFCs, HFCs, or other chemical coolants. Although TEC technology is not widely used for refrigeration yet, in the commercial market one can find a number of products utilizing TECs. For instance, there exists a TEC product designed to cool a bed mattress.

See now the illustrative schematic in Figure 20.4. The materials are dissimilar since one is predominantly n-type and the other p-type semiconductor. In the right image is a thermoelectric generator. If the resistor at the bottom of the top figure would be replaced by a voltmeter, we would get a **thermocouple** that senses temperature changes. In the left image of Figure 20.4, we have a schematic for a thermoelectric cooler. In fact, as suggested by the figure, such a device can be easily switched to a heater by reversing the voltage, hence its label as a heat pump.

A typical thermoelectric device is shown in Figure 20.5. An array of n-type and p-type semiconductor thermoelements sandwiched between two ceramic plates are connected electrically in series and thermally in parallel. In fact, functioning TE devices can be quite small.

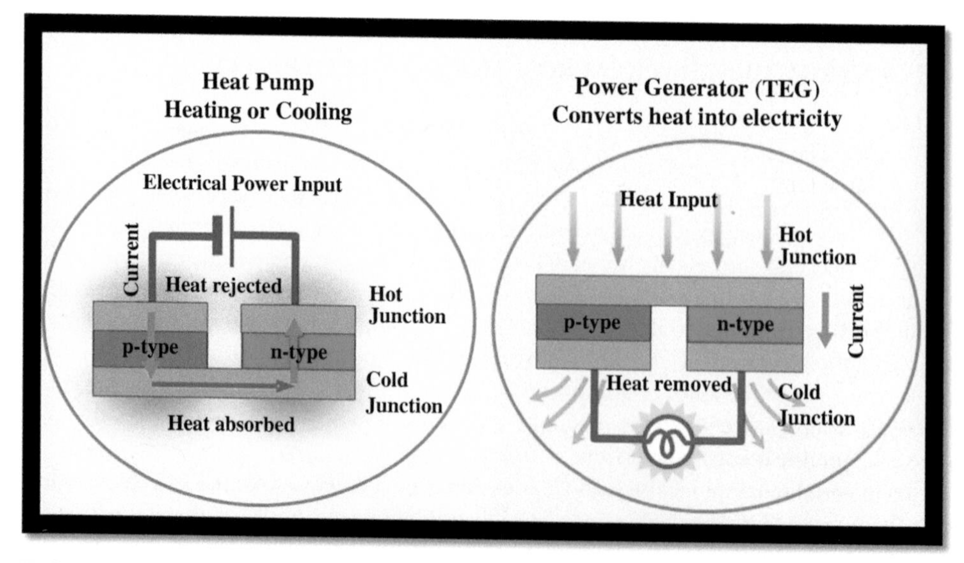

FIGURE 20.4 Schematic illustrating the uses of the Peltier (left) and the Seebeck (right) effects. The thermoelectric materials are n-type and p-type semiconductors.

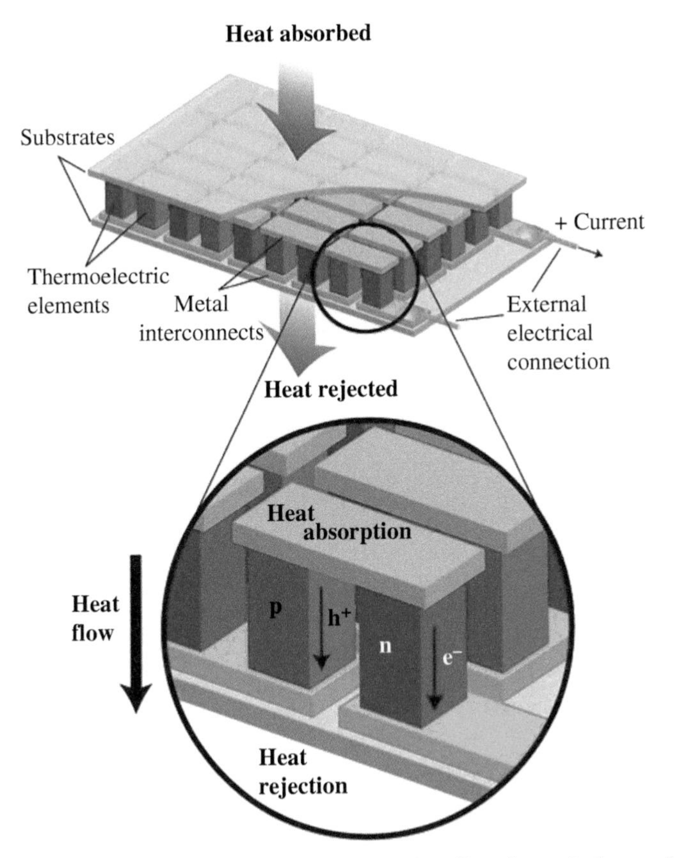

FIGURE 20.5 Thermoelectric module showing the direction of charge flow on both cooling and power generation. *Source*: Reprinted by permission from Macmillan Publishers Ltd [19]: G. Jeffrey Snyder & Eric S. Toberer, *Nature Materials* **2008**, *7*, 105. © 2008.

At this point the reader can ask skeptically: why, given these two wonderful effects, cars running on electricity from car exhausts are not in use? Nor do we see solid state refrigerators. The answer is that TE materials necessarily have to be subjected to repetitive cycles of large temperature change; the materials we know survive only a small number of such cycles. This is the case unless we have vacuum, hence TE devices are in good use on spaceships. Dramatic extension of service lives of TE materials and devices is needed. There is, needless to say, work advancing in this area [18].

20.6 DEGRADATION OF MATERIALS

The topics of corrosion and degradation were discussed already in the previous Chapter since these processes are largely instigated at a materials surface. In some cases degradation of a material or component in service is unwanted; corrosion is a well-known example. However, a larger problem is that created by increasing amounts of materials essentially "thrown out" into the environment: these materials fill garbage dumps, landfills, go into rivers or are simply scattered around. (This is why we see along highways announcements of fines for littering.) What is needed is recycling (discussed in the next section) or at least degradation of materials and components. In the latter case, used materials will return to Nature from which they originated. Degradation in the environment occurs mainly by two categories of chemical reactions:

- **Biodegradation** = biotic degradation: reactions taking place through the action of enzymes and/or chemical decomposition associated with living organisms (bacteria, fungi) or their secretion products
- **Abiotic degradation**: reactions including photo-degradation, oxidation, and hydrolysis

The time scale of these reactions can vary widely. One also has to consider the toxicity of disposed materials. The slowness of degradation poses a large problem worldwide. Many plastics are hardly biodegradable, thus they remain in the environment for many years. Non-biodegradable plastics have been found even on remote islands uninhabited by human beings. The potential impact on marine life is a concern since excess waste gets carried away in the world's oceans and rivers. Given this situation, much work is being done on biodegradable polymers—such as polyesters studied by Diaz, Katsarava, and Puiggali [20]—as one example of many.

Most metals eventually corrode due to oxygen and water exposure. A principal concern with many disposed items—including metals and ceramics—is whether they contain hazardous metals or other compounds that will eventually contaminate soil and water resources through toxic water runoff.

Biologists consider in particular *anthropogenic effects*, that is those originating from human activity. To some extent, Nature takes care of results of such activity. Justyna Ciesielczuk and her colleagues [21] studied a coal waste dump superimposed on a former ordinary dump in Katowice-Wełnowiec in Silesia from 2008 to 2015. The dump was closed in 1996. A number of plants had appeared growing in the dump, apparently attracted by self-heating of organic matter dispersed in the waste. In 2010 a plant called *Atriplex nitens* appeared among the grass and in the summer of that year reached a gigantic height of 2 m. Other plants in the dump have also shown gigantism—not seen elsewhere; likely causes are unusually high local concentrations of CO_2 and of nitrogen compounds. The Silesians conclude, among others things, that [21]:

> Coal waste dumps form unique plant habitats that promote the development of specific plant communities. The plants reflect anthropogenic influences, and the geochemical- and mineralogical characteristics of the coal wastes. Burning parts of dumps make, as the temperature decreases, specific habitats for plants with high requirements for heat, light exposure, moisture, CO_2 and nitrogen compounds.

Separating different categories of rubbish and recyclable materials according to degradability and recyclability can significantly reduce the amount of toxic waste that goes into landfills and increase the reclamation of recyclable materials. Improving—and increasing the prevalence of—these activities on a national and global scale ought to be an important goal.

20.7 RECYCLING

It is clear that waste disposal may cause (costly) air and water pollution and that land allocated for landfills could be more productively utilized. Additionally it is true that recycling offers benefits in terms of materials supply and energy usage. Nevertheless, Gaines

[22] stresses the fact that "not all recycling processes are created equal". Even a process as seemingly simple as recycling a plastic PET bottle entails various alternative routes, each of which has its own impacts on the environment. Recycled material can re-enter the life cycle stream at various points. In the present example, recycled PET could be mixed with virgin material to make new bottles; or recycled PET could be mixed with virgin PET or some other plastic to create an entirely different product—one that may be more or less valuable than the original bottle; or recycled PET might be processed into fibers and used in composite materials; or the same recycled PET could be chemically modified, transforming it into a completely different chemical substance that can be used for synthesis of other materials. Each of these routes is characterized by its own energy cost and environmental impact.

One should always be mindful that in determining the sustainability of any recycling method, there are many factors involved and usually tradeoffs, as well. Presently, a common, useful technique for making evaluations and comparisons between such options is **life-cycle analysis (LCA)**. Life-cycle analysis is a systems approach that considers all stages in the life cycle of a product or service [22]. Life-cycle analysis is very helpful in assessing the benefits and drawbacks for different disposal methods of items that are otherwise typically thrown away. Importantly, the results of LCA for a given product may differ from one country to the next, or even from city to city within a country, since factors such as price and availability of materials, energy, and labor may play a role [22]. Going back to the start, *more products might possibly be designed with the method of final disposition in mind.* While the benefits of recycling are widely recognized, there is no one approach that yields optimal results. **Recycling** might work best for one item, **re-use** for another, and **combustion** for a third. Economic impact, institutional constraints, and consumer preferences also play a role in determining the final outcome [22, 23]. As an exercise, consider the possibilities for re-using, recycling, or combusting (as waste-to-energy) a few commonly disposed of items: paper (newsprint and office paper), soda cans and bottles, and automotive batteries. (For further discussion of life-cycle analysis for these three products, reference the article by Gaines [22].)

20.8 FINAL THOUGHTS

A separate line of human activities related to nature and the environment consists in using natural materials to create products useful for humans. This is again an area of wide-ranging activities. Here we shall provide one example only. Sujitra Wongkasemjit and her coworkers [24] create ethanol from a variety of grasses. Different pretreatments are needed for different grass types. For two separate types of the vetiver grass (*Vetiveria zizanioides*) the yield is 0.14 g of ethanol for 1.0 g of initial grass. The same group has also demonstrated how to make sugar from grasses [25].

The role of Materials Science and Engineering in sustainable development extends even beyond what has been discussed here. It is clear however that material and product design can be tailored to reduce negative impacts on society and the environment. Furthermore, material advancements can sometimes be used to mitigate negative effects produced by certain technologies. In some cases, the challenges for MSE are great. Philip Ball [26] wrote:

> Energy production is the principal source of human pollution, generating wastes that cause global warming, acid rain, urban smog, and radioactive contamination. But the development of new materials for efficient batteries and solar cells now promises a cleaner way to power the world.

FIGURE 20.6 Work for the preservation of our planet—as well as against it.

Decades after these words were penned, there remains much to be done in developing better batteries and solar cells. Furthermore, there may be other avenues—besides only these two—in materials advancements that have the potential to significantly improve sustainable development.

Not everybody is conscious of the need for preservation of our environment—this in spite of much space devoted to the issue in public media. Moreover, we often observe that efforts on this front are being exerted cross-wise, with no real progress achieved. Figure 20.6 provides an illustration of this. Evaluating the costs of solutions for improved sustainability requires care and attention to details. A simple declaration that one's processes or product are sustainable and non-contaminating does not make them so. Furthermore, we are in danger of a vicious cycle in which we manufacture a thing to ameliorate a problem while the manufacturing creates another problem requiring another remedy which though good in itself generates another problem of contamination or sustainability. The cartoon in Figure 20.7 illustrates this dilemma of materials in the environment.

20.9 SELF-ASSESSMENT QUESTIONS

1. Explain the importance of water as a material.
2. Discuss technologies for obtaining energy from sunlight.
3. What is a practical use of the Seebeck effect? Think of an application, one not mentioned in the text, for how you could use this phenomenon to make something that would be useful in your own life.
4. Before a material is thrown away as rubbish it first has a service lifetime. Discuss material degradation in both contexts, highlighting what you think are important aspects for sustainability and safety. Be sure to mention possible conflicts, if any, between what is desired in each context.
5. Describe and distinguish re-use, recycling, and combustion.
6. What is life cycle analysis and why is it important?

FIGURE 20.7 "I guess gray is the new green." Irony in the name: A manufacturing plant intended for drug production actually appears to be generating a health *problem* by polluting air and water, and possibly the drug is intended to treat problems caused by just such pollution! *Source*: Used with permission from Jerry King, Jerry King Cartoons, Inc.

REFERENCES

1. R. LeSar, K.C. Chen & D. Apelian, Teaching sustainable development in materials science and engineering. *MRS Bulletin* **2012**, *37*, 449.
2. O.A. Ogunseitan & J.M. Schoenung, Human health and ecotoxicological considerations in materials selection for sustainable product development. *MRS Bulletin* **2012**, *37*, 356.
3. R. Katsarava & Z. Gomurashvili, Biodegradable polymers composed of naturally occurring α-amino acids, Ch. 5, in *Handbook of Biodegradable Polymers: Synthesis, Characterization and Applications*, editors A. Lendlein & A. Sisson, Wiley-VCH: Weinheim 2011.
4. World Commission on Environment and Development (UN WCED), *Report of the World Commission on Environment and Development: Our Common Future*, Annex to General Assembly document A/42/427, Oxford University Press: Oxford 1987.
5. W. Yu, T. Reitberger, T. Hjertberg, J. Oderkerk, F.R. Costa, V. Englund & U.W. Gedde, *Polymer Degrad. & Stabil.* **2015**, *111*, 1.
6. A.Y. Ku & A.P. Shapiro, The energy-water nexus: Water use trends in sustainable energy and opportunities for materials research and development. *MRS Bulletin* **2012**, *37*, 439.
7. D.S. Ginley & D. Cahen, editors. *Fundamentals of Materials for Energy and Environmental Sustainability*, Cambridge University Press: Cambridge **2011**.
8. S.S. Hecker, M. Englert & M.C. Miller, Nuclear non-proliferation, Ch. 14, in Ref. [7].
9. A.R. Whittington, Polymer Characterization of Medical Devices for Use in Cancer Patients, *POLYCHAR 23 World Forum on Advanced Materials*, University of Nebraska-Lincoln: Lincoln, NE, May 11–15, **2015**.

10. Union of Concerned Scientists, *How Solar Energy Works*, http://www.ucsusa.org/clean_energy/our-energy-choices/renewable-energy/how-solar-energy-works.html#.Vcj1ERNVhBc, accessed August 10, **2015**.
11. D.S. Ginley, R. Collins & D. Cahen, Direct solar energy conversion with photovoltaic devices, Ch 18, in Ref. [7].
12. D.J. Coutinho & R. Mendonça Faria, *Appl. Phys. Letters* **2013**, *103*, 223304.
13. O. Malinkiewicz, A. Yella, Y.H. Lee, G. Minguez Espallargas, M. Graetzel, M.K. Nazeeruddin & H.J. Bolink, *Nature Photonics* **2014**, *8*, 128.
14. C. Roldán-Carmona, O. Malinkiewicz, A. Soriano, G. Minguez Espallargas, A. Garcia, P. Reinecke, T. Kroyer, M.I. Dar, M.K. Nazeeruddin & H.J. Bolink, *Energy & Environ. Sci.* **2014**, *7*, 994.
15. B. Rybtchinski & M.R. Wasielewski, Artificial photosynthesis for solar energy conversion, Ch. 27, in Ref. [7].
16. L. Savage, *Optics and Photonics News*, February **2013**, http://www.osa-opn.org/home/articles/volume_24/february_2013/features/artificial_photosynthesis_saving_solar_energy_for/#.Vc5h2PlVhBc, accessed March 4, 2016.
17. W. Brostow, G. Granowski, N. Hnatchuk, J. Sharp & J.B. White, *J. Mater. Ed.* **2014**, *36*, 175.
18. W. Brostow, T. Datashvili, H.E. Hagg Lobland, T. Hilbig, L. Su, C. Vinado & J.B. White, *J. Mater. Res.* **2012**, *27*, 2930.
19. G.J. Snyder & E.S. Toberer, *Nat. Mater.* **2008**, *7*, 105.
20. A. Diaz, R. Katsarava & J. Puiggali, *Internat. J. Molec. Sci.* **2014**, *15*, 7065.
21. J. Ciesielczuk, A. Czylok, M.J. Fabiańska & M. Misz-Kennan, *Environ. & Socio-Econ. Studies* **2015**, *3* (2), 1.
22. L. Gaines, *MRS Bulletin* **2012**, *37*, 333.
23. V. Goodship, *Introduction to Plastics Recycling*, 2nd Edn., Smithers Rapra: Shrewsbury **2007**.
24. A. Banka, T. Komolwanich & S. Wongkasemjit, *Cellulose* **2015**, *22*, 9.
25. T. Komolwanich, S. Prasertwasu, D. Khumsupan, P. Tatijarern, T. Chaisuwan, A. Luengnaruemitchai & S. Wongkasemjit, *Mater. Res. Innovat.* **2016**, *20*, 259.
26. P. Ball, *Made to Measure: New Materials for the 21st Century*, Princeton University Press: Princeton, NJ **1997**, p. 244.

21

MATERIALS TESTING AND STANDARDS

Hugh Griffith as Monsieur Bonnet: "I know about their so called tests."
Audrey Hepburn as Nicole Bonnet: "Papa, they aren't so called, they are!"
—in the film *How to Steal a Million* directed by William Wyler.

21.1 INTRODUCTION

You might wonder why the topic of testing and standards deserves its own chapter. The motto for this module gives us at least one reason. Monsieur Bonnet and his daughter are talking about testing the authenticity of works of art—in view of an imminent visit in Paris of an expert in such testing from Zurich. Another reason for a separate chapter is the fact that having an understanding of different testing procedures and what information they can provide better enables one to frame appropriate research questions and experiments. Of course, there is room for creativity here and the development of new techniques, but the basic categories of testing are fairly well distinguished. Moreover, the establishment of standards is essential if we want to have valid comparisons, for example in evaluating data or materials from Lab A versus Lab B, and so on.

21.2 STANDARDS AND METRICS

National and international standards such as those of the International Standards Organization (**ISO**) tell us how to do certain things. For instance, if one wants to determine the friction of plastic film and sheeting, there is an ISO standard describing a method to do so. Results obtained can then be compared to results of others obtained by

Materials: Introduction and Applications, First Edition. Witold Brostow and Haley E. Hagg Lobland.

following the same ISO standard. The American Society for Testing and Materials (**ASTM**, *not "for Testing Materials"*) publishes more than thirty volumes of standards. These are not limited to only the materials testing methods discussed so far in this book. The American Society for Testing and Materials publishes standards for atmospheric analysis, cement and concrete, paper and packaging, underground utilities, and wood, just to name a few. Of a somewhat different sort, chemistry standards are published by the International Union of Pure and Applied Chemistry (**IUPAC**) and are also sometimes relevant to MSE.

All industrial nations have bodies that deal with standardization: the British Standards Institution (BSI) in the United Kingdom, the Dansk Standard in Denmark, the Sveriges standardiseringsråd in Sweden, and so on. The American Society for Testing and Materials differs from these, however, because it is not a part of or supported by the government of the country. Thus, the system of standards in the United States is based on voluntary consensus. Separate from the organizations dealing with standardization are those concerned with metrology, primarily with units and a few fundamental standards.

As new properties are discovered or deemed important, a need for new standards may be warranted. We saw this in the previous chapter on materials and the environment. Since sustainability and impact are not behaviors (like tensile strength) that can be directly measured, new metrics for their assessment must be developed.

21.3 TESTING

Consider now the following types of testing procedures:

1. **Routine** tests, usually in industrial testing laboratories, often for verifying manufacturer's specifications or assessing the outcome of production and forming techniques.
2. **Exploratory** tests on new (and existing) materials, the backbone of materials research.
3. **Destructive** tests, usually on parts rather than on whole structures.
4. **Scale model** tests, often on mechanical properties of a small model.
5. **Nondestructive** tests, many, often used for quality control manufacturing.
6. **Probing** tests, also non-destructive, but up to a maximum allowed value of stress, voltage, etc., often to "prove" the capability and performance.
7. **Inspections**, qualitative, may, for instance, utilize X-ray examinations, but are often only visual, as inspecting dimensional accuracy or surface finish.

A given test method, for instance, dynamic mechanical analysis, may be employed in more than one of the above categories. Inspections, particularly **visual inspections**, ought to be more highly regarded. They can be used in conjunction with other testing procedures. Although digital probes are powerful tools for collecting data, a human can observe by visual survey aspects undetected by a mechanical or electronic device. Therefore, a visual survey includes also inspections of equipment and data to ensure proper operating conditions.

It seems that the most famous visual inspection in the literature comes from a book by Mark Twain [1]:

> Aunt Polly placed small trust in such evidence. She went out to see for herself; and she would have been content to find twenty percent of Tom's statement true. When she found the entire fence whitewashed, and not only whitewashed but elaborately coated and recoated, and even a streak added to the ground, her astonishment was almost unspeakable.

The fence in Hannibal, Missouri, that a young Sam Clemens (real name of Mark Twain) truly whitewashed is still there. Indeed, one of us (WB) has seen it there in Hannibal.

Given a material—or device or product consisting of several materials—how do we know it will perform as expected? We can make our assessment by relying on a standard procedure (as described in Section 21.2), which is essentially someone else's assertion that the standard has been met. Alternatively, we can conduct our own test. Although thousands of testing methods exist, one should not be surprised by the need to develop a new one in order to solve a particular research or analysis problem. Nevertheless, searching through extant methods and standard procedures is a good place to begin the process of designing a new method.

21.4 MICROSCOPY TESTING

Owing to tremendous advancements in microscopy during the last half-century, microscopy test methods are becoming more widely accessible. Since these testing procedures are so important, we include here further information on the major classifications of microscopy in Table 21.1. It is evident from this table that large amounts of information can be obtained via microscopy.

Microscopy of polymers provides useful connections to their mechanical properties, as discussed in the books by Michler and Balta Calleja [2]. Likewise one can find resources detailing microscopy methods and data for all the major classes of materials and for some specialized types as well. Different kinds of microscopy are explained in what is already considered a classic textbook by Charles Kittel [3]. Though microscopy techniques continue to be advanced, the book remains quite comprehensive, covering well even newer approaches such as methods of atomic force microscopy (AFM) (discussed by us in Section 19.10) [3].

Returning to the motto of this Chapter... MSE is useful in detection of counterfeit art—a somewhat hidden topic of conversation of Nicole Bonnet with her father. However, much more often MSE has an indispensable role in **art conservation** and **art restoration**: these entail trying to keep a sculpture or a painting as close to its original condition as possible for as long as possible. One of the pioneers of this line of activity was the well-known physicist Michael Faraday who studied damaging effects of environment on works of art. Friedrich Rathgen of the Royal Museums in Berlin is considered by many as the father of modern artwork conservation. He published a book on the topic in 1898, translated into English in 1905 [4]. An institution for the purpose, namely, the International Institute for Conservation of Historic and Artistic Works (IIC), was incorporated in the United Kingdom in 1950.

TABLE 21.1 Summary of Main Forms of Microscopy and Their Features and Limitations

Technique	Detection Mode	Magnification/ Resolution	Features	Requirements	Limitations
Optical Microscopy (OM)	visible light	2,000× magnification	ambient temperature, sample in air or water, relatively cheap and easy maintenance, quick analysis	thin transparent specimens for transmission mode	small depth of the field
Scanning Electron Microscopy (SEM)	electron reflection, with electron energy 10–30 keV	20,000× magnification	relatively easy sample prep, high depth of field, essentially a real-time image scan	conductive samples or conductive gold coatings on samples, operates in a vacuum (except ESEM[a])	no color resolution
Transmission Electron Microscopy (TEM)	electron transmission, electron energy 100–400 keV	atomic resolution, approx. 10× more magnification than SEM	obtains SEM image simultaneously; determines atomic structure, particle size, shape, crystal dislocation; elemental analysis; essentially real time scan	thin samples (approx. 100 nm)	small field of view; expensive; sample prep might alter sample structure; possibility of damage by electron beam
Atomic Force Microscopy (AFM)[b]	laser light, photodiode, and piezoelectric scanner	<0.1 nm height resolution, a few nm lateral resolution	3D surface images, no special sample prep, no vacuum needed, operated in air and liquid, good for surface topography, detection of nano-size surface features, different modes (probes) allow surface property maps (e.g., height, optical absorption, magnetism)		slow image generation; smaller image size than with SEM

[a] Environmental Scanning Electron Microscopy.

[b] Also called Scanning Probe Microscopy. Includes variant modes such as Scanning Tunneling Microscopy, Scanning Capacitance Microscopy, Scanning Thermal Microscopy, and more.

Some of the current methods in artwork conservation are known to us from previous chapters. Szczepanowska, Jha, and Mathia [5] studied stains induced by dematiaceous (black pigmented) fungi on a cellular paper matrix, both on the surface and in the bulk. Such formations are a consequence of biodeterioration. Environmental scanning electron microscopy and imaging in backscattered electron mode (SEM-BSE) was combined with confocal laser scanning microscopy (CLSM) to analyze the damage. The spherical fruiting bodies (perithecia) embedded in paper were not detected by CLSM, but ESEM provided data at and near the surface. Interactions of the fungi with the paper were assessed by 3D visualization using x-ray microtomography at the ESRF (European Synchrotron Radiation Facility) in Grenoble. Szczepanowska and her colleagues note that for such a project, multi-scale analysis is needed. The ultimate purpose of their work is the development of "preservation strategies for cultural heritage, such as historic and artistic works on paper infested by fungi".

21.5 SENSORS IN TESTING

We have mentioned sensors in various locations throughout this book. We can distinguish two kinds of sensors: qualitative and quantitative. **Qualitative sensors** serve as indicators, for instance, to detect the presence of a certain compound in a material or a change in some material property. Thus, sensors for detection of the presence of organic liquid solvents and their vapors mentioned in Section 9.11 belong to this category. Of course, there is in each case a certain minimum concentration of the liquid or vapor that the qualitative sensor can detect. Deformation sensors, discussed in Section 12.2, belong here also. As discussed in Section 12.2, the combination of a nematic monomer liquid crystal and a cholesteric polymer liquid crystal undergoes a color change when stress is applied. The sensor provides notice of a change in stress without providing a specific numeric value.

Quantitative sensors provide additional feedback in the form of numerical values while they can always be used as qualitative sensors. Typically a sensor measures one specific quantity, however, Véronique Michaud and her colleagues are developing sensors that measure *simultaneously* strain and temperature in composites [6, 7]. We have talked about shape memory alloys (SMAs) in Section 12.6. Michaud and coworkers use optical SMA wires, fiber sensors with 150 μm diameter, embedded in laminate composites. The fibers have been subjected to **grating**, that is, providing equidistant parallel lines on their surfaces, resulting in periodic change of the refractive index. Two fiber types are needed, with different core dopants, so that they will react differently to temperature changes. The technique has been applied successfully to unidirectional laminates consisting of Kevlar fibers in an epoxy matrix. The progress of curing the epoxy can thus be followed in terms of both temperature and strain changes.

A process known as **solvent healing** can be induced in some thermoset polymer matrices, subjected to aging, in which crack formation occurs [8]. To achieve that end, such matrices based on an epoxy have to be under-cured since residual monomer is needed for crack healing. Mechanical testing has to be performed to verify that healing has taken place—although possibly strain monitoring, such as by the technique developed by Michaud, could be applied for that purpose.

21.6 SUMMARY

We have already acknowledged that there are situations in which a standard or recommended method cannot be found for the problem of interest. Often such situations involve extreme working conditions, e.g., very low temperatures or high magnetic fields; new or little-known properties; new or little-known materials and device components. We hope that at least some of the readers of this book will come upon such problems—and that they will ultimately solve them.

21.7 SELF-ASSESSMENT QUESTIONS

1. Are visual inspections worthwhile? Why?
2. Do you think standards for testing are necessary? Should standard test methods be used in all experimental testing? In all industrial testing? Explain your answers.
3. Look at the material objects all around you. Find an object that is likely to have undergone some kind of testing before it made its way into your environment. Discuss why and how (i.e., methods) it might have been tested. Do not fail to consider all the stages of its development and commercialization in your explanation.

REFERENCES

1. Mark Twain, *The Adventures of Tom Sawyer* (many editions).
2. G.H. Michler & F.J. Balta-Calleja, *Nano- and Micromechanics of Polymers: Structure Modification and Improvement of Properties*, Hanser: Munich/Cincinnati, OH **2012**; G.H. Michler, *Atlas of Polymer Structures*, Hanser, Munich-Cincinnati **2016**.
3. C. Kittel, with a contribution by P. McEuen, *Introduction to Solid State Physics*, 8th edn., John Wiley & Sons, Inc.: Hoboken, NJ **2005**.
4. F. Rathgen, *The Preservation of Antiquities: A Handbook for Curators*, Cambridge University Press **1905**.
5. H.M. Szczepanowska, D. Jha, & Th.G. Mathia, *J. Anal. At. Spectrom.* **2015**, *30*, 651.
6. H.-J. Yoon, D.M. Costantini, V. Michaud, H.G. Limberger, R.P. Salathé, C.-G. Kim, & C.S. Hong, *J. Intelligent Mater. Systems & Str.* **2006**, *17*, 1059.
7. L.P. Canal, B.D. Manshadi, V. Michaud, J. Botsis, G. Violakis, & H.G. Limberger, Monitoring strain gradients in adhesive composite joints by embedded fibre Bragg grating sensors, in *Sixth International Conference on Composites Testing and Model Identification*, editors O.T. Thomsen, B.F. Sørensen, & C. Berggreen, Aalborg University: Aalborg, Denmark **2013**, p. 91.
8. S. Neuser & V. Michaud, *Polymer Chem.* **2013**, *4*, 4993.

NUMERICAL VALUES OF IMPORTANT PHYSICAL CONSTANTS

Atomic mass constant	m_u	$1.660538921 \cdot 10^{-27}$ kg
Avogadro's number	A_0	$6.022045 \cdot 10^{23}$ mol^{-1}
Boltzmann constant	k	$1.380662 \cdot 10^{-23}$ J $\cdot$ K^{-1}
Electric vacuum permittivity	ε_0	$8.854187 \cdot 10^{-12}$ F $\cdot$ m^{-1}
Electron mass	m_e	$9.109382 \cdot 10^{-31}$ kg
Elementary charge	e	$1.6021766 \cdot 10^{-19}$ C
Faraday constant	F, A_0e	96485.3 C $\cdot$ mol^{-1}
Gas constant	R, A_0k	8.31446 J $\cdot$ mol^{-1} $\cdot$ K^{-1}
Magnetic vacuum permeability	μ_0	$1.256637061 \times 10^{-6}$ N $\cdot$ A^{-2}
Newton's gravitation constant	G	$6.67408 \cdot 10^{-11}$ m^3 kg^{-1} s^{-2}
Planck's constant	h	$6.626070 \cdot 10^{-34}$ J $\cdot$ s
Proton mass	m_p	$1.672621 \cdot 10^{-27}$ kg
Speed of light in vacuum	c	2.99792458×10^8 m $\cdot$ s^{-1}

Materials: Introduction and Applications, First Edition. Witold Brostow and Haley E. Hagg Lobland.

NAME INDEX

Adhikari, Rameshwar, 316
Apelian, Diran, 427
Amstutz, G.C., 56
Aouadi, S.M., 424

Baglin, John E.E., 166
Ball, Philip, 129, 439
Balta Calleja, Francisco J., 445
Banerjee, Rajarshi, 211
Bartholin, Rasmus, 347
Bednorz, J. Georg, 371
Blech, Ilan, 67
Bergmann, G., 215
Bismarck, Alexander, 232, 423
Blümich, Bernhard, 387
Boltzmann, Ludwig, 25
Bonaparte, Emperor Napoleon, 58
Bottino, Marco C., 219
Boubaker, Karem, 376
Bragg, W.L., 70
Bratychak, Michael, 162
Bravais, Auguste, 50
de Broglie, Prince Louis V., 70, 326
Bunsen, Robert, 26
Burgers, Jan, 62
Burriesci, Gaetano, 219

Casona, Alejandro, 241
Castaño, Victor M., 86, 236, 237
Celsius, Anders, 331
Chan, Chin Han, 180
Choi, Hyoung Jin, 284
de Coulomb, Charles-Augustin, 405
Coutinho, D.J., 376, 435
Curie, Pierre, 53, 256, 384
Czochralski, Jan, 109

Datashvili, Tea, 317
Davidovits, Joseph, 203
Debye, Peter J.W., 12, 325
Delaunay, Boris, 75, 79
Demkowicz, Michael J., 109, 116
DiBenedetto, A.T., 94
Dirac, P.A.M., 340
D'Souza, Nandika A., 238
Du, Jincheng, 134, 143, 350

Ehrenfest, Paul, 47
Einstein, Albert, 325, 352, 357

Materials: Introduction and Applications, First Edition. Witold Brostow and Haley E. Hagg Lobland.

Enders, Sabine, 184
Escher, Maurits C., 52, 59

Fahrenheit, D. Gabriel, 331
Faraday, Michael, 445
Fermi, Enrico, 340, 359
Flory, Paul J., 168, 172, 183, 207, 282
Foresti, Daniele, 18
Fourier, Joseph, 70
Fowler, Sir Ralph, 23
Föll, Helmut, 62
Fredro, Count Aleksander, 67
Fresnel, Augustin-Jean, 347

Gardner, Erle Stanley, 289
Gedde, Ulf W., xv, 40, 166, 184, 246, 430
Geim, Andre, 130
Gencel, Osman, 210
Gibbs, Josiah W., 26
von Goethe, Johann Wolfgang, 11
Gorman, B.P., 85
Gorodetsky, Alon, 353
Gotman, Irina, 236
Grünberg, Peter A., 387
Grüneisen, Eduard, 325
Guggenheim, Edward A., 23, 326
Gutmanas, Elazar Y., 236

Hanlon, Roger, 353
Hartmann, Bruce, 323
He, Jiasong, 232
Hedenqvist, Mikael, 184
Heeger, Alan J., 371
Henning, Sven, 316
Hess, Michael, 188, 356
von Helmholtz, Hermann, 26
von Hippel, Arthur, 3
Hnatchuk, Nathalie, 190
Hoffmann, Roald, 163, 168
Humphreys, Colin, 60

Infeld, Leopold, 352
de la Isla, Agustin, 416

Jackovich, Dacia, 187, 270
Janowski, Gregg M., 219
Johnson, William L., 111

Kamerlingh Onnes, Heike, 322, 371
Kassman Rudolphi, Åsa, 411
Katsarava, Ramaz, 236, 428, 438
Keesom, Willem Hendrik, 323
Kelvin, Lord, 331
Kepler, Johannes, 56
Kornfield, Julia, 161
Kozytskiy, A.V., 350
Kraynik, A.M., 82
Kubát, Josef, 300
Kuhlmann-Wilsdorf, Doris, 99
Kulicke, W.-M., 283
Kwolek, Stephanie, 174

Landau, Lev, 26
Landfester, Katarina, 234
Lensen, Marga C., 220, 230
Levinskas, Rimantas, 106
Lind, Jonna, 411
Litt, Morton, 323
Lucas, Elizabete F., 40, 156, 162

Magritte, René, 8
MacDiarmid, Alan G., 371
Malinkiewicz, O., 435
Mano, João F., 77, 217
Massieu, François, 26
Maxwell, James C., 322
McGlashan, Maxwell L., 21, 25, 326
McKittrick, Joanna, 228
Menard, K.P., 40, 302, 324
Mendonça Faria, R., 435
Micciulla, Samantha, 187
Michaud, Véronique, 264, 447
Michler, Goerg H., 86, 182, 445
Mohs, Friedrich, 312
Moodera, J.S., 386
Mott, Sir Nevill F., 368
Mueller, D.W., 85
Mukherjee, Sundeep, 115
Murr, Lawrence, 110
Müller, Alejandro J., 222
Müller, Christian, 370
Müller, K. Alex, 371
Myshkin, Nikolai, 418

Nag, Soumya, 211
Narkis, Moshe, 189, 200, 316, 424

Needleman, Alan, 198
Néel, Louis, 385
Nernst, Walther, 25
Newton, Sir Isaac, 274, 392
Novoselov, Konstantin, 130

Oliveira Jr., O.N., 355, 376
Olszewski, Karol, 322
Onck, P.R., 83
Onsager, Lars, 26

Pahler, Raymond H., 95, 205, 335, 432
Pal, Sagar, 288
Pauli, Wolfgang, 340, 359
Pauling, Linus, 15
Peltier, Jean C.A., 435
Penczek, Stanisław, 176
Petrenko, Viktor, 7
Perez, J.M., 85
Planck, Max, 24, 26
Popescu, M., 351
Prud'homme, Robert E., 222
Purcell, Edward M., 332

Quintanilla, John, 85

Rabi, Isidor I., 379, 387
Rabinowicz, Ernest, 418
Raines, K.S., 128
Raman, Sir C.V., 346
Ramsay, Sir William, 322
Ratner, Buddy D., 214
Rayleigh, Lord, 322
Reidy, R.F., 43, 85
Reinitzer, Friedrich, 242
Rigdahl, Mikael, 301
Riggs, Mark, 392
Roy, Rustum, 47, 125, 134, 166
Rusek, Piotr, 190, 423
Ruskin, John, 45

Samulski, Edward T., 242
Scharf, Thomas W., 350, 400, 420
Schmachtenberg, Ernst, 200
Schönhals, Andreas, 200
Shyshchak, Olena, xix
Seebeck, Thomas J., 435
Seifalian, Alexander, 219
Shechtman, Dan, 67
Shepherd, Nigel D., 350
Shibaev, Petr, 246
Silverstein, Michael S., 189
Simoes, Ricardo, 116, 189
Singh, R.P., 288
Singer, Irwin L., 409
Skłodowska-Curie, Marie, 53
Srinivasan, S.G., 147
Stachurski, Z.H., 80
Stingelin, Natalie, 355, 369

Termonia, Yves, 280
Tribus, Myron, 25
Twain, Mark, v, 445

Ulam, Stanisław M., 183
Uygunoglu, Tayfun, 210
Whittington, Abby, 432

Voevodin, Andrey, 420
Voronoi, Hrihoriy, 73

van der Waals, J.D., 2, 194, 322
Wahl, Kathryn J., 227
Wegner, Gerhard, 187
White, Mary Ann, 339, 345, 377, 387
Windle, A.H., 83
Wongkasemjit, Sujitra, 439
Wróblewski, Zygmunt, 322

Xu, Chunye, 255

Young, Marcus L., 264

Zabinski, J.S., 420
Zakin, Jaques L., 282
Zhia, Z.H., 199
Zhigilei, L.V., 61

SUBJECT INDEX

Note: Page numbers referring to figures are given in **bold** and those referring to tables in *italics*. Page numbers <u>underlined</u> refer to information in the shaded gray boxes.

acid rain, 158, 439
acrystallotropic point, 40
adhesion, 187–188, 208, 220, **221**, 227, 296
aerogel, 85, 183, 232, 329
aging, 81–82, 102, 281
anisotropy, anisotropic, 52, 130, 193, 194, 207, 208, 244, 258, 346
ankylography, 128
antimicrobial materials, 200, 230–231
antiplasticizer, 186
asphalt, asphaltenes,157, 204
azeotrope(s), 35–37, 155
azeotropic system, **35, 36**

Bingham fluid, 248, 249, 274–275, **275**
Bingham plastic *see* Bingham fluid
biomimetics, 227
birefractive *see* birefringence
birefringence, 346–347, **347**, 351
biocompatibility, 114, 212
blood-brain barrier, 235
Boeing 787 Dreamliner, 4, 165
Boltzmann
  constant, 24, 331
  distribution, 332, **333**
bond(s)
  chemical, 11, 12, 120, 178, 179, 182, 339, 401 *see also* intramolecular interactions
  covalent, 12, *18*, 119, 121, 129, 178
  hydrogen, **14**, 14–15, *18*, 32, 161
  ionic, 12, *18*, 257
  metallic, 12, *18*, 94
bone(s),114, 120, 207–208, **212**, *214*, 215–217, 228, <u>230</u>
boron nitride, 132, 363
brittleness, 144, 304–305, 311–312, **311**, **313**, 413, 417, **418**
buckminsterfullerene, 130, **131**

carbon nanotubes, 119–120, 130, **131**, <u>231</u>, 363, 364, 411
cellulose, 205–206, **214**, 229, 232

cement, 139, **142**, 202, 203, 227
cermet, 196–198, *197*, 202
chemical potential, 27, 47, 156
clad metals, 108, 201
coal tar, 154, **157**, 159, **160**, 195
coated metals, 108, 202
collagen, 207, 228, 258
compressibility, 46, 247, 321
copolymer(s), 167, **168**, **171**, 304
corrosion, 106, *107*, *108*, 110, 114–115, 216–217, 399–400, 437
coupling agent, 193, 208
creep, *64*, 82, 109, 147, 196, 197, 199, 299, **299**, 301–302, **302**, 431
crosslinking, 147, 168–170, **169**, 213, 262, 280, 398, 403

debonding, 194
defect(s), 46, 60–65, 99–100, 301, 339, 364–367, 368
degradation, 233, 282, 284, 435–438
Delaunay tessellation, 75, **75**
dendrimer, 180–181, **181**
dendritic phase, 109, 113, **113**
detergent(s), 17, 157, 175–176
devitrification, 81, 133
diamond, *18*, 46, 58, **59**, *101*, 129, *313*, *382*, 405
differential scanning calorimetry (DSC), 40, 304, 323–324, **324**
dipole *see* forces
dislocation(s) *see* defect
distillation, 32–34, 36, 154–155
drag reduction, 281–285
drunk walk problem, 179
dynamic mechanical analysis (DMA), 302–305, **305**

elastomer(s), 170–171, 247, 254–255, **296**
elastomer, thermoplastic, 171
electrical resistivity,199, 358–359, 363
electrochromic materials, 255–256
electroluminescence, 342, 350
electromagnetic spectrum, **336**
electronegativity, 15
electro-optical effect, 348–349
electrorheological fluids, 248, 252–254, **252**, **253**
electrostatic forces or interactions, **13**, 15, 207, 253–254, 401
emulsion, 156–157, 172, 269
emulsion polymerization, 175–176, 234, 285
energy, 23, 26, 48, 161, 428
engineering strain, 290
engineering stress, 290
enthalpy of combustion, *153*, 153, 166
entropy, 24–25
equilibrium, 21, 23
eutectic, 40–42, **41**, **42**
extensive properties, 22

fatigue, **306**, 312
feldspar(s), 127–128, **136**, 138
flocculation, 286
foam(s), 82, **83**, **202**
  aluminum foam(s), 84, 109
  ceramic foam(s), 85
  metal foam(s), 85, 109, 114
  polymeric foam(s), 85, **173**, 187, 291
forces, **13**
  cohesive, 18, 395
  Debye, 12
  dipole-dipole, 13, **13**
  dipole, induced, 12
  dipole, ion-induced, 13, **14**
  dispersion, 12, **13**
  ion-dipole, 13, **14**
  London dispersion *see* dispersion forces
  van der Waals *see* dispersion forces
fracking, 153
free volume, 18, 77, 78, 78
friction, 272, 273, 404, **407**
  dynamic (kinetic), 405
  static, 405

galvanized iron, 108
gel(s), 134, 182–183
Gibbs function, 26, 47, **47**
glass transition, 76–79, 82, 111, 112, 133, 187, 224, 247, 262, 296, 304, 323–324, *324*
glassy metals, 109–116
glazing, 120, 138
graphene, 130, **130**, 363
graphite, 53, **54**, 129–130, 420
gutta-percha, 179

hardness, 99, **99**, 100, *101*, **108**, 120, 132, 197, 312–314, *313*, **314**, **394**, 409
heart valves, 220
Helmholtz function, 26
heterogeneous system, 22
homogeneous system, 22
homopolymer(s), 167
hybrid composites, 200
hydrocracking, 158
hydrogel, 182, 213, 222
hydroxyapatite, 207, 218, 228

ice, 14–15, **338**
immiscible, 22
immunogenicity, 212
impurities, 60
inorganic materials, 119 *see also* minerals
intensive properties, 22
interaction potential, 15, 58
intermolecular interactions, 12, 16
interphase, 187
intramolecular interactions, 11
invisibility, 352–354
iron + carbon phase diagram,105, **105**
isobaric expansivity, 320
isotropy, isotropic, 196, 208, 292

kaolinite, 128, 136, **136**, **137**
Kevlar, 174, **175**, 195

laminate(s), **173**, 192, 200–202
Landau potential, 26
laser(s), 170, 343, 348, 420
lattice energy, 58
lever rule, 40
levitation, 18, 115 *see also* magnetic levitation train
light emitting diode(s) (LEDs), 342
lignin, 205–206, 232
lime glass, 134
lime mortar, 189
liquid(s), 16–17 *see also* surface tension, viscosity
liquid crystal(s), 242–247
lonsdaleite, 132
lubrication, 132, 419
luminescent materials, 349

magnetic levitation (maglev) train(s), 388
magnetic resonance, 387
magnetic susceptibility, 381
magnetoresistance, 387
magnetorheological fluids, 248
megasupramolecules, 151
melt flow index (MFI), 272
MEMS, 420
metamaterials, 353
mica(s), 128, **136**, 136–137
microelectromechanical systems *see* MEMS
mineral(s), 68, 91, 122
mineraloid, 68–69
miscible, miscibility, 22
molecular dynamics simulations, 116, 183–184, 301
Monte Carlo simulations, 183–184
MSE (materials science and engineering) triangle, 5–6, **5**, 11

nacre, 146, 227
negative temperatures, 330–332
nucleation, 48–50
  heterogeneous, 50
  homogeneous, 50
Nylon, 172, **174**, 178, **229**

opal (opalescent), 68–69, **69**, 174, 344
optical illusions, **351**, **352**
organic materials, 119
oxidation, 395

pantographs, 421
Pauli exclusion principle, 340, 359
Peltier effect, 435–436, **436**
percolation, 79
periodic table of the elements, **92**
perovskite, 123
photonic materials, 343
photovoltaic materials, 362, 434
physical interactions *see* intermolecular interactions
piezoelectrics, 256, 373
plasma, 7, 110
plasmon *see* plasma
plasticizer, 186
polyethylene, 167

polymer concrete(s), 204
polymorph, polymorphism, 58
porcelain, 138
porosity, 138
positrons, 357
potential energy, 15
precipitation hardening, 102
primary interactions *see* intramolecular interactions
pyroelectric effect, 256

quasicrystals, 68

radial distribution function, 70
refraction of light, 344, 346
Reynolds number, 270
rheology, 269
rubber, natural, 168, 180, 183

scaffolds, 217, 230
scratch resistance, 411, **412**
secondary interactions *see* intermolecular interactions
Seebeck effect, 435–436, **436**
semiconductor(s), 32, **46**, 109, 340–343, **360**, 363–370
shape memory alloy(s) (SMAs), 63, 105, 260, **261**
shape memory polymers (SMPs), 261, **262**
sialons, 128–129
silica, nanoporous, 143
silk, **173**, **229**
sled test, 406, **407**
sliding wear, 417
slurry, 125, 285
smart materials, 231
smelting, 93
soap(s), 17, 82
solar thermal devices, 432–434
spider silk, 227
spinels, 123, **124**
spherulites, 182, 222
standards, 443
state function, 23
stiffness, 298
strain hardening
  in metals, *64*, 99, 294, **295**
  in polymers, 414, **417**
stress concentration factor, 306
stress relaxation, 299, **299**, **300**, 300–301
superconductivity, 371
supercritical fluid, 31
surface tension, 16–17, **17**, 395, 402
  see also wetting angle
suspension, 172, 269, 285–286
sustainability, 427–428, 439

thermal conductivity, 129, 327–330
thermodynamic(s), 21
thermodynamic potentials, 25
thermodynamic stability, 21, 28
thermogravimetric analysis (TGA), 326
thermoplastic, **170**
thermoset, **170**
thixotropy, 275
tissue engineering, 217, 220, 230
toughness, 298, 310–311
  impact, 310
  tensile, 312
transport properties, 22, 26
tribology, 391–393, 419
tribometer, 406–408, **407**
triple point, 30
turbine engine, **126**

vascularization, 219, 219
Vickers hardness, *101*, 312–314
vinyl(s), **160**, 175, 178, 180
viscoelasticity, 277–280, 302–305
viscosity, 16, 270, 273
Voronoi tessellation or diagram, 83, **74**, 74, **83**, **103**, 361
vulcanization, 168–169

water, 32, 428
wear, 418
wetting angle, 17–18, 401–402
wood, 205
wool, **173**, **229**

xerogel, 85